Spin

ILYA KUPROV

Spin

From Basic Symmetries to Quantum Optimal Control

Springer

ILYA KUPROV
Southampton, UK

ISBN 978-3-031-05606-2 ISBN 978-3-031-05607-9 (eBook)
https://doi.org/10.1007/978-3-031-05607-9

This Springer imprint is published by the registered company Springer Nature Switzerland AG
The registered company address is: Gewerbestrasse 11, 6330 Cham, Switzerland

Preface

It is reasonable to assume that reality is causal, uniform, and isotropic. These assumptions lead to the conservation of energy, linear momentum, and angular momentum. When Lorentz invariance is also assumed, the resulting symmetry group (translations, rotations, inversions, and now space-time boosts) yields only two conserved quantities. One is the invariant mass; the other is a sum of angular momentum and something else, which appears because boost generators commute into rotation generators: special relativity has more ways of rotating things than Newtonian physics. That extra quantity is called spin.

Common interpretations of spin are smoke and mirrors, born of futile attempts to visualise the Lorentz group in three dimensions. I decided here to let the algebra speak for itself, but also to ignore mathematicians, chiefly Cartan, who had generated much fog around spin physics. Illustrations are drawn from magnetic resonance, where real-life applications had served to keep the formalism elegant and clear.

This book breaks with the harmful tradition of taking analytical derivations below matrix level—there are too many papers and books that painstakingly discuss every eigenvector. The same applies to perturbation theories, total spin representations, propagators, … In all those cases, numerical methods in matrix notation are ten lines of text and five lines of *MATLAB*. In this book, I pointedly avoid mathematical spaghetti: derivations only proceed to a point at which a computer can take over.

This is a foundation, not a literature review. If something is not here, then I did not think it fundamental enough given the time available; do let me know, it may appear in the next edition.

Southampton, UK
August 2022

ILYA KUPROV

Contents

Mathematical Background 1

Spin physics derives its mathematical structure from linear algebra and group theory. This chapter gives the minimal necessary introduction to the relevant topics (Fig. 1.1) from the user perspective: there are few derivations, few proofs, and no unnecessary generalisations. In a few tight spots, we appeal to empirical observations and physical intuition.

The examples used in this chapter are drawn from elementary geometry, calculus, and undergraduate quantum mechanics with which the reader is expected to be familiar.

1.1 Sets and Maps

A *set* is a collection of objects, called *elements* or *members* of the set, defined by listing the objects or specifying their common properties. A set $\mathcal{U}$ is called a *subset* of another set $\mathcal{V}$ (denoted $\mathcal{U}\subseteq\mathcal{V}$) if all members of $\mathcal{U}$ are also members of $\mathcal{V}$, and a *proper subset* of $\mathcal{V}$ (denoted $\mathcal{U}\subset\mathcal{V}$) if the two sets are not the same [1]. The following operations will be useful later:

1. *Union* $\mathcal{U}\cup\mathcal{V}$ is a set of all elements that are members of either $\mathcal{U}$, or $\mathcal{V}$, or both.
2. *Intersection* $\mathcal{U}\cap\mathcal{V}$ is the set of all elements that are simultaneously members of $\mathcal{U}$ and $\mathcal{V}$.
3. *Difference* $\mathcal{U}\backslash\mathcal{V}$ is the set of all members of $\mathcal{U}$ that are not members of $\mathcal{V}$.
4. *Cartesian product* $\mathcal{U}\times\mathcal{V}$ is the set of all ordered pairs (u,v) in which $u\in\mathcal{U}$ and $v\in\mathcal{V}$.

A *map* $M:\mathcal{U}\to\mathcal{V}$ from a set $\mathcal{U}$ into a set $\mathcal{V}$ is a relation that associates to every member of $\mathcal{U}$ exactly one member of $\mathcal{V}$. It is defined by a set of ordered pairs $\{u,v\}$,

I. KUPROV, *Spin*,
https://doi.org/10.1007/978-3-031-05607-9_1

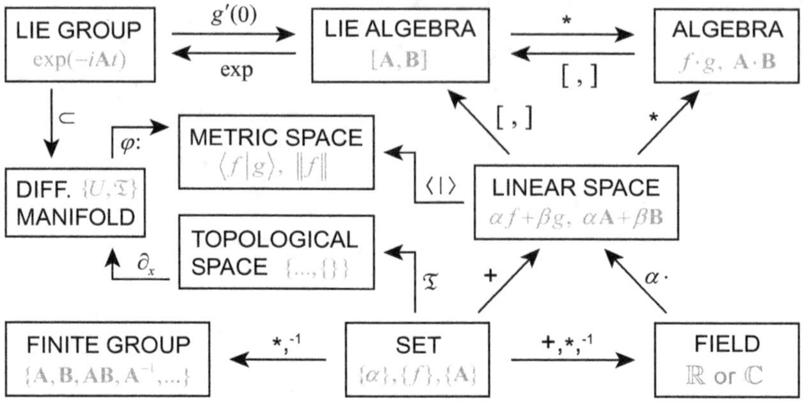

Fig. 1.1. A schematic map of the mathematical infrastructure of quantum theory. Postulating that wavefunctions exist defines their set. It then becomes a space when we add the superposition principle. The field of complex numbers is needed to make superpositions. When coefficients are associated with probabilities, a norm appears which makes the space of wavefunctions a metric space. Operator spaces and groups appear as sets of linear transformations of the wavefunction space. Finite groups govern things like particle permutations, and Lie groups perform continuous transformations such as translations, rotations, and time evolution. Tangent spaces of Lie groups are algebras containing the operators that appear in the equations of motion and conservation laws

such that $u \in \mathcal{U}$ and $v \in \mathcal{V}$, in which for every $u \in \mathcal{U}$ there is exactly one ordered pair of which u is the first element. A map is called:

1. *Injective* if distinct elements of $\mathcal{U}$ are mapped into distinct elements of $\mathcal{V}$.
2. *Surjective* if each element of $\mathcal{V}$ is mapped into by at least one element of $\mathcal{U}$.
3. *Bijective* if the map creates a one-to-one correspondence between elements of $\mathcal{U}$ and $\mathcal{V}$. A map that is simultaneously injective and surjective is bijective.

A map that preserves some additional features of the set (e.g. order or metric) is called a *morphism*. A set is called *closed* under an operation if performing that operation on a member of the set always produces a member of the same set.

1.2 Topological Spaces

Let $\mathcal{X}$ be a set of elements of any nature which we call *points*. Let $\mathcal{T}$ be a function assigning to each $x \in \mathcal{X}$ a collection $\mathcal{T}(x)$ of subsets of $\mathcal{X}$ called *neighbourhoods* of x. $\mathcal{T}$ is called a *neighbourhood topology* of $\mathcal{X}$, and $(\mathcal{X}, \mathcal{T})$ is called a *topological space*, if the following are true [1]:

1. For every point $x \in \mathcal{X}$, $\mathcal{T}(x)$ contains at least one neighbourhood; each neighbourhood in $\mathcal{T}(x)$ contains x.

2. For any two neighbourhoods $\mathcal{U}(x)$ and $\mathcal{V}(x)$ of the same point $x \in \mathcal{X}$, there exists a neighbourhood that is a subset of both $\mathcal{U}(x)$ and $\mathcal{V}(x)$.
3. If a point $y \in \mathcal{X}$ belongs to a neighbourhood $\mathcal{U}(x)$ of some $x \in \mathcal{X}$, there exists a neighbourhood $\mathcal{V}(y)$ that is a subset of $\mathcal{U}(x)$.
4. For any two different points $x, y \in \mathcal{X}$ there exist neighbourhoods $\mathcal{U}(x)$ and $\mathcal{V}(y)$ with no points in common.

A topological space $(\mathcal{X}, \mathcal{T})$ is called *disconnected* if $\mathcal{X}$ can be obtained as a union of two or more non-empty subsets that include neighbourhoods of all of their points, but have no points in common. If that is not the case, the topological space is called *connected* (Fig. 1.2).

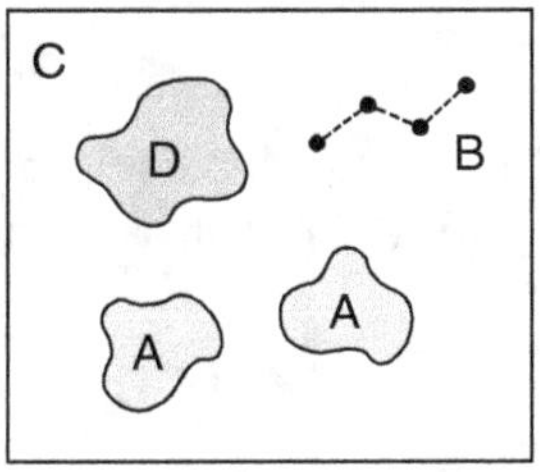

	connected	**path-connected**	**simply connected**
A	no	no	no
B	yes	no	no
C	yes	yes	no
D	yes	yes	yes

Fig. 1.2 A schematic illustration of the concepts of connectedness, path-connectedness, and simple connectedness. The dotted lines indicate that the discrete points are listed as neighbours in the neighbourhood topology

When the set $\mathcal{X}$ is continuous, we can define a *path* from a point $x \in \mathcal{X}$ to a point $y \in \mathcal{X}$ as a continuous function $f : [0, 1] \to \mathcal{X}$ with $f(0) = x$ and $f(1) = y$. A topological space is called *path-connected* if there is a path between any two of its points, and *simply connected* if any path between those points can be continuously transformed into any other path between the same points (Fig. 1.2).

A *covering space* of a topological space $\mathcal{X}$ is a topological space $\mathcal{Y}$ together with a continuous map $p : \mathcal{Y} \to \mathcal{X}$ such that for every $x \in \mathcal{X}$ there exists a neighbourhood $\mathcal{U}(x) \in \mathcal{T}$ that has multiple disjoint preimages in $\mathcal{Y}$, each of which has a one-to-one correspondence of elements with $\mathcal{U}(x)$. The number of preimages is called the *multiplicity* of the cover. A covering space is called a *universal covering space* when it is simply connected. An example is given in Fig. 1.3.

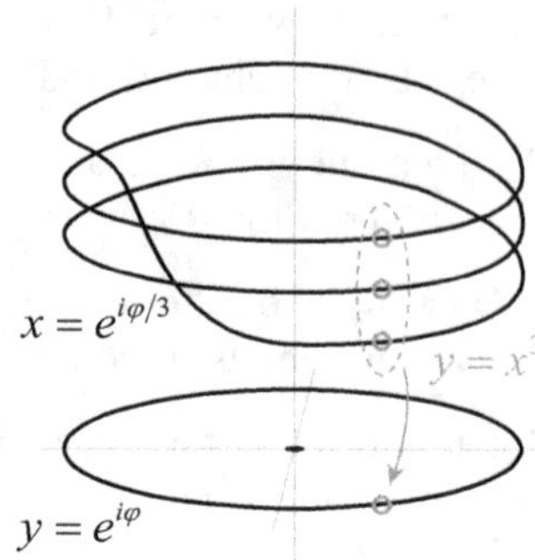

Fig. 1.3 A schematic illustration of the concept of multiple cover: a unit circle in the complex plane is triply covered by the topological space of its cubic roots

1.3 Fields

A *field* is a set $\mathbb{F}$ containing elements $\{\alpha, \beta, \gamma, ...\}$ of any nature and equipped with two operations, called *addition* ("+") and *multiplication* ("·"), that have the following properties [2]:

1. $\mathbb{F}$ is closed under addition and multiplication:

$$\forall \alpha, \beta \in \mathbb{F} \quad \alpha + \beta \in \mathbb{F} \quad \text{and} \quad \alpha \cdot \beta \in \mathbb{F} \tag{1.1}$$

2. Addition and multiplication operations are *associative*:

$$\forall \alpha, \beta, \gamma \in \mathbb{F} \quad (\alpha + \beta) + \gamma = \alpha + (\beta + \gamma) \text{ and } (\alpha \cdot \beta) \cdot \gamma = \alpha \cdot (\beta \cdot \gamma) \tag{1.2}$$

3. Addition and multiplication operations are *commutative*:

$$\forall \alpha, \beta \in \mathbb{F} \quad \alpha + \beta = \beta + \alpha \quad \text{and} \quad \alpha \cdot \beta = \beta \cdot \alpha \tag{1.3}$$

4. $\mathbb{F}$ contains a unique *zero element* and a unique *unit element*:

$$\begin{aligned} \exists! 0 \in \mathbb{F}, \quad \forall \alpha \in \mathbb{F} \quad \alpha + 0 = \alpha \\ \exists! 1 \in \mathbb{F}, \quad \forall \alpha \in \mathbb{F} \quad \alpha \cdot 1 = \alpha \end{aligned} \tag{1.4}$$

5. Each element in $\mathbb{F}$ has a unique *additive inverse* in $\mathbb{F}$:

$$\forall \alpha \in \mathbb{F} \quad \exists!(-\alpha) \in \mathbb{F}, \quad \alpha + (-\alpha) = 0 \tag{1.5}$$

   and a unique *multiplicative inverse* in $\mathbb{F}$:

$$\forall (\alpha \neq 0) \in \mathbb{F} \quad \exists! \alpha^{-1} \in \mathbb{F}, \quad \alpha \cdot \alpha^{-1} = 1 \tag{1.6}$$

6. Multiplication is *distributive* over addition:

$$\forall\alpha, \beta, \gamma \in \mathbb{F} \quad \alpha \cdot (\beta + \gamma) = \alpha \cdot \beta + \alpha \cdot \gamma \tag{1.7}$$

The fields that make appearances in this book are $\mathbb{C}$ (complex numbers), and $\mathbb{R}$ (real numbers). Elements of fields are called *scalars*. A field $\mathbb{F}$ is called *algebraically closed* if every polynomial of a non-zero degree with coefficients taken from $\mathbb{F}$ has at least one root in $\mathbb{F}$. By the fundamental theorem of algebra [332], $\mathbb{C}$ is algebraically closed, but $\mathbb{R}$ is not. This is a physics book—hereinafter, we avoid unnecessary generality and alternate between $\mathbb{C}$ and $\mathbb{R}$ as appropriate.

1.4 Linear Spaces

A *linear space* $\mathcal{V}$ over a field $\mathbb{F}$ is a set containing elements $\{\mathbf{a}, \mathbf{b}, \mathbf{c}, ...\}$ of any nature and equipped with two operations, called *addition* ("+") and *multiplication by a scalar* ("·"), such that [3]:

1. $\mathcal{V}$ is closed under addition:

$$\forall\mathbf{a}, \mathbf{b} \in \mathcal{V} \quad \mathbf{a} + \mathbf{b} \in \mathcal{V} \tag{1.8}$$

2. $\mathcal{V}$ is closed under multiplication by scalars from $\mathbb{F}$:

$$\forall\alpha \in \mathbb{F} \quad \forall\mathbf{a} \in \mathcal{V} \quad \alpha\mathbf{a} \in \mathcal{V} \tag{1.9}$$

3. Addition is *commutative*:

$$\forall\mathbf{a}, \mathbf{b} \in \mathcal{V} \quad \mathbf{a} + \mathbf{b} = \mathbf{b} + \mathbf{a} \tag{1.10}$$

4. Addition is *associative*:

$$\forall\mathbf{a}, \mathbf{b}, \mathbf{c} \in \mathcal{V} \quad \mathbf{a} + (\mathbf{b} + \mathbf{c}) = (\mathbf{a} + \mathbf{b}) + \mathbf{c} \tag{1.11}$$

5. $\mathcal{V}$ contains a unique *zero element*:

$$\exists\mathbf{0} \in \mathcal{V}, \quad \forall\mathbf{a} \in \mathcal{V} \quad \mathbf{a} + \mathbf{0} = \mathbf{a} \tag{1.12}$$

6. Each element in $\mathcal{V}$ has a unique *additive inverse* in $\mathcal{V}$:

$$\forall\mathbf{a} \in \mathcal{V} \quad \exists!(-\mathbf{a}) \in \mathcal{V}, \quad \mathbf{a} + (-\mathbf{a}) = \mathbf{0} \tag{1.13}$$

7. Multiplication by scalars is *associative*:

$$\forall\alpha, \beta \in \mathbb{F} \quad \forall\mathbf{a} \in \mathcal{V} \quad \alpha(\beta\mathbf{a}) = (\alpha\beta)\mathbf{a} \tag{1.14}$$

8. Multiplication by scalars is *distributive* with respect to elements of both $\mathcal{V}$ and $\mathbb{F}$:

$$\begin{aligned} &\forall\alpha \in \mathbb{F} \quad \forall\mathbf{a},\mathbf{b} \in \mathcal{V} \quad \alpha(\mathbf{a}+\mathbf{b}) = \alpha\mathbf{a}+\alpha\mathbf{b} \\ &\forall\alpha,\beta \in \mathbb{F} \quad \forall\mathbf{a} \in \mathcal{V} \quad (\alpha+\beta)\mathbf{a} = \alpha\mathbf{a}+\beta\mathbf{a} \end{aligned} \tag{1.15}$$

A subset of $\mathcal{V}$ that is itself a linear space is called a *subspace* of $\mathcal{V}$. A *sum* $\mathcal{U}+\mathcal{W}$ of subspaces $\mathcal{U}$ and $\mathcal{W}$ of a space $\mathcal{V}$ is the set of all elements of the form $\mathbf{u}+\mathbf{w}$, where $\mathbf{u} \in \mathcal{U}$ and $\mathbf{w} \in \mathcal{W}$. $\mathcal{U}+\mathcal{W}$ is also a subspace of $\mathcal{V}$. This sum is called a *direct sum* and denoted $\mathcal{U} \oplus \mathcal{W}$ when the two subspaces only have the zero element in common.

A *product* of two spaces $\mathcal{U}$ and $\mathcal{V}$ is the set of all ordered pairs of elements in which the first component belongs to $\mathcal{U}$ and the second component belongs to $\mathcal{V}$:

$$\mathcal{U} \times \mathcal{V} = \{(\mathbf{u},\mathbf{v})|\mathbf{u} \in \mathcal{U},\ \mathbf{v} \in \mathcal{V}\} \tag{1.16}$$

If the field $\mathbb{F}$ is the same for both spaces, then addition and multiplication by scalar may be inherited by the product. The resulting set of all linear combinations of pairs of elements is called *direct product*:

$$\mathcal{U} \otimes \mathcal{V} = \left\{ \sum_{nk} \alpha_{nk}(\mathbf{u}_n \otimes \mathbf{v}_k)|\mathbf{u}_n \in \mathcal{U},\ \mathbf{v}_k \in \mathcal{V},\ \alpha_{nk} \in \mathbb{F} \right\} \tag{1.17}$$

where the operation $\otimes$ must be distributive on either side with respect to addition:

$$\begin{aligned} &(\mathbf{u}_1+\mathbf{u}_2) \otimes \mathbf{v} = \mathbf{u}_1 \otimes \mathbf{v}+\mathbf{u}_2 \otimes \mathbf{v} \quad \forall\mathbf{u}_{1,2} \in \mathcal{U} \quad \forall\mathbf{v} \in \mathcal{V} \\ &\mathbf{u}_1 \otimes (\mathbf{v}_1+\mathbf{v}_2) = \mathbf{u}_1 \otimes \mathbf{v}_1+\mathbf{u}_1 \otimes \mathbf{v}_2 \quad \forall\mathbf{u} \in \mathcal{U} \quad \forall\mathbf{v}_{1,2} \in \mathcal{V} \end{aligned} \tag{1.18}$$

and associative on either side with respect to multiplication by scalars:

$$(\alpha\mathbf{u}) \otimes \mathbf{v} = \mathbf{u} \otimes (\alpha\mathbf{v}) = \alpha(\mathbf{u} \otimes \mathbf{v}) \quad \forall\mathbf{u} \in \mathcal{U} \quad \forall\mathbf{v} \in \mathcal{V} \quad \forall\alpha \in \mathbb{F} \tag{1.19}$$

A direct product of two spaces is itself a space.

1.4.1 Inner Product Spaces

Here we narrow the discussion down to the field of complex numbers $\mathbb{C}$; the complex conjugation operation will be denoted by an asterisk: $(a+ib)^* = a - ib$ for real a and b. An *inner product space* is a linear space $\mathcal{V}$ over $\mathbb{C}$ equipped with a map, called *inner product*

$$\langle\cdot\,|\,\cdot\rangle : (\mathcal{V} \times \mathcal{V}) \to \mathbb{C} \tag{1.20}$$

that satisfies the following properties:

1. Conjugate symmetry:

$$\forall \mathbf{a}, \mathbf{b} \in \mathcal{V} \quad \langle \mathbf{a} \mid \mathbf{b} \rangle = \langle \mathbf{b} \mid \mathbf{a} \rangle^* \tag{1.21}$$

   from which it follows that $\langle \mathbf{a} \mid \mathbf{a} \rangle$ is real.

2. Linearity in the second argument:

$$\forall \mathbf{a}, \mathbf{b}, \mathbf{c} \in \mathcal{V} \quad \forall \alpha, \beta \in \mathbb{C} \quad \langle \mathbf{a} \mid \alpha\mathbf{b} + \beta\mathbf{c} \rangle = \alpha\langle \mathbf{a} \mid \mathbf{b} \rangle + \beta\langle \mathbf{a} \mid \mathbf{c} \rangle \tag{1.22}$$

   from which it follows that linearity in the first argument involves conjugation:

$$\forall \mathbf{a}, \mathbf{b}, \mathbf{c} \in \mathcal{V} \quad \forall \alpha, \beta \in \mathbb{C} \quad \langle \alpha\mathbf{a} + \beta\mathbf{b} \mid \mathbf{c} \rangle = \alpha^*\langle \mathbf{a} \mid \mathbf{c} \rangle + \beta^*\langle \mathbf{b} \mid \mathbf{c} \rangle \tag{1.23}$$

3. Positive definiteness:

$$\forall \mathbf{a} \in \mathcal{V} \quad \langle \mathbf{a} \mid \mathbf{a} \rangle \geq 0 \tag{1.24}$$

   such that $\langle \mathbf{a} | \mathbf{a} \rangle$ is only zero when $\mathbf{a} = \mathbf{0}$.

The square root of the inner product of a vector with itself is called *norm*:

$$\|\boldsymbol{a}\| = \sqrt{\langle \boldsymbol{a} | \boldsymbol{a} \rangle} \tag{1.25}$$

Inner product spaces are topological (Sect. 1.2) because norm may be used to define neighbourhoods.

1.4.2 Linear Combinations and Basis Sets

A set of elements $\{\mathbf{a}_1, ..., \mathbf{a}_n\}$ of a space $\mathcal{V}$ over a field $\mathbb{F}$ is called *linearly independent* if

$$\alpha_1\mathbf{a}_1 + \cdots + \alpha_n\mathbf{a}_n = \mathbf{0} \tag{1.26}$$

is only true when all $\alpha_k \in \mathbb{F}$ are zero. The largest number of linearly independent elements one can find in a space is called the *dimension* of the space [3]. The dimension of the sum of two subspaces is equal to the sum of their dimensions minus the dimension of their intersection. The dimension of a direct product space is equal to the product of the dimensions of its components.

An element $\mathbf{b} \in \mathcal{V}$ is called a *linear combination* of elements $\{\mathbf{a}_1, ..., \mathbf{a}_n\}$ if such scalars $\{\alpha_1, ..., \alpha_n\}$ may be found in $\mathbb{F}$ that

$$\alpha_1\mathbf{a}_1 + \cdots + \alpha_n\mathbf{a}_n = \mathbf{b} \tag{1.27}$$

The scalars $\{\alpha_1, \ldots, \alpha_n\}$ are called *expansion coefficients* of the element $\mathbf{b}$ in the set $\{\mathbf{a}_1, \ldots, \mathbf{a}_n\}$. Such expansions are unique if and only if $\{\mathbf{a}_1, \ldots, \mathbf{a}_n\}$ are linearly independent.

A set of elements $\{\mathbf{a}_1, \ldots, \mathbf{a}_n\}$ of $\mathcal{V}$ is called a *basis set* of $\mathcal{V}$ if $\{\mathbf{a}_1, \ldots, \mathbf{a}_n\}$ are linearly independent and any element of $\mathcal{V}$ may be expressed as their linear combination. All basis sets of a given space have the same number of elements, this number is equal to the dimension of $\mathcal{V}$. Once a basis set is chosen, any element of the space is uniquely defined by its expansion coefficients in that basis.

If an inner product (Sect. 1.4.1) is defined in $\mathcal{V}$, it may be used to find the expansion coefficients of any element in the specified basis. If $\mathbf{b} \in \mathcal{V}$ and $\{\mathbf{a}_1, \ldots, \mathbf{a}_n\}$ is a basis set of $\mathcal{V}$, then taking an inner product of both sides of Eq. (1.27) with each $\mathbf{a}_k$ yields

$$\underset{k=1}{\overset{n}{\{}}\ \alpha_1\langle\mathbf{a}_k|\mathbf{a}_1\rangle + \cdots + \alpha_n\langle\mathbf{a}_k|\mathbf{a}_n\rangle = \langle\mathbf{a}_k|\mathbf{b}\rangle \tag{1.28}$$

This is a system of n equations for n unknown expansion coefficients $\{\alpha_1, \ldots, \alpha_n\}$; linear independence of $\{\mathbf{a}_1, \ldots, \mathbf{a}_n\}$ guarantees that it always has a unique solution.

Equation (1.28) is simplified if two further conditions are imposed on the basis set: that all elements have a unit norm (*normalisation*), and that pairwise inner products are zero (*orthogonality*). These conditions may be summarised as $\langle\mathbf{a}_k|\mathbf{a}_m\rangle = \delta_{km}$, where δ_{km} is the *Kronecker symbol* (equal to 1 when $k = m$, and to zero otherwise). With this condition in place, most inner products on the left side of Eq. (1.28) disappear:

$$\alpha_k = \langle\mathbf{a}_k|\mathbf{b}\rangle \tag{1.29}$$

The simplicity of Eq. (1.29) makes orthonormal basis sets popular in practical calculations. In particular, explicit expressions for norms become straightforward. For the popular *2-norm* and its scalar product:

$$\langle\mathbf{a}|\mathbf{b}\rangle = \sum_n a_n^* b_n \quad , \quad \|\mathbf{a}\|_2 = \left(\sum_n |a_n|^2\right)^{1/2} \tag{1.30}$$

More generally, the *p-norm* is defined as follows, with a special case at infinity:

$$\|\mathbf{a}\|_p = \left(\sum_k |a_k|^p\right)^{1/p} , \qquad \|\mathbf{a}\|_\infty = \max_k\{|a_k|\} \tag{1.31}$$

The corresponding expressions for function spaces involve integrals:

$$\langle f|g\rangle = \int_{-\infty}^{\infty} f^*(x)g(x)\mathrm{d}x, \quad \|f\|_2 = \left(\int_{-\infty}^{\infty} |f(x)|^2 \mathrm{d}x\right)^{1/2}$$
$$\|f\|_p = \left(\int_{-\infty}^{\infty} |f(x)|^p \mathrm{d}x\right)^{1/p}, \quad \|f\|_\infty = \max_x\{|f(x)|\} \tag{1.32}$$

In the case of multiple discrete and continuous arguments, all continuous arguments are integrated over their domains, and a summation is performed over the indices of all discrete arguments.

1.4.3 Operators and Superoperators

Let $\mathcal{U}$ and $\mathcal{V}$ be spaces over the same field $\mathbb{F}$, and M be such a map from $\mathcal{U}$ to $\mathcal{V}$ that for any $\alpha \in \mathbb{F}$ and any $\mathbf{a}, \mathbf{b} \in \mathcal{U}$ the following is true:

$$M(\alpha\mathbf{a}) = \alpha M(\mathbf{a}), \quad M(\mathbf{a}+\mathbf{b}) = M(\mathbf{a}) + M(\mathbf{b}) \tag{1.33}$$

Then M is called a *linear operator* (or *homomorphism*), $M(\mathbf{a})$ is called the *image* of $\mathbf{a}$ in $\mathcal{V}$, and $\mathbf{a}$ is called a *preimage* of $M(\mathbf{a})$ in $\mathcal{U}$. An image is always unique, but a pre-image need not be.

A linear operator mapping a space into itself is called a *linear transformation* (or *endomorphism*) of that space. Invertible endomorphisms are called *automorphisms.* If $\mathcal{U}$ and $\mathcal{V}$ are spaces over the same field and M is a linear operator mapping $\mathcal{U}$ into $\mathcal{V}$ in a mutually unique and reversible way, then M is called an *isomorphism* and spaces $\mathcal{U}$ and $\mathcal{V}$ are called *isomorphic*:

$$\mathcal{U} \cong \mathcal{V} \tag{1.34}$$

If addition and multiplication by a scalar from the same field are defined for linear operators:

$$[\alpha M](\mathbf{a}) = \alpha M(\mathbf{a}) \quad [M+N](\mathbf{a}) = M(\mathbf{a}) + N(\mathbf{a}) \tag{1.35}$$

the set of all endomorphisms of a space $\mathcal{V}$ becomes a space itself. This space is denoted $\mathrm{End}(\mathcal{V})$ and called *endomorphism space* of $\mathcal{V}$. A *homomorphism space* $\mathrm{Hom}(\mathcal{V}, \mathcal{W})$ from a space $\mathcal{V}$ to a space $\mathcal{W}$ is defined in a similar way. Automorphisms cannot be made a space because the sum of two invertible maps is not necessarily invertible.

To determine the dimension of End($\mathcal{V}$), consider the action of its elements on a basis set $\{\mathbf{a}_1, ..., \mathbf{a}_n\}$ of $\mathcal{V}$. For every element $\mathbf{a}_k$, its image must itself have an expansion in the same basis set:

$$\left\{ M(\mathbf{a}_k) = \sum_{m=1}^{n} \mu_{km}\mathbf{a}_m \right._{k=1}^{n} \tag{1.36}$$

and the knowledge of the coefficients μ_{km} defines the map completely, because

$$\forall\{\alpha_1, ..., \alpha_n\} \in \mathbb{F} \quad M\left(\sum_{k=1}^{n} \alpha_k \mathbf{a}_k\right) = \sum_{k=1}^{n} \alpha_k M(\mathbf{a}_k) = \sum_{k,m=1}^{n} \alpha_k \mu_{km}\mathbf{a}_m \tag{1.37}$$

There are n^2 of those independent coefficients, and all relations are linear. Therefore, the dimension of End($\mathcal{V}$) is n^2. Later in this book, we will also encounter the n^4-dimensional space End(End($\mathcal{V}$))—the space of *linear superoperators* that transform linear operators that transform $\mathcal{V}$.

1.4.4 Representations of Linear Spaces

Linear spaces of the same dimension over the same field are isomorphic. This means that all n-dimensional linear spaces over a field $\mathbb{F}$ are isomorphic to $\mathbb{F}^n$—we can abstract from the specific nature of the elements of any particular linear space and work instead with their images in $\mathbb{F}^n$, whose elements are vectors. This is an example of a *representation*—a map from one mathematical structure to another that preserves some properties and relations between objects.

When vector representations of linear spaces are used, homomorphisms are represented by matrices. If $A = \{\mathbf{a}_1, ..., \mathbf{a}_N\}$ and $B = \{\mathbf{b}_1, ..., \mathbf{b}_K\}$ are basis sets of spaces $\mathcal{A}$ and $\mathcal{B}$, and $M : \mathcal{A} \to \mathcal{B}$ is a linear operator, then a matrix $\mathbf{M}$ is called a *matrix representation* of M in the pair of basis sets A and B, if

$$\forall \mathbf{a} \in \mathcal{A} \quad \forall \mathbf{b} \in \mathcal{B} \quad M(\mathbf{a}) = \mathbf{b} \quad \Rightarrow \quad \mathbf{M}\boldsymbol{\alpha} = \boldsymbol{\beta} \tag{1.38}$$

where $\boldsymbol{\alpha}$ is the column of expansion coefficients of $\mathbf{a}$ in the basis A, and $\boldsymbol{\beta}$ is the column of expansion coefficients of $\mathbf{b}$ in the basis B:

$$\mathbf{a} = \sum_{n=1}^{N} \alpha_n \mathbf{a}_n \quad , \quad \mathbf{b} = \sum_{k=1}^{K} \beta_k \mathbf{b}_k \tag{1.39}$$

When written out explicitly, Eq. (1.38) becomes

$$\mathbf{M}\boldsymbol{\alpha} = \boldsymbol{\beta} \quad \Leftrightarrow \quad \sum_{n=1}^{N} \mu_{kn}\alpha_n = \beta_k \tag{1.40}$$

where the scalars μ_{kn} are called *matrix elements* of M in the pair of basis sets A and B. When the basis sets are orthonormal, the expression is simple:

$$\mu_{kn} = \langle \mathbf{b}_k | M(\mathbf{a}_n) \rangle \tag{1.41}$$

A particular representation often used in quantum mechanics connects spaces of square-integrable complex functions of real arguments and spaces of complex vectors [4]:

$$f(x) = \sum_{n=1}^{\infty} \alpha_n g_n(x) \quad \Leftrightarrow \quad |f\rangle = \begin{pmatrix} \alpha_1 \\ \alpha_2 \\ \vdots \end{pmatrix}, \quad \alpha_n = \langle g_n(x) | f(x) \rangle \tag{1.42}$$

where $\{g_n(x)\}$ is an orthonormal (with respect to the 2-norm) basis set and x need not be a single argument. This representation maps linear integrodifferential operators into matrices, and therefore maps linear integrodifferential equations into linear algebraic equations that are easier to solve.

1.4.5 Operator Norms and Inner Products

Inner products and norms on linear spaces $\mathcal{A}$ and $\mathcal{B}$ may be used to define an *induced norm* and *induced inner product* on the corresponding homomorphism space $\mathrm{Hom}(\mathcal{A}, \mathcal{B})$:

$$\|M\| = \sup\left\{ \frac{\|M(\mathbf{a})\|_{\mathcal{B}}}{\|\mathbf{a}\|_{\mathcal{A}}} \middle| \mathbf{a} \in \mathcal{A},\ \|\mathbf{a}\|_{\mathcal{A}} \neq 0 \right\} \tag{1.43}$$

The following induced norms are popular for matrix representations of homomorphisms:

$$\begin{aligned} \|\mathbf{M}\|_1 &= \max_k \sum_n |\mu_{nk}| && \text{(maximum absolute column sum)} \\ \|\mathbf{M}\|_2 &= \sigma_{\max}(\mathbf{M}) && \text{(maximum singular value)} \\ \|\mathbf{M}\|_\infty &= \max_n \sum_k |\mu_{nk}| && \text{(maximum absolute row sum)} \end{aligned} \tag{1.44}$$

In the physical sciences context, the 2-norm is usually the quantity of interest, but singular values are computationally expensive. Other norms provide cheaper bounds, for example:

$$\|A\|_2 \leq \sqrt{\|A\|_1 \|A\|_\infty} \tag{1.45}$$

Treating the matrices as Cartesian vectors with two indices yields the *Frobenius metric*:

$$\langle \mathbf{M}|\mathbf{K}\rangle_{\mathrm{F}} = \sum_{pq} \mu_{pq}^* \kappa_{pq} = \mathrm{Tr}\left(\mathbf{M}^\dagger \mathbf{K}\right), \quad \|\mathbf{M}\|_{\mathrm{F}} = \sqrt{\sum_{pq} |\mu_{pq}|^2} \tag{1.46}$$

where μ_{pq} and κ_{pq} are the elements of matrices **M** and **K**, respectively [5].

1.4.6 Representing Matrices with Vectors

In the theories of spin dynamics described later in this book, it is common to see vector representations of wavefunctions transformed by matrix representations of linear operators: a complex vector space $\mathbb{C}^N$ and a matrix space $\mathrm{End}(\mathbb{C}^N)$ of operators acting on it. In spin physics, this structure is colloquially called *Hilbert space formalism*. When the equations of motion are generalised to describe spin system ensembles, we would also encounter linear maps between matrix spaces; such maps are called *superoperators* and the picture is called *Liouville space formalism* [6].

Numerical implementations of Liouville space formalism are simplified when the matrix space $\mathrm{End}(\mathbb{C}^N)$ is treated as a vector space of dimension N^2, and the superoperator space $\mathrm{End}(\mathrm{End}(\mathbb{C}^N))$ as a space of $N^2 \times N^2$ matrices:

$$\mathrm{End}(\mathbb{C}^N) \cong \mathbb{C}^{N^2}, \quad \mathrm{End}(\mathrm{End}(\mathbb{C}^N)) \cong \mathrm{End}\left(\mathbb{C}^{N^2}\right) \tag{1.47}$$

In practical programming, each operator matrix **M** is stretched column-wise into a vector **m**. A tedious index transformation demonstrates that a matrix product **AMB** is then mapped into $[\mathbf{B}^{\mathrm{T}} \otimes \mathbf{A}]\mathbf{m}$. This defines the isomorphism in Eq. (1.47) because the action by a linear superoperator $\mathcal{B} \in \mathrm{End}(\mathrm{End}(\mathbb{C}^N))$ on any matrix $\mathbf{M} \in \mathrm{End}(\mathbb{C}^N)$ can be written as a linear combination of two-sided matrix products:

$$\mathcal{B}(\mathbf{M}) = \sum_{nk} \alpha_{nk} \mathbf{B}_n \mathbf{M} \mathbf{B}_k, \quad \mathbf{B}_n, \mathbf{B}_k \in \mathrm{End}(\mathbb{C}^N), \quad \alpha_{nk} \in \mathbb{C} \tag{1.48}$$

where the set $\{\mathbf{B}_n\}$ is a basis of $\mathrm{End}(\mathbb{C}^N)$. In particular, for commutation superoperators:

$$[\mathbf{A}, \mathbf{M}] = \mathbf{AM} - \mathbf{MA} \quad \rightarrow \quad [\mathbf{1} \otimes \mathbf{A} - \mathbf{A}^{\mathrm{T}} \otimes \mathbf{1}]\mathbf{m} \tag{1.49}$$

where **1** is a unit matrix of the same dimension as **A**, and the transpose *does not* become a Hermitian conjugate for complex matrices.

1.5 Groups and Algebras

A set $\mathcal{G}$ with an associative binary product is called a *semigroup* if it is closed under that product:

$$\forall g, h \in \mathcal{G} \quad gh \in \mathcal{G} \tag{1.50}$$

If additionally, the following relations hold in $\mathcal{G}$:

1. $\exists! e \in \mathcal{G}, \quad \forall g \in \mathcal{G} \quad eg = g$ (there exists a unique unit element)
2. $\forall g \in \mathcal{G} \quad \exists! g^{-1} \in \mathcal{G}, \quad gg^{-1} = e$ (for every element there exists a unique inverse)

then the set is called a *group*. The elements of $\mathcal{G}$ need not commute, but if they do, the group is called *Abelian*. The number of elements in the group is called the *order of the group* and denoted $|\mathcal{G}|$. A subset of $\mathcal{G}$ which is itself a group is called a *subgroup* of $\mathcal{G}$. A subgroup $\mathcal{N}$ is called a *normal subgroup* of $\mathcal{G}$ if $\mathcal{N}$ is invariant under the similarity transformation by the elements of $\mathcal{G}$:

$$\mathcal{N} \triangleleft \mathcal{G} \quad \Leftrightarrow \quad \forall n \in \mathcal{N} \quad \forall g \in \mathcal{G} \quad gng^{-1} \in \mathcal{N} \tag{1.51}$$

Groups $\mathcal{G}$ and $\mathcal{H}$ are called *isomorphic* if there is a one-to-one correspondence $\varphi : \mathcal{G} \to \mathcal{H}$ such that

$$\forall g_1, g_2 \in \mathcal{G} \quad \varphi(g_1 g_2) = \varphi(g_1)\varphi(g_2) \tag{1.52}$$

where the multiplication $g_1 g_2$ takes place in $\mathcal{G}$ and the multiplication $\varphi(g_1)\varphi(g_2)$ takes place in $\mathcal{H}$.

The general notion of group direct product is too complicated for our purposes here, but if $\mathcal{G}$ and $\mathcal{H}$ are normal subgroups of the same group, then their *direct product* is defined as

$$\mathcal{G} \times \mathcal{H} = \{gh | g \in \mathcal{G}, h \in \mathcal{H}\} \tag{1.53}$$

1.5.1 Finite, Discrete, and Continuous Groups

When the elements of a group are operators acting on a normed space $\mathcal{V}$, Eq. (1.43) may be used to establish an induced norm for the elements of the group, and thus to establish a topology. For a given element $g \in \mathcal{G} \subseteq \mathrm{End}(\mathcal{V})$ and any positive $\varepsilon \in \mathbb{R}$, the set of all elements $h \in \mathcal{G}$ such that

$$\|g - h\| < \varepsilon \tag{1.54}$$

Table 1.1 Examples of physical systems conforming to groups of different topologies

Topology	Physical symmetry groups
Finite, discrete	Molecular symmetry groups in chemistry: C_{2v} (water), D_{6h} (benzene), T_d (methane), etc.
Infinite, discrete	Crystallographic groups: $Fd\bar{3}m$ (diamond), P4mm (lead titanate), etc.
Infinite, continuous	Translation group, rotation group, time propagation group, Lorentz group, etc.

is called the *ε-neighbourhood* of g. If every element of the group $\mathcal{G}$ has an ε-neighbourhood that does not contain other elements of $\mathcal{G}$, the group $\mathcal{G}$ is called *discrete*. If a group has a finite number of elements, it is called a *finite* group; otherwise, it is called *infinite* (Table 1.1).

A group $\mathcal{G}$ is called a *continuous* or *Lie group* [7] if it is also a differentiable manifold: if there exists a function $g(t_1, \ldots, t_k)$ of k variables that is continuous within the $|t_i| < \varepsilon$ cube, defines all elements of $\mathcal{G}$ within the ε-neighbourhood of g, and has different values for different parameters $(t_1, \ldots, t_k)$. Such a function is called a *parametrisation* of the group around g. The number of variables k is specific to each group; it is called the *dimension* of the group, and the group itself is called *k-parametric*.

1.5.2 Conjugacy Classes and Centres

Two elements $a, b \in \mathcal{G}$ are called *conjugate* if there is an element $g \in \mathcal{G}$ such that $gag^{-1} = b$. The set of all elements of the form gag^{-1} with g running over the group is called the *conjugacy class* of a:

$$\mathrm{Cl}(a) = \{gag^{-1} | g \in \mathcal{G}\} \tag{1.55}$$

Each element of a group can only belong to one conjugacy class. Two conjugacy classes of a group are either identical or disjoint—groups are *partitioned* into conjugacy classes. The identity element always forms its own class, and the number of classes in Abelian groups is equal to the order of the group. The number of elements in each conjugacy class is a divisor of the group order (Table 1.2).

Table 1.2 Centres and conjugacy classes of common point symmetry groups of molecules

Molecular symmetry group	Conjugacy classes	Centre
C_{3v} (ammonia molecule)	Rotations, reflections, identity	Identity operator
T_d (methane molecule)	120-degree rotations, 180-Degree rotations, 90-degree rotation + inversion, reflections	Identity operator
C_{2v} (water molecule)	Identity, 180-degree rotation, XZ reflection, YZ reflection	Whole group

The *centre* $Z(\mathcal{G})$ of a group $\mathcal{G}$ is the set of elements that commute with every element of $\mathcal{G}$:

$$Z(\mathcal{G}) = \{z \in \mathcal{G} | \forall g \in \mathcal{G},\, zg = gz\} \tag{1.56}$$

The centre is an Abelian subgroup of $\mathcal{G}$. A group with $Z(\mathcal{G}) = \{e\}$ is called *centreless* . It is important to remember that additive operations are not defined on groups. Unless further structure is implied (for example, group action on a space) the definition of commuting elements is strictly as in Eq. (1.56).

1.5.3 Group Actions, Orbits, and Stabilisers

If $\mathcal{G}$ is a group of operators acting on a set $\mathcal{A}$, and the set is closed with respect to that action, then the map $\mathcal{G} : \mathcal{A} \to \mathcal{A}$ is called a *group action* by $\mathcal{G}$ on $\mathcal{A}$. A subset $\mathcal{B} \subseteq \mathcal{A}$ is called *invariant* under the action by $\mathcal{G}$ if no element is taken outside of $\mathcal{B}$ by the action:

$$\forall b \in \mathcal{B} \quad \forall g \in \mathcal{G} \qquad gb \in B \tag{1.57}$$

and *fixed* under the action by $\mathcal{G}$ if all elements are mapped into themselves:

$$\forall b \in \mathcal{B} \quad \forall g \in \mathcal{G} \qquad gb = b \tag{1.58}$$

The *group orbit* $\mathcal{G}a$ of an element $a \in \mathcal{A}$ is the subset of $\mathcal{A}$ spanned by the group action on a:

$$\mathcal{G}a = \{ga | g \in \mathcal{G}\} \tag{1.59}$$

The set of orbits of all elements of $\mathcal{A}$ is a *partition* of $\mathcal{A}$—distinct orbits do not intersect. If $\mathcal{A}$ is a space, then different orbits belong to non-intersecting subspaces of $\mathcal{A}$. A common physical example is time evolution (action by the time propagation group on the initial condition) in the presence of a conservation law: trajectories corresponding to different values of the conserved quantity do not intersect.

The subset of $\mathcal{G}$ leaving a particular element $a \in \mathcal{A}$ unchanged is a subgroup of $\mathcal{G}$ called the *stabiliser subgroup* $\mathcal{G}_a$ of a:

$$\mathcal{G}_a = \{g \in \mathcal{G} | ga = a\} \tag{1.60}$$

The group of rotations in $\mathbb{R}^3$ is a good example: the subgroup of rotations around the Z-axis is the stabiliser subgroup of the vectors directed along that axis. If $\mathcal{G}_a = \{e\}$ for all $a \in \mathcal{A}$, the group action $\mathcal{G} : \mathcal{A} \to \mathcal{A}$ is called *free*, an example is the translation group in one dimension.

1.5.4 Matrix Representations of Groups

The set of all invertible $N \times N$ matrices acting on an N-dimensional vector space is a group under matrix multiplication; this group is called the *general linear group* and denoted $\mathrm{GL}(N)$. A map $\mathbf{P} : \mathcal{G} \to \mathrm{GL}(N)$ is called an N-dimensional *matrix representation* of $\mathcal{G}$ if the map preserves the group operation:

$$\forall g, h \in \mathcal{G} \quad \mathbf{P}(gh) = \mathbf{P}(g)\mathbf{P}(h) \tag{1.61}$$

and the unit operator in $\mathcal{G}$ is mapped into the unit matrix in $\mathrm{GL}(N)$. Matrix representations can be:

- *Faithful:* when different elements of $\mathcal{G}$ are mapped into different elements of $\mathrm{GL}(N)$. An example of a faithful representation is the *regular representation*, which uses the fact that $\mathcal{G}$ is able to act on itself by multiplication. If we define $\Delta_{mk}^{(n)}$ such that $g_n g_m = \Delta_{mk}^{(n)} g_k$, then the matrix corresponding to the regular representation of g_n is $[\mathbf{P}(g_n)]_{mk} = \Delta_{mk}^{(n)}$. The dimension of the regular representation is equal to the order of the group.
- *Unfaithful:* a representation in which two or more elements of $\mathcal{G}$ are mapped into the same element of $\mathrm{GL}(N)$. Properties and relations between elements may be changed as a result. Examples of lossy representations are *scalar representation* where each $g \in \mathcal{G}$ is mapped into the determinant of its faithful representation, and *trivial representation*, in which all elements of $\mathcal{G}$ are mapped into the unit matrix.

Two representations $\mathbf{P}$ and $\mathbf{Q}$ of the same group $\mathcal{G}$ are called *equivalent*, if they have the same dimension, and there is a similarity transformation taking one into the other:

$$\exists \mathbf{S} \in \mathrm{GL}(n), \quad \forall g \in \mathcal{G} \quad \mathbf{Q}(g) = \mathbf{S}\mathbf{P}(g)\mathbf{S}^{-1} \tag{1.62}$$

All representations of finite groups are equivalent to representations with unitary matrices, those are called *unitary representations*. Such representations are preferred in practical work because the orthogonality theorems discussed in the next section become particularly simple.

Let $\mathbf{P}$ and $\mathbf{Q}$ be representations of $\mathcal{G}$ by matrices of dimension N and K, respectively. Because

$$\begin{gathered} \forall \mathbf{A}, \mathbf{C} \in \mathrm{GL}(N) \quad \forall \mathbf{B}, \mathbf{D} \in \mathrm{GL}(K) \\ (\mathbf{A} \oplus \mathbf{B})(\mathbf{C} \oplus \mathbf{D}) = (\mathbf{AC}) \oplus (\mathbf{BD}) \end{gathered} \tag{1.63}$$

matrices of the form $\mathbf{P}(g) \oplus \mathbf{Q}(g)$ are also a representation of $\mathcal{G}$, called the *direct sum of representations* $\mathbf{P}$ and $\mathbf{Q}$. A similar property exists for direct products:

$$\begin{aligned} &\forall \mathbf{A}, \mathbf{C} \in \mathrm{GL}(N) \quad \forall \mathbf{B}, \mathbf{D} \in \mathrm{GL}(K) \\ &(\mathbf{A} \otimes \mathbf{B})(\mathbf{C} \otimes \mathbf{D}) = (\mathbf{AC}) \otimes (\mathbf{BD}) \end{aligned} \tag{1.64}$$

and therefore matrices of the form $\mathbf{P}(g) \otimes \mathbf{Q}(g)$ are also a representation of $\mathcal{G}$, called the *direct product of representations* $\mathbf{P}$ and $\mathbf{Q}$.

A matrix representation $\mathbf{P}$ of a group $\mathcal{G}$ is called *reducible* if it can be cast by a similarity transformation into the following form:

$$\mathbf{P}(g) = \begin{pmatrix} \mathbf{P}_1(g) & \mathbf{X}(g) \\ \mathbf{0} & \mathbf{P}_2(g) \end{pmatrix} \quad \forall g \in \mathcal{G} \tag{1.65}$$

where $\mathbf{P}_1(g)$, $\mathbf{P}_2(g)$ and $\mathbf{X}(g)$ are blocks of the matrix $\mathbf{P}(g)$. A representation is *fully reducible* if it can be transformed into a direct sum of representations of lower dimension:

$$\mathbf{P}(g) = \begin{pmatrix} \mathbf{P}_1(g) & \mathbf{0} & \mathbf{0} \\ \mathbf{0} & \mathbf{P}_2(g) & \mathbf{0} \\ \mathbf{0} & \mathbf{0} & \cdots \end{pmatrix} = \mathbf{P}_1(g) \oplus \mathbf{P}_2(g) \oplus \ldots \tag{1.66}$$

If neither transformation is possible, the representation $\mathbf{P}$ is called *irreducible.* Any representation of a finite group is either irreducible or fully reducible. The number of unique (up to a similarity transformation) irreducible representations of a finite group is equal to the number of conjugacy classes. For infinite groups, if a representation is unitary and reducible, then it is fully reducible.

1.5.5 Orthogonality Theorems and Characters

Matrix elements of irreducible representations satisfy an orthogonality condition called the *great orthogonality theorem*: if $\mathbf{P}_a$ and $\mathbf{P}_b$ are non-equivalent unitary irreps of a group $\mathcal{G}$, then [8]:

$$\frac{\sqrt{N_a N_b}}{|\mathcal{G}|} \sum_{g \in \mathcal{G}} [\mathbf{P}_a(g)]^*_{jk} [\mathbf{P}_b(g)]_{lm} = \delta_{ab} \delta_{jl} \delta_{km} \tag{1.67}$$

where $|\mathcal{G}|$ is the order of the group, and $N_{a,b}$ are the dimensions of the two irreps. Loosely speaking, if the matrices of unitary irreps are stacked like decks of cards on a horizontal table, different vertical columns are orthogonal within each stack, and between the stacks corresponding to non-equivalent irreps. In the case of continuous groups, the sum in Eq. (1.67) becomes an integral.

The requirement for the irreps to be non-equivalent is illustrated by matrix traces, which are invariant under similarity transformations:

$$\mathrm{Tr}\left[\mathbf{SP}(g)\mathbf{S}^{-1}\right] = \mathrm{Tr}\left[\mathbf{S}^{-1}\mathbf{SP}(g)\right] = \mathrm{Tr}[\mathbf{P}(g)] \tag{1.68}$$

Table 1.3 Character table for the permutation group of three identical objects, for example, the three hydrogen atoms in the ammonia molecule

C_{3v} (chem)		E	$C_3(120°)$	$C_3(240°)$	σ_1	σ_2	σ_3
S_3 (math)		[1 2 3]	[3 1 2]	[2 3 1]	[2 1 3]	[1 3 2]	[3 2 1]
	A_1	1	1	1	1	1	1
Irreps	A_2	1	1	1	–1	–1	–1
	E	2	–1	–1	0	0	0
Classes		Identity	Rotations		Reflections/swaps		

where we have used the fact that the trace of a matrix product is invariant under cyclic permutations of the matrices. This trace is called the *character* of the matrix representation $\mathbf{P}(g)$ of an element g:

$$\chi_{\mathbf{P}}(g) = \mathrm{Tr}[\mathbf{P}(g)] \tag{1.69}$$

As Eq. (1.68) demonstrates, equivalent representations of the same group element have equal characters.

An example of the *character table* of a group, listing all non-equivalent irreps, is given in Table 1.3 for the S_3 permutation group, known to chemists as C_{3v}—the symmetry group of the ammonia molecule.

It also follows from Eq. (1.68) and the definition of conjugacy class (Sect. 1.5.2) that representations of all elements in the same class have the same character—this is visible in Table 1.3. After computing the traces of the matrices appearing in the great orthogonality theorem:

$$\sum_{jklm}\left(\frac{\sqrt{N_a N_b}}{|\mathcal{G}|}\sum_{g\in\mathcal{G}}[\mathbf{P}_a(g)]^*_{jk}[\mathbf{P}_b(g)]_{lm}\right)\delta_{jk}\delta_{lm} = \sum_{jklm}\delta_{ab}\delta_{jl}\delta_{km}\delta_{jk}\delta_{lm} \tag{1.70}$$

and simplifying, we conclude that character strings of non-equivalent irreps are orthogonal as vectors:

$$\frac{1}{|\mathcal{G}|}\sum_{g\in\mathcal{G}}\chi^*_{\mathbf{P}_a}(g)\chi_{\mathbf{P}_b}(g) = \delta_{ab} \tag{1.71}$$

This is the *little orthogonality theorem*, it is also evident in Table 1.3.

1.5.6 Algebras and Lie Algebras

A space $\mathfrak{a}$ over a field $\mathbb{F}$ becomes an *algebra* when a binary product $[\cdot,\cdot]:(\mathfrak{a}\times\mathfrak{a}) \to \mathfrak{a}$ is defined such that

1. $\forall \mathbf{A}, \mathbf{B} \in \mathfrak{a} \quad [\mathbf{A}, \mathbf{B}] \in \mathfrak{a}$ (the space is closed under the product)
2. $\forall \mathbf{A}, \mathbf{B}, \mathbf{C} \in \mathfrak{a} \quad [\mathbf{A}+\mathbf{B}, \mathbf{C}] = [\mathbf{A}, \mathbf{C}] + [\mathbf{B}, \mathbf{C}]$ (the product is left-distributive)
3. $\forall \mathbf{A}, \mathbf{B}, \mathbf{C} \in \mathfrak{a} \quad [\mathbf{A}, \mathbf{B}+\mathbf{C}] = [\mathbf{A}, \mathbf{B}] + [\mathbf{A}, \mathbf{C}]$ (the product is right-distributive)
4. $\forall \mathbf{A}, \mathbf{B} \in \mathfrak{a} \quad \forall \alpha, \beta \in \mathbb{F} \quad [\alpha\mathbf{A}, \beta\mathbf{B}] = \alpha\beta[\mathbf{A}, \mathbf{B}]$ (multiplication by scalars is associative)

where the notation for the product (square bracket) has been adapted to our future use of this formalism for matrices. A subspace of $\mathfrak{a}$ that is itself an algebra is called a *subalgebra* of $\mathfrak{a}$.

As we saw in Sect. 1.4, in a given basis set $\{\mathbf{V}_k\}$ of $\mathfrak{a}$, any element has a unique expansion:

$$\mathbf{A} = \sum_n \alpha_n \mathbf{V}_n, \qquad \mathbf{B} = \sum_k \beta_k \mathbf{V}_k \tag{1.72}$$

In particular, this must hold for products of basis elements; this implies that products of basis elements are themselves linear combinations of basis elements:

$$[\mathbf{V}_n, \mathbf{V}_k] = \sum_m c_{nk}^m \mathbf{V}_m \tag{1.73}$$

where c_{nk}^m are called *structure coefficients* or *structure constants*. They are useful because they connect the expansion coefficients of operands to the expansion coefficients of the product:

$$\begin{aligned} [\mathbf{A}, \mathbf{B}] &= \sum_{nk} \alpha_n \beta_k [\mathbf{V}_n, \mathbf{V}_k] = \sum_{nkm} \alpha_n \beta_k c_{nk}^m \mathbf{V}_m \\ &= \sum_m \gamma_m \mathbf{V}_m, \quad \gamma_m = \sum_{nk} \alpha_n c_{nk}^m \beta_k \end{aligned} \tag{1.74}$$

Algebras occur whenever linear combinations and products are simultaneously defined. When the properties of the algebra are narrowed down to only allow a specific kind of binary product, called *Lie bracket* (not necessarily a commutator), with the following properties:

5. $\forall \mathbf{A}, \mathbf{B} \in \mathfrak{a} \quad [\mathbf{A}, \mathbf{B}] = -[\mathbf{B}, \mathbf{A}]$ (product is antisymmetric)
6. $\forall \mathbf{A}, \mathbf{B}, \mathbf{C} \in \mathfrak{a} \quad [\mathbf{A}, [\mathbf{B}, \mathbf{C}]] + [\mathbf{B}, [\mathbf{C}, \mathbf{A}]] + [\mathbf{C}, [\mathbf{A}, \mathbf{B}]] = 0$ (*Jacobi identity* holds)

the algebra is called a *Lie algebra*, as distinct from an *associative algebra*, where instead:

$$\forall \mathbf{A}, \mathbf{B}, \mathbf{C} \in \mathfrak{a} \quad [\mathbf{A}, [\mathbf{B}, \mathbf{C}]] = [[\mathbf{A}, \mathbf{B}], \mathbf{C}] \tag{1.75}$$

An associative algebra with a multiplicative unit element is called *unital*.

1.5.7 Exponential and Tangent Maps

For any Lie algebra $\mathfrak{a}$, the set of exponentials of its elements is a group because:

1. The product of two exponentials of elements of $\mathfrak{a}$ is an exponential of another element of $\mathfrak{a}$ (*Baker-Campbell-Hausdorff formula* [9–11], presented here without proof):

$$\forall \mathbf{A}, \mathbf{B} \in \mathfrak{a} \quad \exp(\mathbf{A}) \exp(\mathbf{B}) = \exp\left(\mathbf{A} + \mathbf{B} + \frac{1}{2}[\mathbf{A}, \mathbf{B}] + \cdots\right) \tag{1.76}$$

   where the exponential on the right contains a linear combination of nested commutators of **A** and **B**, which must be an element of $\mathfrak{a}$. Although the *operator exponential* itself is sometimes defined using the associative product:

$$\exp(\mathbf{A}) = \sum_{n=1}^{\infty} \frac{\mathbf{A}^n}{n!} \tag{1.77}$$

   only Lie brackets actually occur in Eq (1.76) which we here require.

2. The zero element of $\mathfrak{a}$ is exponentiated into the unit element of the group:

$$\forall \mathbf{A} \in \mathfrak{a} \quad \exp(\mathbf{0}) \exp(\mathbf{A}) = \exp(\mathbf{A}) \exp(\mathbf{0}) = \exp(\mathbf{A}) \tag{1.78}$$

3. The exponential of each element of $\mathfrak{a}$ has a multiplicative inverse:

$$\forall \mathbf{A} \in \mathfrak{a} \quad \exists!(-\mathbf{A}) \in \mathfrak{a}, \quad \exp(\mathbf{A}) \exp(-\mathbf{A}) = \exp(\mathbf{0}) \tag{1.79}$$

This relationship between the Lie algebra and the corresponding Lie group is called the *exponential map*. The elements of the basis set of $\mathfrak{a}$ are called *generators* of that group; the number of linearly independent generators is called the *dimension* of the Lie algebra [7]. If the algebra $\mathfrak{a}$ is n-dimensional, the resulting group $\exp(\mathfrak{a})$ is n-parametric.

Given a basis set $\{\mathbf{V}_k\}$ of an n-dimensional algebra $\mathfrak{a}$ and parameters $\boldsymbol{\alpha} \in \mathbb{F}^n$, the group element generated by the exponential map is

$$\mathbf{G}(\boldsymbol{\alpha}) = \exp[\alpha_1 \mathbf{V}_1 + \alpha_2 \mathbf{V}_2 + \cdots] \tag{1.80}$$

When only $\mathbf{G}(\boldsymbol{\alpha})$ is available, the generators $\{\mathbf{V}_k\}$ may be recovered by taking partial derivatives at the unit element of the group:

$$\left.\frac{\partial \mathbf{G}}{\partial \alpha_k}\right|_{\boldsymbol{\alpha}=0} = \lim_{\alpha_k \to 0} \frac{\exp[\alpha_k \mathbf{V}_k] - \exp[\mathbf{0}]}{\alpha_k} = \lim_{\alpha_k \to 0} \frac{\mathbf{1} + \alpha_k \mathbf{V}_k + O(\alpha_k^2) - \mathbf{1}}{\alpha_k} = \mathbf{V}_k \tag{1.81}$$

This is called *tangent map* because the set of such derivatives spans the tangent hyperplane to $\mathbf{G}(\boldsymbol{\alpha})$ at $\boldsymbol{\alpha} = 0$. It also spans the parent space of the original Lie algebra—we can therefore conclude that the Lie algebra generating a Lie group is its *tangent space at identity* [7].

1.5.8 Ideals, Simple and Semisimple Algebras

An *ideal* of a Lie algebra $\mathfrak{a}$ is a subalgebra $\mathfrak{h} \subseteq \mathfrak{a}$ such that $[\mathfrak{a}, \mathfrak{h}] \subseteq \mathfrak{h}$. It follows that the orthogonal complement $\mathfrak{h}^{\perp}$ of $\mathfrak{h}$ in $\mathfrak{a}$ is also an ideal, and therefore $\mathfrak{a} = \mathfrak{h} \oplus \mathfrak{h}^{\perp}$, where the two components of the direct sum are sub-algebras of $\mathfrak{a}$. This may be repeated until there are no non-trivial ideals left, with the conclusion that the algebra splits into a direct sum of its non-trivial ideals. The algebras that cannot be split further are called *simple*. The algebras that are direct sums of simple algebras are called *semisimple*. These designations are inherited by the corresponding Lie groups: a *simple group* is the group whose only normal subgroups (those invariant under conjugation by members of the group) are the trivial subgroup and the group itself [12].

1.5.9 Matrix Representations of Lie Algebras

Let $\mathfrak{a}$ be a Lie algebra and $\mathcal{V}$ a vector space over the same field. A *matrix representation* $\mathbf{P}$ of $\mathfrak{a}$ on $\mathcal{V}$ is a linear map $\mathbf{P} : \mathfrak{a} \to \mathrm{End}(\mathcal{V})$, such that

$$\forall \mathbf{A}, \mathbf{B} \in \mathfrak{a} \qquad \mathbf{P}([\mathbf{A}, \mathbf{B}]) = [\mathbf{P}(\mathbf{A}), \mathbf{P}(\mathbf{B})] \tag{1.82}$$

Faithful and irreducible representations of Lie algebras are defined in the same way as they are for groups (Sect. 1.5.4): a faithful representation maps different elements of $\mathfrak{a}$ into different elements of $\mathrm{End}(\mathcal{V})$, and an irreducible representation is one that cannot be brought into a block-diagonal form by a similarity transformation.

A particular choice of $\mathcal{V}$ is the parent space of the Lie algebra itself, on which it acts by Lie bracket:

$$\forall \mathbf{A}, \mathbf{B} \in \mathfrak{a} \qquad \mathrm{ad}_{\mathbf{A}}(\mathbf{B}) = [\mathbf{A}, \mathbf{B}] \tag{1.83}$$

This action is called *adjoint endomorphism*. It follows from the definition of the Lie bracket that the map $\mathbf{A} \to \mathrm{ad}_{\mathbf{A}}$ is linear, and that the set of elements $\{\mathrm{ad}_{\mathbf{A}} | \mathbf{A} \in \mathfrak{a}\}$ is a Lie algebra itself:

$$\forall \mathbf{A}, \mathbf{B}, \mathbf{C} \in \mathfrak{a} \qquad [\mathrm{ad}_{\mathbf{A}}, \mathrm{ad}_{\mathbf{B}}](\mathbf{C}) = \mathrm{ad}_{[\mathbf{A},\mathbf{B}]}(\mathbf{C}) \tag{1.84}$$

This construction may be used to build a faithful matrix representation of $\mathfrak{a}$, called *adjoint representation*. In a given basis set $\{\mathbf{V}_k\}$, the adjoint action by one basis element on another is determined by the structure coefficients in Eq. (1.73), and thus the matrix elements are

$$\mathrm{ad}_{\mathbf{V}_n}(\mathbf{V}_k) = \sum_m c_{nk}^m \mathbf{V}_m \quad \Rightarrow \quad [\mathbf{P}(\mathrm{ad}_{\mathbf{V}_n})]_{mk} = c_{nk}^m \tag{1.85}$$

For simple Lie algebras, the adjoint representation is irreducible because subrepresentations of $\mathrm{ad}_{\mathfrak{a}}$ would correspond to the ideals of $\mathfrak{a}$, and simple algebras have no non-trivial ideals [12].

1.5.10 Envelopes, Complexifications, and Covers

Although it is possible to formulate some physical theories purely in terms of Lie brackets, ignoring the associative product on the algebras of physical operators is unreasonable. Lie algebras in physics, therefore, come embedded into associative algebras, and their bracket is the *commutator*:

$$[\mathbf{A}, \mathbf{B}] = \mathbf{AB} - \mathbf{BA} \tag{1.86}$$

It is also desirable to have Hermitian representations—observable operators in quantum mechanics are Hermitian. However, a commutator of two Hermitian operators is not, and so physicists define their Lie algebras as closed with respect to $i[\mathbf{A}, \mathbf{B}]$ product, which is Hermitian for Hermitian operands. These decorations lead to the following definitions:

1. A *universal enveloping algebra* of a Lie algebra $\mathfrak{g}$ over a field $\mathbb{F}$ is a pair $\{\mathfrak{U}(\mathfrak{g}), M\}$, such that
   (a) $\mathfrak{U}(\mathfrak{g})$ is a unital associative algebra over $\mathbb{F}$;
   (b) the map $M : \mathfrak{g} \to \mathfrak{U}(\mathfrak{g})$ is linear and preserves the Lie bracket relation
   $$[M(\mathbf{A}), M(\mathbf{B})] = M([\mathbf{A}, \mathbf{B}]); \tag{1.87}$$
   (c) for any associative unital algebra $\mathfrak{a}$ over $\mathbb{F}$, and for any linear map $N : \mathfrak{g} \to \mathfrak{a}$ that preserves the Lie bracket, there exists a unique homomorphism $K : \mathfrak{U}(\mathfrak{g}) \to \mathfrak{a}$ such that N is a superposition of M and K.

   The general process of obtaining $\mathfrak{U}(\mathfrak{g})$ from $\mathfrak{g}$ is complicated, but a shortcut exists in physics where faithful matrix representations are available: $\mathbf{P}(\mathfrak{U}(\mathfrak{g}))$ is obtained by adding the associative matrix product to the list of operations defined on the matrix space of $\mathbf{P}(\mathfrak{g})$. For any Lie algebra over any field, the universal enveloping algebra is unique up to an isomorphism.

2. The *complexification* $\mathfrak{g}_{\mathbb{C}}$ of a Lie algebra $\mathfrak{g}$ over $\mathbb{R}$ is the algebra that is obtained by extending the field from $\mathbb{R}$ to $\mathbb{C}$. This is useful because the representation map is complex-linear:

$$\forall \mathbf{A}, \mathbf{B} \in \mathfrak{g} \qquad \mathbf{P}(\mathbf{A} \pm i\mathbf{B}) = \mathbf{P}(\mathbf{A}) \pm i\mathbf{P}(\mathbf{B}) \tag{1.88}$$

and $\mathbf{P}(\mathfrak{g}_{\mathbb{C}})$ is irreducible if and only if $\mathbf{P}(\mathfrak{g})$ is irreducible. Thus, $\mathbf{P}(\mathfrak{g})$ and $\mathbf{P}(\mathfrak{g}_{\mathbb{C}})$ have the same invariant subspaces, and finding them for $\mathbf{P}(\mathfrak{g}_{\mathbb{C}})$ is sometimes easier. The corresponding group $\exp(\mathfrak{g}_{\mathbb{C}})$ is called the complexification of the group $\exp(\mathfrak{g})$

For simply connected matrix Lie groups, a homomorphism φ of Lie algebras uniquely corresponds to a homomorphism Φ of the corresponding groups:

$$\exp[\varphi(\mathbf{A})] = \Phi[\exp(\mathbf{A})] \tag{1.89}$$

If a Lie group is not simply connected, it may be useful to find another group with an isomorphic algebra that is simply connected. Formally, a *universal cover* of a matrix Lie group $\mathcal{G}$ is a simply connected matrix Lie group $\mathcal{H}$ and a homomorphism (called *covering map*) $\Phi : \mathcal{H} \to \mathcal{G}$, such that the associated Lie algebra homomorphism $\varphi : \mathfrak{h} \to \mathfrak{g}$ is an isomorphism. It follows that, if two algebras of simply connected groups are isomorphic, then their groups also are [12].

Envelopes and complexifications are useful because they are supersets of the Lie algebra: their representations are, therefore, also representations of the algebra. Because these supersets are larger, their irreps may turn out to be reducible for the original algebra. Although a cover is a different Lie group, the isomorphism of algebras requires the irreps of the original algebra to be irreps of the covering algebra. However, the cover algebra may also have irreps of its own that are not irreps of the original algebra.

1.5.11 Cartan Subalgebras, Roots, and Weights

If $\mathfrak{g}$ is a semisimple Lie algebra and $\mathfrak{h} \subseteq \mathfrak{g}$ is its largest Abelian subalgebra, then $\mathfrak{h}$ is called a *Cartan subalgebra* of $\mathfrak{g}$, and its dimension is called *Cartan rank* of $\mathfrak{g}$. An equivalent definition of the Cartan rank is the number of simple components in the semi-simple algebra [13].

In a given matrix representation $\mathbf{P}$ of $\mathfrak{g}$, all matrices in $\mathbf{P}(\mathfrak{h})$ commute, and therefore can be simultaneously diagonalised. Each simultaneous eigenvector $\mathbf{v}_n$ will be associated with multiple eigenvalues, one for each matrix in $\mathbf{P}(\mathfrak{h})$. Those eigenvalues make a vector of their own, called *weight vector*—the notion of eigensystem is generalised:

$$\{\mathbf{P}(\mathbf{H}_1) \quad \mathbf{P}(\mathbf{H}_2) \quad \cdots\}\mathbf{v}_n = \{\mu_{1n} \quad \mu_{2n} \quad \cdots\}\mathbf{v}_n \tag{1.90}$$

where $\{\mathbf{H}_k\}$ is a basis set of $\mathfrak{h}$, and $\{\mu_{1n} \quad \mu_{2n} \quad \cdots\}$ is the corresponding weight vector. Different irreducible representations of a given simple Lie algebra have different highest weights.

A similar structure may be built for matrices that simultaneously commute with all elements of $\mathbf{P}(\mathfrak{h})$:

$$[\{\mathbf{P}(\mathbf{H}_1) \quad \mathbf{P}(\mathbf{H}_2) \quad \cdots\}, \mathbf{A}_n] = \{\alpha_{1n} \quad \alpha_{2n} \quad \cdots\}\mathbf{A}_n \tag{1.91}$$

where $\{\alpha_{1n} \quad \alpha_{2n} \quad \cdots\}$ is called *root vector*. Elaborate analytical procedures exist for constructing and classifying representations of Lie algebras and groups using roots and weights. Those procedures are deliberately omitted from this book which advocates numerical methods (Sect. 2.5.3) instead.

1.5.12 Killing Form and Casimir Elements

A simple Lie algebra $\mathfrak{g}$ has trivial ideals—none of its generators commute with all other generators. However, its universal enveloping algebra $\mathfrak{U}(\mathfrak{g})$ turns out to contain an element that does. Finding it is instructive. If an element $\mathbf{C} \in \mathfrak{U}(\mathfrak{g})$ commutes with $\mathfrak{g}$, it must be in the kernel of the adjoint map:

$$\forall \mathbf{A} \in \mathfrak{g} \quad \mathrm{ad}_{\mathbf{A}}(\mathbf{C}) = 0 \tag{1.92}$$

At the same time, $\mathfrak{U}(\mathfrak{g})$ is spanned by polynomials of the generators of $\mathfrak{g}$, and we have already seen a polynomial function that similarity transformations leave invariant: the inner product (Sect. 1.4). Our next step is therefore to define an inner product on $\mathfrak{g}$.

For matrices, a natural choice is the Frobenius product (Sect. 1.4.5) in the most straightforward faithful matrix representation we can construct—the adjoint representation (Sect. 1.5.9):

$$K(\mathbf{X}, \mathbf{Y}) = \mathrm{Tr}[\mathrm{ad}_{\mathbf{X}}\mathrm{ad}_{\mathbf{Y}}] \tag{1.93}$$

This inner product is called the *Killing form* (after Wilhelm Killing [14]) on $\mathfrak{g}$. It is bilinear, symmetric, and invariant under the adjoint action by $\mathfrak{g}$:

$$\begin{gathered}\forall \mathbf{X}, \mathbf{Y}, \mathbf{Z} \in \mathfrak{g} \quad K(\mathbf{X}, \mathbf{Y}) = K(\mathbf{Y}, \mathbf{X}) \\ K(\mathrm{ad}_{\mathbf{Z}}(\mathbf{X}), \mathrm{ad}_{\mathbf{Z}}(\mathbf{Y})) = K(\mathbf{X}, \mathbf{Y})\end{gathered} \tag{1.94}$$

When the elements of $\mathfrak{g}$ are represented by coefficient vectors in some basis $\{\mathbf{V}_k\}$, the Killing form becomes a metric tensor on the corresponding space. Using Eq. (1.85):

$$\kappa_{ab} = K(\mathbf{V}_a, \mathbf{V}_b) = \mathrm{Tr}[\mathrm{ad}(\mathbf{V}_a)\mathrm{ad}(\mathbf{V}_b)] = \sum_{mk} c_{am}^{k} c_{bk}^{m} \tag{1.95}$$

and it follows that the corresponding norm is conserved. This norm yields the *Casimir element* [15]:

$$\mathbf{C} = \sum_{ab} \mathbf{V}_a \kappa_{ab} \mathbf{V}_b \tag{1.96}$$

that commutes with the whole of $\mathfrak{g}$. In the context of physics, it corresponds to a quantity that remains unchanged under the group—in other words, to a conservation law [261]. Semi-simple Lie groups have multiple Casimir elements, one for each simple component.

1.6 Building Blocks of Spin Physics

This section introduces physically motivated instances of the structures described above. Although they could have been presented in an abstract way, we are compelled to be specific because mathematics and physics communities use a different notation for commutators and the exponential map (Table 1.4).

Physicists' commutator is convenient because **A**, **B**, and **C** can be simultaneously *Hermitian* (equal to their own conjugate-transpose)—in quantum mechanics, operators corresponding to real-valued physical observables are Hermitian. For example, for orbital angular momentum operators:

$$\left[\hat{L}_{\mathrm{X}}, \hat{L}_{\mathrm{Y}}\right] = i\hat{L}_{\mathrm{Z}} \quad \overset{\text{rep.}}{\longrightarrow} \quad [\mathbf{L}_{\mathrm{X}}, \mathbf{L}_{\mathrm{Y}}] = i\mathbf{L}_{\mathrm{Z}} \tag{1.97}$$

where a convention in physics is that hats indicate multiplicative or integrodifferential operators and bold fonts indicate a matrix representation.

Physicists' exponential map comes from the general solution of Schrödinger's equation:

$$\frac{\partial}{\partial t}\psi(t) = -i\hat{H}(t)\psi(t) \quad \Rightarrow \quad \psi(t+\mathrm{d}t) = \exp\left[-i\hat{H}(t)\mathrm{d}t\right]\psi(t) \tag{1.98}$$

where a Hermitian Hamiltonian $\hat{H}(t)$ yields a unitary propagator $\exp\left[-i\hat{H}(t)\mathrm{d}t\right]$, which conserves the 2-norm on the wavefunction space, and therefore conserves the sum of probabilities.

1.6.1 Euclidean and Minkowski Spaces

A linear space $\mathcal{V}$ over $\mathbb{R}$ is called *Euclidean* [16] if it has an inner product $\langle \cdot \mid \cdot \rangle : (\mathcal{V} \times \mathcal{V}) \to \mathbb{R}$ that satisfies the following properties:

Table 1.4 Physicists' and mathematicians' commutator and exponential map

	Mathematics	Physics
Commutator	$[\mathbf{A}, \mathbf{B}] = \mathbf{C}$	$[\mathbf{A}, \mathbf{B}] = i\mathbf{C}$
Exponential map	$\exp[\mathbf{A}]$	$\exp[-i\mathbf{A}]$

1. Symmetry:

$$\forall \mathbf{a}, \mathbf{b} \in \mathcal{V} \quad \langle \mathbf{a} | \mathbf{b} \rangle = \langle \mathbf{b} | \mathbf{a} \rangle \tag{1.99}$$

2. Bilinearity:

$$\forall \mathbf{a}, \mathbf{b}, \mathbf{c} \in \mathcal{V} \quad \forall \alpha, \beta \in \mathbb{R} \quad \begin{array}{c} \langle \mathbf{a} | \alpha \mathbf{b} + \beta \mathbf{c} \rangle = \alpha \langle \mathbf{a} | \mathbf{b} \rangle + \beta \langle \mathbf{a} | \mathbf{c} \rangle \\ \langle \alpha \mathbf{a} + \beta \mathbf{b} | \mathbf{c} \rangle = \alpha \langle \mathbf{a} | \mathbf{c} \rangle + \beta \langle \mathbf{b} | \mathbf{c} \rangle \end{array} \tag{1.100}$$

3. Positive definiteness:

$$\forall \mathbf{a} \in \mathcal{V} \quad \langle \mathbf{a} | \mathbf{a} \rangle \geq 0 \tag{1.101}$$

such that $\langle \mathbf{a} | \mathbf{a} \rangle$ is only zero when $\mathbf{a} = \mathbf{0}$.

Euclidean spaces correspond to the classical geometry where norms are real and non-negative. Time cannot be introduced into this picture in a way that is consistent with experimental observations: Michelson and Morley demonstrated in 1887 that measurement devices in all inertial frames of reference report the same speed of light, even if those frames are moving relative to one another [17]. An explanation proposed by Lorentz in 1892 was that moving objects contract in the direction of travel, and that their clocks slow down as velocity is increased [18]. The corresponding transformations were named *boosts*; observations suggested that boosts are *orthochronous* (preserve the direction of time), and satisfy the properties of a Lie group [7]: closure (two boosts in a sequence are another boost), invertibility (it is possible to undo a boost), the existence of a unique unit boost (corresponding to no change in velocity), and continuity (the dependence on velocity is continuous). Boosts also preserve vector addition operation on space-time, and are therefore linear operators representable by matrices.

If we point the X-axis in the direction of travel and use ct instead of t to make units consistent in space-time vectors, we arrive at the following general form for the boost operation:

$$\begin{bmatrix} ct' \\ x' \end{bmatrix} = \mathbf{K}_\mathrm{X}(v) \begin{bmatrix} ct \\ x \end{bmatrix} \tag{1.102}$$

where $\mathbf{K}_\mathrm{X}(v)$ is a real 2×2 matrix that depends on the velocity v. It must obey the group properties:

$$\begin{aligned} \mathbf{K}_\mathrm{X}(v_1 + v_2) &= \mathbf{K}_\mathrm{X}(v_2)\mathbf{K}_\mathrm{X}(v_1) \\ \mathbf{K}_\mathrm{X}(-v)\mathbf{K}_\mathrm{X}(+v) &= \mathbf{K}_\mathrm{X}(0) = \mathbf{1} \end{aligned} \tag{1.103}$$

These conditions yield a system of equations for the elements of the matrix; the solution is

$$\mathbf{K}_\mathrm{X}(v) = \frac{1}{\sqrt{1 - v^2/c^2}} \begin{bmatrix} 1 & -v/c \\ -v/c & 1 \end{bmatrix} \tag{1.104}$$

Importantly, the quantity this transformation leaves invariant is $x^2 - c^2t^2$, and therefore the metric on space-time is not positive definite (*c.f.* Sect. 1.4.1)—this is not a Euclidean space.

The procedure may be repeated for the other directions to obtain similar expressions for $\mathbf{K}_\mathrm{Y}$ and $\mathbf{K}_\mathrm{Z}$. In three spatial dimensions, the map $\langle\cdot\mid\cdot\rangle : (\mathcal{V}\times\mathcal{V}) \to \mathbb{R}$ that is invariant under boosts is

$$\langle\mathbf{a}|\mathbf{b}\rangle = -a_0b_0 + a_1b_1 + a_2b_2 + a_3b_3 \tag{1.105}$$

where $\mathbf{a}$ and $\mathbf{b}$ are vectors with elements listed in the order $\{ct, x, y, z\}$. This map is symmetric and bilinear, and may be viewed as a kind of inner product:

1. $\forall\mathbf{a},\mathbf{b}\in\mathcal{V} \quad \langle\mathbf{a}|\mathbf{b}\rangle = \langle\mathbf{b}|\mathbf{a}\rangle$
2. $\forall\mathbf{a},\mathbf{b},\mathbf{c}\in\mathcal{V} \quad \forall\alpha,\beta\in\mathbb{R} \quad \langle\alpha\mathbf{a}+\beta\mathbf{b}|c\rangle = \alpha\langle\mathbf{a}|\mathbf{c}\rangle + \beta\langle\mathbf{b}|\mathbf{c}\rangle$

However, the corresponding norm:

$$\|\mathbf{a}\| = \sqrt{\langle\mathbf{a}|\mathbf{a}\rangle} = \sqrt{-a_0^2 + a_1^2 + a_2^2 + a_3^2} \tag{1.106}$$

need not be a real number. This construct is called *Minkowski space* [19], its points are called *events*, and the "distance" between two events, as defined by Eq. (1.106), is invariant under spatial rotations and boosts.

1.6.2 Special Orthogonal Group in Three Dimensions

Orthogonal transformations of an N-dimensional metric vector space over $\mathbb{R}$ are those that preserve the 2-norm and its associated inner product. Such transformations are represented faithfully by *orthogonal matrices*, for which an equivalent definition is that their transpose is equal to their inverse:

$$\forall\mathbf{x},\mathbf{y}\in\mathbb{R}^N \quad \langle\mathbf{x}|\mathbf{O}^\mathrm{T}\mathbf{O}|\mathbf{y}\rangle = \langle\mathbf{x}\mid\mathbf{y}\rangle \quad \Leftrightarrow \quad \mathbf{O}^\mathrm{T} = \mathbf{O}^{-1} \tag{1.107}$$

The superposition of two orthogonal transformations is also orthogonal. Together with the identity transformation, this makes a group (Sect. 1.4.2), called *orthogonal group* and denoted $\mathrm{O}(N)$.

Because $\mathbf{O}^\mathrm{T}\mathbf{O} = \mathbf{1}$ and the matrices are real, the eigenvalues of $\mathbf{O}$ (and therefore the determinant which is their product) can only be equal to ± 1. Since $\det(\mathbf{AB}) = \det(\mathbf{A})\det(\mathbf{B})$, the subset of $\mathrm{O}(N)$ in which the matrices have a determinant of $+1$ is a subgroup, called the *special orthogonal group*:

$$\mathrm{SO}(N) = \{\mathbf{A}\in\mathrm{O}(N)|\det(\mathbf{A}) = 1\} \tag{1.108}$$

Thus, $O(N)$ has two connected components, and the one containing the identity is $SO(N)$. In a three-dimensional Euclidean space, $O(3) = SO(3) \times I$. The elements of $SO(3)$ are *rotations*, and the non-unit element of $I = \{+\mathbf{1}, -\mathbf{1}\}$ performs coordinate *inversion* operation.

$SO(3)$ is a non-Abelian triparametric simple Lie group. Its generators with respect to the $\exp[-i\mathfrak{g}]$ map may be obtained (Sect. 1.5.7) from rotation matrices (corkscrew rule for positive rotation, ISO 31-11) around the three Cartesian axes:

$$\mathbf{R}_X(\varphi) = \begin{pmatrix} 1 & 0 & 0 \\ 0 & +\cos\varphi & -\sin\varphi \\ 0 & +\sin\varphi & +\cos\varphi \end{pmatrix} = \exp[-i\mathbf{L}_X\varphi], \quad \mathbf{L}_X = \begin{pmatrix} 0 & 0 & 0 \\ 0 & 0 & -i \\ 0 & +i & 0 \end{pmatrix}$$
$$\mathbf{R}_Y(\varphi) = \begin{pmatrix} +\cos\varphi & 0 & +\sin\varphi \\ 0 & 1 & 0 \\ -\sin\varphi & 0 & +\cos\varphi \end{pmatrix} = \exp[-i\mathbf{L}_Y\varphi], \quad \mathbf{L}_Y = \begin{pmatrix} 0 & 0 & +i \\ 0 & 0 & 0 \\ -i & 0 & 0 \end{pmatrix}$$
$$\mathbf{R}_Y(\varphi) = \begin{pmatrix} +\cos\varphi & -\sin\varphi & 0 \\ +\sin\varphi & +\cos\varphi & 0 \\ 0 & 0 & 1 \end{pmatrix} = \exp[-i\mathbf{L}_Z\varphi], \quad \mathbf{L}_Z = \begin{pmatrix} 0 & -i & 0 \\ +i & 0 & 0 \\ 0 & 0 & 0 \end{pmatrix} \tag{1.109}$$

The corresponding Lie algebra $\mathfrak{so}(3)$ is spanned by linear combinations of $\{\mathbf{L}_X, \mathbf{L}_Y, \mathbf{L}_Z\}$ with real coefficients. Commutation relations and the Casimir element are obtained by direct calculation:

$$[\mathbf{L}_a, \mathbf{L}_b] = i\varepsilon_{abc}\mathbf{L}_c \qquad \mathbf{C} = \mathbf{L}_X^2 + \mathbf{L}_Y^2 + \mathbf{L}_Z^2 \tag{1.110}$$

where ε_{abc} is the Levi-Civita symbol [331]. The general form of the group element:

$$\mathbf{R}(\mathbf{n}, \varphi) = \exp[-i(n_X\mathbf{L}_X + n_Y\mathbf{L}_Y + n_Z\mathbf{L}_Z)\varphi] \tag{1.111}$$

has the physical meaning of a rotation around the unit vector $\mathbf{n}$ by an angle φ. It follows that groups of uniaxial rotations are uniparametric Abelian subgroups of $SO(3)$.

$SO(3)$ is not simply connected; this may be seen from a geometric interpretation of Eq. (1.111): every rotation may be identified with a vector whose direction defines the rotation axis and the length defines the rotation angle—$SO(3)$ is thus mapped onto a ball of radius π. This ball has a periodic boundary: $+\pi$ and $-\pi$ rotations around the same axis are the same transformation, and thus the opposing points on the surface of the ball correspond to the same rotation. This creates two non-intersecting classes of continuous paths between any two elements of $SO(3)$: one through the volume of the ball, and one that cuts across the surface. Because the entry and the exit points on the surface must stay diametrically opposed, there is no possibility of a continuous transformation between the two classes of paths. We, therefore, conclude that $SO(3)$ is doubly connected.

1.6.2.1 Parametrisation of Rotations

Throughout this book, all three-dimensional rotations are *active*: we always rotate the object, and never the reference frame. The simplest way to specify a rotation unambiguously is to give the corresponding 3×3 rotation matrix. Unfortunately, that was rarely done in the history of physics and engineering—a large number of rotation specification conventions exist, some of them exasperatingly fiddly, usually as a consequence of the generators associated with different parameters not being linearly independent [20].

1.6.2.2 Euler Angles Parametrisation

This is a historically important [21] but treacherous convention that is regularly blamed for airplane and satellite malfunctions. The active rotation convention is as follows:

1. Rotate the object about the *Z*-axis through an angle $\gamma \in [0, 2\pi)$
2. Rotate the object about the *Y*-axis through an angle $\beta \in [0, \pi)$
3. Rotate the object about the *Z*-axis through an angle $\alpha \in [0, 2\pi)$

For column vectors in $\mathbb{R}^3$, the following matrix performs the rotation:

$$\mathbf{R}(\alpha, \beta, \gamma) = \begin{pmatrix} +\cos\alpha & -\sin\alpha & 0 \\ +\sin\alpha & +\cos\alpha & 0 \\ 0 & 0 & 1 \end{pmatrix} \begin{pmatrix} +\cos\beta & 0 & +\sin\beta \\ 0 & 1 & 0 \\ -\sin\beta & 0 & +\cos\beta \end{pmatrix} \begin{pmatrix} +\cos\gamma & -\sin\gamma & 0 \\ +\sin\gamma & +\cos\gamma & 0 \\ 0 & 0 & 1 \end{pmatrix} \tag{1.112}$$

Euler angles should be avoided whenever possible: the transformation from $\{\alpha, \beta, \gamma\}$ into the rotation matrix $\mathbf{R}$ is well defined, but the inverse transformation is not, because the first and the last generator are the same. When $\beta = 0$ or $\beta = \pi$, it is impossible to extract α and γ individually because only $\alpha + \gamma$ is constrained. This creates insidious numerical difficulties.

1.6.2.3 Angle-Axis Parametrisation

Any rotation may be specified as a unit vector $\mathbf{n}$ and a turning angle φ around that vector:

$$\mathbf{R}(\mathbf{n}, \varphi) = \exp[-i(\mathbf{L}_X n_X + \mathbf{L}_Y n_Y + \mathbf{L}_Z n_Z)\varphi] \tag{1.113}$$

The connection to the rotation matrix is:

$$\mathbf{R}(\mathbf{n}, \varphi) = \begin{bmatrix} \cos\varphi + n_X^2(1-\cos\varphi) & n_X n_Y(1-\cos\varphi) - n_Z\sin\varphi & n_X n_Z(1-\cos\varphi) + n_Y\sin\varphi \\ n_Y n_X(1-\cos\varphi) + n_Z\sin\varphi & \cos\varphi + n_Y^2(1-\cos\varphi) & n_Y n_Z(1-\cos\varphi) - n_X\sin\varphi \\ n_Z n_X(1-\cos\varphi) - n_Y\sin\varphi & n_Z n_Y(1-\cos\varphi) + n_X\sin\varphi & \cos\varphi + n_Z^2(1-\cos\varphi) \end{bmatrix} \tag{1.114}$$

Composition of rotations is particularly simple in the angle-axis parametrisation. Given two rotations $(\mathbf{n}_1, \varphi_1)$ and $(\mathbf{n}_2, \varphi_2)$, the rotation $(\mathbf{n}, \varphi)$ corresponding to their superposition is

$$\begin{aligned} \mathbf{a} &= \tan\left(\frac{\varphi_1}{2}\right)\mathbf{n}_1, \quad \mathbf{b} = \tan\left(\frac{\varphi_2}{2}\right)\mathbf{n}_2 \\ \mathbf{c} &= \frac{\mathbf{a}+\mathbf{b}-\mathbf{a}\times\mathbf{b}}{1-\mathbf{a}\cdot\mathbf{b}}, \quad \mathbf{c} = \tan\left(\frac{\varphi}{2}\right)\mathbf{n} \end{aligned} \tag{1.115}$$

When expressing a rotation via parameters is for some reason unavoidable, of the many possible ways of doing that, the angle-axis parametrisation is the least problematic.

1.6.2.4 Irreducible Representations

The smallest faithful irrep of SO(3) is three-dimensional, an example is Eq. (1.109); other irreps are obtained by taking direct products of those matrices and diagonalising the Casimir element of the resulting representation. A tedious calculation demonstrates [22] that irreducible representations have dimension $l(l+1)$ where l is a positive integer, and m runs in integer increments between $-l$ and $+l$:

$$\begin{aligned} &[\![\mathbf{L}^2]\!]^{(l)}_{m,m} = l(l+1), \quad [\![\mathbf{L}_\mathrm{Z}]\!]^{(l)}_{m,m} = m \\ &[\![\mathbf{L}_+]\!]^{(l)}_{m+1,m} = \sqrt{l(l+1)-m(m+1)}, \quad \mathbf{L}_\mathrm{X} = (\mathbf{L}_+ + \mathbf{L}_-)/2 \\ &[\![\mathbf{L}_-]\!]^{(l)}_{m-1,m} = \sqrt{l(l+1)-m(m-1)}, \quad \mathbf{L}_\mathrm{Y} = (\mathbf{L}_+ - \mathbf{L}_-)/2i \end{aligned} \tag{1.116}$$

where the double square brackets indicate a matrix representation. Irreps of finite rotations (called *Wigner D matrices* [23]) are obtained by exponentiating the generators:

$$\begin{aligned} \mathfrak{D}^{(l)}(\alpha,\beta,\gamma) &= e^{-i\mathbf{L}_\mathrm{Z}^{(l)}\alpha} e^{-i\mathbf{L}_\mathrm{Y}^{(l)}\beta} e^{-i\mathbf{L}_\mathrm{Z}^{(l)}\gamma} \\ \mathfrak{D}^{(l)}(\mathbf{n},\varphi) &= e^{-i\left[n_\mathrm{X}\mathbf{L}_\mathrm{X}^{(l)} + n_\mathrm{Y}\mathbf{L}_\mathrm{Y}^{(l)} + n_\mathrm{Z}\mathbf{L}_\mathrm{Z}^{(l)}\right]\varphi} \end{aligned} \tag{1.117}$$

for active ZYZ Euler angles and the angle-axis parametrisation, respectively.

1.6.3 Special Unitary Group in Two Dimensions

Unitary transformations of an N-dimensional metric vector space over $\mathbb{C}$ are those that preserve the norm and the scalar product. They are represented faithfully by a *unitary matrices*, for which an equivalent definition is that their conjugate transpose is equal to the inverse:

$$\forall \mathbf{x},\mathbf{y} \in \mathbb{C}^N \quad \langle \mathbf{x}|\mathbf{U}^\dagger \mathbf{U}|\mathbf{y}\rangle = \langle \mathbf{x} \mid \mathbf{y}\rangle \Leftrightarrow \mathbf{U}^\dagger = \mathbf{U}^{-1} \tag{1.118}$$

The superposition of two unitary transformations is also unitary. Together with the identity transformation, this makes a group (Sect. 1.5), called *unitary group* and denoted $\mathrm{U}(N)$.

Because $\mathbf{U}^\dagger\mathbf{U} = \mathbf{1}$, all eigenvalues of $\mathbf{U}$ must have the form $e^{i\varphi}$ where φ is a real number. Their product, i.e. the determinant of $\mathbf{U}$, must therefore also have that form. Since $\det(\mathbf{AB}) = \det(\mathbf{A})\det(\mathbf{B})$, the subset of $\mathrm{U}(N)$ in which $\det(\mathbf{U}) = 1$ is a subgroup, called *special unitary group*:

$$\mathrm{SU}(N) = \{\mathbf{U} \in \mathrm{U}(N) |\det(\mathbf{U}) = 1\} \tag{1.119}$$

1.6.3.1 Parametrisation

In two dimensions, the definition leads to the following general form for the elements of $\mathrm{SU}(2)$:

$$\mathrm{SU}(2) = \left\{ \left(\begin{matrix} z & w \\ -w^* & z^* \end{matrix} \right) \middle| z, w \in \mathbb{C},\ |z|^2 + |w|^2 = 1 \right\} \tag{1.120}$$

To build generators, we will observe that unitary matrices are complex exponentials of Hermitian ones:

$$\mathbf{H}^\dagger = \mathbf{H} \quad \Leftrightarrow \quad [\exp(-i\mathbf{H})][\exp(-i\mathbf{H})]^\dagger = \mathbf{1} \tag{1.121}$$

and that the unit determinant of $\exp(-i\mathbf{H})$ requires $\mathbf{H}$ to be traceless, because $\det[e^{\mathbf{A}}] = e^{\mathrm{Tr}(\mathbf{A})}$. Therefore, $\mathrm{SU}(2)$ is generated under $\exp[-ig]$ by traceless 2×2 Hermitian matrices.

A complex 2×2 matrix has eight independent real parameters. The Hermitian property requires the imaginary parts of the diagonal elements to be zero, this leaves six parameters. The requirement for zero trace puts a linear constraint on the diagonal elements and leaves five independent parameters. Finally, the requirement for the matrix to be Hermitian means that the off-diagonal elements must be complex conjugates of each other; this leaves three independent real parameters. It follows that $\mathrm{SU}(2)$ is simply connected—the three parameters can be mapped onto a unit sphere in four dimensions:

$$\begin{aligned} z = \alpha + i\beta, \quad & w = \gamma + i\delta, \quad \alpha,\beta,\gamma,\delta \in \mathbb{R} \\ |z|^2 + |w|^2 = 1 \quad & \Leftrightarrow \quad \alpha^2 + \beta^2 + \gamma^2 + \delta^2 = 1 \end{aligned} \tag{1.122}$$

which is connected, path-connected, and compact.

1.6.3.2 Irreducible Representations

Three linearly independent traceless Hermitian 2 × 2 matrices are easily found; a particularly convenient set (chosen for simple commutation relations and sparse direct products) is

$$\mathbf{S}_{\mathrm{X}}=\begin{pmatrix}0 & 1/2\\ 1/2 & 0\end{pmatrix},\quad \mathbf{S}_{\mathrm{Y}}=\begin{pmatrix}0 & -i/2\\ i/2 & 0\end{pmatrix},\quad \mathbf{S}_{\mathrm{Z}}=\begin{pmatrix}1/2 & 0\\ 0 & -1/2\end{pmatrix} \tag{1.123}$$

where Cartesian indices are used in the expectation of the physical meaning that these matrices will acquire later in the book. The corresponding Lie algebra $\mathfrak{su}(2)$ is spanned by linear combinations of $\{\mathbf{S}_{\mathrm{X}},\mathbf{S}_{\mathrm{Y}},\mathbf{S}_{\mathrm{Z}}\}$ with real coefficients. The general form of the group element is

$$\mathbf{U}(\boldsymbol{\omega})=\exp[-i(\omega_{\mathrm{X}}\mathbf{S}_{\mathrm{X}}+\omega_{\mathrm{Y}}\mathbf{S}_{\mathrm{Y}}+\omega_{\mathrm{Z}}\mathbf{S}_{\mathrm{Z}})],\quad \boldsymbol{\omega}\in\mathbb{R}^3 \tag{1.124}$$

Commutation relations and the complex envelope are obtained by a direct calculation:

$$[\mathbf{S}_{\mathrm{X}},\mathbf{S}_{\mathrm{Y}}]=i\mathbf{S}_{\mathrm{Z}},\quad [\mathbf{S}_{\mathrm{Z}},\mathbf{S}_{\mathrm{X}}]=i\mathbf{S}_{\mathrm{Y}},\quad [\mathbf{S}_{\mathrm{Y}},\mathbf{S}_{\mathrm{Z}}]=i\mathbf{S}_{\mathrm{X}} \tag{1.125}$$

$$\mathbf{S}_{+}=\mathbf{S}_{\mathrm{X}}+i\mathbf{S}_{\mathrm{Y}},\quad \mathbf{S}_{-}=\mathbf{S}_{\mathrm{X}}-i\mathbf{S}_{\mathrm{Y}},\quad \mathbf{S}^2=\mathbf{S}_{\mathrm{X}}^2+\mathbf{S}_{\mathrm{Y}}^2+\mathbf{S}_{\mathrm{Z}}^2 \tag{1.126}$$

$$[\mathbf{S}^2,\mathbf{S}_{\mathrm{X,Y,Z}\pm}]=0,\quad [\mathbf{S}_{\mathrm{Z}},\mathbf{S}_{\pm}]=\pm\mathbf{S}_{\pm},\quad [\mathbf{S}_{+},\mathbf{S}_{-}]=2\mathbf{S}_{\mathrm{Z}} \tag{1.127}$$

The smallest faithful representation is two-dimensional, an example is Eq. (1.123). Irreps of other dimensions are obtained by taking Kronecker products of those matrices and diagonalising the Casimir operator. A tedious calculation [22] demonstrates that irreducible representations have dimension $s(s+1)$ where s is a positive integer or half-integer, and m runs in integer increments between $-s$ and $+s$:

$$\begin{aligned}
&[\![\mathbf{S}^2]\!]^{(s)}_{m,m}=s(s+1),\quad [\![\mathbf{S}_{\mathrm{Z}}]\!]^{(l)}_{m,m}=m\\
&[\![\mathbf{S}_{+}]\!]^{(s)}_{m+1,m}=\sqrt{s(s+1)-m(m+1)},\quad \mathbf{S}_{\mathrm{X}}=(\mathbf{S}_{+}+\mathbf{S}_{-})/2\\
&[\![\mathbf{S}_{-}]\!]^{(s)}_{m-1,m}=\sqrt{s(s+1)-m(m-1)},\quad \mathbf{S}_{\mathrm{Y}}=(\mathbf{S}_{+}-\mathbf{S}_{-})/2i
\end{aligned} \tag{1.128}$$

Irreps corresponding to integer values of s are the same as the irreps of $\mathfrak{so}(3)$; representations of group elements are obtained by exponentiating (using physicists' map, Sect. 1.6) linear combinations of generators with real coefficients.

1.6.3.3 Normalisation-Commutation Dilemma

Irreducible representations obtained above are faithful for the Lie algebra, but not for its associative envelope: commutation relations do not depend on the representation dimension, but product relations do—for example, the square of $\mathbf{S}_{\mathrm{Z}}^{(1/2)}$ is a multiple of a unit matrix, but the square of $\mathbf{S}_{\mathrm{Z}}^{(1)}$ is not. This means that any metric would be representation-dependent; the Frobenius norm is a good example:

$$
\begin{aligned}
&\|\mathbf{A}\|_{\mathrm{F}}=\sqrt{\mathrm{Tr}\left[\mathbf{A}^{\dagger}\mathbf{A}\right]} \\
&\left\|\mathbf{S}_{\mathrm{X}}^{(s)}\right\|_{\mathrm{F}}=\left\|\mathbf{S}_{\mathrm{Y}}^{(s)}\right\|_{\mathrm{F}}=\left\|\mathbf{S}_{\mathrm{Z}}^{(s)}\right\|_{\mathrm{F}}=\frac{\sqrt{s(s+1)(2s+1)}}{3}
\end{aligned}
\tag{1.129}
$$

The generators should not be normalised because that would make commutation properties representation dependent, and that is not a good idea: $\mathfrak{su}(2)$ is first and foremost a Lie algebra, and only optionally a metric space. This normalisation-commutation dilemma is delicate: as we shall later see, consistent commutation relations are more important than consistent normalisation. The same applies to $\mathfrak{so}(3)$.

1.6.4 Relationship Between SU(2) and SO(3)

SU(2) and SO(3) both have three real parameters, and it is apparent from Eqs. (1.110) and (1.125) that their generator sets have identical commutation relations:

$$
[\mathbf{S}_a,\mathbf{S}_b]=i\varepsilon_{abc}\mathbf{S}_c \qquad [\mathbf{L}_a,\mathbf{L}_b]=i\varepsilon_{abc}\mathbf{L}_c \tag{1.130}
$$

meaning that $\mathfrak{su}(2)$ and $\mathfrak{so}(3)$ are isomorphic as Lie algebras. However, their Lie groups are not isomorphic—two different elements of SU(2) correspond to each element of SO(3), for example:

$$
\begin{aligned}
\exp[-2\pi i\mathbf{S}_{\mathrm{Y}}]&=\begin{pmatrix}-1&0\\0&-1\end{pmatrix}, \quad \exp[-2\pi i\mathbf{L}_{\mathrm{Y}}]=\begin{pmatrix}1&0&0\\0&1&0\\0&0&1\end{pmatrix} \\
\exp[-4\pi i\mathbf{S}_{\mathrm{Y}}]&=\begin{pmatrix}+1&0\\0&+1\end{pmatrix}, \quad \exp[-4\pi i\mathbf{L}_{\mathrm{Y}}]=\begin{pmatrix}1&0&0\\0&1&0\\0&0&1\end{pmatrix}
\end{aligned}
\tag{1.131}
$$

Thus, the orbits in SU(2) are twice as long as those in SO(3)—in the special unitary group, the parameter goes to 4π in any direction before the group starts repeating itself. This is a double cover (Sect. 1.5.10): the factor group $\mathrm{SU}(2)/\mathrm{S}_2$ is isomorphic to SO(3), where S_2 is the permutation group of two objects. It is sometimes erroneously stated that SU(2) and SO(3) have the same Lie algebra. That is not the case—$\mathfrak{su}(2)$ and $\mathfrak{so}(3)$ are isomorphic as Lie algebras, but they are not identical.

1.7 Linear Time-Invariant Systems

In the early days of electromagnetic spectroscopy, the absorption spectrum of the sample was recorded by measuring absorption at each frequency or wavelength in turn [24]—the so-called slow passage method. The arrival of pulse-acquire Fourier spectroscopy [25] brought improvements in speed and sensitivity. Even though Fourier methods are commonly rationalised using physical arguments, the relationship between pulse and frequency responses of a physical system can be derived from mathematical assumptions. Consider system with an input $x(t)$ and an output $y(t)$:

$$y(t) = \Phi\{x(t)\} \tag{1.132}$$

the system Φ is called *linear time-invariant (LTI)* if:

$$\begin{aligned} \Phi\{\alpha x_k(t) + \beta x_m(t)\} = \alpha\Phi\{x_k(t)\} + \beta\Phi\{x_m(t)\} = \alpha y_k(t) + \beta y_m(t) \\ \Phi\{x_k(t - t_0)\} = y_k(t - t_0) \end{aligned} \tag{1.133}$$

where $x_k(t)$ are input signals, $y_k(t)$ are output signals, and α, β, t_0 are constants. In other words, an LTI system behaves *linearly* with respect to the input signal—a sum of two input signals returns a sum of their corresponding outputs, and multiplying an input by a constant number multiplies the output by the same number. If the input is shifted in time by a constant amount, the system output is shifted by the same amount, but otherwise remains the same—this is called *time invariance*.

If the input function $x(t)$ has a linear expansion in some orthonormal basis set $\{g_k(t)\}$, then:

$$x(t) = \sum_k \chi_k g_k(t), \quad \chi_k = \langle g_k(t) \mid x(t) \rangle = \int_{-\infty}^{\infty} g_k^*(t) x(t) \mathrm{d}t, \tag{1.134}$$

where the star denotes complex conjugation. In the continuous limit, the sum becomes an integral transform with some kernel $g(k, t)$:

$$x(t) = \int_{-\infty}^{\infty} \chi(k) g(k, t) \mathrm{d}k \tag{1.135}$$

The response to an arbitrary input function can then be written in terms of responses to either the individual basis functions or the integral transform kernel:

$$\Phi\{x(t)\} = \sum_k \chi_k \Phi\{g_k(t)\} = \int_{-\infty}^{\infty} \chi(k)\Phi\{g(k,t)\}\mathrm{d}k \tag{1.136}$$

This means that the set of responses to the basis functions defines an LTI system completely—known responses to $\{g_k(t)\}$ or $g(k,t)$ enable straightforward calculation of the response to any input.

1.7.1 Pulse and Frequency Response

Consider the system response to a delta-function, which is here a model for a short and sharp input pulse. By the definition [26] of $\delta(t)$, for any continuous compactly supported function $x(t)$:

$$x(t) = \int_{-\infty}^{\infty} x(\tau)\delta(\tau - t)\mathrm{d}\tau \tag{1.137}$$

When we apply our LTI system to both sides of this expression, it becomes

$$\Phi\{x(t)\} = \int_{-\infty}^{\infty} x(\tau)\Phi\{\delta(\tau - t)\}\mathrm{d}\tau = \int_{-\infty}^{\infty} x(\tau)h(\tau - t)\mathrm{d}\tau \tag{1.138}$$

The function $h(t) = \Phi\{\delta(t)\}$ is called the *pulse response* of the system. The integral in Eq. (1.138) is a convolution of $x(t)$ with $h(t)$ that may be abbreviated as

$$\Phi\{x(t)\} = x(t) * h(t) \tag{1.139}$$

The conclusion is that the pulse response $h(t)$ contains complete information about the LTI black box Φ and allows one to predict its response to an arbitrary input $x(t)$.

Consider now the response of the system to an oscillatory input $\mathrm{e}^{i\omega t}$:

$$\begin{aligned}\Phi\{\mathrm{e}^{i\omega t}\} &= \int_{-\infty}^{\infty} \mathrm{e}^{i\omega\tau}h(\tau - t)\mathrm{d}\tau = \mathrm{e}^{i\omega t}\int_{-\infty}^{\infty} \mathrm{e}^{i\omega(\tau - t)}h(\tau - t)\mathrm{d}\tau \\ &= \mathrm{e}^{i\omega t}\int_{-\infty}^{\infty} h(\tau)\mathrm{e}^{i\omega\tau}\mathrm{d}\tau = H(\omega)\mathrm{e}^{i\omega t}, \quad H(\omega) = \int_{-\infty}^{\infty} h(\tau)\mathrm{e}^{i\omega\tau}\mathrm{d}\tau\end{aligned} \tag{1.140}$$

This demonstrates that oscillatory inputs are eigenfunctions: an LTI system cannot shift frequencies, it can only alter their coefficients. The function $H(\omega)$ is called the *frequency response* of the system; it is the Fourier transform [27] of the pulse response $h(t)$.

1.7.2 Properties of the Fourier Transform

Engineers, physicists, mathematicians, and computer scientists have different conventions on frequency signs and multipliers; here we use the convention that yields unitary transformations (the 2-norm of the function is preserved). For a function $f(t)$, the *forward Fourier transform* [27] is

$$\hat{F}_+ f(t) = \frac{1}{\sqrt{2\pi}} \int_{-\infty}^{\infty} f(t) \mathrm{e}^{-i\omega t} \mathrm{d}t \tag{1.141}$$

and the *backward Fourier transform* is:

$$\hat{F}_- f(t) = \frac{1}{\sqrt{2\pi}} \int_{-\infty}^{\infty} f(t) \mathrm{e}^{i\omega t} \mathrm{d}t \tag{1.142}$$

In the discussion below, we assume that $f(t)$ is absolutely integrable—a sufficient condition for the integrals in Eqs. (1.141) and (1.142) to exist. Both transforms are obviously linear with respect to $f(t)$.

1. Forward and backward Fourier transforms are each other's inverse. This is proven by direct inspection using one of the definitions of delta function:

$$\delta(t) = \frac{1}{2\pi} \int_{-\infty}^{\infty} \mathrm{e}^{i\omega t} \mathrm{d}\omega = \frac{1}{2\pi} \lim_{a \to \infty} \int_{-a}^{a} \mathrm{e}^{i\omega t} \mathrm{d}\omega \tag{1.143}$$

where the integral is understood to mean Cauchy's principal value [334] defined by the limit on the right-hand side. With that in place, we have

$$\begin{aligned} \hat{F}_+ \hat{F}_- f(t) &= \frac{1}{2\pi} \int_{-\infty}^{\infty} \left[\int_{-\infty}^{\infty} f(t') \mathrm{e}^{i\omega t'} \mathrm{d}t' \right] \mathrm{e}^{-i\omega t} \mathrm{d}\omega = \frac{1}{2\pi} \int_{-\infty}^{\infty} f(t') \mathrm{d}t' \int_{-\infty}^{\infty} \mathrm{e}^{i\omega (t'-t)} \mathrm{d}\omega \\ &= \int_{-\infty}^{\infty} f(t') \delta(t'-t) \mathrm{d}t' = f(t), \quad \hat{F}_- \hat{F}_+ f(t) = \cdots = f(t) \end{aligned} \tag{1.144}$$

Most software packages place the zero frequency on the *edge* of the spectrum – commands like MATLAB's `fftshift` rotate the data to put the zero frequency in the centre.

2. Forward and backward Fourier transforms are unitary operators. This is also proven by direct inspection. For the forward transform:

$$\int_{-\infty}^{\infty} \left|\hat{F}_{\pm}f(t)\right|^2 \mathrm{d}\omega = \int_{-\infty}^{\infty} \left[\frac{1}{\sqrt{2\pi}}\int_{-\infty}^{\infty} f(t)\mathrm{e}^{\mp i\omega t}\mathrm{d}t\right]\left[\frac{1}{\sqrt{2\pi}}\int_{-\infty}^{\infty} f^*(t')\mathrm{e}^{\pm i\omega t'}\mathrm{d}t'\right]\mathrm{d}\omega$$
$$= \frac{1}{2\pi}\int_{-\infty}^{\infty}\mathrm{d}t\int_{-\infty}^{\infty}\mathrm{d}t'\left[f(t)f^*(t')\int_{-\infty}^{\infty}\mathrm{e}^{\mp i\omega(t-t')}\mathrm{d}\omega\right]$$
$$= \int_{-\infty}^{\infty}\mathrm{d}t\int_{-\infty}^{\infty}\mathrm{d}t'[f(t)f^*(t')\delta(t'-t)] = \int_{-\infty}^{\infty} f(t)f^*(t)\mathrm{d}t = \int_{-\infty}^{\infty}|f(t)|^2\mathrm{d}t \tag{1.145}$$

and likewise for the backward one. For electromagnetic signals, the integral of the absolute square is proportional to signal power, this is therefore known as the *power theorem*.

3. The Fourier transform of a derivative is

$$\hat{F}_{\pm}f^{(k)}(t) = (\pm i\omega)^k \hat{F}_{\pm}f(t) \tag{1.146}$$

This property is proven by induction—we start with the kth derivative of a function $f(t)$ and perform integration by parts:

$$\hat{F}_{\pm}f^{(k)}(t) = \frac{1}{\sqrt{2\pi}}\int_{-\infty}^{\infty} f^{(k)}(t)\mathrm{e}^{\mp i\omega t}\mathrm{d}t$$
$$= \mathrm{e}^{\mp i\omega t}f^{(k-1)}(t)\Big|_{-\infty}^{\infty} + \frac{\pm i\omega}{\sqrt{2\pi}}\int_{-\infty}^{\infty} f^{(k-1)}(t)\mathrm{e}^{\mp i\omega t}\mathrm{d}t \tag{1.147}$$
$$= \frac{\pm i\omega}{\sqrt{2\pi}}\int_{-\infty}^{\infty} f^{(k-1)}(t)\mathrm{e}^{\mp i\omega t}\mathrm{d}t = (\pm i\omega)\hat{F}_{\pm}f^{(k-1)}(t)$$

where, on the right hand side, the order of the derivative has been reduced by one, and $\pm i\omega$ appeared in front. Repeating this process multiple times yields Eq. (1.146). Note that the proof requires the function and its derivatives to vanish at $\pm\infty$. In the common special case when $f(t<0) = 0$, the relation reads

$$\frac{1}{\sqrt{2\pi}}\int_{0}^{\infty} f^{(k)}(t)\mathrm{e}^{\mp i\omega t}\mathrm{d}t = \pm\frac{i\omega}{\sqrt{2\pi}}\int_{0}^{\infty} f^{(k-1)}(t)\mathrm{e}^{\mp i\omega t}\mathrm{d}t - f^{(k-1)}(0) \tag{1.148}$$

This property is useful for solving initial value problems for ordinary differential equations.

4. Derivatives of a Fourier transform are

$$\left[\hat{F}_{\pm}f(t)\right]^{(k)} = (\mp i)^{k}\hat{F}_{\pm}\left(t^{k}f(t)\right) \tag{1.149}$$

the proof is simple because the derivative here is with respect to ω—it may be taken directly:

$$\begin{aligned}\frac{1}{\sqrt{2\pi}}\int_{-\infty}^{\infty}\left[f(t)\mathrm{e}^{\mp i\omega t}\right]^{(k)}\mathrm{d}t &= \frac{1}{\sqrt{2\pi}}\int_{-\infty}^{\infty} f(t)\left[\mathrm{e}^{\mp i\omega t}\right]^{(k)}\mathrm{d}t \\ &= \frac{1}{\sqrt{2\pi}}\int_{-\infty}^{\infty}(\mp it)^{k}f(t)\mathrm{e}^{\mp i\omega t}\mathrm{d}t\end{aligned} \tag{1.150}$$

5. Fourier transform maps convolution into multiplication:

$$\begin{aligned} f * g = \int_{-\infty}^{\infty} f(\tau)g(t-\tau)\mathrm{d}\tau &= \hat{F}_{-}\left(\hat{F}_{+}[f(t)]\hat{F}_{+}[g(t)]\right) \\ &= \hat{F}_{+}\left(\hat{F}_{-}[f(t)]\hat{F}_{-}[g(t)]\right)\end{aligned} \tag{1.151}$$

This is *convolution theorem*. The proof is lengthy but educational. Writing out the Fourier transforms and simplifying yields

$$\begin{aligned}&\hat{F}_{-}\left(\hat{F}_{+}[f(t)]\hat{F}_{+}[g(t)]\right) \\ &= \frac{1}{2\pi}\int_{-\infty}^{\infty}\left[\int_{-\infty}^{\infty} f(t')\mathrm{e}^{-i\omega t'}\mathrm{d}t'\right]\left[\int_{-\infty}^{\infty} g(t'')\mathrm{e}^{-i\omega t''}\mathrm{d}t''\right]\mathrm{e}^{i\omega t}\mathrm{d}\omega\end{aligned} \tag{1.152}$$

Absorbing the $e^{i\omega t}$ term under the integral in the first set of square brackets produces

$$\hat{F}_-\left(\hat{F}_+[f(t)]\hat{F}_+[g(t)]\right) = \frac{1}{2\pi}\int_{-\infty}^{\infty}\left[\int_{-\infty}^{\infty} f(t')e^{-i\omega(t'-t)}dt'\right]\left[\int_{-\infty}^{\infty} g(t'')e^{-i\omega t''}dt''\right]d\omega \tag{1.153}$$

Shifting the integration variable under the first set of square brackets yields

$$\hat{F}_-\left(\hat{F}_+[f(t)]\hat{F}_+[g(t)]\right) = \frac{1}{2\pi}\int_{-\infty}^{\infty}\left[\int_{-\infty}^{\infty} f(t'+t)e^{-i\omega t'}dt'\right]\left[\int_{-\infty}^{\infty} g(t'')e^{-i\omega t''}dt''\right]d\omega \tag{1.154}$$

Rearranging the integrals to take the ω integral first then yields:

$$\hat{F}_-\left(\hat{F}_+[f(t)]\hat{F}_+[g(t)]\right) = \frac{1}{2\pi}\int_{-\infty}^{\infty}dt'\int_{-\infty}^{\infty}dt''\left[f(t'+t)g(t'')\int_{-\infty}^{\infty} e^{-i\omega(t'+t'')}d\omega\right] \tag{1.155}$$

After we recognise the delta function and use the property given in Equation (1.137), we get

$$\begin{aligned}\hat{F}_-\left(\hat{F}_+[f(t)]\hat{F}_+[g(t)]\right) &= \int_{-\infty}^{\infty}dt'\int_{-\infty}^{\infty}dt'' f(t'+t)g(t'')\delta(t'+t'')\\ &= \int_{-\infty}^{\infty} f(t-t'')g(t'')dt'' = f*g\end{aligned} \tag{1.156}$$

and likewise for $\hat{F}_+\left(\hat{F}_-[f(t)]\hat{F}_-[g(t)]\right)$. An alternative formulation is

$$\hat{F}_\pm(f*g) = \hat{F}_\pm[f(t)]\hat{F}_\pm[g(t)] \tag{1.157}$$

A corollary, called *Wiener-Khinchin theorem* [28], is useful for computing autocorrelation functions:

$$\int_{-\infty}^{\infty} f^*(\tau)f(t+\tau)d\tau = \hat{F}_+\left(\left|\hat{F}_- f(t)\right|^2\right) \tag{1.158}$$

6. When time is shifted in the input function, its Fourier transform is modulated:

$$\begin{aligned}\hat{F}_{\pm}f(t-t_0) &= \frac{1}{\sqrt{2\pi}}\int\limits_{-\infty}^{\infty} f(t-t_0)\mathrm{e}^{\mp i\omega t}\mathrm{d}t \\ &= \frac{1}{\sqrt{2\pi}}\int\limits_{-\infty}^{\infty} f(t)\mathrm{e}^{\mp i\omega(t+t_0)}\mathrm{d}t = \mathrm{e}^{\mp i\omega t_0}\hat{F}_{\pm}f(t)\end{aligned} \tag{1.159}$$

Reciprocally, the Fourier transform of a modulated function ends up being shifted:

$$\begin{aligned}\hat{F}_{\pm}\left[\mathrm{e}^{\mathrm{i}\omega_0 t}f(t)\right] &= \frac{1}{\sqrt{2\pi}}\int\limits_{-\infty}^{\infty} f(t)\mathrm{e}^{\mp \mathrm{i}(\omega\mp\omega_0)t} \\ &= \hat{F}_{\pm}[f(t)](\omega\mp\omega_0)\end{aligned} \tag{1.160}$$

These relations constitute the *modulation theorem* for the Fourier transforms.

1.7.3 Causality and Hilbert Transform

The physical principle of *causality* (in this context, "no output before input") places additional constraints on the response functions of LTI systems. If the excitation pulse arrives at time zero, the response $f(t)$ must vanish for $t<0$, and therefore must not change if we multiply it by the *step function* $h(t)$:

$$\begin{aligned}& f(t)c(t) = f(t)h(t), \\ & h(t) = \begin{Bmatrix} 0 \text{ for } t<0 \\ 1 \text{ for } t>0 \end{Bmatrix} = \int\limits_{-\infty}^{t} \delta(\tau)\mathrm{d}\tau, \quad c(t) = 1\end{aligned} \tag{1.161}$$

In the frequency domain, this means that the corresponding convolutions should be equal, that is,

$$f(\omega) * c(\omega) = f(\omega) * h(\omega) \tag{1.162}$$

The Fourier transform of a constant function is a multiple of delta function:

$$c(\omega) = \frac{1}{\sqrt{2\pi}}\int\limits_{-\infty}^{\infty} c(t)\mathrm{e}^{-i\omega t}\mathrm{d}t = \frac{1}{\sqrt{2\pi}}\int\limits_{-\infty}^{\infty} \mathrm{e}^{-i\omega t}\mathrm{d}t = \sqrt{2\pi}\delta(\omega) \tag{1.163}$$

The Fourier transform of the step function is a subtle matter [335] that is outside the scope of this book—we will not discuss it here and simply state the result:

$$h(\omega) = \frac{1}{\sqrt{2\pi}} \int_{-\infty}^{\infty} h(t) \mathrm{e}^{-\mathrm{i}\omega t} \mathrm{d}t = \frac{1}{\sqrt{2\pi}} \left(\pi\delta(\omega) - \frac{i}{\omega} \right) \tag{1.164}$$

Substitution of these results into Eq. (1.162) yields

$$\sqrt{2\pi} f(\omega) * \delta(\omega) = \frac{1}{\sqrt{2\pi}} f(\omega) * \left(\pi\delta(\omega) - \frac{i}{\omega} \right) \tag{1.165}$$

By the definition of the delta function, $f(\omega) * \delta(\omega) = f(\omega)$. After applying this simplification, we get the following convolution identity (called *Hilbert transform* [29]) relating $f(\omega)$ to itself:

$$f(\omega) = \frac{1}{i\pi} f(\omega) * \frac{1}{\omega} \quad \Leftrightarrow \quad f(\omega) = \frac{1}{i\pi} \int_{-\infty}^{\infty} \frac{f(\nu)}{\nu - \omega} \mathrm{d}\nu \tag{1.166}$$

After separating real and imaginary parts of $f(\omega)$, we arrive at *Kramers–Kronig relations* [30,31]:

$$f_{\mathrm{re}}(\omega) = \frac{1}{\pi} \int_{-\infty}^{\infty} \frac{f_{\mathrm{im}}(\nu)}{\nu - \omega} \mathrm{d}\nu, \quad f_{\mathrm{im}}(\omega) = -\frac{1}{\pi} \int_{-\infty}^{\infty} \frac{f_{\mathrm{re}}(\nu)}{\nu - \omega} \mathrm{d}\nu \tag{1.167}$$

that connect the real and the imaginary parts of the Fourier transform of a causal signal.

What Exactly Is Spin? 2

There are basic actions that preserve the identity of physical systems: translations in space and time, rotations, relativistic boosts, etc. Sets of such actions are groups under superposition; a natural classification of physical systems is therefore by irreducible representations of those groups—different irreps correspond to different patterns of behaviour under fundamental symmetry operations.

In this chapter we use group theory to turn reasonable assumptions about reality into equations of motion and conservation laws. In the Poincare group of special relativity, we find two Casimir elements: one corresponding to the invariant mass, the other to the sum of orbital angular momentum and something else. The extra quantity appears because relativistic boost generators commute into rotation generators—there are more ways of rotating things in the Minkowski space-time of Special Relativity than Newtonian mechanics had in $\mathbb{R}^3$. That extra quantity is retained by point particles—it is called spin.

2.1 Time Translation Group

We assume that reality is knowable. For a finite and isolated physical system, this implies the possibility of a descriptor Ψ that contains complete information about the system state. Empirical observations suggest that reality evolves in time according to some rules. Therefore, given the system state at a time t, the state at some future time $t+\tau$ must be:

$$\Psi(t+\tau) = \hat{U}(t,\tau)\Psi(t) \tag{2.1}$$

where the time propagation operator $\hat{U}$ applies the rules. In the observations so far, those rules appear to be unchanging. Therefore, the propagator can only depend on the time increment τ:

I. KUPROV, *Spin*,
https://doi.org/10.1007/978-3-031-05607-9_2

$$\Psi(t+\tau) = \hat{U}(\tau)\Psi(t) \tag{2.2}$$

Reality looks continuous in time, and velocities are aspects of that reality—it is reasonable to assume that all time derivatives of Ψ are continuous. Therefore, a Taylor series must exist:

$$\Psi(t+\tau) = \sum_{n=0}^{\infty} \frac{\tau^n}{n!} \frac{\partial^n}{\partial t^n} \Psi(t) \tag{2.3}$$

After comparing this with Eq. (2.2), we conclude that

$$\hat{U}(\tau) = \sum_{n=0}^{\infty} \frac{\tau^n}{n!} \frac{\partial^n}{\partial t^n} = \exp\left(\tau \frac{\partial}{\partial t}\right) \quad \Rightarrow \quad \Psi(t+\tau) = \exp\left(\tau \frac{\partial}{\partial t}\right) \Psi(t) \tag{2.4}$$

It follows that the set of all time propagators is a Lie group generated by $\partial/\partial t$. This group is Abelian—all irreducible representations are one-dimensional. The eigenfunctions of the generator are exponentials; we are at liberty (Sect. 1.5.4) to pick them so as to have unitary irreps:

$$\frac{\partial}{\partial t}\left[e^{-i\omega t}\right] = -i\omega\left[e^{-i\omega t}\right] \tag{2.5}$$

where ω is a real number with the physical meaning of oscillation frequency. The irreducible representations of the time propagation generator therefore are:

$$\mathbf{P}_{\omega} : \frac{\partial}{\partial t} \to (-i\omega) \tag{2.6}$$

The system must evolve in time under some representation of the propagator group, which will be a similarity transformation away from a direct sum of irreducible representations. Therefore:

$$\mathbf{P}_{\omega_1 \oplus \omega_2 \oplus \cdots} : \frac{\partial}{\partial t} \to -i\mathbf{H} \tag{2.7}$$

where $\mathbf{H}$ is a matrix whose eigenvalues are the frequencies of the irreducible components. Once a matrix representation is chosen, we may loosely write:

$$\frac{\partial}{\partial t} = -i\mathbf{H} \tag{2.8}$$

The same representation maps the system state descriptor $\Psi(t)$ into a vector which we will denote $|\Psi(t)\rangle$. After inserting Eq. (2.7) into the right side of Eq. (2.4):

$$|\Psi(t+\tau)\rangle = \exp(-i\mathbf{H}\tau)|\Psi(t)\rangle \tag{2.9}$$

After rewriting the exponential as $\exp(-i\mathbf{H}\tau) = \mathbf{1} - i\mathbf{H}\tau + O(\tau^2)$ and rearranging the terms, we obtain:

$$\frac{|\Psi(t+\tau)\rangle - |\Psi(t)\rangle}{\tau} = -i\mathbf{H}|\Psi(t)\rangle + O(\tau) \tag{2.10}$$

In the limit of $\tau \to 0$, the left hand side becomes a derivative, and we obtain the *Schrödinger form* of the equation of motion for the system descriptor:

$$\frac{\partial}{\partial t}|\Psi(t)\rangle = -i\mathbf{H}|\Psi(t)\rangle \tag{2.11}$$

This is not necessarily Schrödinger's actual equation [32], only the general algebraic form that our assumptions about reality require any equation of motion to have. It follows (proof by direct inspection) that a linear combination of solutions is also a solution; this is called *superposition principle*.

Because $\mathbf{H}$ is Hermitian, its exponential in Eq. (2.9) is a unitary operator—the 2-norm of the vector representation of $\Psi(t)$ is conserved. At the same time, the superposition principle allows us to reduce the representation and break the evolution up into irreps that evolve independently. Thus, in the representation that diagonalises $\mathbf{H}$, the absolute square of each element of $|\Psi(t)\rangle$ acquires a physical meaning of the contribution of that irrep to some invariant total. Experimental evidence suggests that this quantity is probability.

2.2 Full Translation Group

Empirical observations suggest that physical reality has three continuous and uniform spatial dimensions, and that the identity of a finite isolated physical system is unchanged by translations, which satisfy the properties of a Lie group. By the same argument as in Eqs. (2.2)–(2.4), translation generators are

$$\hat{T}_0 = \frac{\partial}{\partial ct}, \quad \hat{T}_1 = \frac{\partial}{\partial x}, \quad \hat{T}_2 = \frac{\partial}{\partial y}, \quad \hat{T}_3 = \frac{\partial}{\partial z} \tag{2.12}$$

where c is the proportionality coefficient between units of space and time. These operators commute—space-time translations are, therefore, a four-parametric Abelian Lie group with the following elements:

$$\hat{T}(\mathbf{r}) = \exp\left[\hat{T}_0 r_0 + \hat{T}_1 r_1 + \hat{T}_2 r_2 + \hat{T}_3 r_3\right], \quad \mathbf{r} = \begin{bmatrix} ct & x & y & z \end{bmatrix} \in \mathbb{R} \tag{2.13}$$

The eigenfunctions of the generators are again exponentials:

$$\begin{aligned} \hat{T}_0\left[e^{-i\omega_0 t}\right] &= -i(\omega_0/c)\left[e^{-i\omega_0 t}\right] \\ \hat{T}_k\left[e^{-i\omega_k r_k}\right] &= -i\omega_k\left[e^{-i\omega_k r_k}\right], \quad k \in \{1,2,3\} \end{aligned} \tag{2.14}$$

where the frequencies ω_k identify the irreps; ω_0 is called energy, the frequencies with respect to spatial coordinates are called *linear momenta*. The corresponding *linear momentum operators* are conventionally defined as

$$\hat{p}_0 = i\frac{\partial}{\partial ct}, \quad \hat{p}_1 = -i\frac{\partial}{\partial x}, \quad \hat{p}_2 = -i\frac{\partial}{\partial y}, \quad \hat{p}_3 = -i\frac{\partial}{\partial z} \tag{2.15}$$

The Casimir operator (Sect. 1.5.12) of this group under the Minkowski metric (Sect. 1.6.1):

$$m^2c^2 = \hat{p}_0^2 - \hat{p}_1^2 - \hat{p}_2^2 - \hat{p}_3^2 \tag{2.16}$$

corresponds to a quantity that connects energy with linear momentum and remains unchanged by translations and boosts; we can recognise the *invariant mass*. It must here be multiplied by another instance of the coefficient c to match the units of mass and momentum.

We conclude that irreps of the translation group classify physical systems by energy, linear momenta, and mass. Their operators commute with the time translation generator, they are therefore conserved. When the mass is zero in Eq. (2.16), we obtain what looks like a wave equation:

$$\hat{p}_0^2 = \hat{p}_1^2 + \hat{p}_2^2 + \hat{p}_3^2 \Rightarrow \frac{\partial^2}{\partial t^2} = c^2\left(\frac{\partial^2}{\partial x^2} + \frac{\partial^2}{\partial y^2} + \frac{\partial^2}{\partial z^2}\right) \tag{2.17}$$

It clarifies the physical meaning of the proportionality coefficient c—it is the velocity with which massless waves travel through empty space; in other words, the *speed of light*.

2.3 Rotation Group

Reality appears to be isotropic—a finite isolated physical system may be continuously rotated around any axis in three dimensions, and the rotation preserves its identity. By the same argument as above, for a system descriptor Ψ that depends on a rotation angle φ around a particular axis:

$$\Psi(\varphi+\alpha)=\exp\left[\alpha\frac{\partial}{\partial\varphi}\right]\Psi(\varphi) \tag{2.18}$$

Reality also looks 2π periodic with respect to rotations. This means that, unlike translations, the eigenfunctions of the rotation generator:

$$\frac{\partial}{\partial\varphi}\left[e^{-im\varphi}\right]=-im\left[e^{-im\varphi}\right] \tag{2.19}$$

have discrete frequencies because periodicity must be enforced:

$$\exp[-im(\varphi+2\pi)]=\exp[-im\varphi] \quad \Rightarrow \quad m\in\mathbb{Z} \tag{2.20}$$

Thus, the uniaxial rotation group is Abelian and its irreps are one-dimensional. The integer index m enumerating the irreps is called *orbital angular momentum* around the axis of rotation.

Rotations in three dimensions are superpositions of rotations around the three Cartesian axes. Using the chain rule to translate the angle derivative in Eq. (2.18) into the Cartesian form yields the following generators under the $\exp[-i\mathbf{A}]$ exponential map:

$$\hat{L}_{\mathrm{X}}=-i\left(y\frac{\partial}{\partial z}-z\frac{\partial}{\partial y}\right),\quad \hat{L}_{\mathrm{Y}}=-i\left(z\frac{\partial}{\partial x}-x\frac{\partial}{\partial z}\right),\quad \hat{L}_{\mathrm{Z}}=-i\left(x\frac{\partial}{\partial y}-y\frac{\partial}{\partial x}\right) \tag{2.21}$$

for rotations around the indicated axes. Recalling the linear momenta expressions from Eq. (2.12), we can recognise the non-relativistic orbital angular momentum:

$$\mathbf{L}=\mathbf{r}\times\mathbf{p} \tag{2.22}$$

The commutation relations are obtained from Eq. (2.21) by a tedious direct calculation:

$$[\hat{L}_{\mathrm{X}},\hat{L}_{\mathrm{Y}}]=i\hat{L}_{\mathrm{Z}},\quad [\hat{L}_{\mathrm{Y}},\hat{L}_{\mathrm{Z}}]=i\hat{L}_{\mathrm{X}},\quad [\hat{L}_{\mathrm{Z}},\hat{L}_{\mathrm{X}}]=i\hat{L}_{\mathrm{Y}} \tag{2.23}$$

We now recognise the SO(3) group (Sect. 1.6.2). It commutes with the time translation generator—in a finite and isolated non-relativistic physical system, orbital angular momentum is conserved. Because SO(3) is non-Abelian, the irreps are to be found by diagonalising the Casimir operator:

$$\hat{L}^2=\hat{L}_{\mathrm{X}}^2+\hat{L}_{\mathrm{Y}}^2+\hat{L}_{\mathrm{Z}}^2 \tag{2.24}$$

We can simultaneously diagonalise one of the generators (traditionally $\hat{L}_Z$), but only one because they do not commute with each other. In spherical coordinates:

$$\hat{L}^2 = -\frac{1}{\sin\theta}\frac{\partial}{\partial\theta}\left(\sin\theta\frac{\partial}{\partial\theta}\right) - \frac{1}{\sin^2\theta}\frac{\partial^2}{\partial\varphi^2}, \quad \hat{L}_Z = -i\frac{\partial}{\partial\varphi} \tag{2.25}$$

An arduous exercise in calculus, first published by Pierre-Simon Laplace [33], yields a family of eigenfunctions Laplace called *spherical harmonics*:

$$Y_{l,m}(\theta,\varphi) = \sqrt{\frac{2l+1}{4\pi}\frac{(l-m)!}{(l+m)!}}P_l^m(\cos\theta)e^{im\varphi}, \quad l \in \mathbb{Z}_{\geq 0},\ m = -l,\ldots,l \tag{2.26}$$

where P_l^m are *associated Legendre polynomials* [34]:

$$\begin{aligned} P_l^m(x) &= \frac{(-1)^m}{2^l l!}\left(1-x^2\right)^{m/2}\frac{d^{l+m}}{dx^{l+m}}\left(x^2-1\right)^l \\ P_l^{-m}(x) &= (-1)^m\frac{(l-m)!}{(l+m)!}P_l^m(x), \ m > 0 \end{aligned} \tag{2.27}$$

As intended, spherical harmonics are eigenfunctions of $\hat{L}^2$ and $\hat{L}_Z$:

$$\begin{cases} \hat{L}^2 Y_{l,m}(\theta,\varphi) = l(l+1)Y_{l,m}(\theta,\varphi) \\ \hat{L}_Z Y_{l,m}(\theta,\varphi) = mY_{l,m}(\theta,\varphi) \end{cases} \tag{2.28}$$

where l is called *spherical rank*—this index enumerates irreducible representations of SO(3). Calculating matrix elements of rotation operators in the spherical harmonic basis leads, after another arduous exercise in calculus, to the irreducible representation expressions given in Eqs. (1.116) and (1.117).

2.4 Lorentz Group

The conserved quantities arising from the symmetries analysed so far have been unremarkable: energy, momenta, mass. This changes when relativistic boosts are added to the picture—there are more ways of rotating things in Minkowski space-time than there had been in the Euclidean space. This modifies the corresponding group invariants and leads to the emergence of a new quantity—spin.

2.4.1 Boost Generators

Expressions for Lorentz boost matrices were obtained in Sect. 1.6.1; they may be converted into a more elegant form by setting $v/c = \tanh\phi$, where the hyperbolic angle ϕ is called rapidity:

$$\mathbf{K}_\mathrm{X}(\phi) = \begin{pmatrix} \cosh(\phi) & -\sinh(\phi) \\ -\sinh(\phi) & \cosh(\phi) \end{pmatrix} = \exp\left[-i\begin{pmatrix} 0 & -i \\ -i & 0 \end{pmatrix}\phi\right] \tag{2.29}$$

and likewise for $\mathbf{K}_\mathrm{Y}$ and $\mathbf{K}_\mathrm{Z}$. After comparing this with rotation generators in Eqs. (1.109) and (2.21), and making appropriate variable replacements, we obtain the differential form of *boost generators*:

$$\begin{gathered} \hat{K}_\mathrm{X} = -i\left(x\frac{\partial}{\partial ct} + ct\frac{\partial}{\partial x}\right), \quad \hat{K}_\mathrm{Y} = -i\left(y\frac{\partial}{\partial ct} + ct\frac{\partial}{\partial y}\right), \\ \hat{K}_\mathrm{Z} = -i\left(z\frac{\partial}{\partial ct} + ct\frac{\partial}{\partial z}\right) \end{gathered} \tag{2.30}$$

Unlike Eq. (2.21), these generators have a positive sign in the brackets—boosts are hyperbolic, rather than trigonometric, transformations of space-time. It may be verified by direct inspection that the three boosts on their own do not generate a group—their commutators are rotation generators, for example:

$$\left[\hat{K}_\mathrm{X}, \hat{K}_\mathrm{Y}\right] = -i\hat{L}_\mathrm{Z}, \quad \hat{L}_\mathrm{Z} = -i\left(x\frac{\partial}{\partial y} - y\frac{\partial}{\partial x}\right) \tag{2.31}$$

Another tedious calculation shows that the set of three rotation generators and three boost generators is closed under the commutation operation:

$$\left[\hat{L}_n, \hat{L}_k\right] = i\varepsilon_{nkm}\hat{L}_m, \quad \left[\hat{K}_n, \hat{K}_k\right] = -i\varepsilon_{nkm}\hat{L}_m, \quad \left[\hat{L}_n, \hat{K}_k\right] = i\varepsilon_{nkm}\hat{K}_m \tag{2.32}$$

where ε_{nkm} is the Levi-Civita symbol [333]. The corresponding algebra generates a six-parametric Lie group, called the *Lorentz group* [35], which preserves the Minkowski metric (Sect. 1.6.1). It has three Euclidean and one pseudo-Euclidean component, and is therefore denoted SO(3,1); its algebra is $\mathfrak{so}(3,1)$.

Commutation relations in Eq. (2.32) indicate that rotation and boost generators are not individually Lorentz invariant—they are interconverted by similarity transformations, for example:

$$\begin{aligned} e^{-i\hat{K}_\mathrm{Z}\varphi}\hat{K}_\mathrm{X}e^{+i\hat{K}_\mathrm{Z}\varphi} &= \hat{K}_\mathrm{X}\cosh(\varphi) - \hat{L}_\mathrm{Y}\sinh(\varphi) \\ e^{-i\hat{K}_\mathrm{Z}\varphi}\hat{L}_\mathrm{X}e^{+i\hat{K}_\mathrm{Z}\varphi} &= \hat{L}_\mathrm{X}\cosh(\varphi) + \hat{K}_\mathrm{Y}\sinh(\varphi) \end{aligned} \tag{2.33}$$

Although $\{\hat{L}_X, \hat{L}_Y, \hat{L}_Z\}$ do span a proper sub-algebra of $\mathfrak{so}(3,1)$, their irreps are different from those of $\mathfrak{so}(3)$ because generators of relativistic boosts must also be represented.

2.4.2 Irreps of Lorentz Group

A tedious calculation of the Killing form (Sect. 1.5.12) yields the following Casimir elements:

$$\begin{aligned} \hat{L}^2 - \hat{K}^2 &= \hat{L}_X^2 + \hat{L}_Y^2 + \hat{L}_Z^2 - \hat{K}_X^2 - \hat{K}_Y^2 - \hat{K}_Z^2 \\ \hat{L} \cdot \hat{K} &= \hat{L}_X \hat{K}_X + \hat{L}_Y \hat{K}_Y + \hat{L}_Z \hat{K}_Z \end{aligned} \tag{2.34}$$

The eigenfunction problem is formidable, but an elegant trick [36] involving complexification (Sect. 1.5.10) reduces the task to a known case. If we define the following complex linear combinations

$$\begin{cases} \hat{M}_n = \left(\hat{L}_n + i\hat{K}_n\right)/2 \\ \hat{N}_n = \left(\hat{L}_n - i\hat{K}_n\right)/2 \end{cases} \Leftrightarrow \begin{cases} \hat{L}_n = \hat{M}_n + \hat{N}_n \\ i\hat{K}_n = \hat{M}_n - \hat{N}_n \end{cases}, \quad n \in \{X, Y, Z\} \tag{2.35}$$

and obtain their commutators relations, we find two three-dimensional ideals (Sect. 1.5.8):

$$\left[\hat{M}_n, \hat{M}_k\right] = i\varepsilon_{nkm}\hat{M}_m \ , \quad \left[\hat{N}_n, \hat{N}_k\right] = i\varepsilon_{nkm}\hat{N}_m \ , \quad \left[\hat{M}_n, \hat{N}_k\right] = 0 \tag{2.36}$$

These ideals are spanned by Hermitian matrices and satisfy the commutation relations of both $\mathfrak{so}(3)$ and $\mathfrak{su}(2)$. Because those algebras have different irreducible representations, we must decide exactly which ones we have here. The difference (Sect. 1.6.4) is the period of group orbits: 2π under $\mathfrak{so}(3)$ and 4π under $\mathfrak{su}(2)$. Let us, therefore, look at the orbit period, for example under $\hat{M}_X$:

$$\left[\hat{L}_X, \hat{K}_X\right] = 0 \quad \Rightarrow \quad \exp\left[-i\frac{\hat{L}_X + i\hat{K}_X}{2}\alpha\right] = \exp\left[-i\hat{L}_X\frac{\alpha}{2}\right]\exp\left[\hat{K}_X\frac{\alpha}{2}\right] \tag{2.37}$$

For the rotation, it is clear from Eq. (1.109) that the parameter α must go to 4π for a full revolution to happen. For the boost, the same conclusion follows from Eq. (2.29). Thus, both ideals are $\mathfrak{su}(2)$:

$$\mathfrak{so}(3,1)_{\mathbb{C}} \cong \mathfrak{su}(2)_{\mathbb{C}} \oplus \mathfrak{su}(2)_{\mathbb{C}} \tag{2.38}$$

We already know (Sect. 1.6.3) the irreps of $\mathfrak{su}(2)$. Irreps of a direct sum of two simple Lie algebras are spanned by Kronecker products of the irreps of the components, therefore:

$$\begin{cases} \mathbf{M}_k^{(j_1 j_2)} = \mathbf{S}_k^{(j_1)} \otimes \mathbf{1}^{(j_2)} \\ \mathbf{N}_k^{(j_1 j_2)} = \mathbf{1}^{(j_1)} \otimes \mathbf{S}_k^{(j_2)} \end{cases}, \quad \begin{cases} \mathbf{L}_k^{(j_1 j_2)} = \mathbf{M}_k^{(j_1 j_2)} + \mathbf{N}_k^{(j_1 j_2)} \\ i\mathbf{K}_k^{(j_1 j_2)} = \mathbf{M}_k^{(j_1 j_2)} - \mathbf{N}_k^{(j_1 j_2)} \end{cases} \tag{2.39}$$

where j_1 and j_2 are independent positive integer or half-integer indices labelling the irrep, $\mathbf{S}_k^{(j)}$ are rank j irreps of $\mathfrak{su}(2)$, and $\mathbf{1}^{(j)}$ is a unit matrix of the same dimension as $\mathbf{S}_k^{(j)}$.

In this case, the irreps of the complexification are the same as the irreps of the parent group—direct inspection indicates that the Casimir operators in Eq. (2.34) are multiples of the unit matrix

$$\begin{aligned} \left[\mathbf{L}^2 - \mathbf{K}^2\right]^{(j_1 j_2)} &= 2\left[j_1(j_1+1) + j_2(j_2+1)\right]\mathbf{1}^{(j_1 j_2)} \\ \left[\mathbf{L} \cdot \mathbf{K}\right]^{(j_1 j_2)} &= -i\left[j_1(j_1+1) - j_2(j_2+1)\right]\mathbf{1}^{(j_1 j_2)} \end{aligned} \tag{2.40}$$

indicating that the representations in Eq. (2.39) are irreducible for $\mathfrak{so}(3,1)$.

2.4.3 Irreps of Lorentz Group with Parity

Reality appears to have well-defined properties under $\{x, y, z\}$ coordinate inversion—some quantities (charge, mass, orbital angular momentum) are invariant, and others (position, linear momentum, vector potential) change sign. Because the generators of the Lorentz group have different behaviour under parity—rotations are invariant, and boosts change sign—the empirically established requirement for symmetry or antisymmetry of observables under inversion modifies its representation structure.

Consider the direct product $\mathrm{SO}(3,1) \times \mathrm{P}$ of $\mathrm{SO}(3,1)$ and the *parity group* P, which contains the identity operation $\hat{E}$ and a Cartesian coordinate inversion operation $\hat{I}$:

$$\mathrm{P} = \{\hat{E}, \hat{I}\}, \quad \hat{I}\{x, y, z\} = \{-x, -y, -z\} \tag{2.41}$$

Because inversion changes the sign of boost generators, but not of rotation generators, $\hat{L} \cdot \hat{K}$ is not an invariant of $\mathrm{SO}(3,1) \times \mathrm{P}$, but $\hat{L}^2 - \hat{K}^2$ still is. Thus, the number of Casimir operators changes when inversion is introduced, and we must reconsider the representation structure.

Inversion exchanges $\hat{M}_k$ and $\hat{N}_k$ in Eq. (2.35), and thereby also swaps the two instances of $\mathfrak{su}(2)_{\mathbb{C}}$ in Eq. (2.38). This means that (j_1, j_2) irrep of $\mathrm{SO}(3,1)$ is mapped by inversion into (j_2, j_1). Accordingly, when $j_1 \neq j_2$, irreducible representations of $SO(3,1)\times P$ are direct sums of irreps of $SO(3,1)$ with exchanged indices: $(j_1,j_2)\oplus(j_2,j_1)$. When $j_1=j_2=j$, this yields two redundant identical copies, and the irreducible representation is simply (j,j).

2.4.4 Poincare Group and the Emergence of Spin

The full symmetry group of special relativity must include all translations, rotations, and boosts. This is called *Poincare group* [37], its commutation relations may be obtained by direct inspection:

$$\begin{gathered}\left[\hat{L}_n,\hat{L}_k\right]=i\varepsilon_{nkm}\hat{L}_m,\quad\left[\hat{K}_n,\hat{K}_k\right]=-i\varepsilon_{nkm}\hat{L}_m,\quad\left[\hat{L}_n,\hat{K}_k\right]=i\varepsilon_{nkm}\hat{K}_m\\ \left[\hat{L}_n,\hat{p}_k\right]=i\varepsilon_{nkm}\hat{p}_m,\quad\left[\hat{L}_n,\hat{p}_0\right]=0,\quad\left[\hat{p}_\mu,\hat{p}_\nu\right]=0\\ \left[\hat{K}_n,\hat{p}_k\right]=i\delta_{nk}\hat{p}_0,\quad\left[\hat{K}_n,\hat{p}_0\right]=i\hat{p}_n\\ n,k,m\in\{1,2,3\},\quad\mu,\nu\in\{0,1,2,3\}\end{gathered}\tag{2.42}$$

This group has two Casimir operators—the invariant mass from the translation subgroup (Sect. 2.2):

$$m^2c^2=\hat{p}_0^2-\hat{p}^2,\quad\hat{p}^2=\hat{p}_1^2+\hat{p}_2^2+\hat{p}_3^2\tag{2.43}$$

and the relativistic generalisation of orbital angular momentum $\hat{L}^2-\hat{K}^2$ that has the irreps enumerated by the two half-integer indices in Eq. (2.40). Thus, what is conserved in Minkowski space-time is the sum of orbital angular momentum invariant $\hat{L}^2$ and something else that is contained in $\hat{K}^2$. To maintain a sentimental connection with celestial mechanics, and to retain a modicum of intuition about its behaviour, that quantity is called *spin*.

2.5 Dirac's Equation and Electron Spin

The symmetry arguments presented above are only a framework that classifies isolated physical systems, including elementary particles, by the irreps of the Poincare group. The question of which particle belongs to which irrep is decided empirically; in this book, we deal with electrons and nuclei. The former appear to be elementary, and the internal structure of the latter will be considered in Chap. 3.

2.5.1 Dirac's Equation

The only irrep of $SO(3,1)\times P$ that fits the observed behaviour of electrons (two values with opposite sign [38] of what appears to be non-orbital angular momentum

[39]) is $(1/2,0) \oplus (0,1/2)$. After using Eq. (2.39) and taking the direct sum, we obtain the following matrices:

$$\mathbf{L}_n = \begin{pmatrix} \boldsymbol{\sigma}_n/2 & \mathbf{0} \\ \mathbf{0} & \boldsymbol{\sigma}_n/2 \end{pmatrix}, \quad \mathbf{K}_n = \begin{pmatrix} i\boldsymbol{\sigma}_n/2 & \mathbf{0} \\ \mathbf{0} & -i\boldsymbol{\sigma}_n/2 \end{pmatrix}, \quad n \in \{\mathrm{X,Y,Z}\}$$
$$\boldsymbol{\sigma}_\mathrm{X} = \begin{pmatrix} 0 & 1 \\ 1 & 0 \end{pmatrix}, \quad \boldsymbol{\sigma}_\mathrm{Y} = \begin{pmatrix} 0 & -i \\ i & 0 \end{pmatrix}, \quad \boldsymbol{\sigma}_\mathrm{Z} = \begin{pmatrix} 1 & 0 \\ 0 & -1 \end{pmatrix} \tag{2.44}$$

where $\{\boldsymbol{\sigma}_\mathrm{X}, \boldsymbol{\sigma}_\mathrm{Y}, \boldsymbol{\sigma}_\mathrm{Z}\}$ are *Pauli matrices* [40]. At the same time, the spectrum of values measured experimentally for the linear momentum of free electrons is continuous. We are, therefore, compelled to use a continuous representation for the translation subgroup, and an irreducible matrix representation for the Lorentz subgroup. The two invariance relations become

$$\begin{cases} \hat{p}^2 = \hat{p}_0^2 - m^2c^2 \\ \mathbf{L}^2 - \mathbf{K}^2 = 3/2 \end{cases} \tag{2.45}$$

where the 3/2 comes from Eq. (2.40). After multiplying these equations term by term, we obtain

$$(\mathbf{L}^2 - \mathbf{K}^2)\hat{p}^2 = \frac{3}{2}(\hat{p}_0^2 - m^2c^2) \quad \Rightarrow \quad \begin{pmatrix} \boldsymbol{\sigma}^2\hat{p}^2 & \mathbf{0} \\ \mathbf{0} & \boldsymbol{\sigma}^2\hat{p}^2 \end{pmatrix} = \hat{p}_0^2 - m^2c^2 \tag{2.46}$$

where $\boldsymbol{\sigma}^2 = \boldsymbol{\sigma}_\mathrm{X}^2 + \boldsymbol{\sigma}_\mathrm{Y}^2 + \boldsymbol{\sigma}_\mathrm{Z}^2$. A special property of the matrices spanning the $(1/2,0)\oplus(0,1/2)$ irrep, verifiable by direct inspection, is that $\sigma^2\hat{p}^2 = (\sigma \cdot \hat{\mathbf{p}})^2$, where:

$$\mathbf{p} = \begin{bmatrix} \hat{p}_\mathrm{X} \\ \hat{p}_\mathrm{Y} \\ \hat{p}_\mathrm{Z} \end{bmatrix}, \quad \boldsymbol{\sigma} = \begin{bmatrix} \boldsymbol{\sigma}_\mathrm{X} \\ \boldsymbol{\sigma}_\mathrm{Y} \\ \boldsymbol{\sigma}_\mathrm{Z} \end{bmatrix} \tag{2.47}$$

With these vector operators in place, after a rearrangement:

$$\hat{p}_0^2 = \begin{pmatrix} m^2c^2 + (\boldsymbol{\sigma} \cdot \mathbf{p})^2 & \mathbf{0} \\ \mathbf{0} & m^2c^2 + (\boldsymbol{\sigma} \cdot \mathbf{p})^2 \end{pmatrix} \tag{2.48}$$

After remembering (Sect. 2.2) that $\hat{p}_0 = i\partial/\partial ct$, we obtain

$$\frac{\partial^2}{\partial ct^2} = -\begin{bmatrix} m^2c^2 + (\boldsymbol{\sigma}\cdot\mathbf{p})^2 & \mathbf{0} \\ \mathbf{0} & m^2c^2 + (\boldsymbol{\sigma}\cdot\mathbf{p})^2 \end{bmatrix} \tag{2.49}$$

This equation is second-order with respect to time, but the reasoning in Sect. 2.1 compels us to seek a first-order equation with a traceless Hermitian generator on the right-hand side:

$$\frac{\partial}{\partial t} = -i\mathbf{H}, \qquad \mathbf{H}^2 = \begin{bmatrix} m^2c^4 + c^2(\boldsymbol{\sigma}\cdot\mathbf{p})^2 & \mathbf{0} \\ \mathbf{0} & m^2c^4 + c^2(\boldsymbol{\sigma}\cdot\mathbf{p})^2 \end{bmatrix} \tag{2.50}$$

For the matrices in question, the generator must have the structure:

$$\mathbf{H} = \begin{pmatrix} +\mathbf{a} & i\mathbf{b} \\ i\mathbf{b} & -\mathbf{a} \end{pmatrix} \tag{2.51}$$

where **a** and **b** are Hermitian 2×2 matrices. Placing this into Eq. (2.50) and solving a system of matrix equations yields, after a tedious calculation, several solutions. The conventional one is

$$\frac{\partial}{\partial t} = -i\begin{bmatrix} +mc^2 & c(\boldsymbol{\sigma}\cdot\mathbf{p}) \\ c(\boldsymbol{\sigma}\cdot\mathbf{p}) & -mc^2 \end{bmatrix} \tag{2.52}$$

and the rest differ by minus signs that pronounce on which directions are to be called positive in space and time. This is the *Dirac equation* [41], an accurate description of the dynamics of a free relativistic electron at energies much lower than the gamma photon pair production energy (1.03 meV; above that energy, more sophisticated treatments using quantum field theory must be used).

An optional final adjustment is to bring units of time, frequency, position, and energy into the conventions that humans have historically agreed to use. This introduces the *Planck constant* $\hbar$ that translates between units of frequency and energy [42]:

$$\frac{\partial}{\partial t} = -\frac{i}{\hbar}\begin{bmatrix} +mc^2 & c(\boldsymbol{\sigma}\cdot\mathbf{p}) \\ c(\boldsymbol{\sigma}\cdot\mathbf{p}) & -mc^2 \end{bmatrix} \tag{2.53}$$

and adjusts the definitions of linear and angular momentum operators

$$\hat{L}_{\mathrm{X}} = -i\hbar\left(y\frac{\partial}{\partial z} - z\frac{\partial}{\partial y}\right), \quad \hat{L}_{\mathrm{Y}} = -i\hbar\left(z\frac{\partial}{\partial x} - x\frac{\partial}{\partial z}\right), \quad \hat{L}_{\mathrm{Z}} = -i\hbar\left(x\frac{\partial}{\partial y} - y\frac{\partial}{\partial x}\right)$$

$$\hat{p}_0 = i\hbar\frac{\partial}{\partial ct}, \quad \hat{p}_1 = -i\hbar\frac{\partial}{\partial x}, \quad \hat{p}_2 = -i\hbar\frac{\partial}{\partial y}, \quad \hat{p}_3 = -i\hbar\frac{\partial}{\partial z} \tag{2.54}$$

with the corresponding changes to the commutation relations. It is common to work in angular frequency units for energy. In that convention $\hbar = 1$, and the symbol is then omitted.

2.5.2 Total Angular Momentum and Spin

It may be confirmed by direct inspection that orbital angular momentum operators $\hat{L}_{\mathrm{XYZ}}$ commute with the Casimir invariants $\hat{p}^2$, $\hat{L}^2$, $\partial^2/\partial t^2$ of the Euclidean group, for example:

$$\begin{aligned}\left[\hat{p}^2, \hat{L}_{\mathrm{Z}}\right] &= \left[\hat{p}_{\mathrm{X}}^2 + \hat{p}_{\mathrm{Y}}^2 + \hat{p}_{\mathrm{Z}}^2, x\hat{p}_{\mathrm{Y}} - y\hat{p}_{\mathrm{X}}\right] = \left[\hat{p}_{\mathrm{X}}^2, x\hat{p}_{\mathrm{Y}}\right] - \left[\hat{p}_{\mathrm{Y}}^2, y\hat{p}_{\mathrm{X}}\right] \\ &= \hat{p}_{\mathrm{Y}}\left[\hat{p}_{\mathrm{X}}^2, x\right] - \hat{p}_{\mathrm{X}}\left[\hat{p}_{\mathrm{Y}}^2, y\right] = [\ldots] = 2i(\hat{p}_{\mathrm{X}}\hat{p}_{\mathrm{Y}} - \hat{p}_{\mathrm{Y}}\hat{p}_{\mathrm{X}}) = 0\end{aligned} \tag{2.55}$$

This makes orbital angular momentum a conserved quantity in the Euclidean space. The situation changes in Minkowski space—generators of SO(3) do not commute with the Lorentz group and therefore do not commute with Dirac's evolution generator in Eq. (2.52), which we rewrite in the following form:

$$\mathbf{H} = \boldsymbol{\beta} mc^2 + c\sum_k \boldsymbol{\alpha}_k \hat{p}_k, \quad \boldsymbol{\beta} = \begin{pmatrix} \mathbf{1} & \mathbf{0} \\ \mathbf{0} & -\mathbf{1} \end{pmatrix}, \quad \boldsymbol{\alpha}_k = \begin{pmatrix} \mathbf{0} & \boldsymbol{\sigma}_k \\ \boldsymbol{\sigma}_k & \mathbf{0} \end{pmatrix} \tag{2.56}$$

The following commutators then help us find the new constant of motion:

$$\begin{aligned}\left[\mathbf{H}, \hat{L}_{\mathrm{Z}}\right] &= c\sum_k \left[\boldsymbol{\alpha}_k \hat{p}_k, \hat{L}_{\mathrm{Z}}\right] = c\sum_k \boldsymbol{\alpha}_k \left[\hat{p}_k, \hat{L}_{\mathrm{Z}}\right] = -ic(\boldsymbol{\alpha}_{\mathrm{X}}\hat{p}_{\mathrm{Y}} - \boldsymbol{\alpha}_{\mathrm{Y}}\hat{p}_{\mathrm{X}}) \\ \left[\mathbf{H}, \boldsymbol{\sigma}_{\mathrm{Z}}\right] &= c\sum_k \left[\boldsymbol{\alpha}_k \hat{p}_k, \boldsymbol{\sigma}_{\mathrm{Z}}\right] = c\sum_k \left[\boldsymbol{\alpha}_k, \boldsymbol{\sigma}_{\mathrm{Z}}\right]\hat{p}_k = 2ic(\boldsymbol{\alpha}_{\mathrm{X}}\hat{p}_{\mathrm{Y}} - \boldsymbol{\alpha}_{\mathrm{Y}}\hat{p}_{\mathrm{X}})\end{aligned} \tag{2.57}$$

and likewise for X and Y Cartesian directions. Combining these two equations in such a way as to make the right-hand side vanish reveals that the conserved quantity under Dirac's equation is $\mathbf{J} = \hat{L} + \boldsymbol{\sigma}/2$. This quantity is called *total angular momentum*, and the spin part $\mathbf{S} = \boldsymbol{\sigma}/2$ is now explicit—for an electron, its operators come from the two-dimensional irrep of $\mathfrak{su}(2)$ discussed in Sect. 1.6.3.2:

$$\mathbf{S}_{\mathrm{X}} = \begin{pmatrix} 0 & 1/2 \\ 1/2 & 0 \end{pmatrix}, \quad \mathbf{S}_{\mathrm{Y}} = \begin{pmatrix} 0 & -i/2 \\ i/2 & 0 \end{pmatrix}, \quad \mathbf{S}_{\mathrm{Z}} = \begin{pmatrix} 1/2 & 0 \\ 0 & -1/2 \end{pmatrix} \tag{2.58}$$

This is the consequence of Eq. (2.42) where boost generators commute into rotation generators: there are more ways of rotating things in the Minkowski space-time than there had been in the Euclidean space.

2.5.3 Total Angular Momentum Representation—Numerical

In the previous section, we did not pick any specific irrep for linear and orbital angular momentum operators, keeping them instead in their differential form. Representations of linear momenta are either continuous (free particles) or infinite-dimensional (confined particles); translational quantum dynamics is outside the scope of this book. However, irreps of orbital angular momentum are finite-dimensional (Sect. 1.6.2). Because irreps of $\mathfrak{so}(3)$ are a subset of the irreps of $\mathfrak{su}(2)$, composite systems involving orbital angular momentum and spin are an exercise in representation theory of $\mathfrak{su}(2)$.

Consider a system with two non-interacting subsystems A and B. Any propagator that acts on subsystem A must have the form $\exp\left[-i\mathbf{J}^{(\mathrm{A})}\varphi_{\mathrm{A}}\right]\otimes\mathbf{1}^{(\mathrm{B})}$, where $\mathbf{J}^{(\mathrm{A})}$ is a Hermitian matrix, φ_{A} is a real number, and $\mathbf{1}^{(\mathrm{B})}$ is a unit matrix. Likewise for the operators that act on subsystem B: $\mathbf{1}_{\mathrm{A}}\otimes\exp\left[-i\mathbf{J}^{(\mathrm{B})}\varphi_{\mathrm{B}}\right]$. These operators commute, and therefore the general form of a composite system operator is

$$\begin{aligned}&\left[\exp\left(-i\mathbf{J}^{(\mathrm{A})}\varphi_{\mathrm{A}}\right)\otimes\mathbf{1}^{(\mathrm{B})}\right]\left[\mathbf{1}^{(\mathrm{A})}\otimes\exp\left(-i\mathbf{J}^{(\mathrm{B})}\varphi_{\mathrm{B}}\right)\right]\\&\quad=\exp\left(-i\mathbf{J}^{(\mathrm{A})}\varphi_{\mathrm{A}}\right)\otimes\exp\left(-i\mathbf{J}^{(\mathrm{B})}\varphi_{\mathrm{B}}\right)\end{aligned}\tag{2.59}$$

Direct product representations are reducible; to reduce a rep, we must diagonalise its Casimir element (Sect. 1.5.12). The generators of the group direct product in Eq. (2.59) are

$$\mathbf{J}_k=\mathbf{1}^{(\mathrm{A})}\otimes\mathbf{J}_k^{(\mathrm{B})}+\mathbf{J}_k^{(\mathrm{A})}\otimes\mathbf{1}^{(\mathrm{B})},\quad k\in\{\mathrm{X,Y,Z}\}\tag{2.60}$$

Consider now a specific case of the two subsystems both having $j=1/2$ and therefore the two-dimensional irreps listed in Eq. (2.58) for $\mathbf{J}_{\{\mathrm{X,Y,Z}\}}^{\{\mathrm{A,B}\}}$. Using Eq. (2.60) and taking the sum of squares yields the following Casimir element for the direct product representation:

$$\mathbf{J}^2=\mathbf{J}_{\mathrm{X}}^2+\mathbf{J}_{\mathrm{Y}}^2+\mathbf{J}_{\mathrm{Z}}^2=\begin{pmatrix}2&0&0&0\\0&1&1&0\\0&1&1&0\\0&0&0&2\end{pmatrix}\tag{2.61}$$

When this matrix is diagonalised $\mathbf{J}^2=\mathbf{V}\boldsymbol{\Sigma}\mathbf{V}^\dagger$, we find (on the diagonal of $\boldsymbol{\Sigma}$) one eigenvalue of $j(j+1)=0$ and three eigenvalues of $j(j+1)=2$. After reduction we, therefore, have a one-dimensional irrep with $j=0$, and a three-dimensional irrep with $j=1$. In abbreviated notation:

$$\left(\frac{1}{2}\right) \otimes \left(\frac{1}{2}\right) \cong (0) \oplus (1) \tag{2.62}$$

where the relation is not of identity, but of equivalence—the representations are connected by a unitary transformation accomplished by the eigenvector matrix $\mathbf{V}$.

The example above avoids diagonalising the Casimir element analytically—we rely on numerical linear algebra. That is intentional and recommended because mountains of algebraic spaghetti in thousands of papers are then bypassed. For arbitrary j_1 and j_2, we proceed as follows:

1. Use Eqs. (1.128) to obtain irreducible representations $\mathbf{J}^{\{1,2\}}_{\{X,Y,Z\}}$ for the two subsystems.
2. Use Eq. (2.60) to obtain $\mathbf{J}_{\{X,Y,Z\}}$ and the Casimir element for the composite system.
3. Diagonalise $\mathbf{J}^2$, collect its eigenvectors into sets $\mathbf{V}^{(j)}$ where $j(j+1)$ is the eigenvalue.
4. For each j:
   (a) Project $\mathbf{J}_{\{X,Y,Z\}}$ into the irreducible representation:

$$\mathbf{J}^{(j)}_{\{X,Y,Z\}} = \mathbf{V}^{(j)\dagger}\mathbf{J}_{\{X,Y,Z\}}\mathbf{V}^{(j)} \tag{2.63}$$

   (b) Diagonalise $\mathbf{J}^{(j)}_Z$:

$$\mathbf{J}^{(j)}_Z \to \mathbf{U}\mathbf{J}^{(j)}_Z\mathbf{U}^\dagger, \tag{2.64}$$

   sort the eigenvectors in $\mathbf{U}$ to put the eigenvalues in descending order, and adjust eigenvector phases to ensure that $\mathbf{U}^\dagger\mathbf{J}^{(j)}_X\mathbf{U}$ is real, symmetric, and positive.

   (c) Update irrep projectors to account for the operations performed in (b):

$$\mathbf{V}^{(s)} \to \mathbf{V}^{(s)}\mathbf{U} \tag{2.65}$$

The resulting matrices $\mathbf{J}^{(j)}_{\{X,Y,Z\}} = \mathbf{V}^{(s)\dagger}\mathbf{J}_{\{X,Y,Z\}}\mathbf{V}^{(s)}$ are irreducible representations of $\mathfrak{su}(2)$. They correspond to the possible values of j that the composite system can have. Collectively, these irreducible blocks are called the *total angular momentum representation*; it is just a basis set transformation—often a matter of convenience because the total angular momentum is conserved, and the irreps may sometimes be interpreted physically as non-interacting subspaces of the system state space.

2.5.4 Total Angular Momentum Representation—Analytical

The transformations described in the previous section may be performed analytically. This is rarely efficient in the era of powerful computers, but may be useful for didactic exploration of small systems and in situations where physical insight is obtained by exploring the corresponding conservation laws.

For a direct product of canonical (diagonal $\mathbf{J}_{\mathrm{Z}}$; real, positive, and symmetric $\mathbf{J}_{\mathrm{X}}$) irreducible representations (j_1) and (j_2) of $\mathfrak{su}(2)$, a protracted and technical analysis [22] yields the following reduction:

$$(j_1) \otimes (j_2) \cong \bigoplus_{j=|j_1-j_2|}^{j_1+j_2} (j) \tag{2.66}$$

Given sets of eigenvectors $\{|j_1, m_1\rangle\}$ and $\{|j_2, m_2\rangle\}$ pertaining to the two subsystems, such that:

$$\begin{array}{ll} \mathbf{J}_1^2|j_1,m_1\rangle = j_1(j_1+1)|j_1,m_1\rangle & \mathbf{J}_2^2|j_2,m_2\rangle = j_2(j_2+1)|j_2,m_2\rangle \\ \mathbf{J}_{1\mathrm{Z}}|j_1,m_1\rangle = m_1|j_1,m_1\rangle & , \quad \mathbf{J}_{2\mathrm{Z}}|j_2,m_2\rangle = m_2|j_2,m_2\rangle \end{array} \tag{2.67}$$

the analytical expression for the eigenvectors of the irreducible (j) blocks in Eq. (2.66) is [22]:

$$|j,m\rangle = \sum_{\substack{j_1,j_2,\\ m_1,m_2}} C^{j,m}_{j_1,m_1,j_2,m_2} |j_1,m_1\rangle \otimes |j_2,m_2\rangle \quad , \quad \begin{array}{r} \mathbf{J}^2|j,m\rangle = j(j+1)|j,m\rangle \\ \mathbf{J}_{\mathrm{Z}}|j,m\rangle = m|j,m\rangle \end{array} \tag{2.68}$$

where the expression for *Clebsch-Gordan coefficients* [43]

$$\begin{aligned} C^{j,m}_{j_1,m_1,j_2,m_2} &= \left[\frac{(j_1+j_2-j)!(j_1-j_2+j)!(-j_1+j_2+j)!}{(j_1+j_2+j+1)!}\right]^{1/2} \\ &\times \left[\frac{(2j+1)(j+m)!(j-m)!}{(j_1+m_1)!(j_1-m_1)!(j_2+m_2)!(j_2-m_2)!}\right]^{1/2} \delta_{m,m_1+m_2} \\ &\times \sum_z \frac{(-1)^{j_2+m_2+z}(j+j_2+m_1-z)!(j_1+m_2-m+z)!}{z!(j-j_1+j_2-z)!(j+m_1+m_2-z)!(j_1-j_2-m+z)!} \end{aligned} \tag{2.69}$$

illustrates the point about numerical treatment being preferable. Here, free indices run over all values that produce non-negative numbers under factorials, and factorials of positive half-integers are to be computed using the connection to gamma functions:

$$n! = \Gamma(n+1) = \int_0^\infty e^{-x} x^n \mathrm{d}x \tag{2.70}$$

Due to the abundance of factorials, numerical implementations of Eq. (2.69) must use arbitrary precision arithmetic—standard IEEE754 64-bit floating-point precision is insufficient.

2.5.5 Benefits of the Individual Momentum Representation

Representation reductions discussed in Sects. 2.5.3 and 2.5.4 are only basis set transformations. A reasonable question is whether they are needed at all—we could just as well run simulations using matrices that come straight out of Eq. (2.60). There are computational complexity arguments against reducing direct product representations: the difficulties computing Clebsch-Gordan coefficients, and the fact that matrix diagonalisation is expensive and hard to parallelise. There are also physical reasons: when the total angular momentum arises from combinations of spin and orbital angular momenta of subsystems (in nuclei, *f*-electron shells of lanthanides, etc.), the system may populate the maximal—for the unitary dynamics of a $2j+1$ level quantum system—$\mathfrak{su}(2s+1)$ algebra. In that case, the reduction becomes impractical because $\mathfrak{su}(>2)$ algebras have multiple Casimir elements.

A good example is nuclear quadrupole interaction, discussed in detail in Sect. 3.1.2, where different projections of nuclear spin (in reality, total nuclear angular momentum) are coupled to different spatial charge densities. As a result, nuclear spin Hamiltonian acquires additional terms:

$$\begin{gathered}
\mathbf{H}_{\mathrm{NQI}} = \overbrace{\underbrace{\omega_{\mathrm{X}}\mathbf{S}_{\mathrm{X}} + \omega_{\mathrm{Y}}\mathbf{S}_{\mathrm{Y}} + \omega_{\mathrm{Z}}\mathbf{S}_{\mathrm{Z}}}_{\mathfrak{su}(2)} + \sum_{q=-2}^{2} b_{2q}\mathbf{O}_{2q}}^{\mathfrak{su}(3)}, \qquad \omega_{\mathrm{XYZ}}, b_{2q} \in \mathbb{R} \\
\mathbf{O}_{2,-2} = \left(\mathbf{S}_{+}^{2} - \mathbf{S}_{-}^{2}\right)/2i, \quad \mathbf{O}_{2,+2} = \left(\mathbf{S}_{+}^{2} + \mathbf{S}_{-}^{2}\right)/2, \\
\mathbf{O}_{2,-1} = [\mathbf{S}_{\mathrm{Z}}, \mathbf{S}_{\mathrm{Y}}]_{+}, \quad \mathbf{O}_{2,+1} = [\mathbf{S}_{\mathrm{Z}}, \mathbf{S}_{\mathrm{X}}]_{+}, \quad \mathbf{O}_{2,0} = 3\mathbf{S}_{\mathrm{Z}}^{2} - \mathbf{S}^{2}
\end{gathered} \tag{2.71}$$

The structure of the Zeeman part (first three terms) of this Hamiltonian conforms to $\mathfrak{su}(2)$, but the sum of the Zeeman and the quadrupole interaction parts has eight generators and may be shown to be isomorphic to $\mathfrak{su}(3)$ when the dimension of the irrep is greater than two. This situation is illustrated in Fig. 2.1; the effect of external physical interactions is to push the Hamiltonian out of a particular n-dimensional irrep $\mathbf{P}(n) : \mathfrak{su}(2) \to \mathfrak{gl}(n)$ and into some bigger subset of $\mathfrak{gl}(n)$.

We, therefore, conclude that irreps of $\mathfrak{su}(2)$ are only building blocks for the Hamiltonians of richer physics that lives in their envelopes. The consequences of this empirically motivated extension are profound: an envelope of an n-dimensional irrep of $\mathfrak{su}(2)$ could be isomorphic to the envelope of an n-dimensional irrep of

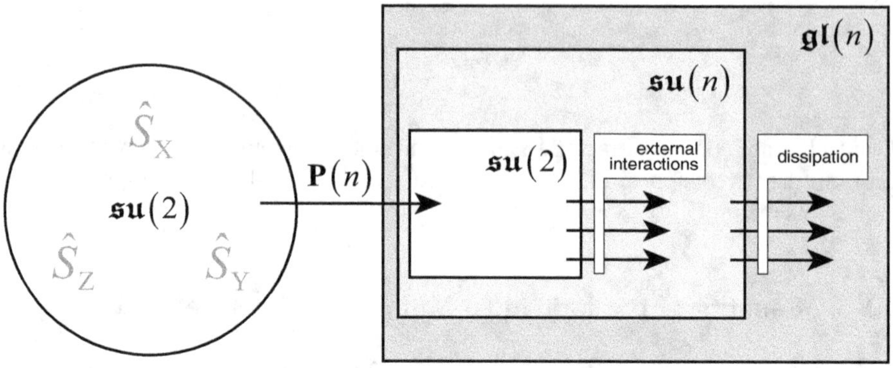

Fig. 2.1 A schematic illustration of how system dynamics escapes from an n-dimensional irrep of su(2) into a larger subset of gl(n). The reason is the presence of external interactions (for example, electrostatic ones) that influence the spatial wavefunctions associated with different total angular momentum states. This introduces additional generators into the dynamics, and the resulting algebra is larger than su(2)

$\mathfrak{su}(n)$: for example, the dynamical algebra of a system of k identical (and identically interacting) spin-1/2 nuclei is still $\mathfrak{su}(2)$, but for a system of k non-identical (and/or differently interacting) spin-1/2 nuclei it may be as large as $\mathfrak{su}(2^k)$ even before dissipative dynamics is considered. In special cases, various sub-algebras of $\mathfrak{su}(2^k)$ may turn up—for example, when permutation symmetry is present, and also when some generators are knocked out by missing interactions.

We must now leave $\mathfrak{su}(2)$ behind: it was useful for setting things up, but no longer applies to situations when spin interacts with orbital angular momentum and the outside world. We will mostly shun the total angular momentum representation: for interacting systems, direct products of spin and orbital angular momentum operators of individual particles are more convenient.

2.6 Weakly Relativistic Limit of Dirac's Equation

For an electron of rest mass m moving at a small fraction of the speed of light, the dominant term in Dirac's Hamiltonian

$$\mathbf{H}=\begin{bmatrix}+mc^2 & \mathbf{0}\\ \mathbf{0} & -mc^2\end{bmatrix}+\begin{bmatrix}\mathbf{0} & c(\boldsymbol{\sigma}\cdot\mathbf{p})\\ c(\boldsymbol{\sigma}\cdot\mathbf{p}) & \mathbf{0}\end{bmatrix}=\mathbf{H}_0+\mathbf{H}_1 \tag{2.72}$$

is the separation between positive and negative energy state pairs contained in $\mathbf{H}_0$. This may be seen from comparing matrix 2-norms (Sect. 1.4.5) of $\mathbf{H}_0$ and $\mathbf{H}_1$:

$$\|\mathbf{H}_0\|=mc^2,\quad \|\mathbf{H}_1\|=c\sqrt{\hat{p}_X^2+\hat{p}_Y^2+\hat{p}_Z^2}=mc\sqrt{\hat{v}_X^2+\hat{v}_Y^2+\hat{v}_Z^2} \tag{2.73}$$

where $\hat{p}_{XYZ} = m\hat{v}_{XYZ}$. For a slow electron, $\|\mathbf{H}_0\| \gg \|\mathbf{H}_1\|$. This allows the mixing that $\mathbf{H}_1$ causes between positive and negative energy eigenspaces of $\mathbf{H}_0$ to be treated approximately. This section introduces one of the many ways to apply this approximation, and explores its consequences. The treatment is simplified—relativistic retardation effects are ignored and interactions unrelated to spin are not discussed.

2.6.1 Zitterbewegung

The presence of negative energy states in the Hamiltonian leads to rapid oscillations in the electron position (*zitterbewegung*, after the German word for fluttering motion) and other observables [44]. To see how these oscillations appear, consider the operators $\mathbf{P}_\pm$ corresponding to the probabilities $p_\pm$ of the electron being found in the positive and the negative energy eigenspaces of $\mathbf{H}_0$:

$$\begin{gathered} p_\pm(t) = \langle \Psi(t) | \mathbf{P}_\pm | \Psi(t) \rangle \\ \mathbf{P}_+ = \begin{bmatrix} \mathbf{1} & \mathbf{0} \\ \mathbf{0} & \mathbf{0} \end{bmatrix}, \quad \mathbf{P}_- = \begin{bmatrix} \mathbf{0} & \mathbf{0} \\ \mathbf{0} & \mathbf{1} \end{bmatrix} \end{gathered} \tag{2.74}$$

For an electron that starts in the positive energy eigenspace:

$$p_+(0) = 1, \quad p_-(0) = 0 \tag{2.75}$$

a tedious calculation yields the following dynamics under the full Hamiltonian from Eq. (2.72):

$$\begin{gathered} \frac{\partial}{\partial t}\langle \mathbf{P}_\pm \rangle = \frac{i}{\hbar}\langle [\mathbf{H}, \mathbf{P}_\pm] \rangle \quad \Rightarrow \quad \begin{cases} p_+(t) = 1 - \varepsilon^2 \sin^2(\omega t) \\ p_-(t) = \varepsilon^2 \sin^2(\omega t) \end{cases} \\ \varepsilon^2 = v^2/(c^2 + v^2), \quad \hbar\omega = mc^2\sqrt{1 + v^2/c^2} \end{gathered} \tag{2.76}$$

The sinusoidal time dependence in $p_-(t)$ shows the electron dipping into the negative energy subspace; for slow electrons, the dips are shallow because $\varepsilon^2 \ll 1$. The frequency $\omega/2\pi$ exceeds 10^{20} Hz—too high to be relevant (and even observable) on the time scale of regular chemistry and physics. This oscillation is, therefore, a good foundation for approximations that account for the presence of the negative energy states without considering them explicitly.

2.6.2 Negative Energy Subspace Elimination

Consider now Dirac's equation for an electron in the presence of an external scalar potential φ and a vector potential $\mathbf{A}$. Lorentz invariance dictates (we refer the reader to relativistic electromagnetism textbooks here) that the following updates be made to the momentum and the energy operators:

$$\begin{aligned} \mathbf{p} &\rightarrow \boldsymbol{\pi} = \mathbf{p} - q\mathbf{A} \\ i\hbar\partial_t &\rightarrow i\hbar\partial_t - q\varphi \end{aligned} \qquad q_{\text{e}} = -1.602176634 \times 10^{-19}\ \text{C} \tag{2.77}$$

resulting in the following equation:

$$i\hbar\frac{\partial}{\partial t}|\Psi\rangle = \begin{bmatrix} q\varphi + mc^2 & c(\boldsymbol{\sigma}\cdot\boldsymbol{\pi}) \\ c(\boldsymbol{\sigma}\cdot\boldsymbol{\pi}) & q\varphi - mc^2 \end{bmatrix}|\Psi\rangle \tag{2.78}$$

The complex phase of the state vector $|\Psi\rangle$ is arbitrary because it cancels in this equation of motion. It also cancels in the expressions for the observable quantities, all of which have the form $\langle\Psi|\mathbf{A}|\Psi\rangle$. Therefore, the replacement $|\Psi\rangle \rightarrow e^{i\omega t}|\Psi\rangle$ is physically inconsequential and corresponds to shifting the diagonal elements of the Hamiltonian in Eq. (2.78) by $\hbar\omega$. This means that the energy reference point is arbitrary, and only energy differences matter.

Let us now expose the positive and the negative energy components explicitly in $|\Psi\rangle$ and shift the energy reference point to have the negative energy subspace at $-2mc^2$:

$$i\hbar\frac{\partial}{\partial t}\begin{pmatrix} |\psi_{\text{L}}\rangle \\ |\psi_{\text{S}}\rangle \end{pmatrix} = \begin{bmatrix} q\varphi & c(\boldsymbol{\sigma}\cdot\boldsymbol{\pi}) \\ c(\boldsymbol{\sigma}\cdot\boldsymbol{\pi}) & q\varphi - 2mc^2 \end{bmatrix}\begin{pmatrix} |\psi_{\text{L}}\rangle \\ |\psi_{\text{S}}\rangle \end{pmatrix} \tag{2.79}$$

where $|\psi_{\text{L}}\rangle$ is the "large" component which we expect to dominate when the evolution begins in the positive energy subspace, and $|\psi_{\text{S}}\rangle$ is the "small" component which we seek to eliminate from the equation of motion. In terms of the two components individually:

$$\begin{cases} i\hbar\partial_t|\psi_{\text{L}}\rangle = q\varphi|\psi_{\text{L}}\rangle + c(\boldsymbol{\sigma}\cdot\boldsymbol{\pi})|\psi_{\text{S}}\rangle \\ i\hbar\partial_t|\psi_{\text{S}}\rangle = c(\boldsymbol{\sigma}\cdot\boldsymbol{\pi})|\psi_{\text{L}}\rangle + \left(q\varphi - 2mc^2\right)|\psi_{\text{S}}\rangle \end{cases} \tag{2.80}$$

Rearranging the second equation (note that φ does not commute with $\boldsymbol{\pi}$) yields

$$|\psi_{\text{S}}\rangle = \frac{1}{2mc^2 - q\varphi}c(\boldsymbol{\sigma}\cdot\boldsymbol{\pi})|\psi_{\text{L}}\rangle - \frac{i\hbar}{2mc^2 - q\varphi}\partial_t|\psi_{\text{S}}\rangle \tag{2.81}$$

When both sides are averaged over the period of zitterbewegung, $\partial_t|\psi_{\text{S}}\rangle$ disappears —this may be seen from Eq. (2.76), where the average of $\partial_t p_-(t)$ is zero. The averaged equation becomes algebraic:

$$|\overline{\psi}_{\text{S}}\rangle = \frac{1}{2mc^2 - q\varphi}c(\boldsymbol{\sigma}\cdot\boldsymbol{\pi})|\overline{\psi}_{\text{L}}\rangle \tag{2.82}$$

Substituting this into the first equation of the system in Eq. (2.80) yields, after a rearrangement:

$$i\hbar\partial_t|\overline{\psi}_{\mathrm{L}}\rangle = \left[q\varphi + \frac{1}{2m}(\boldsymbol{\sigma}\cdot\boldsymbol{\pi})\frac{1}{1 - q\varphi/2mc^2}(\boldsymbol{\sigma}\cdot\boldsymbol{\pi})\right]|\overline{\psi}_{\mathrm{L}}\rangle \tag{2.83}$$

Therefore, the average Hamiltonian in the positive energy subspace is

$$\overline{\mathbf{H}} = q\varphi + \frac{1}{2m}(\boldsymbol{\sigma}\cdot\boldsymbol{\pi})\frac{1}{1 - q\varphi/2mc^2}(\boldsymbol{\sigma}\cdot\boldsymbol{\pi}) \tag{2.84}$$

In weakly relativistic chemical systems, $q\varphi/2mc^2 \ll 1$, and the Taylor series:

$$\frac{1}{1 - q\varphi/2mc^2} = \sum_{n=0}^{\infty}\left(\frac{q\varphi}{2mc^2}\right)^n \tag{2.85}$$

may be truncated at the first term to yield:

$$\overline{\mathbf{H}} = q\varphi + \frac{1}{2m}(\boldsymbol{\sigma}\cdot\boldsymbol{\pi})^2 + \frac{q}{4m^2c^2}(\boldsymbol{\sigma}\cdot\boldsymbol{\pi})\varphi(\boldsymbol{\sigma}\cdot\boldsymbol{\pi}) \tag{2.86}$$

This is an approximate average Hamiltonian governing the dynamics in the positive energy subspace. Its immediate appearance is obscure; the process of casting it into a physically interpretable form is laborious but instructive. The following identities will be useful:

$$(\boldsymbol{\sigma}\cdot\mathbf{a})(\boldsymbol{\sigma}\cdot\mathbf{b}) = \mathbf{a}\cdot\mathbf{b} + i\boldsymbol{\sigma}\cdot(\mathbf{a}\times\mathbf{b}) \quad \textit{Dirac identity} \tag{2.87}$$

$$(\mathbf{a}\times\mathbf{b})\cdot(\mathbf{c}\times\mathbf{d}) = (\mathbf{a}\cdot\mathbf{c})(\mathbf{b}\cdot\mathbf{d}) - (\mathbf{b}\cdot\mathbf{c})(\mathbf{a}\cdot\mathbf{d}) \; \textit{Binet} - \textit{Cauchy identity in 3D} \tag{2.88}$$

$$\nabla\cdot\mathbf{A} = \nabla\cdot\mathbf{A} = 0 \qquad \textit{Coulomb gauge condition} \tag{2.89}$$

$$\nabla\cdot(\psi\mathbf{A}) = \psi(\nabla\cdot\mathbf{A}) + \mathbf{A}\cdot(\nabla\psi) \quad \textit{product rule for divergence} \tag{2.90}$$

In the transformations discussed below, it is important to remember that the wavefunction is implicit in front of each term of the sum in Eq. (2.86)—this modifies product rules for differential operators.

2.6.3 Zeeman Interactions and Langevin Susceptibility

The first term on the right-hand side of Eq. (2.86) is the energy of our electron in the electrostatic potential, such as it may be—e.g. Coulomb interactions with nuclei and other electrons. Interesting things begin with the $(\boldsymbol{\sigma}\cdot\boldsymbol{\pi})^2$ term, which may be reduced with the help of Dirac's identity in Eq. (2.87):

$$\frac{(\boldsymbol{\sigma}\cdot\boldsymbol{\pi})^2}{2m}=\frac{\pi^2}{2m}+\frac{i\boldsymbol{\sigma}\cdot(\boldsymbol{\pi}\times\boldsymbol{\pi})}{2m} \tag{2.91}$$

The first part contains kinetic energy and an interaction between momentum and vector potential:

$$\frac{\pi^2}{2m}=\frac{(\mathbf{p}-q\mathbf{A})^2}{2m}=\frac{\mathbf{p}^2}{2m}-\frac{q}{2m}(\mathbf{p}\cdot\mathbf{A}+\mathbf{A}\cdot\mathbf{p})+\frac{q^2}{2m}\mathbf{A}\cdot\mathbf{A} \tag{2.92}$$

In the special case of a uniform magnetic field **B**:

$$\mathbf{A}=(\mathbf{B}\times\mathbf{r})/2 \tag{2.93}$$

this interaction reduces to a sum of *orbital Zeeman interaction*

$$\begin{aligned}-\frac{q}{2m}(\mathbf{p}\cdot\mathbf{A}+\mathbf{A}\cdot\mathbf{p})&=-\frac{q}{4m}[\mathbf{p}\cdot(\mathbf{B}\times\mathbf{r})+(\mathbf{B}\times\mathbf{r})\cdot\mathbf{p}]\\&=-\frac{q}{2m}(\mathbf{r}\times\mathbf{p})\cdot\mathbf{B}=-\frac{q}{2m}\mathbf{L}\cdot\mathbf{B}\end{aligned} \tag{2.94}$$

and *diamagnetic magnetisability*: and *Langevin susceptibility*:

$$\frac{q^2}{2m}\mathbf{A}\cdot\mathbf{A}=\frac{q^2}{8m}(\mathbf{B}\times\mathbf{r})\cdot(\mathbf{B}\times\mathbf{r})=\frac{q^2}{8m}\mathbf{B}^{\mathrm{T}}\left[\mathbf{r}^{\mathrm{T}}\mathbf{r}-\mathbf{r}\mathbf{r}^{\mathrm{T}}\right]\mathbf{B} \tag{2.95}$$

In the general case, the meaning of the vector potential terms in Eq. (2.92) clarifies when we use the Coulomb gauge condition to rearrange the equation into

$$\frac{\pi^2}{2m}=\frac{\mathbf{p}^2}{2m}+\mathbf{A}\cdot\left(-\frac{q}{m}\mathbf{p}+\frac{q^2}{2m}\mathbf{A}\right) \tag{2.96}$$

The operator in brackets is a sum of momentum current and gauge current operators associated with the electron motion; these currents interact with the vector potential.

To simplify the second term on the right-hand side of Eq. (2.91), we use the following observations:

$$\begin{gathered}\mathbf{p}=-i\hbar\nabla, \quad \nabla\times(\nabla\psi)=0, \quad \mathbf{A}\times\mathbf{A}=0 \\ \nabla\times(\mathbf{A}\psi)=\psi(\nabla\times\mathbf{A})-\mathbf{A}\times(\nabla\psi), \quad \mathbf{B}=\nabla\times\mathbf{A}\end{gathered} \tag{2.97}$$

The resulting Hamiltonian term is called *spin Zeeman interaction*:

$$\begin{aligned}\frac{i\boldsymbol{\sigma}\cdot(\boldsymbol{\pi}\times\boldsymbol{\pi})}{2m} &= -\frac{q\hbar}{2m}\boldsymbol{\sigma}\cdot(\nabla\times\mathbf{A}+\mathbf{A}\times\nabla)=-\frac{q\hbar}{2m}\boldsymbol{\sigma}\cdot\mathbf{B} \\ &= -\frac{q\hbar}{m}\mathbf{S}\cdot\mathbf{B}=\frac{2e\hbar}{2m}\mathbf{S}\cdot\mathbf{B}=g_{\mathrm{E}}\mu_{\mathrm{B}}\mathbf{S}\cdot\mathbf{B}\end{aligned} \tag{2.98}$$

where $\mu_{\mathrm{B}}=e\hbar/2m$ is Bohr magneton, e is the elementary charge (positive by convention) and g_{E} is the electron g-factor, here equal to exactly 2 because radiative corrections are not yet present. Zeeman interaction may be interpreted as a coupling between the magnetic moment associated with the electron spin and the magnetic field. Eq. (2.98) does not require the magnetic field to be uniform.

2.6.4 Spin-Orbit Coupling

In the remaining $(\boldsymbol{\sigma}\cdot\boldsymbol{\pi})\varphi(\boldsymbol{\sigma}\cdot\boldsymbol{\pi})$ term in Eq. (2.86), we will only discuss the dominant $(\boldsymbol{\sigma}\cdot\mathbf{p})\varphi(\boldsymbol{\sigma}\cdot\mathbf{p})$ part—the rest may be found in specialist literature. The momentum operator obeys the product rule:

$$\begin{aligned}\frac{q}{4m^2c^2}(\boldsymbol{\sigma}\cdot\mathbf{p})\varphi(\boldsymbol{\sigma}\cdot\mathbf{p}) &= -\frac{q\hbar^2}{4m^2c^2}(\boldsymbol{\sigma}\cdot\nabla)\varphi(\boldsymbol{\sigma}\cdot\nabla) \\ = -\frac{q\hbar^2}{4m^2c^2}(\boldsymbol{\sigma}\cdot[\nabla\varphi])(\boldsymbol{\sigma}\cdot\nabla) &- \frac{q\hbar^2\varphi}{4m^2c^2}(\boldsymbol{\sigma}\cdot\nabla)(\boldsymbol{\sigma}\cdot\nabla)\end{aligned} \tag{2.99}$$

Using Dirac's identity to transform both terms on the right-hand side yields

$$\begin{aligned}\frac{q}{4m^2c^2}(\boldsymbol{\sigma}\cdot\mathbf{p})\varphi(\boldsymbol{\sigma}\cdot\mathbf{p}) &= \frac{-i\hbar^2q}{4m^2c^2}\boldsymbol{\sigma}\cdot([\nabla\varphi]\times\nabla) \\ +\cdots &= \frac{\hbar q}{4m^2c^2}\boldsymbol{\sigma}\cdot([\nabla\varphi]\times\mathbf{p})+\cdots\end{aligned} \tag{2.100}$$

where [...] indicates spin-independent electrostatic terms that are of no interest to us here. This is *spin-orbit coupling*; it derives its name from the common case of a centre-symmetric potential $\varphi(r)$:

$$\nabla\varphi(r)=\frac{d\varphi(r)}{dr}\frac{\mathbf{r}}{r}, \quad r=\|\mathbf{r}\|_2 \tag{2.101}$$

In that case, we get an inner product between spin and orbital angular momentum:

$$\mathbf{H}^{\text{SO}} = \frac{q\hbar}{4m^2c^2}\frac{1}{r}\frac{d\varphi}{dr}\boldsymbol{\sigma}\cdot(\mathbf{r}\times\mathbf{p}) = \xi(r)\mathbf{S}\cdot\mathbf{L} \tag{2.102}$$

where $\mathbf{S}$ is a vector of $\mathfrak{su}(2)$ generators from Eq (2.58). An example of this coupling in a non-Coulomb potential will make an appearance when we consider nuclear structure (Sect. 3.2.12).

2.6.5 Spinning Charge Analogy

We can establish a sentimental analogy with Maxwell's electrodynamics. Consider a spherical charge q of mass m moving at a constant velocity $\mathbf{v}$ through space with a static electric field $\mathbf{E}$ and a static magnetic field $\mathbf{B}$. In the rest frame of the charge, relativistic electrodynamics has

$$\begin{aligned} \mathbf{E}' &= \gamma(\mathbf{E} - \mathbf{B}\times\mathbf{v}) - (\gamma-1)(\mathbf{E}\cdot\hat{\mathbf{v}})\hat{\mathbf{v}} \\ \mathbf{B}' &= \gamma\left(\mathbf{B} + \frac{\mathbf{E}\times\mathbf{v}}{c^2}\right) - (\gamma-1)(\mathbf{B}\cdot\hat{\mathbf{v}})\hat{\mathbf{v}} \end{aligned} \tag{2.103}$$

where $\gamma = \left(1 - v^2/c^2\right)^{-1/2}$ and $\hat{\mathbf{v}}$ is a unit vector in the direction of $\mathbf{v}$. In the limit of $v \ll c$, this yields

$$\mathbf{B}' \approx \mathbf{B} + \frac{\mathbf{E}\times\mathbf{v}}{c^2} = \mathbf{B} + \frac{\mathbf{E}\times\mathbf{p}}{mc^2} \tag{2.104}$$

If the charge is rotating in its rest frame, the magnetic moment due to the circular currents is

$$\boldsymbol{\mu} = \frac{q}{2m}\mathbf{L} \tag{2.105}$$

where $\mathbf{L}$ is the angular momentum around the centre of mass. The interaction energy E of this magnetic moment with its rest frame magnetic field is

$$E = -\boldsymbol{\mu}\cdot\mathbf{B}' = -\left[\frac{q}{2m}\mathbf{L}\right]\cdot\mathbf{B} - \left[\frac{q}{2m}\mathbf{L}\right]\cdot\frac{\mathbf{E}\times\mathbf{p}}{mc^2} \tag{2.106}$$

The spin-dependent part of the quantum mechanical Hamiltonian (electron Zeeman and electron spin-orbit interaction), we have derived in the previous section may be brought into a similar form using $\mathbf{E} = -\nabla\varphi$ relation from electrostatics:

$$\mathbf{H}^{\mathrm{EZ+ESO}} = -\left[\frac{q\hbar}{2m}\boldsymbol{\sigma}\right]\cdot\mathbf{B} - \left[\frac{q\hbar}{2m}\boldsymbol{\sigma}\right]\cdot\frac{\mathbf{E}\times\mathbf{p}}{2mc^2} \tag{2.107}$$

where the additional factor of 2 in the denominator is the *Thomas half*—the reflection of the fact that the dynamical group of spin is not SO(3) but SU(2) in which the orbits are twice as long (Sect. 1.6.4). Inserting the definition of the spin operators $\mathbf{S} = \boldsymbol{\sigma}/2$ yields

$$\mathbf{H}^{\mathrm{EZ+ESO}} = 2\mu_{\mathrm{B}}\mathbf{S}\cdot\mathbf{B} + \frac{\mu_{\mathrm{B}}}{mc^2}\mathbf{S}\cdot[\mathbf{E}\times\mathbf{p}] \tag{2.108}$$

Within quantum electrodynamics (the voluminous treatment is outside the scope of this book), radiative corrections to the g-factor appear [45]:

$$g_{\mathrm{E}} = 2\left[1 + \frac{\alpha}{2\pi} + O(\alpha^2)\right] \tag{2.109}$$

where $\alpha = (4\pi\varepsilon_0)^{-1}e^2/\hbar c$ is the fine structure constant. The final expression is

$$\mathbf{H}^{\mathrm{EZ+ESO}} = g_{\mathrm{E}}\mu_{\mathrm{B}}\mathbf{S}\cdot\mathbf{B} + \frac{\mu_{\mathrm{B}}}{mc^2}\mathbf{S}\cdot[\mathbf{E}\times\mathbf{p}] \tag{2.110}$$

This Hamiltonian is strictly applicable only to a weakly relativistic single electron in a Maxwellian electromagnetic field. Systems involving multiple interacting electrons and other particles carrying real or effective spin are discussed in Chap. 3.

2.6.6 Spin as a Magnetic Moment

It is now apparent that the weakly relativistic limit of the Dirac Hamiltonian may be cast into a form where electron spin effectively interacts with the external magnetic field and with the magnetic field of the electric current created by its motion. From Eq. (1.110), the magnetic moment operator is

$$\boldsymbol{\mu} = -g_{\mathrm{E}}\mu_{\mathrm{B}}\mathbf{S}, \quad \mathbf{S} = [\mathbf{S}_{\mathrm{X}} \quad \mathbf{S}_{\mathrm{Y}} \quad \mathbf{S}_{\mathrm{Z}}]^{\mathrm{T}} \tag{2.111}$$

From this point, the theory of spin becomes partly empirical—we assume that a system that has a spin may also have a corresponding magnetic moment, and that those magnetic moments interact. This also turns out to be true for nuclear ground states (Sect. 3.2.12).

The general form of the interaction potential may be obtained from algebraic considerations [46]. Firstly, any Taylor-expandable scalar function of spin-1/2 Pauli matrices is reducible to a linear function of the same matrices; this compels interactions between two electron spins to be at most bilinear. Secondly, the magnetic moment vector $\boldsymbol{\mu}$ is axial, but the distance direction vector $\mathbf{n}$ is polar—their inner product $\boldsymbol{\mu} \cdot \mathbf{n}$ would be a pseudoscalar. In the experimental observations so far, no pseudoscalar potential has ever been seen; it is reasonable to make the same assumption here. This means that out of vectors $\boldsymbol{\mu}_1$, $\boldsymbol{\mu}_2$, $\mathbf{n}_{12}$ only two independent interactions can emerge: $\boldsymbol{\mu}_1 \cdot \boldsymbol{\mu}_2$ and $(\boldsymbol{\mu}_1 \cdot \mathbf{n}_{12})(\boldsymbol{\mu}_2 \cdot \mathbf{n}_{12})$.

After a reordering of terms by irreducible representations of the rotation group, we conclude that any interaction potential for a pair of electrons must have the following form:

$$\mathbf{U} = a(r) + b(r)(\mathbf{S}_1 \cdot \mathbf{S}_2) + c(r)[3(\mathbf{S}_1 \cdot \mathbf{n}_{12})(\mathbf{S}_2 \cdot \mathbf{n}_{12}) - (\mathbf{S}_1 \cdot \mathbf{S}_2)] \tag{2.112}$$

where the coordinate parts are obtained by comparison with the expressions obtained in Maxwell's electromagnetism—essentially, they are postulated and tested experimentally. In a classical system containing two point magnetic moments, the vector potential created by $\boldsymbol{\mu}_2$ at the location of $\boldsymbol{\mu}_1$ is

$$\mathbf{A}(\mathbf{r}_1) = \frac{\mu_0}{4\pi} \frac{\boldsymbol{\mu}_2 \times (\mathbf{r}_1 - \mathbf{r}_2)}{|\mathbf{r}_1 - \mathbf{r}_2|^3} \tag{2.113}$$

A laborious exercise in vector calculus then yields [47]

$$\begin{aligned} U_{12} = -\boldsymbol{\mu}_1 \cdot \mathbf{B}(\mathbf{r}_1) &= -\frac{\mu_0}{4\pi} \boldsymbol{\mu}_1 \cdot \left[\nabla_1 \times \frac{\boldsymbol{\mu}_2 \times (\mathbf{r}_1 - \mathbf{r}_2)}{|\mathbf{r}_1 - \mathbf{r}_2|^3} \right] \\ &= -\frac{\mu_0}{4\pi} \boldsymbol{\mu}_1 \cdot \left[3 \frac{(\mathbf{r}_1 - \mathbf{r}_2)(\mathbf{r}_1 - \mathbf{r}_2)^{\mathrm{T}}}{|\mathbf{r}_1 - \mathbf{r}_2|^5} - \frac{\mathbf{1}}{|\mathbf{r}_1 - \mathbf{r}_2|^3} \right] \cdot \boldsymbol{\mu}_2 \\ &\quad - \frac{\mu_0}{4\pi} \frac{8\pi}{3} \delta(\mathbf{r}_1 - \mathbf{r}_2)(\boldsymbol{\mu}_1 \cdot \boldsymbol{\mu}_2) \end{aligned} \tag{2.114}$$

which has the form prescribed by Eq. (2.112), with the $a(r)$ term missing—it corresponds to the interactions that do not depend on spin. Of the remaining terms, the part containing $b(r)$ is called *contact interaction* and the part containing $c(r)$ is called *dipole interaction*. Both parts do, strictly speaking, originate from the interaction between two dipoles, the separation into contact and non-contact part is dictated by mathematical convenience: they belong to different irreducible representations ($l = 0$ for contact and $l = 2$ for dipole) of the rotation group.

2.6.7 Spin Hamiltonian Approximation

Applying the canonical quantisation procedure [26] to Eq. (2.114) would produce a Hamiltonian that acts on spatial and spin degrees of freedom. A common approximation [48] is to assume spatial degrees of freedom to be in the ground state, such as may exist for each spin configuration that we specify. When this approximation applies, the wavefunction factorises

$$|\Psi\rangle = \sum_n a_n |\psi_n(\mathbf{r})\rangle \otimes |s_n\rangle \tag{2.115}$$

where $\mathbf{r}$ are other degrees of freedom, and $|\psi_n(\mathbf{r})\rangle$ is the ground state wavefunction of those degrees of freedom corresponding to the spin state $|s_k\rangle$. Without loss of generality, we can assume spin states to be a complete orthonormal set. Then the Hamiltonian of the system also factorises

$$\hat{H} = \sum_{pq} \hat{H}_{pq} \otimes |s_p\rangle\langle s_q| \tag{2.116}$$

into a sum of Kronecker products of purely spatial operators $\hat{H}_{pq}$ and spin state projectors $|s_p\rangle\langle s_q|$. In the expression for the energy, we can then integrate away the spatial degrees of freedom:

$$\begin{aligned} E &= \langle\Psi|\hat{H}|\Psi\rangle \\ &= \sum_{knpq} a_k^* \langle s_k| \otimes \langle\psi_k(\mathbf{r})| \left[\hat{H}_{pq} \otimes |s_p\rangle\langle s_q|\right] a_n |\psi_n(\mathbf{r})\rangle \otimes |s_n\rangle \\ &= \sum_{knpq} a_k^* a_n \langle s_k \mid s_p\rangle \langle\psi_k(\mathbf{r})|\hat{H}_{pq}|\psi_n(\mathbf{r})\rangle \langle s_q \mid s_n\rangle \\ &= \sum_{kn} a_k^* a_n \langle\psi_k(\mathbf{r})|\hat{H}_{kn}|\psi_n(\mathbf{r})\rangle = \sum_{kn} a_k^* h_{kn} a_n \end{aligned} \tag{2.117}$$

What remains looks like another inner product. The state vector still contains the original coefficients from Eq. (2.115) but the presence of spatial degrees of freedom is now implicit:

$$E = \langle S|\mathbf{H}_\mathrm{S}|S\rangle \quad |S\rangle = \sum_n a_n |s_n\rangle \tag{2.118}$$

The matrix $\mathbf{H}_\mathrm{S}$ with elements h_{kn} is called *spin Hamiltonian*. It is a significant approximation that can break down in systems (such as lanthanide complexes), where multiple electron spins interact strongly enough for the energies of spatial and spin parts of the full Hamiltonian to become comparable.

2.6.8 Energy Derivative Formalism

Equations (2.98) and (2.112) suggest that, after spatial degrees of freedom had been integrated away, the general form of a spin Hamiltonian in the individual momentum representation is

$$\mathbf{H} = \sum_{k} \mathbf{S}^{(k)} \cdot \mathbf{Z}_k \cdot \mathbf{B} + \sum_{n \neq k} \mathbf{S}^{(n)} \cdot \mathbf{A}_{nk} \cdot \mathbf{S}^{(n)} + \cdots \tag{2.119}$$

where the sum is over spin pairs, $\mathbf{S}^{(k)} = \begin{bmatrix} \mathbf{S}_\mathrm{X}^{(k)} & \mathbf{S}_\mathrm{Y}^{(k)} & \mathbf{S}_\mathrm{Z}^{(k)} \end{bmatrix}^\mathrm{T}$ are vectors of spin operators from Eq. (2.58), $\mathbf{Z}_k$ are 3×3 matrices called *Zeeman tensors*, and $\mathbf{A}_{nk}$ are 3×3 matrices called *interaction tensors*. Equation (2.119) may be viewed as linear and bilinear terms in some Taylor expansion with respect to spin. Further terms arising for effective "spins" such as nuclei and *f*-electron shells are dealt with in Chap. 3.

We have assumed in Sect. 2.6.6 that spins interact through their magnetic moments; obtaining $\mathbf{Z}_k$ and $\mathbf{A}_{nk}$ must therefore involve differentiating the total molecular energy with respect to those magnetic moments and the external field. The first derivative of the *n*th eigenvalue (assumed non-degenerate) of a Hamiltonian that depends continuously on a real parameter α is given by the (improperly named [49]) *Hellmann-Feynman theorem*:

$$\begin{aligned} \frac{\partial E_n}{\partial \alpha} &= \frac{\partial}{\partial \alpha} \langle \Psi_n | \hat{H} | \Psi_n \rangle = \left\langle \frac{\partial \Psi_n}{\partial \alpha} \right| \hat{H} | \Psi_n \rangle + \langle \Psi_n | \frac{\partial \hat{H}}{\partial \alpha} | \Psi_n \rangle + \langle \Psi_n | \hat{H} \left| \frac{\partial \Psi_n}{\partial \alpha} \right\rangle \\ &= \langle \Psi_n | \frac{\partial \hat{H}}{\partial \alpha} | \Psi_n \rangle + E \left\langle \frac{\partial \Psi_n}{\partial \alpha} \middle| \Psi_n \right\rangle + E \left\langle \Psi_n \middle| \frac{\partial \Psi_n}{\partial \alpha} \right\rangle \\ &= \langle \Psi_n | \frac{\partial \hat{H}}{\partial \alpha} | \Psi_n \rangle + E \frac{\partial \langle \Psi_n | \Psi_n \rangle}{\partial \alpha} \quad \Rightarrow \quad \frac{\partial E_n}{\partial \alpha} = \langle \Psi_n | \frac{\partial \hat{H}}{\partial \alpha} | \Psi_n \rangle \end{aligned} \tag{2.120}$$

where we have used the fact that $\langle \Psi_n | \Psi_n \rangle = 1$. When the eigenfunction $|\psi_n\rangle$ is known, this first derivative of the energy is computable directly, but the second derivatives are not—derivatives of the eigenfunction make an appearance when the product rule is applied:

$$\frac{\partial^2 E_n}{\partial \alpha \partial \beta} = \left\langle \frac{\partial \Psi_n}{\partial \beta} \middle| \frac{\partial \hat{H}}{\partial \alpha} \right| \Psi_n \rangle + \langle \Psi_n | \frac{\partial^2 \hat{H}}{\partial \alpha \partial \beta} | \Psi_n \rangle + \langle \Psi_n | \frac{\partial \hat{H}}{\partial \alpha} \left| \frac{\partial \Psi_n}{\partial \beta} \right\rangle \tag{2.121}$$

However, because the eigensystem of the Hermitian operator $\hat{H}$ is an orthonormal basis, we can express the derivative as a linear combination of those eigenfunctions:

$$\left| \frac{\partial \Psi_n}{\partial \beta} \right\rangle = \sum_{k \neq n} b_k^{(n)} | \Psi_k \rangle \tag{2.122}$$

in which the coefficient $b_n^{(n)}$ becomes zero when we assume that the parameter β is such that $\hat{H}'_\beta$ is Hermitian. Other coefficients are obtained by differentiating both sides of $\hat{H}|\Psi_n\rangle = E_n|\Psi_n\rangle$:

$$\frac{\partial \hat{H}}{\partial \beta}|\Psi_n\rangle + \hat{H}\left|\frac{\partial \Psi_n}{\partial \beta}\right\rangle = \frac{\partial E_n}{\partial \beta}|\Psi_n\rangle + E_n\left|\frac{\partial \Psi_n}{\partial \beta}\right\rangle \tag{2.123}$$

After we substitute the expansion in Eq. (2.122) and rearrange the terms, we obtain

$$\left(\frac{\partial \hat{H}}{\partial \beta} - \frac{\partial E_n}{\partial \beta}\right)|\Psi_n\rangle = \sum_{k \neq n} b_k^{(n)}(E_n - E_k)|\Psi_k\rangle \tag{2.124}$$

After multiplying this by $\langle \Psi_m|$ with $m \neq n$ from the left and simplifying:

$$b_m^{(n)} = \frac{\langle \Psi_m|\hat{H}'_\beta|\Psi_n\rangle}{E_n - E_m} \quad \Rightarrow \quad \left|\frac{\partial \Psi_n}{\partial \beta}\right\rangle = \sum_{k \neq n} \frac{\langle \Psi_k|\hat{H}'_\beta|\Psi_n\rangle}{E_n - E_k}|\Psi_k\rangle \tag{2.125}$$

After this is placed into Eqs. (2.121) and further simplifications are applied:

$$\frac{\partial^2 E_n}{\partial \alpha \partial \beta} = \langle \Psi_n|\hat{H}''_{\alpha\beta}|\Psi_n\rangle + \sum_{k \neq n} \frac{\langle \Psi_n|\hat{H}'_\alpha|\Psi_k\rangle\langle \Psi_k|\hat{H}'_\beta|\Psi_n\rangle + c.c.}{E_n - E_k} \tag{2.126}$$

Note that this is not perturbation theory; Eqs. (2.120)–(2.126) are exact. The approximation is instead contained in the assumptions we had made when we stipulated Eq. (2.119). Note also that the expression for the second derivative in this approach requires the energy spectrum to be non-degenerate.

2.6.9 Thermal Corrections

The approach described above is valid for each individual eigenstate. That tends to be a good approximation in small molecules with non-degenerate ground states, but not in systems (e.g. transition metal complexes) where the ground state is degenerate and/or thermally accessible excitations are present. In those cases, the correct quantity to differentiate is *Helmholtz free energy* [50]:

$$F = -\beta^{-1}\ln Z, \quad Z = \mathrm{Tr}\left[\exp\left(-\beta \hat{H}\right)\right], \quad \beta = 1/k_B T \tag{2.127}$$

Derivatives with respect to any parameters x and y are then related by the chain rule to the corresponding derivatives of the Hamiltonian:

$$
\begin{aligned}
F'_x &= \mathrm{Tr}\left[\hat{H}'_x \hat{\rho}_{\mathrm{eq}}\right], \quad \hat{\rho}_{\mathrm{eq}} = \exp\left(-\beta \hat{H}\right) / \mathrm{Tr}\left[\exp\left(-\beta \hat{H}\right)\right] \\
F''_{xy} &= \mathrm{Tr}\left[\left(\hat{H}''_{xy} - \hat{H}'_x \hat{H}'_y\right) \hat{\rho}_{\mathrm{eq}}\right] + \beta\, \mathrm{Tr}\left[\hat{H}'_x \hat{\rho}_{\mathrm{eq}}\right] \mathrm{Tr}\left[\hat{H}'_y \hat{\rho}_{\mathrm{eq}}\right]
\end{aligned}
\tag{2.128}
$$

where we have introduced the thermal equilibrium density matrix $\hat{\rho}_{\mathrm{eq}}$. These expressions are harder to compute than ground state derivatives using Hellmann-Feynman approach, but they are free of denominator singularities and also have the advantage of capturing the temperature dependence of spin Hamiltonian parameters.

3 Bestiary of Spin Hamiltonians

When the system is in the ground state with respect to spatial degrees of freedom, spin dynamics may be approximately segregated in both electronic and nuclear structure theory. The liberty in the choice of approximations makes such treatments less rigorous than Chap. 2 had been: we paint a fig leaf of plausible derivations over empirical expressions. The presentation is simplified—we skip the thorny technicalities of evaluating spatial integrals and focus on the properties of the resulting spin Hamiltonians.

3.1 Physical Side

This section offers a brief introduction to magnetic interactions within nuclear and electronic structure theories. Both subjects are vast and highly developed [51–53]; they are only interesting to us here insofar as they provide, explain, or illustrate the parameters of spin Hamiltonians. Practical details of *ab initio* calculation of those parameters may be found in the specialist literature cited in the subsections below. We assume that the reader is familiar with the electronic structure theory of the hydrogen atom and electrostatic aspects of molecular quantum mechanics [54].

3.1.1 Nuclear Spin and Magnetic Moment

Not even the potential is currently known for the strong nuclear force, but there is experimental evidence that centrally symmetric mean-field (MF) approximations—where each nucleon is assumed to move in some effective average potential created by other nucleons—are qualitatively correct [51]. The corresponding non-relativistic single-nucleon Hamiltonian is

I. KUPROV, *Spin*,
https://doi.org/10.1007/978-3-031-05607-9_3

$$\hat{H}=\frac{\hat{p}^2}{2m}+V_{\mathrm{MF}}(r)=-\frac{\hbar^2}{2m}\nabla^2+V_{\mathrm{MF}}(r) \tag{3.1}$$

More sophisticated and accurate models do exist [336], but we must not drift into the theory of nuclear structure—our objective here is to understand the nature of nuclear spin, magnetic dipole, and electric quadrupole; we will not venture beyond the case of a single unpaired nucleon.

Because the potential is centrally symmetric, the group-theoretical classification of nuclear orbitals is broadly similar to the classification of electron orbitals in the hydrogen atom [54]. After spherical coordinate variable separation, the radial Hamiltonian acquires the following form:

$$\hat{H}_{\mathrm{R}}=-\frac{\hbar^2}{2m}\left[\frac{1}{r^2}\frac{\partial}{\partial r}\left(r^2\frac{\partial}{\partial r}\right)-\frac{l(l+1)}{r^2}\right]+V_{\mathrm{MF}}(r) \tag{3.2}$$

where l is a non-negative integer enumerating the irreducible representations of the rotation group (Sect. 1.6.2.4) obeyed by the angular part (spherical harmonics, same as the hydrogen atom). Electron scattering experiments indicate that the following *Woods-Saxon potential* [55] yields qualitatively correct results:

$$V_{\mathrm{MF}}(r)=-\frac{V_0}{1+\mathrm{e}^{(r-r_0)/a}} \tag{3.3}$$

where V_0 is the potential depth (experimentally, 40.50 + 0.13A MeV, where A is the nuclear mass number), a is the surface diffuseness ($\sim$0.65 fm), and r_0 is the radius of the potential well ($\sim 1.27A^{1/3}$ fm). Woods-Saxon potential changes the orbital classification compared to the hydrogen atom [54]: although the angular wavefunctions are still spherical harmonics with the same alphabetic (*s*, *p*, *d*, *f*,...) notation, their maximum rank l is no longer bounded by the radial quantum number. Orbital energy order is determined by: (a) assumptions about the mean-field potential; (b) assumptions about inter-nucleon interactions. It is customary to ignore the nuclear equivalent of the principal quantum number: energy levels are numbered in the order of appearance—the first *d*-type orbital from the bottom is 1*d*, and the second one is 2*d*, etc. For light nuclei, the resulting orbital filling order is

$$1s,\ 1p,\ 1d,\ 2s,\ 1f,\ 2p,\ 1g,\ 2d,\ 3s,\ \ldots \tag{3.4}$$

Coulomb repulsion between protons ($\sim$1.44 meV for a proton pair at a distance of 1 fm) and spin-orbit coupling ($\sim$5 meV for ^{15}N and ^{17}O, scaling with $(2l+1)A^{-2/3}$) perturb these energy levels, but they are a convenient basis set. Spin-orbit coupling is stronger than Coulomb interaction in nuclei; its Hamiltonian is the same as the one obtained (Sect. 2.6.4) for all moving fermions in Eq. (2.102):

$$\mathbf{H}_{\mathrm{SO}}=-\frac{\hbar}{4m^2c^2}\frac{1}{r}\frac{\mathrm{d}V_{\mathrm{MF}}}{\mathrm{d}r}\boldsymbol{\sigma}\cdot(\mathbf{r}\times\mathbf{p})=V_{\mathrm{SO}}(r)\mathbf{S}\cdot\mathbf{L} \tag{3.5}$$

The sign of this interaction is opposite to what it had been for an electron orbiting a nucleus—unlike in the hydrogen atom, parallel orientations of angular momentum and spin are *lower* in energy than anti-parallel ones. It is easy to see that $\mathbf{S}\cdot\mathbf{L}$ commutes (and therefore shares eigenvectors) with $\mathbf{J}^2 = (\mathbf{L}+\mathbf{S})^2$. A neat way to label those eigenvectors is therefore by eigenvalues of the total angular momentum operator $\mathbf{J}^2$. Choosing a specific orbital quantum number l and a specific spin (here only $s = 1/2$) pins down the matrix representation of the spin-orbit coupling operator:

$$\mathbf{S}\cdot\mathbf{L} = \mathbf{S}_\mathrm{X}\otimes\mathbf{L}_\mathrm{X} + \mathbf{S}_\mathrm{Y}\otimes\mathbf{L}_\mathrm{Y} + \mathbf{S}_\mathrm{Z}\otimes\mathbf{L}_\mathrm{Z} \tag{3.6}$$

where the matrices come from Eqs. (1.116) and (1.123). Its diagonalisation yields the following eigensystem:

$$\mathbf{S}\cdot\mathbf{L}|j\rangle = \frac{1}{2}\left[j(j+1) - l(l+1) - s(s+1)\right]|j\rangle, \quad j = |l-s|,\ldots,l+s \tag{3.7}$$

The resulting nuclear orbital notation convention is similar to the hydrogen atom: first. the appearance order number n, then the letter indicating the orbital quantum number l; the lower index indicates the total angular momentum j, and the upper index in brackets is the maximum occupation number, equal to the number of projections that the total angular momentum can have: $2j+1$:

$$1s_{1/2}^{(2)},\ 2p_{3/2}^{(4)},\ 2p_{1/2}^{(2)}, 3d_{5/2}^{(6)}, 3s_{1/2}^{(2)},\ 3d_{3/2}^{(4)},\ \ldots \tag{3.8}$$

Two copies of this level structure are present—one for protons and one for neutrons; they differ in energy because Coulomb interaction modifies the proton mean-field potential. The difference in the occupation order between proton and neutron orbitals only appears for $n \geq 5$.

What is colloquially called *nuclear spin* is actually the total angular momentum j of the ground state. For ^{13}C, we have an unpaired neutron in the $2p_{1/2}$ neutron orbital:

$$\begin{array}{lll} ^{13}\mathrm{C} & \text{protons:} & 1s_{1/2}^{(2)},\ 2p_{3/2}^{(4)} \\ & \text{neutrons:} & 1s_{1/2}^{(2)},\ 2p_{3/2}^{(4)},\ 2p_{1/2}^{(1)} \end{array} \Rightarrow j = 1/2 \tag{3.9}$$

^{15}N has an unpaired proton in the $2p_{1/2}$ proton orbital:

$$\begin{array}{lll} ^{15}\mathrm{N} & \text{protons:} & 1s_{1/2}^{(2)},\ 2p_{3/2}^{(4)},\ 2p_{1/2}^{(1)} \\ & \text{neutrons:} & 1s_{1/2}^{(2)},\ 2p_{3/2}^{(4)},\ 2p_{1/2}^{(2)} \end{array} \Rightarrow j = 1/2 \tag{3.10}$$

and ^{17}O derives its nuclear spin 5/2 from the unpaired neutron in $3d_{5/2}$:

$$\begin{array}{lll} ^{17}\mathrm{O} & \text{protons:} & 1s_{1/2}^{(2)},\ 2p_{3/2}^{(4)},\ 2p_{1/2}^{(2)} \\ & \text{neutrons:} & 1s_{1/2}^{(2)},\ 2p_{3/2}^{(4)},\ 2p_{1/2}^{(2)},\ 3d_{5/2}^{(1)} \end{array} \Rightarrow j = 5/2 \tag{3.11}$$

With multiple unpaired nucleons, the best way to proceed is by numerical self-consistent field treatments [56]; those are mathematically similar to electronic structure theories [53], but are beyond the scope of this book.

For an arbitrary nuclear state, the magnetic moment $\boldsymbol{\mu}$ is not parallel to either $\mathbf{L}$, or $\mathbf{S}$, or even to $\mathbf{J} = \mathbf{L} + \mathbf{S}$, because orbital and spin g-factors of nucleons are different. Proton and neutron g_S are different because their "spin" is also the total angular momentum of a system of quarks [57]:

$$\boldsymbol{\mu} = \mu_N (g_L \mathbf{L} + g_S \mathbf{S}), \quad \begin{cases} p: & g_L = 1,\ g_S = +5.586 \\ n: & g_L = 0,\ g_S = -3.826 \end{cases} \tag{3.12}$$

This operator does not commute with $\mathbf{J}^2$ and therefore a ground state nucleus is not in a pure state with respect to the magnetic moment. However, at room temperature, nuclear excited states are not populated; we must, therefore, project the operator $\boldsymbol{\mu}$ into the subspace spanned by $|j, m\rangle$ wavefunctions of the ground state multiplet because nothing else is dynamically accessible. By definition then, this projection of $\boldsymbol{\mu}$ is a linear combination $\{\mathrm{J_X}, \mathrm{J_Y}, \mathrm{J_Z}\}$. A tedious calculation [23, 58] yields

$$\boldsymbol{\mu} \to g_J \mu_N \mathbf{J}, \quad g_J = \frac{1}{2j(j+1)} \begin{Bmatrix} g_L [j(j+1) + l(l+1) - s(s+1)] + \\ g_S [j(j+1) - l(l+1) + s(s+1)] \end{Bmatrix} \tag{3.13}$$

Thus, in the ground state, the magnetic moment operator ends up being proportional to the total angular momentum operator and the *nuclear magnetogyric ratio* $\gamma = g_J \mu_N$ is born. Equation (3.13) explains how some nuclei (for example, ^{15}N) acquire their classically impossible negative magnetogyric ratios.

In practical terrestrial conditions, nuclear excited states are not chemically accessible (ΔE upwards of 10 keV, with the exception of ^{229}Th at 8.28 ± 0.17 eV), and relativistic nuclear velocities are uncommon. Thus, a fixed effective nuclear "spin" is a good approximation. The largest half-integer spin for a reasonably stable nucleus is 9/2 (^{73}Ge, ^{83}Kr, etc.), and the largest integer one is 7 (^{176}Lu, $t_{1/2} = 3.78 \times 10^{10}$ years).

3.1.2 Nuclear Electric Quadrupole Moment

Some nuclei are not spherical—for example, the ground state of ^{224}Ra is mango-shaped [59]. In particular, different projection states of the ground state multiplet $|j, m\rangle$ can have different charge distributions in their spatial parts and therefore interact differently with external electric fields. This makes their energy a function of m, and therefore creates a term in the spin Hamiltonian through the mechanism which we will now discuss.

We start with the *multipole expansion* of the electrostatic energy operator. Consider the Taylor expansion of the external electrostatic potential:

$$\begin{aligned}\varphi(\mathbf{r}) &= \varphi(0) + \frac{\partial\varphi(0)}{\partial\mathbf{r}}\cdot\mathbf{r} + \frac{1}{2}\mathbf{r}^{\mathrm{T}}\cdot\frac{\partial^2\varphi(0)}{\partial\mathbf{r}^{\mathrm{T}}\partial\mathbf{r}}\cdot\mathbf{r} + \ldots \\ &= \varphi(0) + \frac{\partial\varphi(0)}{\partial\mathbf{r}}\cdot\mathbf{r} + \frac{1}{2}\mathrm{Tr}\left[\frac{\partial^2\varphi(0)}{\partial\mathbf{r}^{\mathrm{T}}\partial\mathbf{r}}\cdot\left(\mathbf{r}\mathbf{r}^{\mathrm{T}}\right)\right] + \ldots\end{aligned} \tag{3.14}$$

The second derivatives matrix $\partial^2\varphi/\partial\mathbf{r}^{\mathrm{T}}\partial\mathbf{r}$ is traceless by *Laplace's law* ($\nabla^2\varphi = 0$ for the external electrostatic potential), meaning that the trace of $\mathbf{r}\mathbf{r}^{\mathrm{T}}$ also does not influence the result and may be removed: $\mathbf{r}\mathbf{r}^{\mathrm{T}} \to \mathbf{r}\mathbf{r}^{\mathrm{T}} - \mathbf{1}r^2/3$, yielding

$$\varphi(\mathbf{r}) = \varphi(0) + \frac{\partial\varphi(0)}{\partial\mathbf{r}}\cdot\mathbf{r} + \frac{1}{6}\mathrm{Tr}\left[\frac{\partial^2\varphi(0)}{\partial\mathbf{r}^{\mathrm{T}}\partial\mathbf{r}}\cdot\left(3\mathbf{r}\mathbf{r}^{\mathrm{T}} - \mathbf{1}r^2\right)\right] + \ldots \tag{3.15}$$

A nucleus with a charge density $\rho(\mathbf{r})$ would have the energy $E = \int \rho(\mathbf{r})\varphi(\mathbf{r})d^3\mathbf{r}$ in this potential. To simplify the notation, we will introduce the *total charge* q, the *electric dipole vector* $\mathbf{d}$ and the *electric quadrupole tensor* $\mathbf{Q}$ of that nucleus:

$$q = \int \rho(\mathbf{r})d^3\mathbf{r},\ \mathbf{d} = \int \mathbf{r}\rho(\mathbf{r})d^3\mathbf{r},\ \mathbf{Q} = \int \rho(\mathbf{r})\left(3\mathbf{r}\mathbf{r}^{\mathrm{T}} - \mathbf{1}r^2\right)d^3\mathbf{r} \tag{3.16}$$

The quantity $\mathbf{Q}/e$ has the dimension of area; for nuclei, it is commonly quoted in *barns* (1 barn = 100 fm^2, the cross-section area of ^{238}U with respect to thermal neutron capture is 2.7 barn). The energy is

$$E = q\varphi + \frac{\partial\varphi}{\partial\mathbf{r}}\cdot\mathbf{d} + \frac{1}{6}\mathrm{Tr}\left[\frac{\partial^2\varphi}{\partial\mathbf{r}^{\mathrm{T}}\partial\mathbf{r}}\cdot\mathbf{Q}\right] + \ldots \tag{3.17}$$

The charge term has no directional dependence, the electric dipole moment $\mathbf{d}$ is zero because the nuclear ground state wavefunction must be a parity eigenfunction:

$$\Psi(\mathbf{r}) = \pm\Psi(-\mathbf{r}) \ \Rightarrow\ \rho(\mathbf{r}) = \rho(-\mathbf{r}) \tag{3.18}$$

and the contribution from electric multipoles higher than quadrupole is insignificant. We are, therefore, left with the quadrupole as the interesting term:

$$\begin{gathered}\hat{H}_{\mathrm{Q}} = \frac{1}{6}\sum_{nk}\hat{Q}_{nk}V_{nk}, \quad V_{nk} = \frac{\partial^2\varphi}{\partial r_n\partial r_k} \\ \hat{\mathbf{Q}} = 3\mathbf{r}\mathbf{r}^{\mathrm{T}} - \mathbf{1}r^2\end{gathered} \tag{3.19}$$

Because nuclear excited state populations are negligible at terrestrial temperatures, this operator must be projected into the subspace spanned by $|j,m\rangle$ wavefunctions of the ground state term—the same procedure as the one we had used for the magnetic moment. Another tedious calculation [23, 58] yields the following effective Hamiltonian:

$$\mathbf{H}_{\mathrm{Q}} = \frac{eQ}{6j(2j-1)\hbar}\sum_{n,k} V_{nk}\left(\frac{3}{2}[\mathbf{J}_n,\mathbf{J}_k]_+ + \delta_{nk}\mathbf{J}^2\right) \tag{3.20}$$

where eQ is the *nuclear quadrupole moment* and $[\cdot,\cdot]_+$ is an anticommutator. Once this expression is substituted into the Hamiltonian in Eq. (3.19), it becomes an operator acting on the total nuclear angular momentum. When $\mathbf{J}$ is eventually renamed "nuclear spin", this interaction becomes a part of the nuclear spin Hamiltonian. Section 3.3 details its historical parametrisations.

3.1.3 Electronic Structure Derivative Table

To obtain spin Hamiltonian parameters that depend on the electronic structure, we must use the energy derivative formalism described in Sects. 2.6.8–2.6.9 and calculate the derivatives of the electronic structure energy with respect to the external magnetic field and the magnetic moments of electrons and nuclei. Although the formalism is quantum mechanical, the differentiation variables are classical; nuclei are treated as point magnetic dipoles.

The contributions to the vector potential from a uniform (a reasonable assumption on a molecular scale) external magnetic field $\mathbf{B}$ and from a point magnetic dipole $\boldsymbol{\mu}$ are

$$\mathbf{A}_{\mathrm{ext}}(\mathbf{r}) = \frac{\mathbf{B}\times(\mathbf{r}-\mathbf{r}_{\mathrm{g}})}{2}, \quad \mathbf{A}_{\mathrm{dip}}(\mathbf{r}) = \frac{\mu_0}{4\pi}\frac{\boldsymbol{\mu}\times(\mathbf{r}-\mathbf{r}_{\mu})}{|\mathbf{r}-\mathbf{r}_{\mu}|^3} \tag{3.21}$$

where $\mathbf{r}_{\mathrm{g}}$ is the location of the *gauge origin*—an arbitrary point in space that does not influence physical observables (which depend on derivatives of $\mathbf{A}$ with respect to $\mathbf{r}$) but does cause technical problems [60] when numerical calculations are performed using incomplete basis sets. After these expressions are inserted into the gauge-invariant kinetic energy term obtained in Sect. 3.1.2:

$$\frac{\boldsymbol{\pi}^2}{2m_{\mathrm{e}}} = \frac{(\mathbf{p}+e\mathbf{A})^2}{2m_{\mathrm{e}}} = \frac{\mathbf{p}^2}{2m_{\mathrm{e}}} + \frac{e}{2m_{\mathrm{e}}}(\mathbf{p}\cdot\mathbf{A}+\mathbf{A}\cdot\mathbf{p}) + \frac{e^2}{2m_{\mathrm{e}}}\mathbf{A}\cdot\mathbf{A} \tag{3.22}$$

and sums are taken over electrons and nuclei, a protracted exercise in vector calculus yields the expressions collected in Table 1.4, which use the following fundamental constants:

$$\alpha = \frac{1}{4\pi\varepsilon_0}\frac{e^2}{\hbar c} = \frac{\mu_0}{4\pi}\frac{e^2 c}{\hbar} = 0.007\ 297\ 352\ 569$$
$$\mu_B = \frac{e\hbar}{2m_e} = 9.274\ 010\ 078 \times 10^{-24}\ \text{J/T} \tag{3.23}$$

and spin operators from Eq. (2.58). The expressions in Table 3.1 are at most quadratic in the differentiation variables, and therefore the application of the Hellmann-Feynman theorem (Sect. 2.6.8) is straightforward, even if the subsequent evaluation of the constituent integrals is not. Note that $\delta(\mathbf{r})$ has a dimension—inverse cubic metres.

I have taken the liberty of giving the entries of Table 3.1 more descriptive names and indices than those commonly used in electronic structure theory books. This is because the historically tolerated use of "diamagnetic" and "paramagnetic" terminology in this context is not precise.

3.1.4 Spin-Independent Susceptibility

In the context of molecular quantum mechanics, *magnetic susceptibility tensor* $\boldsymbol{\chi}$ is a real symmetric 3 × 3 matrix connecting the external magnetic field $\mathbf{B}$ with the magnetic moment $\boldsymbol{\mu}$ that it induces:

$$\boldsymbol{\mu} = \mu_0^{-1}\boldsymbol{\chi}\mathbf{B} \tag{3.24}$$

With this approximation in place, it follows from the expression for the energy of the induced dipole that the molecular energy is a quadratic function of the external magnetic field:

$$E = -\boldsymbol{\mu}^{\mathrm{T}}\mathbf{B} \quad \Rightarrow \quad E = -\mu_0^{-1}\mathbf{B}^{\mathrm{T}}\boldsymbol{\chi}\mathbf{B} \quad \Rightarrow \quad \mu_0^{-1}\boldsymbol{\chi} = -\frac{\partial^2 E}{\partial\mathbf{B}\partial\mathbf{B}^{\mathrm{T}}} \tag{3.25}$$

and therefore $\boldsymbol{\chi}$ is the second derivative of the energy with respect to the external field. Inspecting the Hamiltonian in Eq. (2.91) with the vector potential of the uniform external field from Eq. (3.21) yields two spin-independent terms with non-zero first and second derivatives with respect to the magnetic field:

$$\mathbf{B}^{\mathrm{T}} \cdot \hat{H}^{\mathrm{LS}} \cdot \mathbf{B} = \frac{e^2}{8m_e}\sum_i \mathbf{B}^{\mathrm{T}}\left(\mathbf{r}_{ig}^{\mathrm{T}}\mathbf{r}_{ig} - \mathbf{r}_{ig}\mathbf{r}_{ig}^{\mathrm{T}}\right)\mathbf{B}$$
$$\hat{H}^{\mathrm{OZ}} \cdot \mathbf{B} = \frac{e}{2m_e}\sum_i \left(\mathbf{r}_{ig} \times \mathbf{p}_i\right) \cdot \mathbf{B} \tag{3.26}$$

These are derived in Sect. 2.6.3; applying the second derivative expression in Eq. (2.126) yields

Table 3.1 Magnetic perturbation operators of relevance in weakly relativistic electronic structure theory. Electrons are numbered by i and j indices, nuclei by n and k indices, and g index stands for the gauge origin. $\mathbf{S}$ indicates a vector of dimensionless $\mathfrak{su}(2)$ generators from Eq. (2.58), $\mathbf{p}$ indicates a vector of linear momentum operators, $\mathbf{B}$ is the external magnetic field vector, Z_n is a dimensionless integer specifying charge of the n-th nucleus

Type	Energy operator	Label
Linear in electron spin	$\mathbf{S}_i \cdot \hat{H}_i^{\mathrm{EZ}} \cdot \mathbf{B} = g_{\mathrm{e}}\mu_{\mathrm{B}}\mathbf{S}_i \cdot \mathbf{B}$	Electron Zeeman
	$\mathbf{S}_i \cdot \hat{H}_{ij}^{\mathrm{ES}} \cdot \mathbf{B} = \frac{\mu_0}{4\pi}\frac{g_{\mathrm{e}}\mu_{\mathrm{B}}^2 e}{2\hbar}\mathbf{S}_i \cdot \left(\frac{\mathbf{r}_{ig}^{\mathrm{T}}\mathbf{r}_{ij}-\mathbf{r}_{ig}\mathbf{r}_{ij}^{\mathrm{T}}}{r_{ij}^3}\right) \cdot \mathbf{B}$	Electron Shielding
	$\mathbf{S}_i \cdot \hat{H}^{\mathrm{ESO}}$: $\mathbf{S}_i \cdot \hat{H}_{ni}^{\mathrm{OON}} = \frac{\mu_0}{4\pi}\frac{g_{\mathrm{e}}\mu_{\mathrm{B}}^2 Z_n}{\hbar}\mathbf{S}_i \cdot \frac{\mathbf{r}_{in}\times\mathbf{p}_i}{r_{in}^3}$	Electron Spin-Orbit... ...with Own Orbit around a Nucleus
	$\mathbf{S}_i \cdot \hat{H}_{ij}^{\mathrm{OOE}} = -\frac{\mu_0}{4\pi}\frac{g_{\mathrm{e}}\mu_{\mathrm{B}}^2}{\hbar}\mathbf{S}_i \cdot \frac{\mathbf{r}_{ij}\times\mathbf{p}_i}{r_{ij}^3}$	...with Own Orbit around an Electron
	$\mathbf{S}_i \cdot \hat{H}_{ij}^{\mathrm{AEO}} = -\frac{\mu_0}{4\pi}\frac{2g_{\mathrm{e}}\mu_{\mathrm{B}}^2}{\hbar}\mathbf{S}_i \cdot \frac{\mathbf{r}_{ij}\times\mathbf{p}_j}{r_{ij}^3}$	...with Another Electron's Orbit
Bilinear in electron spin	$\mathbf{S}_i \cdot \hat{H}_{ij}^{\mathrm{EEC}} \cdot \mathbf{S}_j = -\frac{8\pi}{3}\frac{\mu_0 g_{\mathrm{e}}^2\mu_{\mathrm{B}}^2}{4\pi}\delta(\mathbf{r}_{ij})\mathbf{S}_i \cdot \mathbf{S}_j$	Electron-Electron Contact
	$\mathbf{S}_i \cdot \hat{H}_{ij}^{\mathrm{EED}} \cdot \mathbf{S}_j = \frac{\mu_0 g_{\mathrm{e}}^2\mu_{\mathrm{B}}^2}{4\pi}\mathbf{S}_i \cdot \left(\frac{\mathbf{r}_{ij}^{\mathrm{T}}\mathbf{r}_{ij}-3\mathbf{r}_{ij}\mathbf{r}_{ij}^{\mathrm{T}}}{r_{ij}^5}\right) \cdot \mathbf{S}_j$	Electron-Electron Dipolar
Linear in nuclear spin	$\mathbf{S}_n \cdot \hat{H}_n^{\mathrm{NZ}} \cdot \mathbf{B} = -g_n\mu_{\mathrm{N}}\mathbf{S}_n \cdot \mathbf{B}$	Nuclear Zeeman
	$\mathbf{S}_n \cdot \hat{H}_{ni}^{\mathrm{NS}} \cdot \mathbf{B} = -\frac{\mu_0}{4\pi}\frac{g_n\mu_{\mathrm{N}}\mu_{\mathrm{B}}e}{2\hbar}\times \mathbf{S}_n \cdot \left(\frac{\mathbf{r}_{ig}^{\mathrm{T}}\mathbf{r}_{ni}-\mathbf{r}_{ig}\mathbf{r}_{ni}^{\mathrm{T}}}{r_{ni}^3}\right) \cdot \mathbf{B}$	Nuclear Shielding
	$\mathbf{S}_n \cdot \hat{H}_{ni}^{\mathrm{NSO}} = \frac{\mu_0}{4\pi}\frac{2g_n\mu_{\mathrm{N}}\mu_{\mathrm{B}}}{\hbar}\mathbf{S}_n \cdot \frac{\mathbf{r}_{ni}\times\mathbf{p}_i}{r_{ni}^3}$	Nuclear Spin with electron Orbit
Bilinear in nuclear spin	$\mathbf{S}_n \cdot \hat{H}_{nki}^{\mathrm{NSS}} \cdot \mathbf{S}_k = \left(\frac{\mu_0}{4\pi}\right)^2\frac{g_n g_k\mu_{\mathrm{N}}^2 e^2}{2m_{\mathrm{e}}}\times \mathbf{S}_n \cdot \left(\frac{\mathbf{r}_{ni}^{\mathrm{T}}\mathbf{r}_{ki}-\mathbf{r}_{ki}\mathbf{r}_{ni}^{\mathrm{T}}}{r_{ni}^3 r_{ki}^3}\right) \cdot \mathbf{S}_k$	electron-mediated Nuclear Spin-Spin
Bilinear in electron and nuclear spin	$\mathbf{S}_n \cdot \hat{H}_{ni}^{\mathrm{NED}} \cdot \mathbf{S}_i = -\frac{\mu_0 g_{\mathrm{e}}\mu_{\mathrm{B}}g_n\mu_{\mathrm{N}}}{4\pi}\times \mathbf{S}_n \cdot \left(\frac{\mathbf{r}_{ni}^{\mathrm{T}}\mathbf{r}_{ni}-3\mathbf{r}_{ni}\mathbf{r}_{ni}^{\mathrm{T}}}{r_{ni}^5}\right) \cdot \mathbf{S}_i$	Nuclear-Electron Dipolar
	$\mathbf{S}_n \cdot \hat{H}_{ni}^{\mathrm{NEC}} \cdot \mathbf{S}_i = \frac{8\pi}{3}\frac{\mu_0 g_{\mathrm{e}}\mu_{\mathrm{B}}g_n\mu_{\mathrm{N}}}{4\pi}\delta(\mathbf{r}_{ni})\mathbf{S}_n \cdot \mathbf{S}_i$	Nuclear-Electron Contact
Spin-independent	$\mathbf{B}^{\mathrm{T}} \cdot \hat{H}_i^{\mathrm{LS}} \cdot \mathbf{B} = \frac{e^2}{8m_{\mathrm{e}}}\mathbf{B}^{\mathrm{T}}\left[\mathbf{r}_{ig}^{\mathrm{T}}\mathbf{r}_{ig}-\mathbf{r}_{ig}\mathbf{r}_{ig}^{\mathrm{T}}\right]\mathbf{B}$	Langevin Susceptibility
	$\hat{H}_i^{\mathrm{OZ}} \cdot \mathbf{B} = \frac{e}{2m_{\mathrm{e}}}\left(\mathbf{r}_{ig}\times\mathbf{p}_i\right) \cdot \mathbf{B}$	Orbital Zeeman

$$\mu_0^{-1}\chi = -\langle\Psi_0|\hat{H}^{\mathrm{LS}}|\Psi_0\rangle - 2\sum_{m>0}\frac{\left|\langle\Psi_m|\hat{H}^{\mathrm{OZ}}|\Psi_0\rangle\right|^2}{E_0 - E_m} \tag{3.27}$$

The first term on the right-hand side is *Langevin susceptibility* [61], the second is *Van Vleck susceptibility* [62]. Both are temperature-independent; thermodynamic treatments [63] that lead to temperature-dependent parts are outside the scope of this book.

3.1.5 Hyperfine Coupling

Hyperfine coupling is conventionally defined in magnetic resonance spectroscopy [48] as a bilinear interaction between electron and nuclear spin with the following general spin Hamiltonian:

$$\mathbf{H} = \mathbf{S}_{\mathrm{N}} \cdot \mathbf{A} \cdot \mathbf{S}_{\mathrm{E}} \tag{3.28}$$

where $\mathbf{A}$ is the *hyperfine coupling tensor*; its expression may be obtained by differentiating the Hamiltonian in Eq. (2.91) with respect to nuclear and electron magnetic moments entering the vector potential in Eq. (3.21) and then using the proportionality (Sect. 2.6.6 and Sect. 3.1.1) between magnetic moment and spin operators:

$$\boldsymbol{\mu}_{\mathrm{e}} = -g_{\mathrm{e}}\mu_{\mathrm{B}}\mathbf{S}_{\mathrm{e}}, \quad \boldsymbol{\mu}_{\mathrm{n}} = g_{\mathrm{n}}\mu_{\mathrm{N}}\mathbf{S}_{\mathrm{n}} \tag{3.29}$$

When the dipole vector potential is substituted into Eq. (2.92) and tedious simplifications are applied, the following terms containing nuclear and electron spin operators appear:

$$\mathbf{S}_n \cdot \hat{H}_{ni}^{\mathrm{NEC}} \cdot \mathbf{S}_i = \frac{8\pi}{3}\frac{\mu_0}{4\pi} g_{\mathrm{e}}\mu_{\mathrm{B}} g_n \mu_{\mathrm{N}} \delta(\mathbf{r}_{ni}) \mathbf{S}_n \cdot \mathbf{S}_i \tag{3.30}$$

$$\mathbf{S}_n \cdot \hat{H}_{ni}^{\mathrm{NED}} \cdot \mathbf{S}_i = -\frac{\mu_0}{4\pi} g_{\mathrm{e}}\mu_{\mathrm{B}} g_n \mu_{\mathrm{N}} \mathbf{S}_n \cdot \left(\frac{\mathbf{r}_{ni}^{\mathrm{T}}\mathbf{r}_{ni} - 3\mathbf{r}_{ni}\mathbf{r}_{ni}^{\mathrm{T}}}{r_{ni}^5}\right) \cdot \mathbf{S}_i \tag{3.31}$$

These *nuclear-electron contact* and *nuclear-electron dipole* terms were derived in Sect. 2.6.6, they enter the second derivative part of Eq. (2.126):

$$\mathbf{A} = \langle\Psi_0|\hat{H}^{\mathrm{NEC}} + \hat{H}^{\mathrm{NED}}|\Psi_0\rangle + 2\mathrm{Re}\sum_{m>0}\frac{\langle\Psi_0|\hat{H}^{\mathrm{ESO}}|\Psi_m\rangle\langle\Psi_m|\hat{H}^{\mathrm{NSO}}|\Psi_0\rangle}{E_0 - E_m} \tag{3.32}$$

When the spin-orbit terms are negligible, the interaction has a simple physical interpretation—the isotropic part $a_{\mathrm{iso}}\mathbf{S}_{\mathrm{n}} \cdot \mathbf{S}_{\mathrm{e}}$ arises from contact electron spin density at the nucleus (*contact hyperfine interaction*) and the anisotropic part $\mathbf{S}_{\mathrm{n}} \cdot \mathbf{A}_{\mathrm{aniso}} \cdot \mathbf{S}_{\mathrm{e}}$ contains a real symmetric 3×3 matrix coming from a classical dipolar coupling integral over the electron spin density (*dipolar hyperfine interaction*):

$$\begin{aligned} a_{\text{iso}}^{(n)} &= \frac{8\pi}{3}\frac{\mu_0}{4\pi} g_{\text{e}}\mu_{\text{B}} g_n \mu_{\text{N}} \rho_{\text{spin}}(\mathbf{r}_n) \\ \mathbf{A}_{\text{aniso}}^{(n)} &= -\frac{\mu_0}{4\pi} g_{\text{e}}\mu_{\text{B}} g_n \mu_{\text{N}} \int \left(\frac{(\mathbf{r}-\mathbf{r}_n)^{\text{T}}(\mathbf{r}-\mathbf{r}_n) - 3(\mathbf{r}-\mathbf{r}_n)(\mathbf{r}-\mathbf{r}_n)^{\text{T}}}{|\mathbf{r}-\mathbf{r}_n|^5} \right) \rho_{\text{spin}}(\mathbf{r}) \text{d}V \end{aligned} \tag{3.33}$$

Although the kernel of the integral is both axial and traceless, dipolar hyperfine interaction tensor need not be axial: differently oriented axialities can sum up to give non-zero rhombicity.

The spin-orbit coupling terms in Eq. 3.32 are partly empirical—although the coupling itself does follow from Dirac's equation (Sect. 2.6.4), it must be amended to include radiative corrections to the electron g-tensor (Sect. 2.6.5) and extended to multi-particle systems. As a result, four types of spin-orbit coupling terms appear—the coupling of electron spin to own orbital angular momentum around a nucleus:

$$\mathbf{S}_i \cdot \hat{H}_{ni}^{\text{OON}} = \frac{\mu_0}{4\pi} \frac{g_{\text{e}}\mu_{\text{B}}^2 Z_n}{\hbar} \mathbf{S}_i \cdot \frac{\mathbf{r}_{in} \times \mathbf{p}_i}{r_{in}^3} \tag{3.34}$$

but also to own orbit around other electrons (because all electrostatic potential terms in Eq. (2.102) must be accounted for):

$$\mathbf{S}_i \cdot \hat{H}_{ij}^{\text{OOE}} = -\frac{\mu_0}{4\pi} \frac{g_{\text{e}}\mu_{\text{B}}^2}{\hbar} \mathbf{S}_i \cdot \frac{\mathbf{r}_{ij} \times \mathbf{p}_i}{r_{ij}^3} \tag{3.35}$$

On empirical grounds, we introduce a coupling to other electrons' orbital motion:

$$\mathbf{S}_i \cdot \hat{H}_{ij}^{\text{AEO}} = -\frac{\mu_0}{4\pi} \frac{2g_{\text{e}}\mu_{\text{B}}^2}{\hbar} \mathbf{S}_i \cdot \frac{\mathbf{r}_{ij} \times \mathbf{p}_j}{r_{ij}^3} \tag{3.36}$$

and a coupling of nuclear spin to orbital motion of each electron:

$$\mathbf{S}_n \cdot \hat{H}_{ni}^{\text{NSO}} = \frac{\mu_0}{4\pi} \frac{2g_n\mu_{\text{N}}\mu_{\text{B}}}{\hbar} \mathbf{S}_n \cdot \frac{\mathbf{r}_{ni} \times \mathbf{p}_i}{r_{ni}^3} \tag{3.37}$$

The $\hat{H}^{\text{ESO}}$ term in Eq. 3.32 is a sum of Eqs. (3.34)-(3.36), including sums over electrons (i and j indices) and nuclei (n index). Spin-orbit contribution to hyperfine coupling is important in systems with small excitation energies (they appear in the denominator) and heavy nuclei—due to the approximate Z_n^4 dependence of spin-orbit coupling on the nuclear charge. Because AEO and NSO terms originate in Eq. (2.91) rather than Eq. (2.99), there is no Thomas half.

For localised unpaired electrons (e.g. those of lanthanide ions and nitroxide radicals) at distances over 15 Angstrom from the nucleus, contact and spin-orbit contributions are negligible and the dipolar contribution may be simplified using the point dipole approximation:

$$a_{\text{iso}}^{(n)} = 0, \quad \mathbf{A}_{\text{aniso}}^{(n)} = -\frac{\mu_0}{4\pi}\mu_{\text{B}}g_n\mu_{\text{N}}(\mathbf{1}-\boldsymbol{\sigma}_n)\left(\frac{\mathbf{r}_{ni}^{\text{T}}\mathbf{r}_{ni} - 3\mathbf{r}_{ni}\mathbf{r}_{ni}^{\text{T}}}{r_{ni}^5}\right)\mathbf{g}_{\text{e}} \tag{3.38}$$

This is the limit of classical magnetic dipole coupling; magnetic moments must, therefore, be corrected to account for nuclear and electron shielding tensors discussed in the next section.

3.1.6 Electron and Nuclear Shielding

Zeeman interaction Hamiltonians derived in Sect. 2.6.6 for the electron and Sect. 3.1.1 for nuclei only hold for free electrons and nuclei in vacuum. When they are hosted by a molecule, electronic structure introduces corrections, for which historical conventions stipulate the following expressions

$$\mathbf{H}_{\text{EZ}} = \mu_{\text{B}}\mathbf{S}\cdot\mathbf{g}\cdot\mathbf{B}, \quad \mathbf{H}_{\text{NZ}} = -\mu_{\text{N}}g_n\mathbf{S}\cdot(\mathbf{1}-\boldsymbol{\sigma})\cdot\mathbf{B} \tag{3.39}$$

that are designed to look similar to the corresponding Hamiltonians of the free electron and a bare nucleus in vacuum, respectively. These are definitions of *g-tensor* $\mathbf{g}$ and *chemical shielding tensor* $\boldsymbol{\sigma}$. Another exercise of parsing Eq. (2.91) for terms containing electron magnetic moment, nuclear magnetic moment, and the magnetic field yields, in addition to ESO, NZO, and OZ terms already seen above, the following:

$$\mathbf{S}_i\cdot\hat{H}_{ij}^{\text{ES}}\cdot\mathbf{B} = \frac{\mu_0}{4\pi}\frac{g_{\text{e}}\mu_{\text{B}}^2 e}{2\hbar}\mathbf{S}_i\cdot\left(\frac{\mathbf{r}_{ig}^{\text{T}}\mathbf{r}_{ij} - \mathbf{r}_{ig}\mathbf{r}_{ij}^{\text{T}}}{r_{ij}^3}\right)\cdot\mathbf{B} \tag{3.40}$$

$$\mathbf{S}_n\cdot\hat{H}_{ni}^{\text{NS}}\cdot\mathbf{B} = \frac{\mu_0}{4\pi}\frac{g_n\mu_{\text{N}}\mu_{\text{B}}e}{2\hbar}\mathbf{S}_n\cdot\left(\frac{\mathbf{r}_{ig}^{\text{T}}\mathbf{r}_{ni} - \mathbf{r}_{ig}\mathbf{r}_{ni}^{\text{T}}}{r_{ni}^3}\right)\cdot\mathbf{B} \tag{3.41}$$

Application of the derivative formalism in Eq. (2.126) then yields [64–66]

$$\begin{aligned}\mu_{\text{B}}\mathbf{g} &= \langle\Psi_0|\hat{H}^{\text{EZ}} + \hat{H}^{\text{ES}}|\Psi_0\rangle + 2\text{Re}\sum_{m>0}\frac{\langle\Psi_0|\hat{H}^{\text{ESO}}|\Psi_m\rangle\langle\Psi_m|\hat{H}^{\text{OZ}}|\Psi_0\rangle}{E_0 - E_m}\\ -\mu_{\text{N}}g_n(\mathbf{1}-\sigma) &= \langle\Psi_0|\hat{H}^{\text{NZ}} + \hat{H}^{\text{NS}}|\Psi_0\rangle + 2\text{Re}\sum_{m>0}\frac{\langle\Psi_0|\hat{H}^{\text{NSO}}|\Psi_m\rangle\langle\Psi_m|\hat{H}^{\text{OZ}}|\Psi_0\rangle}{E_0 - E_m}\end{aligned} \tag{3.42}$$

Neither $\mathbf{g}$ nor $\boldsymbol{\sigma}$ is a symmetric matrix, although the antisymmetry is rarely significant (an exception is susceptibility shielding described in Sect. 3.1.7). Situations when electron "spin" is an effective quantity arising from a combination of spin and orbital angular momentum (commonly the case in complexes of *d*- and *f*-elements) are covered in Sect. 3.1.10.2 which deals with zero-field splitting.

For historical reasons [67], chemists define the experimental *chemical shift* as a fractional Zeeman frequency difference between the nucleus in question and the reference nucleus:

$$\delta_{\text{sample}} = \frac{\nu_{\text{sample}} - \nu_{\text{reference}}}{\nu_{\text{reference}}} \tag{3.43}$$

with the reasonably accurate assumption that this quantity does not depend on the magnetic field. Common external reference substances are tetramethylsilane in deuterated chloroform (^{1}H and ^{13}C), liquid ammonia (^{15} N), and 85% H_3PO_4 in water (^{31}P).

3.1.7 Nuclear Shielding by Susceptibility

With the exception of Gd^{3+}, electron spin relaxation times in lanthanide complexes are in the femtoseconds—faster than molecular rotation in liquids and much faster than nuclear spin dynamics in common magnetic fields. As far as nuclear shielding is concerned, lanthanide ions in such complexes can therefore be viewed as classical magnetic susceptibility centres.

Consider a point magnetic dipole $\boldsymbol{\mu}_\text{A}$ at a distance $\mathbf{r}$ from a point magnetic susceptibility $\boldsymbol{\chi} \ll 1$. In an external magnetic field $\mathbf{B}$, the susceptibility acquires a magnetic moment:

$$\boldsymbol{\mu}_\text{B} = \boldsymbol{\chi}\mathbf{B}/\mu_0 \tag{3.44}$$

The energy of the dipole $\boldsymbol{\mu}_\text{A}$ includes interactions with $\mathbf{B}$ and $\boldsymbol{\mu}_\text{B}$:

$$E = -\boldsymbol{\mu}_\text{A}^\text{T}\mathbf{B} - \boldsymbol{\mu}_\text{A}^\text{T}\mathbf{D}\boldsymbol{\mu}_\text{B}, \quad \mathbf{D} = \frac{\mu_0}{4\pi}\left[3\frac{\mathbf{r}\mathbf{r}^\text{T}}{r^5} - \frac{\mathbf{1}}{r^3}\right] \tag{3.45}$$

After using Eq. (3.44) to replace $\boldsymbol{\mu}_\text{B}$ and taking the magnetic field out of the bracket, we obtain

$$E = -\boldsymbol{\mu}_\text{A}^\text{T}(\mathbf{1} + \mathbf{D}\boldsymbol{\chi}/\mu_0)\mathbf{B} \tag{3.46}$$

This looks like an interaction between $\boldsymbol{\mu}_\text{A}$ and $\mathbf{B}$, but with an amplitude correction term, called *dipolar shielding tensor* that has the same mathematical form as chemical shielding in Eq. 3.39:

$$\begin{gathered} E = -\boldsymbol{\mu}_\text{A}^\text{T}(\mathbf{1} - \boldsymbol{\sigma}_\text{DS})\mathbf{B} \\ \boldsymbol{\sigma}_\text{DS} = -\mathbf{D}\boldsymbol{\chi}/\mu_0 = -\frac{1}{4\pi}\left[3\frac{\mathbf{r}\mathbf{r}^\text{T}}{r^5} - \frac{\mathbf{1}}{r^3}\right]\boldsymbol{\chi} \end{gathered} \tag{3.47}$$

Both $\mathbf{D}$ and $\boldsymbol{\chi}$ are symmetric, but their product need not be—dipolar shielding tensors do in general have significant antisymmetric components. When the conventions are switched from shielding to shift, the isotropic part of $\boldsymbol{\sigma}_{\mathrm{DS}}$ gives rise to *pseudocontact shift*:

$$\delta_{\mathrm{PCS}} = -\frac{1}{3}\mathrm{Tr}[\boldsymbol{\sigma}_{\mathrm{DS}}] = \frac{1}{12\pi}\mathrm{Tr}\left\{\left[3\frac{\mathbf{r}\mathbf{r}^{\mathrm{T}}}{r^5} - \frac{\mathbf{1}}{r^3}\right]\boldsymbol{\chi}\right\} \tag{3.48}$$

The name proceeds from a related quantity called *contact chemical shift*, which appears in the situation when the nucleus is close enough to an unpaired electron that contact hyperfine interaction is also present. In that case:

$$E_{\mathrm{n}} = -\boldsymbol{\mu}_{\mathrm{n}}(\mathbf{1}-\boldsymbol{\sigma})\mathbf{B} + \frac{\boldsymbol{\mu}_{\mathrm{n}}\cdot\mathbf{A}\cdot\boldsymbol{\mu}_{\mathrm{e}}}{\gamma_{\mathrm{e}}\gamma_{\mathrm{n}}\hbar} = -\boldsymbol{\mu}_{\mathrm{n}}(\mathbf{1}-\boldsymbol{\sigma})\mathbf{B} + \frac{\boldsymbol{\mu}_{\mathrm{n}}\cdot\mathbf{A}\cdot\boldsymbol{\chi}\cdot\mathbf{B}}{\mu_0\gamma_{\mathrm{e}}\gamma_{\mathrm{n}}\hbar} \tag{3.49}$$

where $\gamma_{\mathrm{e.n}}$ are magnetogyric ratios of the electron and the nucleus in the units of (rad/s)/Tesla, and the units of $\mathbf{A}$ are the conventional (for computational spin dynamics) rad/s. After using the same derivative expression for the chemical shielding and converting from shielding to shift, we obtain the following corrections to the chemical shift tensor and to the isotropic chemical shift:

$$\boldsymbol{\delta}_{\mathrm{CS}} + \boldsymbol{\delta}_{\mathrm{DS}} = -\frac{\mathbf{A}\cdot\boldsymbol{\chi}}{\mu_0\gamma_{\mathrm{e}}\gamma_{\mathrm{n}}\hbar} \quad\Rightarrow\quad \delta_{\mathrm{CS}} + \delta_{\mathrm{PCS}} = -\frac{1}{3}\mathrm{Tr}\left[\frac{\mathbf{A}\cdot\boldsymbol{\chi}}{\mu_0\gamma_{\mathrm{e}}\gamma_{\mathrm{n}}\hbar}\right] \tag{3.50}$$

For a real or effective (in *d*- and *f*-element complexes where orbital contributions exist) electron spin with an isotropic *g*-tensor at high temperature, the contact contribution becomes [68]

$$\delta_{\mathrm{CS}} = \frac{g_{\mathrm{e}}\mu_{\mathrm{B}}}{g_{\mathrm{n}}\mu_{\mathrm{N}}}\frac{S(S+1)h}{3k_{\mathrm{B}}T}a_{\mathrm{iso}}, \quad a_{\mathrm{iso}} = \tfrac{1}{3}\mathrm{Tr}(\mathbf{A}) \tag{3.51}$$

3.1.8 Inter-nuclear Dipolar Interaction

Nuclei are at least five orders of magnitude smaller than molecules; their interactions with each other may be treated using the classical point dipole approximation:

$$E_{\mathrm{DD}} = -\frac{\mu_0}{4\pi}\frac{3(\boldsymbol{\mu}_1\cdot\mathbf{r}_{12})(\mathbf{r}_{12}\cdot\boldsymbol{\mu}_2) - r_{12}^2(\boldsymbol{\mu}_1\cdot\boldsymbol{\mu}_2)}{r_{12}^5} \tag{3.52}$$

where $\mathbf{r}_{12}$ is the distance vector between the nuclei. The spin Hamiltonian is then obtained by replacing the magnetic dipole vectors with the corresponding vectors of spin operators:

$$\boldsymbol{\mu}_k = \gamma_k\mathbf{S}_k \tag{3.53}$$

where γ_k are the magnetogyric ratios. The dipolar spin Hamiltonian in angular frequency units then becomes

$$\mathbf{H}_{\mathrm{DD}} = -\frac{\mu_0}{4\pi}\frac{\gamma_1\gamma_2\hbar}{r^5}\left(3(\mathbf{S}_1\cdot\mathbf{r}_{12})(\mathbf{r}_{12}\cdot\mathbf{S}_2) - r_{12}^2(\mathbf{S}_1\cdot\mathbf{S}_2)\right) = -\frac{\mu_0}{4\pi}\frac{\gamma_1\gamma_2\hbar}{r^5}$$
$$\times\begin{pmatrix}\mathbf{S}_{\mathrm{X}}^{(1)}\\ \mathbf{S}_{\mathrm{Y}}^{(1)}\\ \mathbf{S}_{\mathrm{Z}}^{(1)}\end{pmatrix}^{\mathrm{T}}\begin{pmatrix}3(x_1-x_2)^2-r_{12}^2 & 3(x_1-x_2)(y_1-y_2) & 3(x_1-x_2)(z_1-z_2)\\ 3(y_1-y_2)(x_1-x_2) & 3(y_1-y_2)^2-r_{12}^2 & 3(y_1-y_2)(z_1-z_2)\\ 3(z_1-z_2)(x_1-x_2) & 3(z_1-z_2)(y_1-y_2) & 3(z_1-z_2)^2-r_{12}^2\end{pmatrix}\begin{pmatrix}\mathbf{S}_{\mathrm{X}}^{(2)}\\ \mathbf{S}_{\mathrm{Y}}^{(2)}\\ \mathbf{S}_{\mathrm{Z}}^{(2)}\end{pmatrix} \tag{3.54}$$

By direct inspection, the trace of the matrix in Eq. (3.54) is zero. Diagonalisation would demonstrate that the rhombicity is also zero. Inter-nuclear dipole interaction is therefore traceless and axial.

3.1.9 Inter-nuclear *J*-coupling

The direct interaction of two nuclear magnetic dipoles was obtained in the previous section; here we deal with the contribution mediated by the electronic structure:

$$\mathbf{K}_{ab} = \left.\frac{\partial^2 E}{\partial\boldsymbol{\mu}_a\partial\boldsymbol{\mu}_b^{\mathrm{T}}}\right|_{\boldsymbol{\mu}_a=\boldsymbol{\mu}_b=0} \tag{3.55}$$

This *reduced spin-spin coupling tensor* $\mathbf{K}_{ab}$ is isotope-independent, but only used in electronic structure theory; magnetic resonance spectroscopy measures indirect spin-spin couplings as a splitting (historically, in units of Hz) between lines of experimental NMR spectra, corresponding to

$$\mathbf{J}_{ab} = h\frac{\gamma_a}{2\pi}\frac{\gamma_b}{2\pi}\mathbf{K}_{ab}, \quad J_{ab} = \mathrm{Tr}[\mathbf{J}_{ab}]/3 \tag{3.56}$$

where $\gamma_{a,b} = \hbar^{-1}g_{a,b}\mu_{\mathrm{N}}$ are nuclear magnetogyric ratios with the units of (rad/s)/Tesla—this quantity does depend on the isotope. The anisotropic part of $\mathbf{J}_{ab}$ is undetectable in solids and inconsequential in liquids; only the *scalar spin-spin coupling* J_{ab} is usually quoted in the literature. From Eq. (2.126),

$$\mathbf{K}_{\mathrm{ab}} = \langle\Psi_0|\frac{\partial^2\hat{H}}{\partial\boldsymbol{\mu}_a\partial\boldsymbol{\mu}_b^{\mathrm{T}}}|\Psi_0\rangle + \sum_{k>0}\frac{\langle\Psi_n|\partial\hat{H}/\partial\boldsymbol{\mu}_a|\Psi_k\rangle\langle\Psi_k|\partial\hat{H}/\partial\boldsymbol{\mu}_b^{\mathrm{T}}|\Psi_n\rangle + c.c.}{E_0 - E_k} \tag{3.57}$$

Inserting the point dipole vector potentials from Eq. (3.21) and identifying the terms (Table 3.1) that have non-zero contributions to those derivatives yields [69, 70]

$$h\mathbf{J}_{ab} = \langle \Psi_0 | \hat{H}_{ab}^{\mathrm{NSS}} | \Psi_0 \rangle + 2\mathrm{Re} \sum_{m>0} \frac{\langle \Psi_0 | \hat{H}_a^{\mathrm{NSO}} + \hat{H}_a^{\mathrm{NEC}} + \hat{H}_a^{\mathrm{NED}} | \Psi_m \rangle \langle \Psi_m | \hat{H}_b^{\mathrm{NSO}} + \hat{H}_b^{\mathrm{NEC}} + \hat{H}_b^{\mathrm{NED}} | \Psi_0 \rangle}{E_0 - E_m} \tag{3.58}$$

where the principal contribution to the sum over the excited states is usually from the contact coupling.

3.1.10 Bilinear Inter-electron Couplings

The classification of inter-electron spin interactions proceeds from the turbulent history of that subject. Here, we specify the general mathematical form that any bilinear inter-electron spin interaction must have, and then look at how that form was filled by researchers who did not always consider elegance a priority. Mathematically, any bilinear coupling $\boldsymbol{\mu}_1 \cdot \mathbf{C} \cdot \boldsymbol{\mu}_2$ between two magnetic moment vectors $\boldsymbol{\mu}_1$ and $\boldsymbol{\mu}_2$ must have the following expansion:

$$\begin{gathered} \boldsymbol{\mu}_1 \cdot \mathbf{C} \cdot \boldsymbol{\mu}_2 = a(\boldsymbol{\mu}_1 \cdot \boldsymbol{\mu}_2) + \mathbf{d} \cdot (\boldsymbol{\mu}_1 \times \boldsymbol{\mu}_2) + \boldsymbol{\mu}_1 \cdot \mathbf{A} \cdot \boldsymbol{\mu}_2 \\ a = \mathrm{Tr}(\mathbf{C})/3, \quad \mathbf{A} = \left(\mathbf{C} + \mathbf{C}^{\mathrm{T}}\right)/2 - \mathbf{1}a \\ \mathbf{d} = \left(c_{\mathrm{YZ}} - c_{\mathrm{ZY}} \quad c_{\mathrm{ZX}} - c_{\mathrm{XZ}} \quad c_{\mathrm{XY}} - c_{\mathrm{YX}} \right)^{\mathrm{T}} \Big/ 2 \end{gathered} \tag{3.59}$$

The first term on the right-hand side is called *isotropic* (because it is invariant under rotations), the second term is *antisymmetric* (because it changes sign when $\boldsymbol{\mu}_1$ and $\boldsymbol{\mu}_2$ are permuted), and the third term is *symmetric* (because it does not change sign when $\boldsymbol{\mu}_1$ and $\boldsymbol{\mu}_2$ are permuted).

3.1.10.1 Isotropic Exchange Coupling

The term *exchange coupling* was introduced by Heisenberg [71] who had observed that magnetic properties of systems with localised unpaired electrons (transition metal clusters, organic biradicals, complexes with multiple metal ions, etc.) can be well described by the following spin Hamiltonian:

$$\mathbf{H}_{\mathrm{EX}} = -\sum_{n \neq k} J_{nk} \mathbf{S}^{(n)} \cdot \mathbf{S}^{(k)} = -2 \sum_{n>k} J_{nk} \mathbf{S}^{(n)} \cdot \mathbf{S}^{(k)} \tag{3.60}$$

where $\mathbf{S}^{(n)} = \left\{ \mathbf{S}_{\mathrm{X}}^{(n)}, \mathbf{S}_{\mathrm{Y}}^{(n)}, \mathbf{S}_{\mathrm{Z}}^{(n)} \right\}$ are electron spin operators. The double summation (hence a factor of 2 in the second equation) and the negative sign are also historical. Because J is positive in metallic iron (a ferromagnetic material) and negative in metallic chromium (an antiferromagnetic material), couplings with $J > 0$ and $J < 0$ are historically called *ferromagnetic* and *antiferromagnetic*. These physically motivated names are the recommended way to report signs of exchange couplings because spin Hamiltonian sign conventions differ in the literature.

The term *exchange* comes from one of the mechanisms giving rise to the Hamiltonian in Eq. 3.60. Consider a pair of electrons with spatial wavefunctions $\{\varphi_1(\mathbf{r}), \varphi_2(\mathbf{r})\}$ and spin states $\{\alpha, \beta\}$. Permutation symmetric (S) and antisymmetric (A) product combinations for the spatial part are

$$\begin{aligned} \varphi_{\text{sym}}(\mathbf{r}_1, \mathbf{r}_2) &= \frac{1}{\sqrt{2}}[\varphi_1(\mathbf{r}_1)\varphi_2(\mathbf{r}_2) + \varphi_2(\mathbf{r}_1)\varphi_1(\mathbf{r}_2)] \\ \varphi_{\text{anti}}(\mathbf{r}_1, \mathbf{r}_2) &= \frac{1}{\sqrt{2}}[\varphi_1(\mathbf{r}_1)\varphi_2(\mathbf{r}_2) - \varphi_2(\mathbf{r}_1)\varphi_1(\mathbf{r}_2)] \end{aligned} \tag{3.61}$$

For the spin part, the three permutation symmetric states are collectively called *triplet* and the antisymmetric state is called *singlet*:

$$\begin{gathered} |T_+\rangle = |\alpha\alpha\rangle, \quad |T_0\rangle = \frac{1}{\sqrt{2}}[|\alpha\beta\rangle + |\beta\alpha\rangle], \quad |T_-\rangle = |\beta\beta\rangle \\ |S\rangle = \frac{1}{\sqrt{2}}[|\alpha\beta\rangle - |\beta\alpha\rangle] \end{gathered} \tag{3.62}$$

Fermion wavefunctions must be antisymmetric with respect to permutation of any two particles; we must therefore construct such products of spatial and spin parts as would overall be antisymmetric:

$$\begin{aligned} |\Psi_{\text{S}}\rangle &= \varphi_{\text{sym}}(\mathbf{r}_1, \mathbf{r}_2) \otimes |S\rangle \\ |\Psi_{\text{T}_k}\rangle &= \varphi_{\text{anti}}(\mathbf{r}_1, \mathbf{r}_2) \otimes |T_k\rangle, \quad k \in \{+, 0, -\} \end{aligned} \tag{3.63}$$

The symmetry of the spin state is now unambiguously linked to the symmetry of the spatial wavefunction; this makes spin states of different symmetries have different electronic structure energies. For Coulomb repulsion $\langle \Psi | r_{12}^{-1} | \Psi \rangle$, straightforward substitution and rearrangement yields

$$\begin{aligned} \langle \Psi_{\text{S}} | \frac{1}{r_{12}} | \Psi_{\text{S}} \rangle &= C + J, \quad \langle \Psi_{\text{T}_k} | \frac{1}{r_{12}} | \Psi_{\text{T}_k} \rangle = C - J \\ C &= \int |\varphi_1(\mathbf{r}_1)|^2 \frac{1}{r_{12}} |\varphi_2(\mathbf{r}_2)|^2 \mathrm{d}^3\mathbf{r}_1 \mathrm{d}^3\mathbf{r}_2 \\ J &= \int \varphi_1(\mathbf{r}_1)\varphi_1^*(\mathbf{r}_2) \frac{1}{r_{12}} \varphi_2(\mathbf{r}_2)\varphi_2^*(\mathbf{r}_1) \mathrm{d}^3\mathbf{r}_1 \mathrm{d}^3\mathbf{r}_2 \end{aligned} \tag{3.64}$$

The two energies are the same in the classical part C, which is just Coulomb repulsion between two clouds of charge density $|\varphi_1(\mathbf{r}_1)|^2$ and $|\varphi_2(\mathbf{r}_2)|^2$. However, they are different in the *exchange integral* part J, called so because particle labels are exchanged on either side of r_{12}^{-1} so that the product is not a classical probability density. For this term to be non-zero, there must be overlap between $\varphi_1(\mathbf{r})$ and $\varphi_2(\mathbf{r})$, otherwise the products $\varphi_1(\mathbf{r}_1)\varphi_2^*(\mathbf{r}_1)$ and $\varphi_2(\mathbf{r}_2)\varphi_1^*(\mathbf{r}_2)$ would vanish.

Now that the relative energies of the four spin states are known, the effective spin Hamiltonian may be obtained using projector expansion:

$$\mathbf{H}_{\mathrm{EX}} = (C+J)|S\rangle\langle S| + (C-J)(|T_+\rangle\langle T_+| + |T_0\rangle\langle T_0| + |T_-\rangle\langle T_-|) \tag{3.65}$$

Written in terms of the projection operators of individual spins, this becomes

$$\mathbf{H}_{\mathrm{EX}} = \begin{pmatrix} |\alpha\alpha\rangle \\ |\alpha\beta\rangle \\ |\beta\alpha\rangle \\ |\beta\beta\rangle \end{pmatrix}^{\mathrm{T}} \cdot \begin{pmatrix} C-J & 0 & 0 & 0 \\ 0 & C & -J & 0 \\ 0 & -J & C & 0 \\ 0 & 0 & 0 & C-J \end{pmatrix} \cdot \begin{pmatrix} \langle\alpha\alpha| \\ \langle\alpha\beta| \\ \langle\beta\alpha| \\ \langle\beta\beta| \end{pmatrix} \tag{3.66}$$

Direct inspection shows that this is equal to the following combination of spin operators:

$$\mathbf{H}_{\mathrm{EX}} = \mathbf{1}(C - J/2) - 2J\mathbf{S}^{(1)} \cdot \mathbf{S}^{(2)} \tag{3.67}$$

The energy offset operator $\mathbf{1}(C - J/2)$ commutes with everything and does not influence the dynamics prescribed by the Liouville - von Neumann equation. It may be dropped, at which point we obtain an equation of the same general form as Eq. (3.60). This electrostatic term is called the *direct exchange*.

Indirect mechanisms contributing to isotropic exchange coupling are complicated [72]; when no simplifying assumptions are available, the matter is best approached by treating J as a parameter of an effective spin Hamiltonian responsible for different energies of states with different total spin. Consider two localised spin centres A and B with fixed expectation values of $\mathbf{S}^2$: $S_{\mathrm{A}}(S_{\mathrm{A}}+1)$ and $S_{\mathrm{B}}(S_{\mathrm{B}}+1)$ respectively. The total spin operator for the composite system is then:

$$\mathbf{S}^2 = \mathbf{S}_{\mathrm{A}}^2 + \mathbf{S}_{\mathrm{B}}^2 + 2\mathbf{S}_{\mathrm{A}} \cdot \mathbf{S}_{\mathrm{B}} \;\Rightarrow\; 2\mathbf{S}_{\mathrm{A}} \cdot \mathbf{S}_{\mathrm{B}} = \mathbf{S}^2 - \mathbf{S}_{\mathrm{A}}^2 - \mathbf{S}_{\mathrm{B}}^2 \tag{3.68}$$

In a system where spin is a good quantum number, the energy operator eigenstates have specific total spin. For any two energy eigenstates Ψ_1 and Ψ_2 that differ in the total spin:

$$\begin{aligned} E_1 &= -J_{\mathrm{AB}}\left(\langle\mathbf{S}^2\rangle_1 - \langle\mathbf{S}_{\mathrm{A}}^2\rangle - \langle\mathbf{S}_{\mathrm{B}}^2\rangle\right) \\ E_2 &= -J_{\mathrm{AB}}\left(\langle\mathbf{S}^2\rangle_2 - \langle\mathbf{S}_{\mathrm{A}}^2\rangle - \langle\mathbf{S}_{\mathrm{B}}^2\rangle\right) \end{aligned} \tag{3.69}$$

When these equations are subtracted, $\langle\mathbf{S}_{\mathrm{A}}^2\rangle$ and $\langle\mathbf{S}_{\mathrm{B}}^2\rangle$ cancel, and we obtain *Yamaguchi equation* [73]:

$$J_{\mathrm{AB}} = -\frac{E_1 - E_2}{\langle\mathbf{S}^2\rangle_1 - \langle\mathbf{S}^2\rangle_2} \tag{3.70}$$

For a system with two localised unpaired electrons, J is equal to half the singlet-triplet gap. At the time of writing, computational chemists have no rigorous procedure for computing exchange couplings in systems with more than two electron spin centres, but Eq. 3.60 usually fits experimental data well.

3.1.10.2 Antisymmetric Exchange Coupling

The energy derivative formalism in Sect. 2.6.8 demonstrates that exchange coupling does have the antisymmetric component that was mentioned as a possibility in Eq. 3.59. This may be seen from the derivative expression for the bilinear inter-electron coupling

$$g_e^2 \mu_B^2 \mathbf{C}_{a,b} = \langle \Psi_0 | \hat{H}_{a,b}^{\mathrm{EEC}} + \hat{H}_{a,b}^{\mathrm{EED}} | \Psi_0 \rangle + 2\mathrm{Re} \sum_{m>0} \frac{\langle \Psi_0 | \hat{H}_a^{\mathrm{ESO}} | \Psi_m \rangle \langle \Psi_m | \hat{H}_b^{\mathrm{ESO}} | \Psi_0 \rangle}{E_0 - E_m} \tag{3.71}$$

which contains a product of matrix elements of two electron spin-orbit operators that may involve different electrons. Those may be rearranged, using vector calculus identities, as follows:

$$(\mathbf{S}_a \cdot \mathbf{L}_p)(\mathbf{L}_q \cdot \mathbf{S}_b) = (\mathbf{S}_b \cdot \mathbf{L}_p)(\mathbf{S}_a \cdot \mathbf{L}_q) + (\mathbf{L}_p \times \mathbf{L}_q)(\mathbf{S}_a \times \mathbf{S}_b) \tag{3.72}$$

The second term on the right-hand side is responsible for the *antisymmetric exchange coupling*, sometimes called *Dzyaloshinskii-Moriya interaction* [74, 75]. Orbital angular momentum operators disappear when spatial integrals are taken to leave the following contribution to the spin Hamiltonian:

$$\mathbf{H}_{nk}^{\mathrm{DM}} = \mathbf{d}_{nk} \cdot (\mathbf{S}_n \times \mathbf{S}_k) \tag{3.73}$$

where $\mathbf{d}$ is the *antisymmetric exchange coupling vector* between spins n and k.

3.1.10.3 Notes on Zero-Field Splitting

Zero-field splitting is a vast topic [76, 77] that this book cannot cover; here we look at ZFS spin Hamiltonians, but make no attempt to derive them. For *d*- and *f*-element ions, the reason is illustrated in Fig. 3.1—by the time we reach the ligand field splitting that is responsible for ZFS in those systems, the "spin" is no longer just spin: much like it happened with the nuclear structure in Sect. 3.1.1, spin-orbit coupling comes into play and we must consider the total angular momentum. This is because spin projection is not a good quantum number in the presence of spin-orbit coupling, but the total angular momentum projection.

In the phenomenological ZFS Hamiltonian, the splitting mechanism is the same as the one discussed for NQI in Sect. 3.1.2: different projection states of the multiplet have different spatial wavefunctions that interact differently with the electric fields, for example, from the ligands that surround the metal ion. ZFS spin

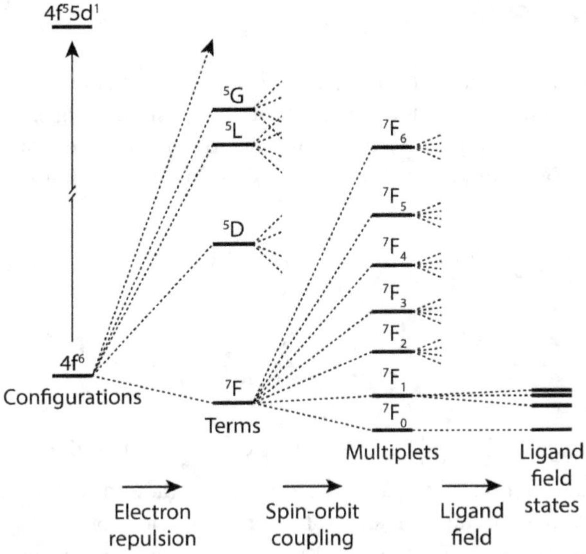

Fig. 3.1 Hierarchy of energy-level splittings in a Eu^{3+} complex (adapted from [78], energy-level positioning is schematic and not to scale). From left to right: **electronic configuration** classification that proceeds from irreducible representations of the rotation group of the spherically symmetric Coulomb potential of the nuclear charge; **term splitting** that proceeds from inter-electron repulsion; **multiplet splitting** caused by the spin-orbit coupling that extends the classification from spin and orbital angular momentum to the total angular momentum; **ligand field splitting** wherein different total angular momentum projection states have different energy as a result of being associated with different spatial wavefunctions that interact with external electric fields

Hamiltonian in such systems is essentially a matrix Taylor series, conventionally reordered so as to have an expansion in Stevens operators (historically dominant but not recommended, see Sect. 3.2.3) or spherical tensor operators (Sect. 3.2.2, recommended)

$$\mathbf{H}_{\text{ZFS}} = \sum_{k,q} B_k^q \mathbf{O}_k^q = \sum_l \sum_{m=-l}^{l} a_{lm} \mathbf{T}_{lm} \tag{3.74}$$

that refer to the total angular momentum of the multiplet. The general problem of calculating the coefficients from first principles is difficult [79], but they may be fitted to experimental data.

In simpler systems (e.g. organic radicals), zero-field splitting is a quadratic coupling tensor that may be obtained [80] by taking the second derivative of the molecular energy with respect to electron magnetic moments in the total electron momentum representation:

$$\mathbf{H}_{\text{ZFS}} = \mathbf{S} \cdot \mathbf{Z} \cdot \mathbf{S} \tag{3.75}$$

where $\mathbf{Z}$ is a real symmetric 3×3 matrix called *zero-field splitting tensor*.

3.2 Algebraic Side

After the electronic structure is dropped into the ground state (or a finite temperature ensemble) and integrated away (Sects. 3.1.3–3.1.10.3), and nuclei are assigned their spins, magnetogyric ratios, and quadrupole tensors (Sects. 3.1.1–3.1.2), the spin Hamiltonian of a molecular system acquires the following general form:

$$\begin{aligned}\mathbf{H} = &\sum_{n} \mathbf{S}^{(n)} \cdot \mathbf{Z}_n \cdot \mathbf{B} + \sum_{n>k} \mathbf{S}^{(n)} \cdot \mathbf{A}_{n,k} \cdot \mathbf{S}^{(k)} \\ &+ \sum_{n} \mathbf{S}^{(n)} \cdot \mathbf{Q}_n \cdot \mathbf{S}^{(n)} + \sum_{n} \sum_{l=4,6,\dots} \sum_{m=-l}^{l} a_{lm}^{(n)} \mathbf{T}_{lm}^{(n)}\end{aligned} \tag{3.76}$$

where $\mathbf{S}^{(n)} = \left\{ \mathbf{S}_\mathrm{X}^{(n)} \quad \mathbf{S}_\mathrm{Y}^{(n)} \quad \mathbf{S}_\mathrm{Z}^{(n)} \right\}$ are spin operators of the n-th electron or nucleus, $\mathbf{B}$ is the external magnetic field, $\mathbf{Z}_n$ are Zeeman interaction tensors stemming from chemical shielding (for nuclei) and g-tensors (for electrons), $\mathbf{A}_{n,k}$ are spin-spin coupling tensors arising from the multitude of mechanisms discussed in Sect. 3.1 (including antisymmetric coupling terms because the matrix $\mathbf{A}_{n,k}$ need not be symmetric), $\mathbf{Q}_n$ are quadrupolar (for nuclei) and second-rank zero-field splitting (for electrons) tensors, and the triple sum is higher rank electron zero-field splitting (Sect. 3.1.10.3) expressed as a linear combination of irreducible spherical tensor operators $\mathbf{T}_{lm}^{(n)}$ (Sect. 3.2.2) of rank l and projection m.

This Hamiltonian has general algebraic properties dictated by the mathematical nature of each interaction; it also carries much historical baggage on units and specification conventions. This section provides an overview of the mathematical side, and of the more advisable specification conventions.

3.2.1 Interaction Classification

At the level of the spin Hamiltonian, there is no algebraic difference between nuclear and electron spin, although only the latter, and only sometimes, is actually spin in the Dirac sense (Chap 2). Physical mechanisms vary (Sect 3.1), but the resulting terms can only have a few generic algebraic forms:

1. **Linear in spin**: coupling to external vectors, such as magnetic field and orbital angular momentum. Nuclear Zeeman interaction, electron Zeeman interaction, spin-rotation coupling, and spin-orbit coupling belong to this type. The general form is exemplified by Zeeman interaction:

$$\mathbf{H}_Z = \mathbf{S} \cdot \mathbf{Z} \cdot \mathbf{B} = \begin{pmatrix} \mathbf{S}_X & \mathbf{S}_Y & \mathbf{S}_Z \end{pmatrix} \begin{pmatrix} z_{XX} & z_{XY} & z_{XZ} \\ z_{YX} & z_{YY} & z_{YZ} \\ z_{ZX} & z_{ZY} & z_{ZZ} \end{pmatrix} \begin{pmatrix} B_X \\ B_Y \\ B_Z \end{pmatrix} \tag{3.77}$$

where **S** is a vector of spin operators from Sect. 1.6.3.2, **B** is the magnetic field vector, and **Z** is the Zeeman interaction tensor.

2. **Bilinear in spin**: coupling between spins—*J*-coupling, dipolar coupling, exchange interaction, and hyperfine coupling belong to this type. For the example of hyperfine coupling:

$$\mathbf{H}_{\text{HFC}} = \mathbf{S}^{(\text{N})} \cdot \mathbf{A} \cdot \mathbf{S}^{(\text{E})} = \begin{pmatrix} \mathbf{S}_X^{(\text{N})} & \mathbf{S}_Y^{(\text{N})} & \mathbf{S}_Z^{(\text{N})} \end{pmatrix} \begin{pmatrix} a_{XX} & a_{XY} & a_{XZ} \\ a_{YX} & a_{YY} & a_{YZ} \\ a_{ZX} & a_{ZY} & a_{ZZ} \end{pmatrix} \begin{pmatrix} \mathbf{S}_X^{(\text{E})} \\ \mathbf{S}_Y^{(\text{E})} \\ \mathbf{S}_Y^{(\text{E})} \end{pmatrix} \tag{3.78}$$

where $\mathbf{S}^{(\text{N,E})}$ are vectors of spin operators from Sect. 1.6.3.2 for the nucleus and the electron, and **A** is the hyperfine coupling tensor.

3. **Quadratic and higher order**: as described in Sect. 3.1, these are caused indirectly by other interactions, but manifest in the spin Hamiltonian as a coupling between a spin and itself, for example, nuclear quadrupolar interaction and electron zero-field splitting. For NQI:

$$\mathbf{H}_Q = \mathbf{S} \cdot \mathbf{Q} \cdot \mathbf{S} = \begin{pmatrix} \mathbf{S}_X & \mathbf{S}_Y & \mathbf{S}_Z \end{pmatrix} \begin{pmatrix} q_{XX} & q_{XY} & q_{XZ} \\ q_{YX} & q_{YY} & q_{YZ} \\ q_{ZX} & q_{ZY} & q_{ZZ} \end{pmatrix} \begin{pmatrix} \mathbf{S}_X \\ \mathbf{S}_Y \\ \mathbf{S}_Z \end{pmatrix} \tag{3.79}$$

where **S** is a vector of spin operators from Sect. 1.6.3.2 and **Q** is the quadrupolar coupling tensor. Only ZFS (Sect. 3.1.10.3) can have terms of spherical rank higher than two; they are commonly expressed as a linear combination of irreducible spherical tensor operators introduced in Sect. 3.2.2:

$$\mathbf{H}_{\text{ZFS}} = \sum_{l=2,4,\ldots} \sum_{m=-l}^{l} a_{lm} \mathbf{T}_{lm} \tag{3.80}$$

ZFS occurs in spin Hamiltonians of *d*- and *f*-element systems where spin-orbit coupling and ligand field splittings are so weak as to require additional terms for the spin Hamiltonian approximation to remain valid (Section 3.1.10.3).

A common operation applied to spin Hamiltonians is the spatial rotation of the molecule that the spins belong to. Under an instantaneous molecular rotation in physical space, the projections of the spins on the laboratory frame axes remain the same, but the interaction tensors—being functions of spatial interactions within the electronic structure theory—are rotated. We will now set up a systematic formalism for that.

3.2.2 Irreducible Spherical Tensors

Irreducible representations of the 3D rotation group were introduced in Sect. 1.6.2; they are spanned by spherical harmonics Y_{lm}. In that basis set, the matrices representing rotation operations are Wigner D matrices with elements $\mathfrak{D}^{(l)}_{m',m}(\Omega)$, where Ω is some parametrisation of the rotation group. A rotation $\hat{R}(\Omega)$ mixes spherical harmonics of the same rank:

$$\hat{R}(\Omega)Y_{lm} = \sum_{m'=-l}^{l} Y_{lm'}\mathfrak{D}^{(l)}_{m',m}(\Omega) \tag{3.81}$$

Spin operators $\mathbf{T}_{lm}$ that have the same property are called *irreducible spherical tensor (IST) operators*:

$$\hat{R}(\Omega)\mathbf{T}_{lm} = \sum_{m'=-l}^{l} \mathbf{T}_{lm'}\mathfrak{D}^{(l)}_{m',m}(\Omega) \tag{3.82}$$

They are obtained by taking a Cartesian expression for Y_{lm} and replacing products of $\{x, y, z\}$ with symmetrised products of $\{\mathbf{S}_\mathrm{X}, \mathbf{S}_\mathrm{Y}, \mathbf{S}_\mathrm{Z}\}$. A numerically friendly procedure is

1. For each spin, generate $\mathbf{S}_+$ and $\mathbf{S}_-$ matrices as described in Sect. 1.6.3.
2. For a specific rank l, the irreducible spherical operator with the largest projection is

$$\mathbf{T}_{ll} = (-1)^l 2^{-l/2}\mathbf{S}_+^l \tag{3.83}$$

3. ISTs with other projection numbers are obtained by sequential lowering:

$$\mathbf{T}_{l,m-1} = \frac{[\mathbf{S}_-, \mathbf{T}_{lm}]}{\sqrt{l(l+1) - m(m-1)}} \tag{3.84}$$

4. For *multi-spin ISTs*, the multiplication process is inherited from spherical harmonics:

$$\mathbf{T}_{LM} = \sum_{l_{1,2},m_{1,2}} C^{L,M}_{l_1,m_1,l_2,m_2}\left[\mathbf{T}_{l_1m_1} \otimes \mathbf{T}_{l_2m_2}\right] \tag{3.85}$$

where $\mathbf{T}_{l_1m_1}$ and $\mathbf{T}_{l_2m_2}$ act on different spins, and $C^{L,M}_{l_1,m_1,l_2,m_2}$ are Clebsch-Gordan coefficients [43, 81].

Expressions via Cartesian spin projection operators are the same for all spin quantum numbers. First and second rank single-spin ISTs are

$$\begin{aligned} \mathbf{T}^{(\mathrm{S})}_{1,0} &= \mathbf{S}_{\mathrm{Z}}, \quad \mathbf{T}^{(\mathrm{S})}_{1,\pm1} = \mp\frac{1}{\sqrt{2}}\mathbf{S}_{\pm} \\ \mathbf{T}^{(\mathrm{S})}_{2,\pm2} &= +\frac{1}{2}\mathbf{S}^2_{\pm}, \quad \mathbf{T}^{(\mathrm{S})}_{2,\pm1} = \mp\frac{1}{2}(\mathbf{S}_{\mathrm{Z}}\mathbf{S}_{\pm} + \mathbf{S}_{\pm}\mathbf{S}_{\mathrm{Z}}) \\ \mathbf{T}^{(\mathrm{S})}_{2,0} &= +\sqrt{\frac{2}{3}}\left(\mathbf{S}^2_{\mathrm{Z}} - \frac{1}{4}(\mathbf{S}_{+}\mathbf{S}_{-} + \mathbf{S}_{-}\mathbf{S}_{+})\right) \end{aligned} \tag{3.86}$$

These operators are not, and should not be, normalised (Sect. 1.6.3.3). Their coefficients are inherited from the commutation relations of $\mathfrak{su}(2)$ and propagated through the projections by Eq. (3.84). Two-spin irreducible spherical tensor operators up to second spherical rank are

$$\begin{aligned} \mathbf{T}^{(\mathrm{L,S})}_{0,0} &= \mathbf{L}_{\mathrm{X}}\mathbf{S}_{\mathrm{X}} + \mathbf{L}_{\mathrm{Y}}\mathbf{S}_{\mathrm{Y}} + \mathbf{L}_{\mathrm{Z}}\mathbf{S}_{\mathrm{Z}} \\ \mathbf{T}^{(\mathrm{L,S})}_{1,\pm1} &= -\frac{1}{2}(\mathbf{L}_{\pm}\mathbf{S}_{\mathrm{Z}} - \mathbf{L}_{\mathrm{Z}}\mathbf{S}_{\pm}), \quad \mathbf{T}^{(\mathrm{L,S})}_{1,0} = -\frac{1}{2\sqrt{2}}(\mathbf{L}_{+}\mathbf{S}_{-} - \mathbf{L}_{-}\mathbf{S}_{+}) \\ \mathbf{T}^{(\mathrm{L,S})}_{2,\pm2} &= +\frac{1}{2}\mathbf{L}_{\pm}\mathbf{S}_{\pm}, \quad \mathbf{T}^{(\mathrm{L,S})}_{2,\pm1} = \mp\frac{1}{2}(\mathbf{L}_{\mathrm{Z}}\mathbf{S}_{\pm} + \mathbf{L}_{\pm}\mathbf{S}_{\mathrm{Z}}) \\ \mathbf{T}^{(\mathrm{L,S})}_{2,0} &= +\sqrt{\frac{2}{3}}\left(\mathbf{L}_{\mathrm{Z}}\mathbf{S}_{\mathrm{Z}} - \frac{1}{4}(\mathbf{L}_{+}\mathbf{S}_{-} + \mathbf{L}_{-}\mathbf{S}_{+})\right) \end{aligned} \tag{3.87}$$

Higher rank ISTs for ZFS Hamiltonians are best generated numerically using Eqs. (3.83), (3.84), (3.85). In practical magnetic resonance simulations, the best operator basis is the direct products of single-spin ISTs.

Useful commutation properties of single-spin ISTs are

$$\begin{aligned} \left[\mathbf{S}_{\mathrm{Z}}, \mathbf{T}^{(\mathrm{S})}_{lm}\right] &= m\mathbf{T}^{(\mathrm{S})}_{lm} \quad \left[\mathbf{S}^2, \mathbf{T}^{(\mathrm{S})}_{lm}\right] = 0 \\ \left[\mathbf{S}_{-}, \mathbf{T}^{(\mathrm{S})}_{lm}\right] &= \sqrt{l(l+1) - m(m-1)}\,\mathbf{T}^{(\mathrm{S})}_{l,m-1} \\ \left[\mathbf{S}_{+}, \mathbf{T}^{(\mathrm{S})}_{lm}\right] &= \sqrt{l(l+1) - m(m+1)}\,\mathbf{T}^{(\mathrm{S})}_{l,m+1} \end{aligned} \tag{3.88}$$

This is an adjoint representation of SO(3): the raising and lowering relations are the same as those in Eq. (1.116), except the action is performed by commutation superoperators.

3.2.3 Stevens Operators

Zero-field splitting has unfortunate historical baggage—it is sometimes reported as coefficients B_k^q in front of operators $\mathbf{O}_k^q$ that a careless postdoc had contrived [82]:

$$\mathbf{H}_{\mathrm{ZFS}} = \sum_{k,q} B_k^q \mathbf{O}_k^q \tag{3.89}$$

and for which no systematic expression exists—they are linear combinations of spin operators reminiscent of the combinations of spherical harmonics used in chemistry textbooks to make atomic orbitals—the intention presumably was to keep the coefficients real. In Tables 3.2 and 3.3.

$$[\mathbf{A},\mathbf{B}]_+ = \mathbf{AB}+\mathbf{BA}, \quad s = S(S+1), \quad c_+ = 1/2, \quad c_- = 1/2i \tag{3.90}$$

Outside historical comparisons, the use of Stevens operators is discouraged—mathematically and physically consistent irreducible spherical tensors should be used instead. Up to the second rank, the translation between Stevens operators and irreducible spherical tensors is straightforward (Table 3.2). For higher ranks (Table 3.3), the translation is best approached numerically, by calculating Frobenius scalar products between the corresponding sets of matrices; this is implemented in *Spinach*.

3.2.4 Hamiltonian Rotations

Consider a static distribution of molecular orientations, or a rotation slow enough to keep the electronic structure in the ground state. In the absence of high-rank ZFS, the simplest way of rotating a spin Hamiltonian is to apply a Cartesian rotation to the interaction tensors in Eq. (3.76), for example:

Table 3.2 Second rank Stevens operators and their connection to irreducible spherical tensor operators

k	q	Stevens operator $\mathbf{O}_k^q$	IST form
2	−2	$(\mathbf{S}_+^2 - \mathbf{S}_-^2)/2i$	$i\left(\mathbf{T}_{2,-2}^{(\mathrm{S})} - \mathbf{T}_{2,2}^{(\mathrm{S})}\right)$
	− 1	$[\mathbf{S}_\mathrm{Z}, \mathbf{S}_+ - \mathbf{S}_-]_+/2i$	$i\left(\mathbf{T}_{2,-1}^{(\mathrm{S})} + \mathbf{T}_{2,1}^{(\mathrm{S})}\right)/2$
	0	$3\mathbf{S}_\mathrm{Z}^2 - \mathbf{S}^2$	$\sqrt{6}\mathbf{T}_{2,0}^{(\mathrm{S})}$
	+ 1	$[\mathbf{S}_\mathrm{Z}, \mathbf{S}_+ + \mathbf{S}_-]_+/2$	$\left(\mathbf{T}_{2,-1}^{(\mathrm{S})} - \mathbf{T}_{2,1}^{(\mathrm{S})}\right)/2$
	+ 2	$(\mathbf{S}_+^2 + \mathbf{S}_-^2)/2$	$\mathbf{T}_{2,-2}^{(\mathrm{S})} + \mathbf{T}_{2,2}^{(\mathrm{S})}$

$$\mathbf{A} \rightarrow \mathbf{R} \cdot \mathbf{A} \cdot \mathbf{R}^{-1} \tag{3.91}$$

where $\mathbf{R}$ is a rotation matrix from Eqs. (1.112–1.115). However, this is logistically inconvenient—once the tensors are rotated, the Hamiltonian in Eq. (3.76) must be recomputed. Important methods—for example, rotational relaxation theories in Chap. 6—are inelegant and hard to use with Cartesian rotations.

A more efficient description uses irreducible representations of the rotation group (Sect. 1.6.2). When Eq. 3.76 is rewritten via ISTs from Sect. 3.3.2, its components become

$$\begin{aligned}
\mathbf{H}_{\mathrm{LS}} &= a_{0,0}^{(\mathrm{L,S})}\mathbf{T}_{0,0}^{(\mathrm{L,S})} + \sum_{m=-1}^{1} a_{1,m}^{(\mathrm{L,S})}\mathbf{T}_{1,m}^{(\mathrm{L,S})} + \sum_{m=-2}^{2} a_{2,m}^{(\mathrm{L,S})}\mathbf{T}_{2,m}^{(\mathrm{L,S})} \\
\mathbf{H}_{\mathrm{BS}} &= a_{0,0}^{(\mathrm{B,S})}\mathbf{T}_{0,0}^{(\mathrm{B,S})} + \sum_{m=-1}^{1} a_{1,m}^{(\mathrm{B,S})}\mathbf{T}_{1,m}^{(\mathrm{B,S})} + \sum_{m=-2}^{2} a_{2,m}^{(\mathrm{B,S})}\mathbf{T}_{2,m}^{(\mathrm{B,S})} \\
\mathbf{H}_{\mathrm{SS}} &= \sum_{l=2,4,\ldots}\sum_{m=-l}^{l} a_{l,m}^{(\mathrm{S})}\mathbf{T}_{l,m}^{(\mathrm{S})}
\end{aligned} \tag{3.92}$$

where $\mathbf{H}_{\mathrm{LS}}$ contains couplings between spins L and S, $\mathbf{H}_{\mathrm{BS}}$ indicates a coupling between spin S and the magnetic field, and $\mathbf{H}_{\mathrm{SS}}$ contains effective quadratic and higher order interactions of spin S (nuclear quadrupolar interactions and zero-field splitting). Higher order ZFS terms are commonly quoted directly in this form (or may be translated into it from Stevens coefficients, Sect. 3.3.3); the translation between 3 × 3 interaction matrices in Eq. (3.76) and the expansion coefficients $a_{l,m}$ in Eq. 3.92 is given in Tables 3.3 and 3.4.

Table 3.3 Higher-rank Stevens operators

k	q	Stevens operator $\mathbf{O}_k^q$
4	0	$35\mathbf{S}_\mathrm{Z}^4 - (30s-25)\mathbf{S}_\mathrm{Z}^2 + (3s^2-6s)\mathbf{1}$
	±1	$c_\pm\left[7\mathbf{S}_\mathrm{Z}^3 - (3s+1)\mathbf{S}_\mathrm{Z}, \mathbf{S}_+ \pm \mathbf{S}_-\right]_+/2$
	±2	$c_\pm\left[7\mathbf{S}_\mathrm{Z}^2 - (s+5)\mathbf{1}, \mathbf{S}_+^2 \pm \mathbf{S}_-^2\right]_+/2$
	±3	$c_\pm\left[\mathbf{S}_\mathrm{Z}, \mathbf{S}_+^3 \pm \mathbf{S}_-^3\right]_+/2$
	±4	$c_\pm\left(\mathbf{S}_+^4 \pm \mathbf{S}_-^4\right)$
6	0	$231\mathbf{S}_\mathrm{Z}^6 - (315s-735)\mathbf{S}_\mathrm{Z}^4 + (105s^2-525s+294)\mathbf{S}_\mathrm{Z}^2 - (5s^3-40s^2+60s)\mathbf{1}$
	±1	$c_\pm\left[33\mathbf{S}_\mathrm{Z}^5 - (30s-15)\mathbf{S}_\mathrm{Z}^3 + (5s^2-10s+12)\mathbf{S}_\mathrm{Z}, \mathbf{S}_+ \pm \mathbf{S}_-\right]_+/2$
	±2	$c_\pm\left[33\mathbf{S}_\mathrm{Z}^4 - (18s+123)\mathbf{S}_\mathrm{Z}^2 + (s^2+10s+102)\mathbf{1}, \mathbf{S}_+^2 \pm \mathbf{S}_-^2\right]_+/2$
	±3	$c_\pm\left[11\mathbf{S}_\mathrm{Z}^3 - (3s+59)\mathbf{S}_\mathrm{Z}, \mathbf{S}_+^3 \pm \mathbf{S}_-^3\right]_+/2$
	±4	$c_\pm\left[11\mathbf{S}_\mathrm{Z}^2 - (s+38)\mathbf{1}, \mathbf{S}_+^4 \pm \mathbf{S}_-^4\right]_+/2$
	±5	$c_\pm\left[\mathbf{S}_\mathrm{Z}, \mathbf{S}_+^5 \pm \mathbf{S}_-^5\right]_+/2$
	±6	$c_\pm\left(\mathbf{S}_+^6 \pm \mathbf{S}_-^6\right)$

Table 3.4 Relations between the components of the Cartesian interaction tensors in Eqs. (3.77), (3.78), (3.79) and their irreducible spherical tensor expansion coefficients

(l,m)	$a_{l,m}$	(l,m)	$a_{l,m}$
$(0,0)$	$(a_{XX}+a_{YY}+a_{ZZ})/3$	$(2,+2)$	$+\frac{1}{2}(a_{XX}-a_{YY}-i(a_{XY}+a_{YX}))$
$(1,+1)$	$-\frac{1}{2}(a_{ZX}-a_{XZ}-i(a_{ZY}-a_{YZ}))$	$(2,+1)$	$-\frac{1}{2}(a_{XZ}+a_{ZX}-i(a_{YZ}+a_{ZY}))$
$(1,0)$	$+\frac{i}{\sqrt{2}}(a_{XY}-a_{YX})$	$(2,0)$	$+\frac{1}{\sqrt{6}}(2a_{ZZ}-(a_{XX}+a_{YY}))$
$(1,-1)$	$-\frac{1}{2}(a_{ZX}-a_{XZ}+i(a_{ZY}-a_{YZ}))$	$(2,-1)$	$+\frac{1}{2}(a_{XZ}+a_{ZX}+i(a_{YZ}+a_{ZY}))$
		$(2,-2)$	$+\frac{1}{2}(a_{XX}-a_{YY}+i(a_{XY}+a_{YX}))$

When the interaction tensor is diagonal, the transformation is particularly simple:

$$\begin{aligned} a_{XX}\mathbf{L}_X\mathbf{S}_X+a_{YY}\mathbf{L}_Y\mathbf{S}_Y+a_{ZZ}\mathbf{L}_Z\mathbf{S}_Z &= \frac{a_{XX}+a_{YY}+a_{ZZ}}{3}\mathbf{T}_{0,0}^{(L,S)} \\ &+ \frac{2a_{ZZ}-(a_{XX}+a_{YY})}{\sqrt{6}}\mathbf{T}_{2,0}^{(L,S)}+\frac{a_{XX}-a_{YY}}{2}\left[\mathbf{T}_{2,-2}^{(L,S)}+\mathbf{T}_{2,2}^{(L,S)}\right] \end{aligned} \tag{3.93}$$

In rigid molecules, the rotational transformation rule for ISTs in Eq. 3.82 is inherited by the full spin system Hamiltonian with Wigner D matrices coming from Eq. (1.117):

$$\mathbf{H}(\Omega)=\mathbf{H}_{\text{iso}}+\sum_{l>0}\sum_{k,m=-l}^{l}\mathfrak{D}_{km}^{(l)}(\Omega)\mathbf{Q}_{km}^{(l)} \tag{3.94}$$

where $\mathbf{Q}_{km}^{(l)}$ (called *rotational basis operators*) are linear combinations of ISTs. Their explicit form:

$$\mathbf{Q}_{km}^{(l)}=\sum_{S}a_{l,m}^{(B,S)}\mathbf{T}_{l,k}^{(B,S)}+\sum_{L>S}a_{l,m}^{(L,S)}\mathbf{T}_{l,k}^{(L,S)}+\sum_{S}a_{l,m}^{(S)}\mathbf{T}_{l,k}^{(S)} \tag{3.95}$$

comes out of a lengthy calculation wherein the form stipulated by Eq. 3.92 is substituted for every term of the Hamiltonian in Eq. (3.76), and a rotation is applied [83]. The expressions for ISTs and spherical tensor coefficients are tabulated above, and Wigner rotation matrices are defined by Eq. (1.117).

In summary, spin Hamiltonian rotations should be set up in the following way:

1. Get all interactions into the 3×3 Cartesian matrix form stipulated by Eq. 3.76. If higher rank zero-field splitting is present, translate it into irreducible spherical tensor coefficients.

2. Translate the 3×3 Cartesian matrices into spherical tensor parameters $a_{l,m}$ using the relations given in Table. 3.4.

3. Compute the isotropic Hamiltonian $\mathbf{H}_{\text{iso}}$ and the rotational basis operators $\mathbf{Q}_{km}^{(l)}$. If at all possible, avoid using Euler angles to parametrise rotations.
4. The spin system Hamiltonian at any orientation relative to the frame of reference in which the original tensors had been specified is given by Eq. (3.94).

In practical spin dynamics simulations, this is the only way of setting up rotations that a software engineer would not later come to regret. In particular, it avoids re-generating coupling operators at every orientation when powder patterns are computed, simplifies the description of magic angle spinning, and also improves the logistics of relaxation theories in Chap. 6—no matter how complicated the rotation sequence is, the Hamiltonian always has the following form:

$$\mathbf{H} = \mathbf{H}_{\text{iso}} + \sum_{l>0} \sum_{k,m=-l}^{l} \left[\cdots \mathfrak{D}_{\text{B}}^{(l)} \mathfrak{D}_{\text{A}}^{(l)} \right]_{km} \mathbf{Q}_{km} \tag{3.96}$$

where the lower indices on the Wigner matrices refer to the individual rotations in the sequence.

3.2.5 Rotational Invariants

From the algebraic perspective, a 3×3 interaction tensor has three rotation invariants:

$$\text{I}_{\mathbf{A}} = \text{Tr}(\mathbf{A}), \quad \text{II}_{\mathbf{A}} = \frac{1}{2}\left(\text{Tr}[\mathbf{A}]^2 - \text{Tr}\left[\mathbf{A}^2\right]\right), \quad \text{III}_{\mathbf{A}} = \det(\mathbf{A}) \tag{3.97}$$

Of these, $\text{I}_{\mathbf{A}}$ determines the isotropic part and survives the rotational averaging, $\text{II}_{\mathbf{A}}$ determines relaxation properties in liquids and powder patterns in solids, and $\text{III}_{\mathbf{A}}$ does not appear to be used in spin physics. The following relation links the three invariants:

$$\mathbf{A}^3 - \text{I}_{\mathbf{A}}\mathbf{A}^2 + \text{II}_{\mathbf{A}}\mathbf{A} - \text{III}_{\mathbf{A}}\mathbf{1} = 0 \tag{3.98}$$

We will also encounter the following invariants in spin relaxation theories (Chap. 6):

$$\begin{aligned} \Delta_{\mathbf{A}}^2 &= a_{\text{XX}}^2 + a_{\text{YY}}^2 + a_{\text{ZZ}}^2 - a_{\text{XX}}a_{\text{YY}} - a_{\text{XX}}a_{\text{ZZ}} - a_{\text{YY}}a_{\text{ZZ}} \\ &\quad + \frac{3}{4}\left[(a_{\text{XY}} + a_{\text{YX}})^2 + (a_{\text{XZ}} + a_{\text{ZX}})^2 + (a_{\text{YZ}} + a_{\text{ZY}})^2\right] \\ \Lambda_{\mathbf{A}}^2 &= (a_{\text{XY}} - a_{\text{YX}})^2 + (a_{\text{XZ}} - a_{\text{ZX}})^2 + (a_{\text{YZ}} - a_{\text{ZY}})^2 \end{aligned} \tag{3.99}$$

From the physical perspective, $\Delta_{\mathbf{A}}^2$ is the squared modulation depth of the second spherical rank component of the interaction under rotational diffusion, and $\Lambda_{\mathbf{A}}^2$ is the squared modulation depth of the first rank component. Mathematically, both quantities satisfy the definition of a norm (Sect. 1.4.5); the corresponding inner products are therefore obtained using polarisation relations:

$$\aleph_{\mathbf{A},\mathbf{B}} = \frac{\Delta^2_{\mathbf{A}+\mathbf{B}} - \Delta^2_{\mathbf{A}-\mathbf{B}}}{4}, \quad \daleth_{\mathbf{A},\mathbf{B}} = \frac{\Lambda^2_{\mathbf{A}+\mathbf{B}} - \Lambda^2_{\mathbf{A}-\mathbf{B}}}{4} \tag{3.100}$$

These quantities will make an appearance in cross-correlated relaxation processes (Chap. 6)

3.3 Historical Conventions

Spin interactions should ideally be reported as 3 × 3 matrices sitting between spin operator vectors in Eq. (3.76), or as coefficients in front of irreducible spherical tensor operators. This section lists some historical reporting conventions; those are problematic, but unfortunately popular. Some of them ignore the possible presence of first spherical rank components—they must never be used, for example, for reporting paramagnetic shielding tensors (Sect. 3.1.7) that have significant antisymmetric contributions.

3.3.1 Eigenvalue Order

A symmetric 3 × 3 interaction tensor **A** with eigenvalues $\{a_{\mathrm{XX}}, a_{\mathrm{YY}}, a_{\mathrm{ZZ}}\}$ is called.

- *isotropic* if $a_{\mathrm{XX}} = a_{\mathrm{YY}} = a_{\mathrm{ZZ}}$
- *axial* if $a_{\mathrm{XX}} = a_{\mathrm{YY}} \neq a_{\mathrm{ZZ}}$
- *rhombic* if $a_{\mathrm{XX}} \neq a_{\mathrm{YY}} \neq a_{\mathrm{ZZ}}$
- *traceless* if $a_{\mathrm{XX}} + a_{\mathrm{YY}} + a_{\mathrm{ZZ}} = 0$

where the eigenvalues are labelled in such a way as to have [84,85]:

$$\begin{array}{ll} |a_{\mathrm{ZZ}} - a_{\mathrm{iso}}| \geq |a_{\mathrm{XX}} - a_{\mathrm{iso}}| \geq |a_{\mathrm{YY}} - a_{\mathrm{iso}}| & \text{Haeberlen order} \\ \text{or} & \\ a_{\mathrm{XX}} \leq a_{\mathrm{YY}} \leq a_{\mathrm{ZZ}} & \text{Mehring order} \end{array} \tag{3.101}$$

where $a_{\mathrm{iso}} = (a_{\mathrm{XX}} + a_{\mathrm{YY}} + a_{\mathrm{ZZ}})/3$. This prescription leads to the axis label switching problem: while the physical properties of the system are continuous and differentiable functions of the elements of **A**, axis labels are not. This creates insidious numerical issues during data fitting—error functionals that depend on axis labels are not, in general, continuous or differentiable at label switching points.

3.3.2 Eigenvalue Reporting

The following eigenvalue combinations are historically significant:

1. Isotropic + anisotropy + asymmetry [84]

$$\begin{aligned} &\begin{cases} \Delta = a_{ZZ} - (a_{XX} + a_{YY})/2 \\ \text{or} \quad \delta = a_{ZZ} - a_{\text{iso}} \end{cases} && \text{anisotropy} \\ &\eta = \frac{3}{2}\frac{a_{YY} - a_{XX}}{\Delta} = \frac{a_{YY} - a_{XX}}{\delta} && \text{asymmetry} \end{aligned} \tag{3.102}$$

$$\begin{aligned} a_{XX} &= a_{\text{iso}} - \Delta(1+\eta)/3 = a_{\text{iso}} - \delta(1+\eta)/2 \\ a_{YY} &= a_{\text{iso}} - \Delta(1-\eta)/3 = a_{\text{iso}} - \delta(1-\eta)/2 \\ a_{ZZ} &= a_{\text{iso}} + 2\Delta/3 = a_{\text{iso}} + \delta \end{aligned} \tag{3.103}$$

with eigenvalues in Haeberlen order. The inverse relations in Eq. (3.103) are single-valued, but numerical implementations of the forward relations in Eq. (3.102) require punctilious housekeeping because axis labels can switch from one fitting iteration to the next.

2. Isotropic + span + skew [337]

$$\begin{aligned} \Omega &= a_{ZZ} - a_{XX} && \text{span} \\ \kappa &= 3(a_{YY} - a_{\text{iso}})/\Omega && \text{skew} \end{aligned} \tag{3.104}$$

$$\begin{aligned} a_{XX} = a_{\text{iso}} - \Omega(3+\kappa)/6, \quad a_{YY} = a_{\text{iso}} + \Omega\kappa/3 \\ a_{ZZ} = a_{\text{iso}} + \Omega(3-\kappa)/6 \end{aligned} \tag{3.105}$$

with eigenvalues in Mehring order. These parameters give a physically intuitive picture of nuclear shielding and electron g-tensor powder patterns. The same numerical problems are present as those described for the anisotropy + asymmetry convention.

3. Isotropic + axiality + rhombicity

$$\begin{aligned} \Delta &= 2a_{ZZ} - (a_{XX} + a_{YY}) && \text{axiality} \\ \delta &= a_{YY} - a_{XX} && \text{rhombicity} \\ a_{XX} &= a_{\text{iso}} - (\Delta + 3\delta)/6 \\ a_{YY} &= a_{\text{iso}} - (\Delta - 3\delta)/6 \\ a_{ZZ} &= a_{\text{iso}} + \Delta/3 \end{aligned} \tag{3.106}$$

with eigenvalues in Mehring order. There are no complications with denominators here, but the axis label switching problem is still present because eigenvalues are ordered explicitly.

4. Nuclear quadrupole interaction parameters
The specification convention for NQI proceeds from its origin (Section 3.2.13) in the interaction between the nuclear quadrupole moment and the electric field gradient:

$$C = \frac{e^2 q_{\mathrm{ZZ}} Q}{\hbar} \quad \eta = \frac{q_{\mathrm{XX}} - q_{\mathrm{YY}}}{q_{\mathrm{ZZ}}} \tag{3.107}$$

where $\{eq_{\mathrm{XX}}, eq_{\mathrm{YY}}, eq_{\mathrm{ZZ}}\}$ are the eigenvalues of the EFG tensor, e is the elementary charge, eQ is the quadrupole moment of the nucleus, and the eigenvalues are in Haeberlen order, meaning that eq_{ZZ} is the one with the largest absolute value because the NQI tensor is traceless. The units of C here are rad/s; η is dimensionless.

Although the electric field gradient is not traceless (by Poisson's equation [86], its trace is proportional to the local charge density), its isotropic part contributes a multiple of a unit matrix

$$\mathbf{S}_{\mathrm{X}}^2 + \mathbf{S}_{\mathrm{Y}}^2 + \mathbf{S}_{\mathrm{Z}}^2 = \mathbf{S}^2 = S(S+1)\mathbf{1} \tag{3.108}$$

to the spin Hamiltonian, and for that reason is dropped from consideration. In the principal axis frame using the parameters introduced above, we then have

$$\begin{aligned} \mathbf{H}_{\mathrm{NQI}} &= -\frac{C}{4S(2S-1)}\left[3\mathbf{S}_{\mathrm{Z}}^2 - \mathbf{S}^2 + \eta\left(\mathbf{S}_{\mathrm{X}}^2 - \mathbf{S}_{\mathrm{Y}}^2\right)\right] \\ \mathbf{H}_{\mathrm{NQI}} &= \frac{\sqrt{6}C}{4S(2S-1)}\mathbf{T}_{2,0}^{(\mathrm{S})} + \frac{\eta C}{4S(2S-1)}\left[\mathbf{T}_{2,-2}^{(\mathrm{S})} + \mathbf{T}_{2,2}^{(\mathrm{S})}\right] \end{aligned} \tag{3.109}$$

The prefactor $C = e^2 q_{\mathrm{ZZ}} Q/\hbar$ in this equation is called the *quadrupole coupling constant*. In matrix notation, the nuclear *quadrupolar interaction tensor* in the principal axis frame is

$$\mathbf{Q} = \frac{C}{4S(2S-1)}\begin{pmatrix} -(1-\eta) & 0 & 0 \\ 0 & -(1+\eta) & 0 \\ 0 & 0 & 2 \end{pmatrix} \tag{3.110}$$

and thus C is a measure of overall amplitude (here proportional to anisotropy because the tensor is traceless) and η is a measure of asymmetry.

5. Second rank zero-field splitting parameters
For historical reasons, second rank ZFS eigenvalues have their own parametrisation

$$\begin{gathered} D=\frac{3}{2}a_{\mathrm{ZZ}}, \quad E=\frac{1}{2}(a_{\mathrm{XX}}-a_{\mathrm{YY}}) \\ a_{\mathrm{XX}}=-D/3+E, \quad a_{\mathrm{YY}}=-D/3-E, \quad a_{\mathrm{ZZ}}=2D/3 \end{gathered} \tag{3.111}$$

with eigenvalues $\{a_{\mathrm{XX}}, a_{\mathrm{YY}}, a_{\mathrm{ZZ}}\}$ in Mehring order. ZFS tensor is traceless for the same reason as the NQI tensor; in the principal axis frame, the Hamiltonian is

$$\mathbf{H}_{\mathrm{ZFS}}=D\left(\mathbf{S}_{\mathrm{Z}}^{2}-\mathbf{S}^{2}/3\right)+E\left(\mathbf{S}_{\mathrm{X}}^{2}-\mathbf{S}_{\mathrm{Y}}^{2}\right) \tag{3.112}$$

In terms of B_2^q parameters (Sect. 3.2.3): $D=3B_2^0$ and $E=B_2^2$.

3.3.3 Practical Considerations

The parameters above, along with the spatial orientation information, are usually obtained by numerical fitting of simulations to experimental data. In finite precision machine arithmetic, the process is a veritable minefield. The following problems may arise:

1. Some matrices either cannot be diagonalised, or cannot be diagonalised over the real field, or cannot be diagonalised uniquely. Unless elaborate constraints are imposed to ensure that interaction tensors are physically sensible at every point in the optimisation, the algorithm would either throw an error, or abscond into complex numbers.
2. Some definitions in Sect. 3.3.2 can yield 0/0 indeterminacies in finite precision arithmetic. When a numerical fitting algorithm runs into those points, it collapses into NaN values.
3. Axis label switching generates singularities and branch cuts in fitting error surfaces. This rules out entire families of fitting algorithms and means that only the methods that do not rely on derivative information (e.g. simplex) work reliably.
4. Some interactions (e.g. chemical shielding) have non-symmetric matrices, meaning that their eigenvectors are not orthogonal, and the transformation between their eigenframe and the molecular frame of reference is not a rotation.
5. Even when eigenvectors of a 3×3 matrix are real and orthonormal, their direction is arbitrary, meaning that there are eight equivalent eigenvector matrices. Of those, four are rotations, meaning that there are four equivalent sets of rotation parameters connecting any two orientations of the tensor. The other four are superpositions of a rotation and an inversion.

6. Even if the rotation matrix is somehow obtained, a numerically stable procedure for converting it into Euler angles cannot exist; this is discussed in Sect. 1.6.2.

For these reasons, specifying spin interactions as combinations of eigenvalues and orientation parameters is not recommended. They should be specified as 3×3 matrices entering the Hamiltonian in Eq. (3.76), or as coefficients in front of irreducible spherical tensor operators.

3.3.4 Visualisation of Interactions

There are two convenient visual representations of amplitude and orientation of spin interaction tensors relative to the molecular frame of reference:

1. **Ellipsoid plots**: a sphere is drawn at an appropriate location (at the nucleus in the case of nucleus-centred interactions, such as CSA / HFC / NQI, and at the spin density centroid for the g-tensor and other electron spin interactions). The sphere is scaled by the three eigenvalues of the interaction tensor in the directions of the three corresponding eigenvectors. A set of axes is drawn inside the resulting ellipsoid, with red axes for positive eigenvalues and blue axes for the negative ones. This method assumes that the eigenvectors are orthogonal, and is therefore only applicable to interactions that have symmetric matrices (Fig. 3.2, left panel; Fig. 3.3, left panel).

2. **Spherical harmonic plots**: the interaction is translated into the spherical tensor convention, and the spherical tensor coefficients are used to multiply the corresponding spherical harmonics. A spherical plot of the resulting function is generated at the nucleus in the case of nucleus-centred interactions, and at the spin density centroid for the g-tensor and other electron spin interactions. This representation has the advantages of being mathematically faithful and being able to represent interactions of spherical ranks higher than two (Fig. 3.2, right panel; Fig. 3.3, right panel).

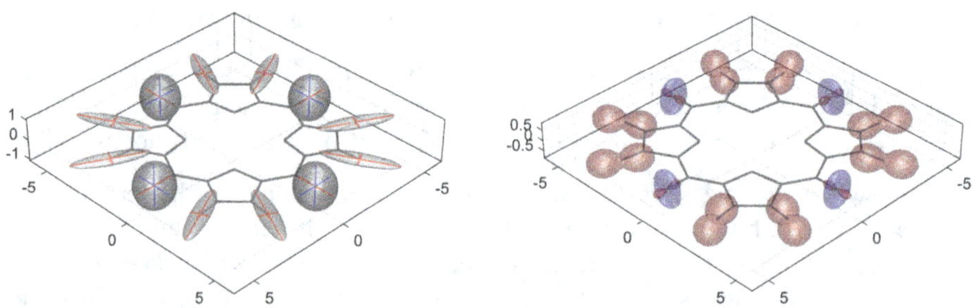

Fig. 3.2 Ellipsoid (left) and spherical harmonic (right) plots of hyperfine coupling tensors in a Cu(II) porphyrin complex. The plots were generated by Spinach from a DFT estimate (B3LYP/def2-SVP in vacuum) by ORCA

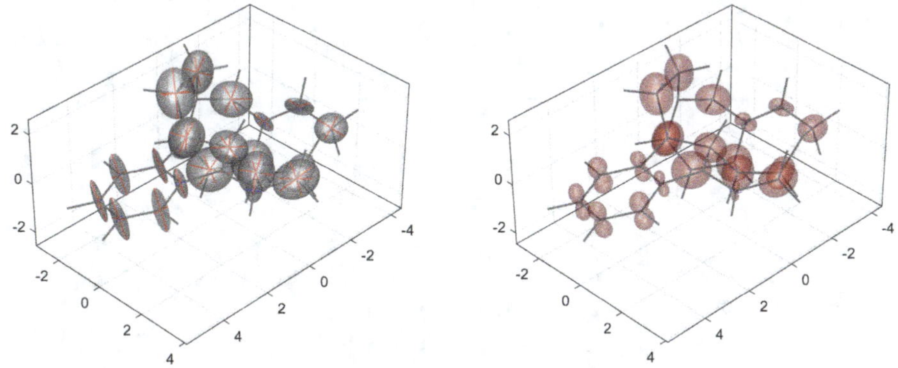

Fig. 3.3 Ellipsoid (left) and spherical harmonic (right) plots of chemical shielding tensors in strychnine. The plots were generated by Spinach [87] from a DFT estimate (B3LYP/cc-pVDZ in vacuum) by Gaussian [88]

Because the interactions themselves belong to abstract spin operator algebras, details of their visualisation in $\mathbb{R}^3$ are a matter of convenience.

Coherent Spin Dynamics

4

From the physical perspective, popular descriptions of spin dynamics follow two general themes. In the *frequency-domain approach*, the Hamiltonian is diagonalised and the properties are obtained from the matrix elements of the various operators in the Hamiltonian eigen-basis. In the *time-domain approach*, the dynamics of the system from a given initial state is calculated forward in time using an equation of motion, and expectation values are computed for the quantities of interest.

At the level of the equation of motion, *wavefunction formalism* using Schrödinger's equation [32] is appropriate for idealised and isolated spin systems in the absence of ensemble averaging, energy dissipation, or incoherent spatial dynamics. Description of ensemble dynamics and decoherence requires the more general *density operator formalism* and its corresponding Liouville–von Neumann equation of motion [89]. An isomorphism that uses superoperators and state vectors [6] is called *Liouville space formalism* [90]; its algebraic structure makes it convenient to add dissipative dynamics, hydrodynamics, diffusion, and chemical kinetics (Fig. 4.1).

4.1 Wavefunction Formalism

This section builds the methodological framework for time-domain simulations; we consider simple cases first, and then use them as building blocks for the more general descriptions. The starting point is Schrödinger's equation [32] for the wavefunction ψ with a time-independent Hamiltonian operator $\hat{H}$:

$$\frac{\partial}{\partial t}\psi(t) = -i\hat{H}\psi(t) \tag{4.1}$$

with an initial condition $\psi(0)$ specified at time zero. Using the Taylor series [91] around $t = 0$, we obtain

I. KUPROV, *Spin*,
https://doi.org/10.1007/978-3-031-05607-9_4

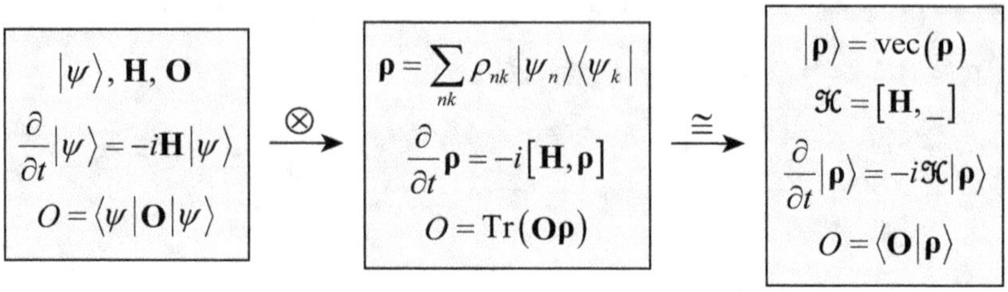

Fig. 4.1 Mathematical relationship between the three commonly used levels of time-domain description of spin dynamics. The need to handle dissipative ensembles mandates the product map from the wavefunction formalism (left, "wavefunction space") into the density operator formalism (middle, "Hilbert space") that can handle thermodynamic ensembles and relaxation at the cost of the state descriptor becoming a matrix. An isomorphism into a vector space, accomplished by stretching that matrix into a vector, then produces a formalism (right, "Liouville space") that resembles the wavefunction formalism, but can be conveniently extended to accommodate (Chap. 5) diffusion, hydrodynamics, and chemical kinetics

$$\psi(t) = \psi(0) + \left.\frac{\partial\psi}{\partial t}\right|_{t=0} t + \frac{1}{2}\left.\frac{\partial^2\psi}{\partial t^2}\right|_{t=0} t^2 + \frac{1}{6}\left.\frac{\partial^3\psi}{\partial t^3}\right|_{t=0} t^3 + \ldots \tag{4.2}$$

Because the Hamiltonian is here assumed to be time-independent, all time derivatives may be obtained by repeatedly differentiating Schrödinger's equation:

$$\begin{aligned} \left.\frac{\partial\psi}{\partial t}\right|_{t=0} &= -i\hat{H}\psi(0) \\ \left.\frac{\partial^2\psi}{\partial t^2}\right|_{t=0} &= \left.\frac{\partial}{\partial t}\left(-i\hat{H}\psi\right)\right|_{t=0} = \left.-i\hat{H}\frac{\partial\psi}{\partial t}\right|_{t=0} = \left(-i\hat{H}\right)^2\psi(0) \\ \left.\frac{\partial^3\psi}{\partial t^3}\right|_{t=0} &= \ldots = \left(-i\hat{H}\right)^3\psi(0), \text{ etc.} \end{aligned} \tag{4.3}$$

When these are placed into Eq. (4.2), the Taylor expansion becomes

$$\begin{aligned} \psi(t) &= \psi(0) + \left(-i\hat{H}t\right)\psi(0) + \frac{1}{2}\left(-i\hat{H}t\right)^2\psi(0) + \ldots = \\ &= \left[\sum_{n=0}^{\infty} \frac{\left(-i\hat{H}t\right)^n}{n!}\right]\psi(0) = \exp\left(-i\hat{H}t\right)\psi(0) \end{aligned} \tag{4.4}$$

and the *exponential propagator* (Sect. 1.5.7) makes an appearance (practical methods for computing matrix exponentials are discussed in Sect. 4.9). It is in practice more convenient to reformulate Eq. (4.4) to take a time step from some previous point t to the next point $t + \Delta t$:

$$\psi(t+\Delta t) = \exp\left(-i\hat{H}\Delta t\right)\psi(t) \tag{4.5}$$

When the Hamiltonian is time dependent, the simplest way to proceed is to slice time into infinitesimal intervals and use Eq. (4.5) in each interval:

$$\frac{\partial\psi}{\partial t} = -i\hat{H}(t)\psi \quad \Rightarrow \quad \psi(t+\mathrm{d}t) = \exp\left(-i\hat{H}(t)\mathrm{d}t\right)\psi(t) \tag{4.6}$$

However, this is still a differential equation that we must solve. Although we will not use the formal analytical solution directly (numerical methods discussed in Sect. 4.9 dominate practical work), it will later serve as the starting point for various approximations. Integration of both sides of Eq. (4.6) yields

$$\psi(t) = \psi(0) - i\int_0^t \hat{H}(t_1)\psi(t_1)\mathrm{d}t_1 \tag{4.7}$$

This may be substituted back into the right-hand side of Schrödinger's equation, and the equation integrated again. After this procedure is applied multiple times, we obtain

$$\begin{aligned}\psi(t) &= \left[1 + (-i)\int_0^t \mathrm{d}t_1\hat{H}(t_1) + (-i)^2\int_0^t \mathrm{d}t_1\int_0^{t_1}\mathrm{d}t_2\hat{H}(t_1)\hat{H}(t_2) + \ldots\right]\psi(0) \\ &= \left[\sum_{n=0}^{\infty}\hat{U}_n(t)\right]\psi(0)\hat{U}_n(t) = (-i)^n\int_0^t \mathrm{d}t_1\int_0^{t_1}\mathrm{d}t_2\cdots\int_0^{t_{n-1}}\mathrm{d}t_n\hat{H}(t_1)\hat{H}(t_2)\cdots\hat{H}(t_n)\end{aligned} \tag{4.8}$$

This is *Dyson series* [92]. Upper integration limits are ordered as $t \geq t_1 \geq t_2 \geq \ldots \geq t_n$; this prevents us from splitting up chained integrals. Outside of simple special cases, this expression is incomputable.

We would be able to proceed if we find a way to make the product $\hat{H}(t_1)\hat{H}(t_2)\cdots\hat{H}(t_n)$ invariant under permutations of $\{t_1, t_2, \ldots, t_n\}$. This is because, for any integrable function $K(t_1, t_2, \ldots, t_n)$ that is invariant under permutations of its arguments, the following relation holds:

$$\int_0^t \mathrm{d}t_1\int_0^{t_1}\mathrm{d}t_2\ldots\int_0^{t_{n-1}}\mathrm{d}t_n K(t_1, t_2, \ldots, t_n) = \frac{1}{n!}\int_0^t \mathrm{d}t_1\int_0^t \mathrm{d}t_2\ldots\int_0^t \mathrm{d}t_n K(t_1, t_2, \ldots, t_n) \tag{4.9}$$

We will do this by introducing a *time-ordering operator* $\overleftarrow{T}$ that reorders times in $\hat{H}(t_1)\hat{H}(t_2)\cdots\hat{H}(t_n)$ in such a way as to preserve the descending sequence. This does not have any effect on Eq. (4.8) because the times are already in order there, but it allows us to apply Eq. (4.9):

$$\hat{U}_n(t) = \frac{(-i)^n}{n!} \int_0^t \mathrm{d}t_1 \int_0^t \mathrm{d}t_2 ... \int_0^t \mathrm{d}t_n \overleftarrow{T}\{\hat{H}(t_1)\hat{H}(t_2)...\hat{H}(t_n)\} \tag{4.10}$$

The time-ordering operator is linear, and therefore:

$$\hat{U}_n(t) = \frac{1}{n!}\overleftarrow{T}\left[(-i)^n \int_0^t \mathrm{d}t_1 \int_0^t \mathrm{d}t_2 ... \int_0^t \mathrm{d}t_n \hat{H}(t_1)\hat{H}(t_2)...\hat{H}(t_n)\right] \tag{4.11}$$

At this point, the chained integrals split up—we now have powers of the same integral:

$$\hat{U}_n(t) = \frac{1}{n!}\overleftarrow{T}\left[-i \int_0^t \hat{H}(t')\mathrm{d}t'\right]^n \tag{4.12}$$

After placing this result into the Dyson series in Eq. (4.8), we obtain

$$\psi(t) = \overleftarrow{T}\left[\sum_{n=0}^{\infty} \frac{1}{n!}\left(-i \int_0^t \hat{H}(t')\mathrm{d}t'\right)^n\right]\psi(0) \tag{4.13}$$

where an operator exponential may again be recognised:

$$\psi(t) = \overleftarrow{\exp}\left(-i \int_0^t \hat{H}(t)\mathrm{d}t\right)\psi(0) \tag{4.14}$$

The arrow indicates a *time-ordered matrix exponential* [92] that is best interpreted in the sense of Eq. (4.6), where infinitesimal time evolution operators are applied, one after another, to the initial condition.

4.1.1 Example: Spin Precession

Consider an isotropically shielded spin in a static magnetic field directed along the Z-axis of the laboratory frame. The equation of motion for the wavefunction has a time-independent Hamiltonian:

$$\mathbf{H} = \omega \mathbf{S}_{\mathrm{Z}}, \quad \frac{\partial}{\partial t}|\psi(t)\rangle = -i\omega \mathbf{S}_{\mathrm{Z}}|\psi(t)\rangle, \quad \mathbf{S}_{\mathrm{Z}} = \begin{pmatrix} +1/2 & 0 \\ 0 & -1/2 \end{pmatrix} \tag{4.15}$$

where ω is the spin precession frequency and $|\psi(t)\rangle$ is a vector representation of $\psi(t)$. The general solution comes from Eq. (4.5):

$$|\psi(t)\rangle = \mathbf{P}(t)|\psi(0)\rangle, \quad \mathbf{P}(t) = \exp(-i\omega \mathbf{S}_{\mathrm{Z}} t) \tag{4.16}$$

The propagator is obtained from its Taylor series definition in Eq. (4.4):

$$\mathbf{P}(t) = \sum_{n=0}^{\infty} \frac{(-i\omega t)^n}{n!} \mathbf{S}_{\mathrm{Z}}^n = \ldots = \begin{pmatrix} \mathrm{e}^{-i\omega t/2} & 0 \\ 0 & \mathrm{e}^{+i\omega t/2} \end{pmatrix} \tag{4.17}$$

When the initial condition corresponds to the spin directed along the Z-axis:

$$|\psi(0)\rangle = \begin{pmatrix} 1 \\ 0 \end{pmatrix} \quad \Rightarrow \quad |\psi(t)\rangle = \begin{pmatrix} \mathrm{e}^{-i\omega t/2} & 0 \\ 0 & \mathrm{e}^{+i\omega t/2} \end{pmatrix} \begin{pmatrix} 1 \\ 0 \end{pmatrix} = \begin{pmatrix} \mathrm{e}^{-i\omega t/2} \\ 0 \end{pmatrix} \tag{4.18}$$

only the phase of the wavefunction is oscillating—the observables stay put

$$\begin{aligned} \langle \psi(t)|\mathbf{S}_{\mathrm{X}}|\psi(t)\rangle &= \ldots = 0 \\ \langle \psi(t)|\mathbf{S}_{\mathrm{Y}}|\psi(t)\rangle &= \ldots = 0 \\ \langle \psi(t)|\mathbf{S}_{\mathrm{Z}}|\psi(t)\rangle &= 1/2 \end{aligned} \tag{4.19}$$

This is to be expected because this initial condition is an eigenvector of the Hamiltonian in Eq. (4.15). When the initial state corresponds instead to the spin directed along the X-axis:

$$\begin{gathered} |\psi(0)\rangle = \frac{1}{\sqrt{2}} \begin{pmatrix} 1 \\ 1 \end{pmatrix}, \qquad \langle \psi(0)|\mathbf{S}_{\mathrm{X}}|\psi(0)\rangle = 1/2 \\ \langle \psi(0)|\mathbf{S}_{\mathrm{Y}}|\psi(0)\rangle = \langle \psi(0)|\mathbf{S}_{\mathrm{Z}}|\psi(0)\rangle = 0 \end{gathered} \tag{4.20}$$

we see precession in the XY plane [93]:

$$
\begin{aligned}
&|\psi(t)\rangle = \frac{1}{\sqrt{2}}\begin{pmatrix} e^{-i\omega t/2} & 0 \\ 0 & e^{+i\omega t/2}\end{pmatrix}\begin{pmatrix}1\\1\end{pmatrix} = \frac{1}{\sqrt{2}}\begin{pmatrix} e^{-i\omega t/2} \\ e^{+i\omega t/2}\end{pmatrix} \\
&\Downarrow \\
&\langle\psi(t)|\mathbf{S}_\mathrm{Z}|\psi(t)\rangle = 0 \\
&\langle\psi(t)|\mathbf{S}_\mathrm{X}|\psi(t)\rangle = \ldots = +\frac{1}{2}\cos(\omega t) \\
&\langle\psi(t)|\mathbf{S}_\mathrm{Y}|\psi(t)\rangle = \ldots = -\frac{1}{2}\sin(\omega t)
\end{aligned}
\tag{4.21}
$$

This is the simplest example of a time-domain simulation in magnetic resonance: the time evolution generator **H** is exponentiated to obtain the propagator, which is then applied to the initial condition to evolve the system forward in time.

4.1.2 Example: Bloch Equations

For an isotropically shielded spin in a (possibly time-dependent) magnetic field **B**, the Hamiltonian is

$$
\mathbf{H} = -\gamma(B_\mathrm{X}\mathbf{S}_\mathrm{X} + B_\mathrm{Y}\mathbf{S}_\mathrm{Y} + B_\mathrm{Z}\mathbf{S}_\mathrm{Z})
\tag{4.22}
$$

where γ is the magnetogyric ratio. Using Heisenberg's equation of motion, we obtain the following equations of motion for the observables corresponding to the three Cartesian spin operators:

$$
\frac{\partial}{\partial t}\langle\mathbf{O}\rangle = i\langle[\mathbf{H},\mathbf{O}]\rangle \quad\Rightarrow\quad
\begin{aligned}
\frac{\partial}{\partial t}\langle\mathbf{S}_\mathrm{X}\rangle &= -i\gamma\langle[B_\mathrm{Y}\mathbf{S}_\mathrm{Y} + B_\mathrm{Z}\mathbf{S}_\mathrm{Z}, \mathbf{S}_\mathrm{X}]\rangle \\
\frac{\partial}{\partial t}\langle\mathbf{S}_\mathrm{Y}\rangle &= -i\gamma\langle[B_\mathrm{X}\mathbf{S}_\mathrm{X} + B_\mathrm{Z}\mathbf{S}_\mathrm{Z}, \mathbf{S}_\mathrm{Y}]\rangle \\
\frac{\partial}{\partial t}\langle\mathbf{S}_\mathrm{Z}\rangle &= -i\gamma\langle[B_\mathrm{X}\mathbf{S}_\mathrm{X} + B_\mathrm{Y}\mathbf{S}_\mathrm{Y}, \mathbf{S}_\mathrm{Z}]\rangle
\end{aligned}
\tag{4.23}
$$

where zero commutators, such as $[\mathbf{S}_\mathrm{X}, \mathbf{S}_\mathrm{X}]$, have been dropped. After working through the remaining commutators (Sect. 1.6.3), we find

$$
\begin{aligned}
\frac{\partial}{\partial t}\langle\mathbf{S}_\mathrm{X}\rangle &= \gamma(B_\mathrm{Z}\langle\mathbf{S}_\mathrm{Y}\rangle - B_\mathrm{Y}\langle\mathbf{S}_\mathrm{Z}\rangle) \\
\frac{\partial}{\partial t}\langle\mathbf{S}_\mathrm{Y}\rangle &= \gamma(B_\mathrm{X}\langle\mathbf{S}_\mathrm{Z}\rangle - B_\mathrm{Z}\langle\mathbf{S}_\mathrm{X}\rangle) \\
\frac{\partial}{\partial t}\langle\mathbf{S}_\mathrm{Z}\rangle &= \gamma(B_\mathrm{Y}\langle\mathbf{S}_\mathrm{X}\rangle - B_\mathrm{X}\langle\mathbf{S}_\mathrm{Y}\rangle)
\end{aligned}
\tag{4.24}
$$

where a vector cross product is now visible:

$$\frac{\partial}{\partial t}\langle \mathbf{S}(t)\rangle = \boldsymbol{\omega}(t) \times \langle \mathbf{S}(t)\rangle \qquad \boldsymbol{\omega}(t) = -\gamma \mathbf{B}(t) \tag{4.25}$$

These are *Bloch equations* for the precession of spin around a magnetic field [94]—so far, without dissipative (Chap. 6) and diffusive (Chap. 5) parts.

4.2 Density Operator Formalism

Consider an ensemble average $\overline{O}$ of an observable quantity O corresponding to a quantum mechanical operator $\mathbf{O}$. Because matrix trace is linear and invariant under cyclic permutations of arguments:

$$\overline{\langle\psi|\mathbf{O}|\psi\rangle} = \overline{\mathrm{Tr}(\langle\psi|\mathbf{O}|\psi\rangle)} = \overline{\mathrm{Tr}(|\psi\rangle\langle\psi|\mathbf{O})} = \mathrm{Tr}\left(\overline{|\psi\rangle\langle\psi|}\mathbf{O}\right) \tag{4.26}$$

It is apparent that the descriptor carrying the information about the ensemble average is not produced by averaging the wavefunction (that average would be zero because the wavefunction phase is arbitrary), but rather by averaging the projector $|\psi\rangle\langle\psi|$ in which phase multipliers cancel:

$$\left|e^{i\varphi}\psi\right\rangle\left\langle e^{i\varphi}\psi\right| = e^{i\varphi}e^{-i\varphi}|\psi\rangle\langle\psi| = |\psi\rangle\langle\psi| \tag{4.27}$$

The quantity $\boldsymbol{\rho} = |\psi\rangle\langle\psi|$ is called *density operator* [89] or *density matrix* (not to be confused with density functional and density matrix in electronic structure theory, those are different things). It is an operator because it acts on a wavefunction and returns another wavefunction:

$$\boldsymbol{\rho}|\varphi\rangle = |\psi\rangle\langle\psi\,|\,\varphi\rangle = a|\psi\rangle, \quad a = \langle\psi\,|\,\varphi\rangle \tag{4.28}$$

It is also Hermitian and *idempotent*—its positive integer powers are equal to itself:

$$\boldsymbol{\rho}^2 = |\psi\rangle\langle\psi\,|\,\psi\rangle\langle\psi| = |\psi\rangle\langle\psi| = \boldsymbol{\rho} \tag{4.29}$$

The physical meaning of $\boldsymbol{\rho}$ follows from the expressions for its matrix elements. The diagonal elements

$$\langle n|\boldsymbol{\rho}|n\rangle = \langle n\,|\,\psi\rangle\langle\psi\,|\,n\rangle = |c_n|^2 = p_n \tag{4.30}$$

correspond to the probability of finding the system in a state $|n\rangle$. It follows that, for a single quantum system, the trace of the density matrix is the sum of all probabilities, and therefore equal to 1. The off-diagonal elements indicate the presence of a superposition in the wavefunction:

$$\begin{gathered} |\psi\rangle = \ldots + c_n|n\rangle + c_k|k\rangle + \ldots \\ \Downarrow \\ \langle n|\boldsymbol{\rho}|k\rangle = \langle n \mid \psi\rangle\langle\psi \mid k\rangle = c_n c_k^* \end{gathered} \tag{4.31}$$

When the ensemble average is applied, the trace of $\boldsymbol{\rho}$ indicates the *purity* of the ensemble: when the wavefunctions of the individual systems are identical up to a phase (*pure ensemble*), the trace of the ensemble average density matrix is equal to 1. However, when individual systems have different wavefunctions, the trace of the average density matrix is smaller than 1 (*mixed ensemble*). In the average density matrix, the off-diagonal elements $\overline{c_n c_k^*}$ acquire the meaning of correlation coefficients between wavefunction components across the ensemble; they are colloquially called *coherences*.

4.2.1 Liouville–Von Neumann Equation

The equation of motion (called *Liouville–von Neumann equation* [89]) for the density operator is obtained from its definition and Schrödinger's equation using the product rule:

$$\begin{gathered} \frac{\partial\boldsymbol{\rho}}{\partial t} = \frac{\partial}{\partial t}(|\psi\rangle\langle\psi|) = \left(\frac{\partial}{\partial t}|\psi\rangle\right)\langle\psi| + |\psi\rangle\left(\frac{\partial}{\partial t}\langle\psi|\right) \\ = -i\mathbf{H}|\psi\rangle\langle\psi| + i|\psi\rangle\langle\psi|\mathbf{H} = -i\mathbf{H}\rho + i\rho\mathbf{H} = -i[\mathbf{H},\boldsymbol{\rho}] \\ \Downarrow \\ \frac{\partial\boldsymbol{\rho}}{\partial t} = -i[\mathbf{H},\boldsymbol{\rho}] \end{gathered} \tag{4.32}$$

The solutions are inherited from Schrödinger's equation: two copies of Eqs. (4.5) or (4.14) are multiplied together to assemble $\boldsymbol{\rho} = |\psi\rangle\langle\psi|$. With a time-independent Hamiltonian:

$$\begin{gathered} |\psi(t+\Delta t)\rangle = \exp(-i\mathbf{H}\Delta t)|\psi(t)\rangle \\ \Downarrow \\ \boldsymbol{\rho}(t+\Delta t) = \exp(-i\mathbf{H}\Delta t)\boldsymbol{\rho}(t)\exp(+i\mathbf{H}\Delta t) \end{gathered} \tag{4.33}$$

and with a time-dependent Hamiltonian:

$$\begin{gathered}\boldsymbol{\rho}(t+\mathrm{d}t)=\exp(-i\mathbf{H}(t)\mathrm{d}t)\boldsymbol{\rho}(t)\exp(+i\mathbf{H}(t)\mathrm{d}t)\\ \Downarrow\\ \boldsymbol{\rho}(t)=\overleftarrow{\exp}\left(-i\int_0^t\mathbf{H}(t)\mathrm{d}t\right)\boldsymbol{\rho}(0)\overrightarrow{\exp}\left(+i\int_0^t\mathbf{H}(t)\mathrm{d}t\right)\end{gathered}\tag{4.34}$$

Efficient numerical methods for evaluating these expressions are discussed in Sect. 4.9.

4.2.2 Calculation of Observables

Using the cyclic permutation property of the matrix trace, we obtain

$$\langle O\rangle=\langle\psi|\mathbf{O}|\psi\rangle=\mathrm{Tr}(\langle\psi|\mathbf{O}|\psi\rangle)=\mathrm{Tr}(\mathbf{O}|\psi\rangle\langle\psi|)=\mathrm{Tr}(\mathbf{O}\boldsymbol{\rho})\tag{4.35}$$

Strictly speaking, this is not a Frobenius inner product of two matrices (the conjugate is missing, see Sect. 1.4.5), but this is equivalent to an inner product because both the observable operator and the density matrix are Hermitian. Care must be taken here when density operator formalism is abused (Sect. 9.4) to include non-Hermitian observables and density matrices.

The equation of motion for observables is identical to the one obtained from Schrödinger's equation:

$$\begin{aligned}\frac{\partial}{\partial t}\langle O\rangle&=\mathrm{Tr}\left(\mathbf{O}\frac{\partial}{\partial t}\boldsymbol{\rho}\right)=\mathrm{Tr}(-i\mathbf{O}[\mathbf{H},\boldsymbol{\rho}])=-i\mathrm{Tr}(\mathbf{O}\mathbf{H}\boldsymbol{\rho}-\mathbf{O}\boldsymbol{\rho}\mathbf{H})\\&=-i\mathrm{Tr}(\mathbf{O}\mathbf{H}\boldsymbol{\rho}-\mathbf{H}\mathbf{O}\boldsymbol{\rho})=i\mathrm{Tr}([\mathbf{H},\mathbf{O}]\boldsymbol{\rho})=i\langle[\mathbf{H},\mathbf{O}]\rangle\end{aligned}\tag{4.36}$$

Thus, for individual quantum systems and pure ensembles, density operator formalism is equivalent to wavefunction formalism.

Liouville–von Neumann equation conserves the trace of the density matrix:

$$\frac{\partial}{\partial t}\mathrm{Tr}(\boldsymbol{\rho})=-i\mathrm{Tr}([\mathbf{H},\boldsymbol{\rho}])=-i\mathrm{Tr}(\mathbf{H}\boldsymbol{\rho}-\boldsymbol{\rho}\mathbf{H})=0\tag{4.37}$$

Because matrix representations of generators of special unitary groups are traceless (Sect. 1.6.3), the trace of $\boldsymbol{\rho}$ does not affect physical observables:

$$\mathrm{Tr}\{\mathbf{O}(\boldsymbol{\rho}-k\mathbf{1})\}=\mathrm{Tr}(\mathbf{O}\boldsymbol{\rho})+k\mathrm{Tr}(\mathbf{O}\mathbf{1})=\mathrm{Tr}(\mathbf{O}\boldsymbol{\rho})\tag{4.38}$$

The contribution from the unit matrix will become important when we consider dissipative dynamics (Chap. 6), but for unitary dynamics discussed here, an appropriate multiple of the unit matrix can always be subtracted out to make $\boldsymbol{\rho}$ traceless.

4.2.3 Spin State Classification

Equation (4.35) has a useful corollary—the expectation values of a complete orthonormal basis set of Hermitian operators are also the expansion coefficients of the density matrix in that basis set. For example, if a spin-½ ensemble is polarised along the Z-axis in the Zeeman basis:

$$|\psi\rangle = |\alpha\rangle = \begin{pmatrix} 1 & 0 \end{pmatrix}^{\mathrm{T}} \tag{4.39}$$

then the corresponding density matrix is

$$\boldsymbol{\rho} = |\psi\rangle\langle\psi| = |\alpha\rangle\langle\alpha| = \begin{pmatrix} 1 \\ 0 \end{pmatrix}\begin{pmatrix} 1 & 0 \end{pmatrix} = \begin{pmatrix} 1 & 0 \\ 0 & 0 \end{pmatrix} = \frac{\mathbf{1}}{2} + \mathbf{S}_{\mathrm{Z}} \tag{4.40}$$

This creates an equivalence between a spin operator and a spin state, and leads to the common practice of using spin operators and density matrices interchangeably: when we say that the ensemble is "in the $\mathbf{S}_{\mathrm{Z}}$ state", this means that its density operator contains a mixture of $\mathbf{S}_{\mathrm{Z}}$ and the unit operator.

Spin states may then be classified by the physical meaning of the corresponding operators. A popular classification identifies (the direct product signs are often skipped for brevity, and a unit matrix is implied on any spin that is not explicitly mentioned in the operator expression):

1. ***Longitudinal Single-Spin Orders***: $\mathbf{S}_{\mathrm{Z}}^{(1)}, \mathbf{S}_{\mathrm{Z}}^{(2)}$, etc. These correspond to population differences between Zeeman energy levels that are one spin flip away from each other. They are also known as *longitudinal magnetisation states*.
2. ***Longitudinal Multi-Spin Orders***: $\mathbf{S}_{\mathrm{Z}}^{(1)} \otimes \mathbf{S}_{\mathrm{Z}}^{(2)}, \mathbf{S}_{\mathrm{Z}}^{(1)} \otimes \mathbf{S}_{\mathrm{Z}}^{(2)} \otimes \mathbf{S}_{\mathrm{Z}}^{(3)}$, etc. These also correspond to population differences across levels connected by single-spin flips, but the sign of the difference depends on the state that other spins have. Also known as *longitudinal correlation states*.
3. ***Transverse Single-Spin Orders***: $\mathbf{S}_{\mathrm{X}}^{(1)}, \mathbf{S}_{\mathrm{Y}}^{(2)}, \mathbf{S}_{+}^{(1)}$, etc. These are off-diagonal terms in the density operator corresponding to observable *transverse magnetisation states*. Some of these operators are non-Hermitian, they appear when corners are being cut (Sect. 9.3.3) in the numerical simulation of phase cycles and quadrature detection schemes.
4. ***Transverse Multi-Spin Orders***: $\mathbf{S}_{\mathrm{X}}^{(1)} \otimes \mathbf{S}_{\mathrm{Y}}^{(2)}$, $\mathbf{S}_{+}^{(1)} \otimes \mathbf{S}_{-}^{(2)}$, etc. These are off-diagonal terms in the density operator that correspond to correlated patterns

of transverse magnetisation of different spins. Such statistical correlations do not directly correspond to observable magnetisation, but they may evolve into other states that do.

5. ***Mixed Spin Orders***: $\mathbf{S}_Z^{(1)} \otimes \mathbf{S}_+^{(2)}$, etc. These do not have a systematic classification and correspond to correlations between the longitudinal and transverse magnetisation of different spins. They become prominent in relaxation-driven experiments (Chap. 6) where the width of the spectral line of a particular spin may depend on the state of nearby spins.

Two broader categories of spin states have historical names:

1. ***Coherences***: a spin state $\boldsymbol{\rho}$ having the following property under the commutation action by the total $\mathbf{S}_Z$ spin projection operator

$$[\mathbf{S}_Z, \boldsymbol{\rho}] = k\boldsymbol{\rho}, \quad \mathbf{S}_Z = \sum_n \mathbf{S}_Z^{(n)} \tag{4.41}$$

is called k-quantum coherence. These states are not Hermitian; the important ones are single-quantum coherences $\mathbf{S}_+$ and $\mathbf{S}_-$ because they correspond to observable transverse magnetization in quadrature-detected magnetic resonance experiments.

An equivalent definition via the exponential action by $\mathbf{S}_Z$ is more physically interpretable. For example, a double-quantum coherence between two spins evolves, under the Zeeman Hamiltonian, at the sum of their Zeeman frequencies:

$$\mathrm{e}^{-i\left(\omega_1 \mathbf{S}_Z^{(1)} + \omega_2 \mathbf{S}_Z^{(2)}\right)t} \boldsymbol{\rho} \mathrm{e}^{+i\left(\omega_1 \mathbf{S}_Z^{(1)} + \omega_2 \mathbf{S}_Z^{(2)}\right)t} = \mathrm{e}^{-i(\omega_1 + \omega_1)t} \boldsymbol{\rho} \tag{4.42}$$

A common example in protein NMR spectroscopy is $\mathbf{H}_+\mathbf{N}_+$ (symbols refer to atom types) in the amide group.

2. ***Correlations***: any direct product of single-spin operators involving k non-unit spin operators is called a *k-spin correlation* because it describes a statistically correlated state of those spins.

Practical analysis of spin system trajectories (Sect. 8.5) is often performed in terms of coherences and correlations that the system is passing through as it evolves.

4.2.4 Superoperators and Liouville Space

Having a commutator in the equation of motion and a double-sided matrix multiplication in the propagation operation (Sect. 4.2.1) is logistically inconvenient. For this reason, density operator formalism is sometimes cast [6] into the adjoint representation (Sect. 1.5.9):

$$\frac{\partial \boldsymbol{\rho}}{\partial t} = -i[\mathbf{H}, \boldsymbol{\rho}] \quad \Rightarrow \quad \frac{\partial \boldsymbol{\rho}}{\partial t} = -i\mathcal{H}\boldsymbol{\rho}$$
$$\mathcal{H}\boldsymbol{\rho} = [\mathbf{H}, \boldsymbol{\rho}] = \mathbf{H}\boldsymbol{\rho} - \boldsymbol{\rho}\mathbf{H} \tag{4.43}$$

where $\mathcal{H}$ is called *commutation superoperator*. Adapting Eqs. (1.48) and (1.49) to this specific case, we get:

$$\boldsymbol{\rho} \to \text{vec}(\boldsymbol{\rho}) \quad \Rightarrow \quad \mathbf{A}\boldsymbol{\rho}\mathbf{B} \to \left(\mathbf{B}^{\mathrm{T}} \otimes \mathbf{A}\right)\text{vec}(\boldsymbol{\rho})$$
$$[\mathbf{H}, \boldsymbol{\rho}] = \mathbf{H}\boldsymbol{\rho} - \boldsymbol{\rho}\mathbf{H} \to \left(\mathbf{1} \otimes \mathbf{H} - \mathbf{H}^{\mathrm{T}} \otimes \mathbf{1}\right)\text{vec}(\boldsymbol{\rho}) \tag{4.44}$$

where the transpose remains (does not become Hermitian conjugate) for complex matrices and **1** is a unit matrix of the same dimension as **H**. Density matrix vectorisation operation is normally not written out explicitly because there is no ambiguity: the representation is determined by the object that acts on the density matrix.

We will also later need *left and right product superoperators*:

$$\mathcal{H}^{(\mathrm{L})}\boldsymbol{\rho} = \mathbf{H}\rho, \quad \mathcal{H}^{(\mathrm{R})}\boldsymbol{\rho} = \rho\mathbf{H} \quad \Rightarrow \quad \mathcal{H} = \mathcal{H}^{(\mathrm{L})} - \mathcal{H}^{(\mathrm{R})} \tag{4.45}$$

This representation is colloquially called *Liouville space* [90]; it makes Eqs. (4.32) and (4.33) neater:

$$\frac{\partial}{\partial t}\boldsymbol{\rho}(t) = -i\mathcal{H}\boldsymbol{\rho}(t) \quad \Rightarrow \quad \boldsymbol{\rho}(t) = \exp(-i\mathcal{H}t)\boldsymbol{\rho}(0)$$
$$\frac{\partial}{\partial t}\boldsymbol{\rho}(t) = -i\mathcal{H}(t)\boldsymbol{\rho}(t) \quad \Rightarrow \quad \boldsymbol{\rho}(t) = \overleftarrow{\exp}\left(-i\int_0^t \mathcal{H}(t)dt\right)\boldsymbol{\rho}(0) \tag{4.46}$$

because they now resemble Schrödinger's equation and its solution, but still support dissipative ensemble dynamics, as well as spatial and chemical processes discussed in Chaps. 5 and 6.

4.2.5 Treatment of Composite Systems

It follows from the representation structure of the time translation group (Sect. 2.1) that a system AB composed of two non-interacting subsystems A and B evolves under the direct product of the corresponding representations of the time evolution operator. Thus, the Hamiltonian of the composite system is

$$\mathbf{H}^{(\mathrm{AB})} = \mathbf{H}^{(\mathrm{A})} \otimes \mathbf{1}^{(\mathrm{B})} + \mathbf{1}^{(\mathrm{A})} \otimes \mathbf{H}^{(\mathrm{B})} \tag{4.47}$$

where $\mathbf{1}^{(\mathrm{A,B})}$ are unit matrices of the dimension matching the dimension of the indicated Hamiltonian. Individual spin operators are extended in the same way, for example:

$$\mathbf{S}_\mathrm{X}^{(\mathrm{A})} = \mathbf{S}_\mathrm{X} \otimes \mathbf{1}^{(\mathrm{B})}, \quad \mathbf{S}_\mathrm{X}^{(\mathrm{B})} = \mathbf{1}^{(\mathrm{A})} \otimes \mathbf{S}_\mathrm{X} \tag{4.48}$$

This creates operators that act each on its own spin, and with a unit matrix on other spins. The wavefunction is then a Kronecker product of the wavefunctions of the seffecubsystems:

$$\left|\psi^{(\mathrm{AB})}\right\rangle = \left|\psi^{(\mathrm{A})}\right\rangle \otimes \left|\psi^{(\mathrm{B})}\right\rangle \tag{4.49}$$

The following properties of Kronecker products are now relevant:

$$\begin{gathered}(\mathbf{A} \otimes \mathbf{B})(\mathbf{C} \otimes \mathbf{D}) = (\mathbf{AC}) \otimes (\mathbf{BD}) \\ \mathrm{Tr}(\mathbf{A} \otimes \mathbf{B}) = \mathrm{Tr}(\mathbf{A})\mathrm{Tr}(\mathbf{B})\end{gathered} \tag{4.50}$$

For the density matrix of the composite system, they yield

$$\begin{aligned}\boldsymbol{\rho}^{(\mathrm{AB})} &= \left|\psi^{(\mathrm{AB})}\right\rangle\left\langle\psi^{(\mathrm{AB})}\right| = \left(\left|\psi^{(\mathrm{A})}\right\rangle \otimes \left|\psi^{(\mathrm{B})}\right\rangle\right)\left(\left\langle\psi^{(\mathrm{A})}\right| \otimes \left\langle\psi^{(\mathrm{A})}\right|\right) \\ &= \left(\left|\psi^{(\mathrm{A})}\right\rangle\left\langle\psi^{(\mathrm{A})}\right|\right) \otimes \left(\left|\psi^{(\mathrm{B})}\right\rangle\left\langle\psi^{(\mathrm{B})}\right|\right) = \boldsymbol{\rho}^{(\mathrm{A})} \otimes \boldsymbol{\rho}^{(\mathrm{B})}\end{aligned} \tag{4.51}$$

This procedure is similarly extended to systems with more than two independent subsystems.

4.2.6 Frequency-Domain Solution

When the Hamiltonian is time independent and dissipative processes are present, it is possible to obtain the frequency spectrum of an observable without computing its time trajectory. This is useful when only a narrow frequency window is of interest, for example in quadrupolar overtone NMR spectroscopy. Let $\mathcal{H}$ be the Hamiltonian commutation superoperator and $\mathcal{R}$ be the negative definite relaxation superoperator (Chap. 6). In the dissipative equation of motion:

$$\boldsymbol{\rho}'(t) = -i(\mathcal{H} + i\mathcal{R})\boldsymbol{\rho}(t) \tag{4.52}$$

we will make a replacement $\boldsymbol{\rho}(t) = \boldsymbol{\sigma}(t) + \boldsymbol{\rho}_\mathrm{eq}$ to isolate the thermal equilibrium state $\boldsymbol{\rho}_\mathrm{eq} = \boldsymbol{\rho}(\infty)$. The definition of thermal equilibrium (Chap. 6) implies that

$$(\mathcal{H}+i\mathcal{R})\boldsymbol{\rho}_{\text{eq}}=0 \tag{4.53}$$

and therefore the equation of motion becomes

$$\boldsymbol{\sigma}'(t)=-i(\mathcal{H}+i\mathcal{R})\boldsymbol{\sigma}(t) \tag{4.54}$$

This helps because $\boldsymbol{\sigma}(\infty)=0$, and the Fourier transform of this equation is uncomplicated:

$$\int_0^\infty \boldsymbol{\sigma}'(t)e^{-i\omega t}\mathrm{d}t=-i(\mathcal{H}+i\mathcal{R})\int_0^\infty \boldsymbol{\sigma}(t)\mathrm{e}^{-i\omega t}\mathrm{d}t \tag{4.55}$$

After integrating the left-hand side by parts, we obtain the Fourier image of Eq. (4.54):

$$\begin{aligned} i\omega\boldsymbol{\sigma}(\omega)-\boldsymbol{\sigma}_0&=-i(\mathcal{H}+i\mathcal{R})\boldsymbol{\sigma}(\omega)\\ \boldsymbol{\sigma}(\omega)&=\int_0^\infty \boldsymbol{\sigma}(t)\mathrm{e}^{-i\omega t}\mathrm{d}t \end{aligned} \tag{4.56}$$

where $\boldsymbol{\sigma}_0=\boldsymbol{\rho}_0-\boldsymbol{\rho}_{\text{eq}}$ is the initial condition. This is now a linear equation for $\boldsymbol{\sigma}(\omega)$ that may be solved using standard numerical methods (Sect. 9.3.4) that avoid inverting the matrix in the round brackets, and compute the following inverse-times-vector directly and efficiently:

$$\boldsymbol{\sigma}(\omega)=-i(\mathcal{H}+i\mathcal{R}+\omega\mathbf{1})^{-1}\boldsymbol{\sigma}_0 \tag{4.57}$$

where $\mathbf{1}$ is a unit matrix of the same dimension as the evolution generator $\mathcal{H}$ and the dissipation generator $\mathcal{R}$. Because the difference between $\boldsymbol{\sigma}(t)$ and $\boldsymbol{\rho}(t)$ is a constant, $\boldsymbol{\sigma}(\omega)$ is only different from $\boldsymbol{\rho}(\omega)$ at zero frequency. Thus, for any Hermitian observable state $\mathbf{O}$ in Liouville space:

$$O(\omega)=-i\langle\mathbf{O}|(\mathcal{H}+i\mathcal{R}+\omega\mathbf{1})^{-1}|\boldsymbol{\sigma}_0\rangle,\quad \omega\neq 0 \tag{4.58}$$

Note that this formalism is not equivalent to frequency-swept spectroscopy where the instrument applies time-dependent external fields—Eq. (4.57) is only a Fourier image of the system trajectory under a time-independent evolution generator in Eq. (4.52).

4.3 Effective Hamiltonians

Strong and boring interactions (Zeeman, NQI, ZFS) can dominate spin Hamiltonians in which the smaller terms—usually spin–spin couplings—are actually the interesting ones. The presence of large static interactions also makes it inconvenient and numerically problematic to calculate the trajectory

$$\begin{gathered} G(\boldsymbol{\rho}_0)=\{\boldsymbol{\rho}_0,\,\mathcal{P}\boldsymbol{\rho}_0,\,\mathcal{P}^2\boldsymbol{\rho}_0,\dots\} \\ \mathcal{P}=\exp(-i\mathcal{H}\Delta t),\quad \Delta t\leq\frac{\pi}{2}\|\mathcal{H}\|_2^{-1} \end{gathered} \tag{4.59}$$

because the discrete time step Δt is very small on the time scale of the simulation, which then necessitates an impractically large number of time steps. This section discusses several ways of eliminating rapid overall dynamics from the picture in order to focus on the weaker interactions of interest.

4.3.1 Interaction Representation

To mitigate the timescale problem, we split the Hamiltonian into the "large, boring, and time-independent" part $\mathbf{H}_0$ and the "small, interesting, and possibly time-dependent" part $\mathbf{H}_1(t)$, and consider a substitution into which the $\mathbf{H}_0$ dynamics is baked explicitly

$$\boldsymbol{\rho}(t)=\mathrm{e}^{-i\mathcal{H}_0 t}\boldsymbol{\sigma}(t) \tag{4.60}$$

in the hope that the remaining dynamics contained in $\boldsymbol{\sigma}(t)$ would be easier to deal with. Placing this substitution into the Liouville–von Neumann equation with $\mathcal{H}(t)=\mathcal{H}_0+\mathcal{H}_1(t)$ yields

$$\frac{\partial}{\partial t}\left(\mathrm{e}^{-i\mathcal{H}_0 t}\boldsymbol{\sigma}(t)\right)=-i(\mathcal{H}_0+\mathcal{H}_1(t))\mathrm{e}^{-i\mathcal{H}_0 t}\boldsymbol{\sigma}(t) \tag{4.61}$$

After using the product rule on the left-hand side and making simplifications, we obtain

$$\frac{\partial}{\partial t}\boldsymbol{\sigma}(t)=-i\mathcal{H}_1^{\mathrm{R}}(t)\boldsymbol{\sigma}(t),\quad \mathcal{H}_1^{\mathrm{R}}(t)=\mathrm{e}^{+i\mathcal{H}_0 t}\mathcal{H}_1(t)\mathrm{e}^{-i\mathcal{H}_0 t} \tag{4.62}$$

A similar calculation using the Hilbert space commutator form of the LvN equation yields:

$$\frac{\partial}{\partial t}\boldsymbol{\sigma}(t) = -i\left[\mathbf{H}_1^{\mathrm{R}}(t), \boldsymbol{\sigma}(t)\right], \quad \begin{cases} \boldsymbol{\rho}(t) = \mathrm{e}^{-i\mathbf{H}_0 t}\boldsymbol{\sigma}(t)\mathrm{e}^{+i\mathbf{H}_0 t} \\ \mathbf{H}_1^{\mathrm{R}}(t) = \mathrm{e}^{+i\mathbf{H}_0 t}\mathbf{H}_1(t)\mathrm{e}^{-i\mathbf{H}_0 t} \end{cases} \tag{4.63}$$

Equations (4.62) and (4.63) do not look like a simplification until we note two things. Firstly, the norm of the evolution generator, previously dominated by the larger term $\mathcal{H}_0$, is now equal to the norm of the smaller term $\mathcal{H}_1(t)$. This is most easily seen for the Frobenius norm (Sect. 1.4.5):

$$\left\|\mathcal{H}_1^{\mathrm{R}}(t)\right\|_{\mathrm{F}}^2 = \mathrm{Tr}\left(\mathrm{e}^{+i\mathcal{H}_0 t}\mathcal{H}_1(t)\mathrm{e}^{-i\mathcal{H}_0 t}\mathrm{e}^{+i\mathcal{H}_0 t}\mathcal{H}_1(t)\mathrm{e}^{-i\mathcal{H}_0 t}\right) = \left\|\mathcal{H}_1(t)\right\|_{\mathrm{F}}^2 \tag{4.64}$$

This will have a positive effect on the accuracy of the approximations that we will later derive. Secondly, because (by our assumption) the "interesting" dynamics under $\mathcal{H}_1(t)$ is much slower than the dynamics under $\mathcal{H}_0$, major simplifications will be obtained below by averaging $\mathcal{H}_1^{\mathrm{R}}(t)$ over the period of $\mathcal{H}_0$.

The transformations in Eqs. (4.60)–(4.63) are called *interaction representation transformations* [95]. When $\mathbf{H}_0$ is Zeeman interaction, they are historically called *rotating frame transformations* [96]—this is because $\mathfrak{su}(2)$ is isomorphic to $\mathfrak{so}(3)$, and $\mathfrak{so}(3)$ generates the rotation group (Sect. 1.6.4). In this context, the original Schrödinger representation of the problem is called *laboratory frame picture*.

4.3.2 Matrix Logarithm Method

In practical magnetic resonance experiments, the objective is to accomplish a specific transformation of the state space—that is, to perform an action by some effective Hamiltonian. A pertinent question is, therefore: *what is the time-independent Hamiltonian* $\overline{\mathbf{H}}$ *that would produce the same effect as a time-dependent Hamiltonian* $\mathbf{H}(t)$ *if it acts for the same period of time?*

$$\exp\left(-i\overline{\mathbf{H}}T\right) = \overleftarrow{\exp}\left(-i\int_0^T \mathbf{H}(t)\mathrm{d}t\right) \tag{4.65}$$

The time T in this context is either the duration of the experiment, or the period of some periodic process: electromagnetic irradiation, Larmor precession, sample spinning, or pulse sequence.

Efficient numerical methods have recently emerged [97] that answer this question directly, by computing the *principal value* (the one with the smallest norm) of the logarithm of the propagator matrix:

$$\overline{\mathbf{H}} = \frac{i}{T}\widehat{ln}\left(\overleftarrow{\exp}\left(-i\int_0^T \mathbf{H}(t)\mathrm{d}t\right)\right) \tag{4.66}$$

where the halo over the logarithm indicates the principal value. In practice, the logarithm is computed using the inverse scaling and squaring method that essentially carries out numerical matrix exponentiation (Sect. 4.9.5) in reverse.

As an example, consider the task of finding the effective Hamiltonian over the period of the rotating frame when the "small and interesting" part $\mathbf{H}_1$ is time independent. This is often the first stage in setting up magnetic resonance simulations. The laboratory frame solution is

$$\boldsymbol{\rho}(t) = \exp(-i(\mathbf{H}_0 + \mathbf{H}_1)t)\boldsymbol{\rho}(0)\exp(+i(\mathbf{H}_0 + \mathbf{H}_1)t) \tag{4.67}$$

After the definition of the rotating frame $\boldsymbol{\sigma}(t) = \exp(+i\mathbf{H}_0 t)\boldsymbol{\rho}(t)\exp(-i\mathbf{H}_0 t)$ from Eq. (4.63) is used, we obtain the following evolution rule for the rotating frame density matrix:

$$\begin{aligned} &\boldsymbol{\sigma}(t) = \mathbf{U}(t)\boldsymbol{\sigma}(0)\mathbf{U}^\dagger(t), \quad \boldsymbol{\sigma}(0) = \boldsymbol{\rho}(0) \\ &\mathbf{U}(t) = \exp(+i\mathbf{H}_0 t)\exp(-i(\mathbf{H}_0 + \mathbf{H}_1)t) \end{aligned} \tag{4.68}$$

We are, therefore, looking for an effective Hamiltonian $\overline{\mathbf{H}}$ that would generate the rotating frame propagator $\mathbf{U}(T)$ at some specific point T in time:

$$\begin{gathered} \exp(-i\overline{\mathbf{H}}T) - \exp(+i\mathbf{H}_0 T)\exp(-i(\mathbf{H}_0 + \mathbf{H}_1)T) \\ \Downarrow \\ \overline{\mathbf{H}} - \frac{i}{T}\widehat{ln}[\exp(+i\mathbf{H}_0 T)\exp(-i(\mathbf{H}_0 + \mathbf{H}_1)T)] \end{gathered} \tag{4.69}$$

When T is the period of evolution under $\mathbf{H}_0$, the propagator $\exp(+i\mathbf{H}_0 T)$ is a unit matrix and thus:

$$\overline{\mathbf{H}} = \frac{i}{T}\widehat{ln}[\exp(-i(\mathbf{H}_0 + \mathbf{H}_1)T)] \tag{4.70}$$

in which we cannot cancel the exponential and the logarithm—the principal value of the logarithm need not be equal to the matrix that has gone into the exponential.

As numerical simulations go, Eq. (4.66) in general and Eq. (4.70), in particular, are the end of the story. However, there is considerable history around analytical expressions for effective Hamiltonians that we need to cover because they provide intellectual insight into the mechanics of spin dynamics experiments.

4.3.3 Baker-Campbell-Hausdorff Formula

An important question arising in the context of sequential application of matrix exponentials is about combining them. The general expansion of $\exp\{\mathbf{A}\}\exp\{\mathbf{B}\} = \exp\{\mathbf{C}\}$, where $\mathbf{C}$ is a linear combination of nested commutators of $\mathbf{A}$ and $\mathbf{B}$, is called the *Baker-Campbell-Hausdorff formula* [9–11]. The neatest expression for $\mathbf{C}$ was found by Dynkin [98], who combined the Taylor series for matrix exponential and logarithm:

$$\mathrm{e}^{\mathbf{A}}\mathrm{e}^{\mathbf{B}} = \sum_{p,q=0}^{\infty} \frac{\mathbf{A}^p\mathbf{B}^q}{p!q!}, \quad \ln(\mathbf{X}) = \sum_{k=1}^{\infty} (-1)^{k-1}\frac{1}{k}(\mathbf{X}-\mathbf{1})^k \tag{4.71}$$

and accounted for the fact that $\mathbf{A}$ and $\mathbf{B}$ do not commute:

$$\mathbf{C} = \ln\{\mathrm{e}^{\mathbf{A}}\mathrm{e}^{\mathbf{B}}\} = \sum_{k=1}^{\infty} (-1)^{k-1}\frac{1}{k}\sum \frac{\mathbf{A}^{p_1}\mathbf{B}^{q_1}\mathbf{A}^{p_2}\mathbf{B}^{q_2}\cdots\mathbf{A}^{p_k}\mathbf{B}^{q_k}}{p_1!q_1!p_2!q_2!\cdots p_k!q_k!} \tag{4.72}$$

where the inner sum is over all k-tuples of pairs of non-negative integers such that $p_i + q_i > 0$. This is an inconvenient expression because it breaks the Lie algebra by making use of associative products. Dynkin found a way of rearranging the inner sum into commutators:

$$\mathbf{C} = \sum_{k=1}^{\infty} (-1)^{k-1}\frac{1}{k}\sum \frac{[\mathbf{A}^{p_1}\mathbf{B}^{q_1}\mathbf{A}^{p_2}\mathbf{B}^{q_2}\cdots\mathbf{A}^{p_k}\mathbf{B}^{q_k}]}{\left(\sum_{n=1}^{k}(p_n+q_n)\right)\prod_{n=1}^{k} p_n!q_n!} \tag{4.73}$$

where the square bracket denotes the right-nested commutator:

$$\begin{aligned}&[\mathbf{A}^{p_1}\mathbf{B}^{q_1}\cdots\mathbf{A}^{p_k}\mathbf{B}^{q_k}] \\ &= \underbrace{[\mathbf{A},[\mathbf{A},\ldots[\mathbf{A},}_{p_1}\underbrace{[\mathbf{B},[\mathbf{B},\ldots[\mathbf{B},}_{q_1}\ldots\underbrace{[\mathbf{A},[\mathbf{A},\ldots[\mathbf{A},}_{p_k}\underbrace{[\mathbf{B},[\mathbf{B},\ldots\mathbf{B}]]}_{q_k}\ldots]]\end{aligned} \tag{4.74}$$

and the inner sum runs the same way as in Eq. (4.72). This is an awkward expression, but a computer may be tasked with collecting terms with the same matrix power [99], so that

$$\mathbf{C} = \sum_{m=1}^{\infty} \mathbf{C}_m(\mathbf{A},\mathbf{B}) \tag{4.75}$$

where $\mathbf{C}_m(\mathbf{A},\mathbf{B})$ are homogeneous Lie polynomials of degree m in $\mathbf{A}$ and $\mathbf{B}$. The first few terms are

$$\mathbf{C}_1 = \mathbf{A} + \mathbf{B},\ \mathbf{C}_2 = \frac{1}{2}[\mathbf{A}, \mathbf{B}]$$
$$\mathbf{C}_3 = \frac{1}{12}[\mathbf{A}, [\mathbf{A}, \mathbf{B}]] + \frac{1}{12}[\mathbf{B}, [\mathbf{B}, \mathbf{A}]] \tag{4.76}$$
$$\mathbf{C}_4 = \frac{1}{24}[\mathbf{A}, [\mathbf{B}, [\mathbf{B}, \mathbf{A}]]]$$

BCH series in the form of Eq. (4.75) always converges when $\|\mathbf{A}\|_2 + \|\mathbf{B}\|_2 < \pi$ ([99], Theorem 3.2), and may converge for larger values of $\|\mathbf{A}\|_2$ and $\|\mathbf{B}\|_2$ in special cases [99].

4.3.4 Zassenhaus Formula

The reverse problem of splitting a matrix exponential leads to the *Zassenhaus formula* (never published by its author, but referred to by Magnus [100]):

$$\exp(\mathbf{A} + \mathbf{B}) = \exp(\mathbf{A}) \exp(\mathbf{B}) \overrightarrow{\prod_{n=2..\infty}} \exp(\mathbf{Z}_n) \tag{4.77}$$

where the matrices $\mathbf{Z}_n$ are obtained by differentiating both sides of the formal expansion

$$\exp(\lambda(\mathbf{A} + \mathbf{B})) = \exp(\lambda\mathbf{A}) \exp(\lambda\mathbf{B}) \overrightarrow{\prod_{n=2..\infty}} \exp(\lambda^n \mathbf{Z}_n) \tag{4.78}$$

with respect to λ and then setting $\lambda = 1$—in practice, this is done using symbolic processing software [101]. The first few terms are

$$\begin{aligned} \mathbf{Z}_2 &= \frac{1}{2}[\mathbf{B}, \mathbf{A}], \quad \mathbf{Z}_3 = \frac{1}{3}[\mathbf{Z}_2, \mathbf{A} + 2\mathbf{B}], \\ \mathbf{Z}_4 &= \frac{1}{12}[[\mathbf{Z}_2, \mathbf{A}], \mathbf{A}] + \frac{1}{4}[[\mathbf{Z}_2, \mathbf{A} + \mathbf{B}], \mathbf{B}] \end{aligned} \tag{4.79}$$

The product series in Eq. (4.77) always converges when $\|\mathbf{A}\|_2 + \|\mathbf{B}\|_2 < 1$, and may converge in other locations for which the criteria are more complicated [101].

4.3.5 Directional Taylor Expansion

Returning now to the quantum mechanical context and notation, consider again the situation where the Hamiltonian is a sum of a "large boring" term $\mathbf{H}_0$ and the "small interesting" term $\mathbf{H}_1$. Interaction representation transformation in Eq. (4.70) is then a special case of a perturbation problem:

$$\overline{\mathbf{H}} = \frac{i}{T} \widehat{ln}[\exp(-i(\mathbf{H}_0 + \alpha \mathbf{H}_1)T)] \tag{4.80}$$

where the evolution generator $\mathbf{H}_0$ is perturbed by a step α in the direction $\mathbf{H}_1$, and T is again the period of $\exp(\pm i\mathbf{H}_0 t)$. By our assumption, $\|\mathbf{H}_1\| \ll \|\mathbf{H}_0\|$, and therefore a Taylor expansion is called for, with respect to the step length parameter α:

$$\overline{\mathbf{H}}(\alpha) = \frac{i}{T} \widehat{ln}\left(e^{-i(\mathbf{H}_0 + \alpha \mathbf{H}_1)T}\right) = \frac{i}{T} \sum_{n=1}^{\infty} \frac{\partial^n}{\partial \alpha^n} \widehat{ln}\left(e^{-i(\mathbf{H}_0 + \alpha \mathbf{H}_1)T}\right)\Bigg|_{\alpha=0} \frac{\alpha^n}{n!} \tag{4.81}$$

The logarithm disappears after the first differentiation. After setting $\alpha = 1$, we obtain simple product rule expressions for the perturbative corrections to the effective Hamiltonian in which nested commutators do not occur, and the number of matrix terms is linear with respect to the approximation order [102]:

$$\overline{\mathbf{H}} = \frac{i}{T} \sum_{n=1}^{\infty} \frac{1}{n} \sum_{k=1}^{n} \frac{\mathbf{D}_{n-k}^{\dagger} \mathbf{D}_k}{(k-1)!(n-k)!}, \quad \mathbf{D}_k = \frac{\partial^k e^{-i(\mathbf{H}_0 + \alpha \mathbf{H}_1)T}}{\partial \alpha^k}\Bigg|_{\alpha=0} \tag{4.82}$$

The directional derivatives $\mathbf{D}_k$ of the matrix exponential are most conveniently computed (Sect. 8.2.4) using auxiliary matrix methods. The first few terms are

$$\begin{aligned} \overline{\mathbf{H}}^{(1)} &= \frac{i}{T} \mathbf{D}_0^{\dagger} \mathbf{D}_1 \\ \overline{\mathbf{H}}^{(2)} &= \frac{i}{2T} \left(\mathbf{D}_1^{\dagger} \mathbf{D}_1 + \mathbf{D}_0^{\dagger} \mathbf{D}_2\right) \\ \overline{\mathbf{H}}^{(3)} &= \frac{i}{6T} \left(\mathbf{D}_2^{\dagger} \mathbf{D}_1 + 2\mathbf{D}_1^{\dagger} \mathbf{D}_2 + \mathbf{D}_0^{\dagger} \mathbf{D}_3\right) \end{aligned} \tag{4.83}$$

Although it works best for small values of $\|\mathbf{H}_1\|$, this expansion has an infinite convergence radius. It is useful when the matrix logarithm in Eq. (4.70) is too expensive and infinite order expansion is not required.

4.3.6 Magnus and Fer Expansions

Consider the general case of a linear equation of motion

$$\mathbf{y}'(t) = \mathbf{A}(t)\mathbf{y}(t) \tag{4.84}$$

with a complex-valued state vector $\mathbf{y}(t)$, a complex-valued evolution generator $\mathbf{A}(t)$, and real-valued time. One form of the solution was proposed by Wilhelm Magnus in 1954 [100]:

$$\mathbf{y}(t) = \exp(\boldsymbol{\Omega}(t))\mathbf{y}(0) = \exp(\boldsymbol{\Omega}_1(t) + \boldsymbol{\Omega}_2(t) + \ldots)\mathbf{y}(0) \tag{4.85}$$

and another by Francis Fer in 1958 [103]:

$$\mathbf{y}(t) = \exp(\mathbf{F}_1(t))\exp(\mathbf{F}_2(t)) \cdots \mathbf{y}(0) \tag{4.86}$$

Magnus expansion converges in the sense that the norm of $\boldsymbol{\Omega}_n(t)$ gets closer to zero as n is increased, and Fer expansion converges in the sense that the residual propagator $\mathbf{U}_n(t)$ in

$$\mathbf{U}(t) = \exp[\mathbf{F}_1(t)]\exp[\mathbf{F}_2(t)]...\exp[\mathbf{F}_n(t)]\mathbf{U}_n(t) \tag{4.87}$$

gets closer to the unit matrix. Truncated forms of both expansions respect the Lie algebraic structure of the problem: the evolution remains symplectic in classical mechanics, and unitary in quantum mechanics. For specific physical systems, generators of different orders in Magnus and Fer expansions may be interpreted (Sect. 4.3.8) as effective interactions with a physical meaning.

Substitution of the Fer expansion into Eq. (4.84) yields, after a lengthy exercise in calculus [103], the following recursive expression for the individual terms:

$$\begin{aligned} \mathbf{F}_n(t) &= \int_0^t \mathbf{A}_{n-1}(s)\mathrm{d}s, \quad \mathbf{A}_0(t) = \mathbf{A}(t) \\ \mathbf{A}_n(t) &= \sum_{k=1}^{\infty} \frac{(-1)^k k}{(k+1)!}[\mathbf{F}_n(t), \mathbf{A}_{n-1}(t)]_k \end{aligned} \tag{4.88}$$

where $[\mathbf{X}, \mathbf{Y}]_k$ denotes a k-fold nested commutator $[\mathbf{X}, ...[\mathbf{X}, \mathbf{Y}]]$. Another lengthy derivation [104], yields the following differential equation for the Magnus generator:

$$\frac{\mathrm{d}}{\mathrm{d}t}\boldsymbol{\Omega}(t) = \mathbf{A}(t) + \sum_{k=1}^{\infty} \frac{(-1)^k B_k}{k!}[\boldsymbol{\Omega}(t), \mathbf{A}(t)]_k \tag{4.89}$$

where B_k are Bernoulli's numbers. This expression yields useful product quadratures (Sect. 4.9); when it is partitioned into terms involving different powers of $\mathbf{A}(t)$, a recursive expression is obtained for the individual terms of the Magnus expansion in Eq. (4.85):

$$\begin{aligned}
\boldsymbol{\Omega}_1(t) &= \int_0^t \mathbf{A}(s)\mathrm{d}s, & \boldsymbol{\Omega}_n(t) &= \sum_{j=1}^{n-1} \frac{B_j}{j!} \int_0^t \mathbf{S}_n^{(j)}(s)\mathrm{d}s \\
\mathbf{S}_n^{(j)}(t) &= \sum_{m=1}^{n-j} \left[\boldsymbol{\Omega}_m(t), \mathbf{S}_{n-m}^{(j-1)}(t)\right], & & 2 \le j \le n-1 \\
\mathbf{S}_n^{(1)}(t) &= [\boldsymbol{\Omega}_{n-1}(t), \mathbf{A}(t)], & \mathbf{S}_n^{(n-1)}(t) &= [\boldsymbol{\Omega}_1(t), \mathbf{A}(t)]_{n-1}
\end{aligned} \tag{4.90}$$

This expansion converges [104] when $\int_0^t \|\mathbf{A}(s)\|_2 ds < \pi$; the first three terms are

$$\begin{aligned}
\boldsymbol{\Omega}_1(t) &= \int_0^t \mathbf{A}(t_1)\mathrm{d}t_1, \quad \boldsymbol{\Omega}_2(t) = \frac{1}{2}\int_0^t \mathrm{d}t_1 \int_0^{t_1} \mathrm{d}t_2 [\mathbf{A}(t_1), \mathbf{A}(t_2)] \\
\boldsymbol{\Omega}_3(t) &= \frac{1}{6}\int_0^t \mathrm{d}t_1 \int_0^{t_1} \mathrm{d}t_2 \int_0^{t_3} \mathrm{d}t_3 ([\mathbf{A}(t_1), [\mathbf{A}(t_2), \mathbf{A}(t_3)]] + [\mathbf{A}(t_3), [\mathbf{A}(t_2), \mathbf{A}(t_1)]])
\end{aligned} \tag{4.91}$$

and the general expression for the subsequent terms is

$$\boldsymbol{\Omega}_n(t) = \sum_{j=1}^{n=1} \frac{B_j}{j!} \sum_{\substack{k_1+\ldots+k_j=n-1 \\ k_1 \ge 1,\ldots,k_j \ge 1}} \int_0^t \left[\boldsymbol{\Omega}_{k_1}(s), \left[\boldsymbol{\Omega}_{k_2}(s), \left[\ldots, \left[\boldsymbol{\Omega}_{k_j}(s), \mathbf{A}(s)\right]\right]\right]\ldots\right]\mathrm{d}s \tag{4.92}$$

4.3.7 Combinations and Corollaries

Other relations may be obtained by combining those discussed above. This section contains, in no particular order, some results that are useful in the spin dynamics context.

1. Baker-Campbell-Hausdorff series may be written in the following form:

$$\ln\left(\mathrm{e}^{\mathbf{A}}\mathrm{e}^{\mathbf{B}}\right) = \mathbf{A} + \mathbf{B} - \int_0^1 \mathrm{d}t \sum_{n=1}^{\infty} \frac{1}{n(n+1)} \left(\mathbf{1} - \mathrm{e}^{\mathrm{ad}_{\mathbf{A}}} \mathrm{e}^{t\,\mathrm{ad}_{\mathbf{B}}}\right)^n \mathbf{B} \tag{4.93}$$

This is useful for generating high-order BCH expansions numerically.

2. A propagator $\exp(\mathbf{A})$ may be castled with another propagator $\exp(\mathbf{B})$ using

$$\begin{aligned}&\exp\{\mathbf{A}\}\exp\{\mathbf{B}\}=\\&\exp\left\{\mathbf{B}+[\mathbf{A},\mathbf{B}]+\frac{1}{2!}[\mathbf{A},[\mathbf{A},\mathbf{B}]]+\frac{1}{3!}[\mathbf{A},[\mathbf{A},[\mathbf{A},\mathbf{B}]]]+\cdots\right\}\exp\{\mathbf{A}\}\end{aligned} \tag{4.94}$$

where the series has an infinite convergence radius.

3. In finite-dimensional Lie algebras, nested commutators eventually start repeating; this allows some BCH expansions to be summed up into closed forms. When $[\mathbf{A},\mathbf{B}]=u\mathbf{A}+v\mathbf{B}+c\mathbf{1}$:

$$\begin{aligned}\ln\left(\mathrm{e}^{\mathbf{A}}\mathrm{e}^{\mathbf{B}}\right)&=\mathbf{A}+\mathbf{B}+f(u,v)[\mathbf{A},\mathbf{B}]\\f(u,v)&=\frac{1}{\mathrm{e}^{-u}-\mathrm{e}^{-v}}\left(\frac{1-\mathrm{e}^{-u}}{u}-\frac{1-\mathrm{e}^{-v}}{v}\right)\end{aligned} \tag{4.95}$$

and the relation in Eq. (4.94) becomes [105]

$$\begin{aligned}\exp\{\mathbf{A}\}\exp\{\mathbf{B}\}&=\exp\left\{\mathbf{B}+v^{-1}(1-\mathrm{e}^{-v})[\mathbf{A},\mathbf{B}]\right\}\exp\{\mathbf{A}\}\\\exp\{\mathbf{B}\}\exp\{\mathbf{A}\}&=\exp\left\{\mathbf{A}+u^{-1}(\mathrm{e}^{u}-1)[\mathbf{A},\mathbf{B}]\right\}\exp\{\mathbf{B}\}\end{aligned} \tag{4.96}$$

This is useful when dealing with the interference of Zeeman and interaction Hamiltonians.

4. The following trace relations may be obtained from the BCH formula:

$$\mathrm{Tr}\left[\ln\left(\mathrm{e}^{\mathbf{A}}\mathrm{e}^{\mathbf{B}}\right)\right]=\mathrm{Tr}[\mathbf{A}]+\mathrm{Tr}[\mathbf{B}] \tag{4.97}$$

$$\mathrm{Tr}\left[e^{\mathbf{A}+\mathbf{B}}\right]\leq\mathrm{Tr}\left[e^{\mathbf{A}}e^{\mathbf{B}}\right]\quad\text{(Golden - Thompson inequality)} \tag{4.98}$$

They are useful for testing numerical implementations.

5. The following expansion for the adjoint map

$$\mathrm{e}^{\mathbf{A}}\mathbf{B}\mathrm{e}^{-\mathbf{A}}=\mathbf{B}+[\mathbf{A},\mathbf{B}]+\frac{1}{2!}[\mathbf{A},[\mathbf{A},\mathbf{B}]]+\frac{1}{3!}[\mathbf{A},[\mathbf{A},[\mathbf{A},\mathbf{B}]]]+\cdots \tag{4.99}$$

will come useful when we calculate (Section 8.2.4) directional derivatives of matrix exponentials.

6. The expression connecting an element of a group direct product to its generator

$$\exp\left[\mathbf{A}\otimes\mathbf{1}^{\mathrm{B}}+\mathbf{1}^{\mathrm{A}}\otimes\mathbf{B}\right]=\exp[\mathbf{A}]\otimes\exp[\mathbf{B}] \tag{4.100}$$

has already been used in Section 2.5, and will be used again when we separate system and bath degrees of freedom in various relaxation theories (Chap. 6).

4.3.8 Average Hamiltonian Theory

Consider a system obeying Schrödinger's equation with a time-dependent Hamiltonian $\mathbf{H}(t)$. The *average Hamiltonian* $\overline{\mathbf{H}}$ over a time interval t is a constant evolution generator that produces the same propagator at time t as the time-ordered exponential [106]:

$$\exp(-i\overline{\mathbf{H}}t) = \overleftarrow{\exp}\left(-i\int_0^t \mathbf{H}(t_1)\mathrm{d}t_1\right) \tag{4.101}$$

A numerical route using matrix logarithms has already been discussed in Sect. 4.3.2:

$$\overline{\mathbf{H}} = \frac{i}{t}\widehat{\ln}\left\{\overleftarrow{\exp}\left[-i\int_0^t \mathbf{H}(t_1)\mathrm{d}t_1\right]\right\} \tag{4.102}$$

However, analytical insights are lost in purely numerical evaluations. A more intellectually satisfying route is to use Magnus expansion (Sect. 4.3.6) because its terms may have physical interpretations:

$$\begin{aligned}
&\overline{\mathbf{H}}^{(0)} = \mathbf{0}, \quad \overline{\mathbf{H}}^{(1)} = \frac{1}{t}\int_0^t \mathbf{H}(t_1)dt_1, \quad \overline{\mathbf{H}}^{(2)} = -\frac{i}{2t}\int_0^t dt_1\int_0^{t_1}[\mathbf{H}(t_1),\mathbf{H}(t_2)]dt_2 \\
&\overline{\mathbf{H}}^{(3)} = -\frac{1}{6t}\int_0^t \mathrm{d}t_1\int_0^{t_1}\mathrm{d}t_2\int_0^{t_2}([\mathbf{H}(t_1),[\mathbf{H}(t_2),\mathbf{H}(t_3)]] + [[\mathbf{H}(t_1),\mathbf{H}(t_2)],\mathbf{H}(t_3)])\mathrm{d}t_3 \\
&\overline{\mathbf{H}}^{(4)} = \frac{i}{12t}\int_0^t \mathrm{d}t_1\int_0^{t_1}\mathrm{d}t_2\int_0^{t_2}\mathrm{d}t_3\int_0^{t_3}\mathrm{d}t_4\begin{pmatrix}[[[\mathbf{H}(t_1),\mathbf{H}(t_2)],\mathbf{H}(t_3)],\mathbf{H}(t_4)] + \\ [\mathbf{H}(t_1),[[\mathbf{H}(t_2),\mathbf{H}(t_3)],\mathbf{H}(t_4)]] + \\ [\mathbf{H}(t_1),[\mathbf{H}(t_2),[\mathbf{H}(t_3),\mathbf{H}(t_4)]]] + \\ [\mathbf{H}(t_2),[\mathbf{H}(t_3),[\mathbf{H}(t_4),\mathbf{H}(t_1)]]]\end{pmatrix}
\end{aligned} \tag{4.103}$$

The average Hamiltonian $\overline{\mathbf{H}} = \overline{\mathbf{H}}^{(1)} + \overline{\mathbf{H}}^{(2)} + \ldots$ satisfies Eq. (4.101) at time t, but not necessarily at other times. When the Hamiltonian is piecewise-constant (e.g. composite pulses) or the time dependence has a short expansion in scalar functions and constant operators (e.g. magic angle spinning):

$$\mathbf{H}(t) = f_1(t)\mathbf{H}_1 + f_2(t)\mathbf{H}_2 + \ldots \tag{4.104}$$

the integrals in Eq. (4.103) are analytical and commutators are straightforward. Machine algebra systems that support piecewise-analytical functions, such as *Mathematica*, are recommended.

4.3.8.1 Zeeman Rotating Frames

A good example of average Hamiltonian theory treatment is the Zeeman rotating frame transformation in magnetic resonance, where the reference Hamiltonian is simple, and the interactions have short spherical tensor expansions. An interacting system of two spin-1/2 particles has the following Hamiltonian:

$$\mathbf{H} = \omega_{\mathrm{L}}\mathbf{L}_{\mathrm{Z}} + \omega_{\mathrm{S}}\mathbf{S}_{\mathrm{Z}} + \mathbf{L}\cdot\mathbf{A}\cdot\mathbf{S} \tag{4.105}$$

where $\mathbf{A}$ is a real, symmetric, and possibly time-dependent 3×3 matrix. In high-field magnetic resonance, the first two terms dominate the Hamiltonian; a popular partitioning is

$$\begin{aligned}\mathbf{H}_0 &= \omega_{\mathrm{ref}}^{(\mathrm{L})}\mathbf{L}_{\mathrm{Z}} + \omega_{\mathrm{ref}}^{(\mathrm{S})}\mathbf{S}_{\mathrm{Z}} \\ \mathbf{H}_1 &= \omega_{\mathrm{off}}^{(\mathrm{L})}\mathbf{L}_{\mathrm{Z}} + \omega_{\mathrm{off}}^{(\mathrm{S})}\mathbf{S}_{\mathrm{Z}} + \mathbf{L}\cdot\mathbf{A}\cdot\mathbf{S} \\ &= \omega_{\mathrm{off}}^{(\mathrm{L})}\mathbf{L}_{\mathrm{Z}} + \omega_{\mathrm{off}}^{(\mathrm{S})}\mathbf{S}_{\mathrm{Z}} + a_{\mathrm{iso}}\mathbf{L}\cdot\mathbf{S} + \sum_{m=-2}^{2} a_m \mathbf{T}_{2,m}^{(\mathrm{L,S})}\end{aligned} \tag{4.106}$$

where $\omega_{\mathrm{ref}}^{(\mathrm{L,S})}$ are large reference frequencies (for example, 600 MHz for protons in a 14.1 T magnet), $\omega_{\mathrm{off}}^{(\mathrm{L,S})}$ are smaller offset frequencies (for example, due to chemical shielding), a_{iso} is the isotropic part of the interaction. Irreducible spherical tensor operators $\mathbf{T}_{2,m}^{(\mathrm{L,S})}$ and their coefficients are defined in Sect. 3.3.2; because they originate in the rotation group, they are eigen operators of the adjoint action by $\mathbf{L}_{\mathrm{Z}}$ and $\mathbf{L}_{\mathrm{Z}} + \mathbf{S}_{\mathrm{Z}}$. The relevant commutators are:

$$\begin{gathered}[\mathbf{L}_{\mathrm{Z}}, \mathbf{L}_{\mathrm{Z}}\mathbf{S}_{\mathrm{Z}}] = [\mathbf{S}_{\mathrm{Z}}, \mathbf{L}_{\mathrm{Z}}\mathbf{S}_{\mathrm{Z}}] = 0 \\ [\mathbf{L}_{\mathrm{Z}} + \mathbf{S}_{\mathrm{Z}}, \mathbf{L}_{\mathrm{X}}\mathbf{S}_{\mathrm{X}} + \mathbf{L}_{\mathrm{Y}}\mathbf{S}_{\mathrm{Y}} + \mathbf{L}_{\mathrm{Z}}\mathbf{S}_{\mathrm{Z}}] = 0 \\ \left[\mathbf{S}_{\mathrm{Z}}, \mathbf{T}_{l,m}^{(\mathrm{S})}\right] = m\mathbf{T}_{l,m}^{(\mathrm{S})}, \ \left[\mathbf{L}_{\mathrm{Z}} + \mathbf{S}_{\mathrm{Z}}, \mathbf{T}_{l,m}^{(\mathrm{L,S})}\right] = m\mathbf{T}_{l,m}^{(\mathrm{L,S})}\end{gathered} \tag{4.107}$$

where the rotating frame generators are highlighted in blue. Application of Eq. (4.99) then results in the following rotating frame transformations:

$$\begin{gathered}e^{i\omega\mathbf{S}_{\mathrm{Z}}t}\mathbf{T}_{lm}^{(\mathrm{S})}e^{-i\omega\mathbf{S}_{\mathrm{Z}}t} = e^{im\omega t}\mathbf{T}_{lm}^{(\mathrm{S})} \\ e^{i\omega(\mathbf{L}_{\mathrm{Z}} + \mathbf{S}_{\mathrm{Z}})t}\mathbf{T}_{lm}^{(\mathrm{L,S})}e^{-i\omega(\mathbf{L}_{\mathrm{Z}} + \mathbf{S}_{\mathrm{Z}})t} = e^{im\omega t}\mathbf{T}_{lm}^{(\mathrm{L,S})}\end{gathered} \tag{4.108}$$

where the right-hand sides only contain scalar functions of time; this makes the integrals in Eq. (4.103) easy to evaluate. We will now present a simple example of this approach.

4.3.8.2 Secular Coupling

Consider first a situation where the two spins in Eq. (4.106) are of the same type, for example, ^{1}H in nuclear magnetic resonance. Then the reference frequencies are the same: $\omega_{\text{ref}} = -\gamma B_0$ (typically hundreds of MHz), and the offset frequencies $\omega_{\text{off}}^{(\text{L,S})}$ come from chemical shifts (typically below 100 kHz). Spin–spin couplings in NMR systems are also below 25 kHz. We, therefore, partition the Hamiltonian so that $\mathbf{H}_0$ contains the common reference frequency, and $\mathbf{H}_1$ contains the offsets and the interaction:

$$\begin{aligned} &\mathbf{H} = \mathbf{H}_0 + \mathbf{H}_1, \quad \mathbf{H}_0 = \omega_{\text{ref}}(\mathbf{L}_\text{Z} + \mathbf{S}_\text{Z}) \\ &\mathbf{H}_1 = \omega_{\text{off}}^{(\text{L})}\mathbf{L}_\text{Z} + \omega_{\text{off}}^{(\text{S})}\mathbf{S}_\text{Z} + a_{\text{iso}}\mathbf{L}\cdot\mathbf{S} + \sum_{m=-2}^{2} a_m \mathbf{T}_{2,m}^{(\text{L,S})} \end{aligned} \tag{4.109}$$

Interaction representation transformation with respect to $\mathbf{H}_0$ is applied using Eq. (4.108):

$$\begin{aligned} \mathbf{H}_1^{\text{R}}(t) &= e^{i\omega_{\text{ref}}(\mathbf{L}_\text{Z}+\mathbf{S}_\text{Z})t}\left(\omega_{\text{off}}^{(\text{L})}\mathbf{L}_\text{Z} + \omega_{\text{off}}^{(\text{L})}\mathbf{S}_\text{Z} + a_{\text{iso}}\mathbf{L}\cdot\mathbf{S} + \sum_{m=-2}^{2} a_{2,m}^{(\text{L,S})}\mathbf{T}_{2,m}^{(\text{L,S})}\right)e^{-i\omega_{\text{ref}}(\mathbf{L}_\text{Z}+\mathbf{S}_\text{Z})t} \\ &= \left(\omega_{\text{off}}^{(\text{L})}\mathbf{L}_\text{Z} + \omega_{\text{off}}^{(\text{S})}\mathbf{S}_\text{Z} + a_{\text{iso}}\mathbf{L}\cdot\mathbf{S} + a_{2,0}^{(\text{L,S})}\mathbf{T}_{2,0}^{(\text{L,S})}\right) + \sum_{m\neq 0} a_{2,m}^{(\text{L,S})} e^{im\omega_{\text{ref}}t}\mathbf{T}_{2,m}^{(\text{L,S})} \end{aligned} \tag{4.110}$$

When the average Hamiltonian is computed over the period $T = 2\pi/\omega_{\text{ref}}$ of $\mathbf{H}_0$, the first-order term in Eq. (4.103) is just the content of the big round brackets in the second line of Eq. (4.110):

$$\overline{\mathbf{H}}_1^{\text{R}(1)} = \omega_{\text{off}}^{(\text{L})}\mathbf{L}_\text{Z} + \omega_{\text{off}}^{(\text{S})}\mathbf{S}_\text{Z} + a_{\text{iso}}\mathbf{L}\cdot\mathbf{S} + a_{2,0}^{(\text{L,S})}\mathbf{T}_{2,0}^{(\text{L,S})} \tag{4.111}$$

because the oscillatory terms integrate to zero. For this approximation to be valid, firstly the lenient overall convergence condition underneath Eq. (4.90) must be satisfied:

$$\int_0^T \left\|\mathbf{H}_1^{\text{R}}(t)\right\|_2 \text{d}t \le T\|\mathbf{H}_1\|_2 < \pi \quad \Rightarrow \quad \|\mathbf{H}_1\|_2 < \omega_{\text{ref}}/2 \tag{4.112}$$

and then the stricter condition that the norm of the second-order term is negligible. Because the second-order term may be expensive, a cheap upper bound is recommended, for example:

$$\begin{aligned}\left\|\overline{\mathbf{H}}_1^{\mathrm{R}(2)}\right\|_2 &\leq \frac{1}{2T}\int_0^T dt_1 \int_0^{t_1} \left\|\left[\mathbf{H}_1^{\mathrm{R}}(t_1), \mathbf{H}_1^{\mathrm{R}}(t_2)\right]\right\|_2 \mathrm{d}t_2 \leq \frac{T}{2}\|\mathbf{H}_1\|_2^2 \\ &= \pi\omega_{\mathrm{ref}}^{-1}\|\mathbf{H}_1\|_2^2 \leq \pi\omega_{\mathrm{ref}}^{-1}\|\mathbf{H}_1\|_1\|\mathbf{H}_1\|_\infty\end{aligned} \tag{4.113}$$

Typical orders of magnitude in high-field ^{1}H NMR are GHz for ω_{ref} and up to 25 kHz for $\|\mathbf{H}_1\|_2$, with the instrumental resolution around 1 Hz. The estimate confirms that the first-order term is sufficient in this case.

The approximation wherein only those Hamiltonian terms that are stationary under the interaction representation are retained is called *secular approximation* [107]; the interaction terms remaining in $\overline{\mathbf{H}}_1^{\mathrm{R}(1)}$ are called *secular terms*. This approximation is useful because: (a) the secular Hamiltonian is simpler than the original interaction Hamiltonian in Eq. (4.109) had been; (b) the dynamics generated by the secular Hamiltonian in the rotating frame is much slower than the dynamics under the laboratory frame Hamiltonian had been—we are only seeing the relative motion of the spins in isolation from their rapid Zeeman precession.

4.3.8.3 Pseudosecular Coupling

Consider now a more subtle case where only one of the two Zeeman interactions is much stronger than the spin–spin coupling. This can happen in ESR spectroscopy: in a 0.33 T magnet, proton Zeeman frequency is only 14 MHz—comparable to hyperfine interactions and therefore not appropriate for inclusion into $\mathbf{H}_0$. In the case of an electron-nuclear system with both spins 1/2, the $\mathbf{H}_0$ part of the Hamiltonian is then only the Zeeman interaction of the electron spin:

$$\begin{aligned}\mathbf{H} &= \mathbf{H}_0 + \mathbf{H}_1, \quad \mathbf{H}_0 = \omega_{\mathrm{ref}}\mathbf{E}_\mathrm{Z} \\ \mathbf{H}_1 &= \omega_{\mathrm{off}}\mathbf{E}_\mathrm{Z} + \omega_\mathrm{N}\mathbf{N}_\mathrm{Z} + a_{\mathrm{iso}}\mathbf{E}\cdot\mathbf{N} + \sum_{m=-2}^{2} a_{2,m}^{(\mathrm{E,N})}\mathbf{T}_{2,m}^{(\mathrm{E,N})}\end{aligned} \tag{4.114}$$

where $\mathbf{E}$ and $\mathbf{N}$ operators refer to electron and nuclear spin, $\omega_\mathrm{E} = \omega_{\mathrm{ref}} + \omega_{\mathrm{off}}$ and ω_N are electron and nuclear Zeeman frequencies, a_{iso} is the isotropic hyperfine coupling, and $a_{2,m}^{(\mathrm{E,N})}$ are the irreducible components of the anisotropic hyperfine coupling.

To make progress, we will now use the $\pm$ form of the isotropic hyperfine interaction term $\mathbf{L}_\mathrm{X}\mathbf{S}_\mathrm{X} + \mathbf{L}_\mathrm{Y}\mathbf{S}_\mathrm{Y} = \frac{1}{2}(\mathbf{L}_+\mathbf{S}_- + \mathbf{L}_-\mathbf{S}_+)$, and also write the irreducible spherical tensor operators (Sect. 3.3.2) out explicitly:

$$\begin{aligned}\mathbf{H}_1 &= \omega_{\text{off}}\mathbf{E}_Z + \omega_N\mathbf{N}_Z + a_{\text{iso}}\mathbf{E}_Z\mathbf{N}_Z + \frac{1}{2}a_{\text{iso}}(\mathbf{E}_+\mathbf{N}_- + \mathbf{E}_-\mathbf{N}_+)\\ &+ \frac{1}{2}a_{2,-2}^{(\text{E,N})}\mathbf{E}_-\mathbf{N}_- + \frac{1}{2}a_{2,2}^{(\text{E,N})}\mathbf{E}_+\mathbf{N}_+\\ &+ \frac{1}{2}a_{2,-1}^{(\text{E,N})}(\mathbf{E}_Z\mathbf{N}_- + \mathbf{E}_-\mathbf{N}_Z) - \frac{1}{2}a_{2,1}^{(\text{E,N})}(\mathbf{E}_Z\mathbf{N}_+ + \mathbf{E}_+\mathbf{N}_Z)\\ &+ a_{2,0}^{(\text{E,N})}\sqrt{\frac{2}{3}}\left(\mathbf{E}_Z\mathbf{N}_Z - \frac{1}{2}(\mathbf{E}_+\mathbf{N}_- + \mathbf{E}_-\mathbf{N}_+)\right)\end{aligned} \tag{4.115}$$

Raising and lowering operators have the following commutation relations with $\mathbf{S}_Z$:

$$[\mathbf{S}_Z, \mathbf{S}_\pm] = \pm\mathbf{S}_\pm \quad \Rightarrow \quad e^{i\omega\mathbf{S}_Z t}\mathbf{S}_\pm e^{-i\omega\mathbf{S}_Z t} = e^{\pm i\omega t}\mathbf{S}_\pm \tag{4.116}$$

An interaction representation transformation with respect to $\mathbf{H}_0 = \omega_{\text{ref}}\mathbf{E}_Z$ then yields

$$\begin{aligned}\mathbf{H}_1^{\text{R}}(t) &= \omega_{\text{off}}\mathbf{E}_Z + \omega_N\mathbf{N}_Z + a_{\text{iso}}\mathbf{E}_Z\mathbf{N}_Z + \frac{1}{2}a_{\text{iso}}\left(\mathrm{e}^{i\omega_{\text{ref}}t}\mathbf{E}_+\mathbf{N}_- + \mathrm{e}^{-i\omega_{\text{ref}}t}\mathbf{E}_-\mathbf{N}_+\right)\\ &+ \frac{1}{2}a_{2,-2}^{(\text{E,N})}\mathrm{e}^{-i\omega_{\text{ref}}t}\mathbf{E}_-\mathbf{N}_- + \frac{1}{2}a_{2,2}^{(\text{E,N})}\mathrm{e}^{i\omega_{\text{ref}}t}\mathbf{E}_+\mathbf{N}_+\\ &- \frac{1}{2}a_{2,1}^{(\text{E,N})}\left(\mathbf{E}_Z\mathbf{N}_+ + \mathrm{e}^{i\omega_{\text{ref}}t}\mathbf{E}_+\mathbf{N}_Z\right)\\ &+ \frac{1}{2}a_{2,-1}^{(\text{E,N})}\left(\mathbf{E}_Z\mathbf{N}_- + \mathrm{e}^{-i\omega_{\text{ref}}t}\mathbf{E}_-\mathbf{N}_Z\right)\\ &+ a_{2,0}^{(\text{E,N})}\sqrt{\frac{2}{3}}\left(\mathbf{E}_Z\mathbf{N}_Z - \frac{1}{4}\left(\mathrm{e}^{i\omega_{\text{ref}}t}\mathbf{E}_+\mathbf{N}_- + \mathrm{e}^{-i\omega_{\text{ref}}t}\mathbf{E}_-\mathbf{N}_+\right)\right)\end{aligned} \tag{4.117}$$

Taking the first-order average Hamiltonian theory integral in Eq. (4.103) over the period of ω_{ref} destroys the oscillating terms:

$$\begin{aligned}\overline{\mathbf{H}}_1^{\text{R}(1)} &= \omega_{\text{off}}\mathbf{E}_Z + \omega_N\mathbf{N}_Z + a_{\text{iso}}\mathbf{E}_Z\mathbf{N}_Z\\ &- \frac{1}{2}a_{2,1}^{(\text{E,N})}\mathbf{E}_Z\mathbf{N}_+ + \frac{1}{2}a_{2,-1}^{(\text{E,N})}\mathbf{E}_Z\mathbf{N}_- + a_{2,0}^{(\text{E,N})}\sqrt{\frac{2}{3}}\mathbf{E}_Z\mathbf{N}_Z\end{aligned} \tag{4.118}$$

and the requirement for the Hamiltonian to be Hermitian further dictates that $a_{-1} = -a_1$, so that

$$\overline{\mathbf{H}}_1^{\text{R}(1)} = \omega_{\text{off}}\mathbf{E}_Z + \omega_N\mathbf{N}_Z + \left(\sqrt{\frac{2}{3}}a_{2,0}^{(\text{E,N})} + a_{\text{iso}}\right)\mathbf{E}_Z\mathbf{N}_Z + a_{2,-1}^{(\text{E,N})}\mathbf{E}_Z\mathbf{N}_X \tag{4.119}$$

This form is useful in ESEEM, ENDOR, and DNP simulations; the term $\mathbf{E}_Z\mathbf{N}_X$ containing the transverse nuclear operator is called *pseudosecular* [108]. The same validity conditions apply as in Sect. 4.3.8.2; the reference frequency here is free electron Zeeman frequency.

4.3.8.4 Weak Coupling

The secular average Hamiltonian in Eq. (4.111) assumed that the spins belonged to the same type of particle (for example, both ^{1}H) and their Zeeman frequencies $\omega^{(L,S)}$ were, therefore, similar. Let us now consider the situation when the particles are different (for example, ^{1}H and ^{13}C). In a high-field magnet (>10 T), the difference between $\omega^{(L)}$ and $\omega^{(S)}$ would be hundreds of MHz. The Hamiltonian is

$$\begin{gathered}\mathbf{H}=\mathbf{H}_0+\mathbf{H}_1,\quad \mathbf{H}_0=\omega_{\mathrm{ref}}^{(L)}\mathbf{L}_Z+\omega_{\mathrm{ref}}^{(S)}\mathbf{S}_Z\\ \mathbf{H}_1=\omega_{\mathrm{off}}^{(L)}\mathbf{L}_Z+\omega_{\mathrm{off}}^{(S)}\mathbf{S}_Z+a_{\mathrm{iso}}\mathbf{L}\cdot\mathbf{S}+\sum_{m=-2}^{2}a_{2,m}^{(L,S)}\mathbf{T}_{2,m}^{(L,S)}\end{gathered}\tag{4.120}$$

where the presence of the reference and the offset parts again reflects experimental realities: the main chemistry-independent Zeeman frequency $\omega_{\mathrm{ref}}^{(L,S)}$ of the bare nucleus in vacuum and the electronic structure correction ω_{off}. In our present case, $\omega_{\mathrm{ref}}^{(L)}$ and $\omega_{\mathrm{ref}}^{(S)}$ are very different.

Because $\mathbf{L}_Z$ and $\mathbf{S}_Z$ commute, so do interaction representation transformations with respect to $\omega_{\mathrm{ref}}^{(L)}\mathbf{L}_Z$ and $\omega_{\mathrm{ref}}^{(S)}\mathbf{S}_Z$—they can be performed in any order. Noting the following commutation relations

$$\begin{gathered}[\mathbf{L}_Z,\mathbf{L}_Z\mathbf{S}_Z]=[\mathbf{S}_Z,\mathbf{L}_Z\mathbf{S}_Z]=0\\ \begin{cases}[\mathbf{L}_Z,\mathbf{L}_+\mathbf{S}_-]=+\mathbf{L}_+\mathbf{S}_-\\ [\mathbf{S}_Z,\mathbf{L}_+\mathbf{S}_-]=-\mathbf{L}_+\mathbf{S}_-\end{cases},\quad \begin{cases}[\mathbf{L}_Z,\mathbf{L}_-\mathbf{S}_+]=-\mathbf{L}_-\mathbf{S}_+\\ [\mathbf{S}_Z,\mathbf{L}_-\mathbf{S}_+]=+\mathbf{L}_-\mathbf{S}_+\end{cases}\end{gathered}\tag{4.121}$$

and going through the same procedures as in Sects. 4.3.8.2, 4.3.8.3 yields the following average Hamiltonian:

$$\overline{\mathbf{H}}_1^{R(1)}=\omega_{\mathrm{off}}^{(L)}\mathbf{L}_Z+\omega_{\mathrm{off}}^{(S)}\mathbf{S}_Z+\left(\sqrt{\frac{2}{3}}a_{2,0}^{(L,S)}+a_{\mathrm{iso}}\right)\mathbf{L}_Z\mathbf{S}_Z$$

The simple ZZ coupling term here is called *weak coupling* [109]. The same validity conditions apply as in Sect. 4.3.8.2, but with respect to the reference frequencies of each spin individually.

4.3.8.5 Monochromatic Irradiation

A popular way of manipulating spin systems is to apply oscillating magnetic fields close to a Zeeman transition frequency. In the case of a cosine modulation of an external magnetic field B_X applied on the X-axis of the laboratory frame of reference:

$$\mathbf{H}(t)=\omega_{\mathrm{ref}}^{(\mathrm{S})}\mathbf{S}_{\mathrm{Z}}+a_{\mathrm{ext}}\cos(\omega_{\mathrm{ext}}t)\mathbf{S}_{\mathrm{X}}+[\ldots] \tag{4.122}$$

where $\omega_{\mathrm{ref}}^{(\mathrm{S})}=-\gamma_{\mathrm{S}}B_{\mathrm{Z}}$ is the reference Zeeman frequency of the spin, $a_{\mathrm{ext}}=-\gamma_{\mathrm{S}}B_{\mathrm{X}}$ is the modulation depth of the Zeeman interaction with the oscillating external field, ω_{ext} is the frequency of that field, and $[\ldots]$ may contain other interactions and offsets.

In high-field magnetic resonance, this is an inconvenient Hamiltonian—because ω_{ext} is close to $\omega_{\mathrm{ref}}^{(\mathrm{S})}$, impractically short time steps may be needed in the solution of the equation of motion. The average Hamiltonian theory provides a workaround; consider an interaction representation with respect to $\omega_{\mathrm{ext}}\mathbf{S}_{\mathrm{Z}}$:

$$\begin{gathered}\mathbf{H}(t)=\mathbf{H}_0+\mathbf{H}_1(t),\quad \mathbf{H}_0=\omega_{\mathrm{ext}}\mathbf{S}_{\mathrm{Z}}\\ \mathbf{H}_1(t)=\left(\omega_{\mathrm{ref}}^{(\mathrm{S})}-\omega_{\mathrm{ext}}\right)\mathbf{S}_{\mathrm{Z}}+a_{\mathrm{ext}}\cos(\omega_{\mathrm{ext}}t)\mathbf{S}_{\mathrm{X}}+[\ldots]\end{gathered} \tag{4.123}$$

In the rotating frame Hamiltonian, the first term is unchanged because it commutes with $\mathbf{S}_{\mathrm{Z}}$ and the rest is transformed using Eq. (4.63):

$$\mathbf{H}_1^{\mathrm{R}}(t)=\left(\omega_{\mathrm{ref}}^{(\mathrm{S})}-\omega_{\mathrm{ext}}\right)\mathbf{S}_{\mathrm{Z}}+a_{\mathrm{ext}}\cos(\omega_{\mathrm{ext}}t)e^{+i\omega_{\mathrm{ext}}\mathbf{S}_{\mathrm{Z}}t}\mathbf{S}_{\mathrm{X}}e^{-i\omega_{\mathrm{ext}}\mathbf{S}_{\mathrm{Z}}t}+[\ldots]^{\mathrm{R}}(t) \tag{4.124}$$

After Eq. (4.99) is applied, the second term acquires a simple trigonometric form and something along the lines of Sects. 4.3.8.2–4.3.8.4 happens to the interactions in the square bracket:

$$\begin{gathered}a_{\mathrm{ext}}\cos(\omega_{\mathrm{ext}}t)\mathrm{e}^{+i\omega_{\mathrm{ext}}\mathbf{S}_{\mathrm{Z}}t}\mathbf{S}_{\mathrm{X}}\mathrm{e}^{-i\omega_{\mathrm{ext}}\mathbf{S}_{\mathrm{Z}}t}\\ =a_{\mathrm{ext}}\cos(\omega_{\mathrm{ext}}t)(\mathbf{S}_{\mathrm{X}}\cos(\omega_{\mathrm{ext}}t)+\mathbf{S}_{\mathrm{Y}}\sin(\omega_{\mathrm{ext}}t))\end{gathered} \tag{4.125}$$

$$[\ldots]^{\mathrm{R}}(t)=\mathrm{e}^{+i\omega_{\mathrm{ext}}\mathbf{S}_{\mathrm{Z}}t}[\ldots]\mathrm{e}^{-i\omega_{\mathrm{ext}}\mathbf{S}_{\mathrm{Z}}t} \tag{4.126}$$

After trigonometric simplifications are applied:

$$\mathbf{H}_1^{\mathrm{R}}(t)=\left(\omega_{\mathrm{ref}}^{(\mathrm{S})}-\omega_{\mathrm{ext}}\right)\mathbf{S}_{\mathrm{Z}}+\frac{a_{\mathrm{ext}}}{2}\mathbf{S}_{\mathrm{X}}+\frac{a_{\mathrm{ext}}}{2}(\mathbf{S}_{\mathrm{X}}\cos(2\omega_{\mathrm{ext}}t)+\mathbf{S}_{\mathrm{Y}}\sin(2\omega_{\mathrm{ext}}t))+[\ldots]^{\mathrm{R}}(t) \tag{4.127}$$

it becomes clear that, to first order in the average Hamiltonian theory, the average of the trigonometric terms over the period of ω_{ext} is zero:

$$\overline{\mathbf{H}}_1^{\mathrm{R}(1)}=\left(\omega_{\mathrm{ref}}^{(\mathrm{S})}-\omega_{\mathrm{ext}}\right)\mathbf{S}_{\mathrm{Z}}+\frac{a_{\mathrm{X}}}{2}\mathbf{S}_{\mathrm{X}}+\overline{[\ldots]}^{\mathrm{R}(1)} \tag{4.128}$$

and the content of $\overline{[\ldots]}^{\mathrm{R}(1)}$ has already been looked at in Sects. 4.3.8.2–4.3.8.4 where it was also much simplified. Thus, to first order in the average Hamiltonian theory, we simply have an effective static transverse magnetic field in the rotating frame. This fact is the foundation of the pulse sequence analysis methods in high-field magnetic resonance spectroscopy (Sect. 4.7). Note that the treatment above would break down in the presence of rapid time dependence in a_{ext} and/or ω_{ext}: when either of them changes significantly over the period of the rotating frame, laboratory frame simulations must be used.

4.3.8.6 Decoupling and Recoupling

A common task in quantum dynamics is creating such time-dependent Hamiltonians as would—when processed using the average Hamiltonian theory—destroy an existing interaction (*decoupling* [110]) or re-introduce a previously destroyed one (*e.g. rotational echo* [111] and rotary resonance [112] recoupling).

Simple decoupling examples are *continuous-wave decoupling*, wherein the time dependence is created by radiofrequency irradiation of the signal of one of the two interacting spins [110]:

$$\begin{aligned} \mathbf{H}(t) &= e^{-i\omega_{\text{RF}} L_{\text{Y}} t}(\mathbf{L}_{\text{Z}}\mathbf{S}_{\text{Z}})e^{+i\omega_{\text{RF}} L_{\text{Y}} t} \\ &= (\mathbf{L}_{\text{Z}}\cos(\omega_{\text{RF}} t) + \mathbf{L}_{\text{X}}\sin(\omega_{\text{RF}} t))\mathbf{S}_{\text{Z}} \quad \Rightarrow \quad \overline{\mathbf{H}}^{(1)} = \mathbf{0} \end{aligned} \tag{4.129}$$

and *composite pulse decoupling*, wherein a piecewise-constant Hamiltonian

$$\mathbf{H}(t_k \le t < t_{k+1}) = \mathbf{H}_k, \quad \tau_k = t_{k+1} - t_k \tag{4.130}$$

is designed so that the piecewise-constant versions of Eqs. (4.103):

$$\overline{\mathbf{H}}^{(1)} = \frac{1}{T}\sum_{k=1}^{N}\mathbf{H}_k\tau_k, \quad \overline{\mathbf{H}}^{(2)} = -\frac{i}{2T}\sum_{k=2}^{N}\sum_{m=1}^{k-1}[\mathbf{H}_k, \mathbf{H}_m]\tau_k\tau_m, \text{ etc.} \tag{4.131}$$

have the unwanted interaction terms destroyed [113]. A simple example of recoupling is *dipolar recoupling*, wherein the secular part of the dipole–dipole interaction Hamiltonian (which yields a zero first-order average under rapid magic angle spinning)

$$\begin{gathered} \mathbf{H}_{\text{DD}}(t) \propto \mathfrak{D}_{0,0}^{(2)}(\mathbf{n}, \omega_{\text{MAS}} t)\mathbf{T}_0^{(2)}(L, S) \\ \mathbf{n} = \frac{1}{\sqrt{3}}[1 \quad 1 \quad 1]^{\text{T}}, \quad \int_0^{2\pi} \mathfrak{D}_{0,0}^{(2)}(\mathbf{n}, \varphi)\text{d}\varphi = 0 \end{gathered} \tag{4.132}$$

is selectively re-introduced into the average Hamiltonian when suitable radiofrequency events are added [111]. The detailed treatment is voluminous [114] and will not be presented here.

4.3.9 Suzuki-Trotter Decomposition

A popular extension of Zassenhaus formula (Sect. 4.3.4) splits the Hamiltonian into parts that have analytical exponentials—for example, the pure momentum Hamiltonian $\hat{p}^2/2m$ and the pure potential energy Hamiltonian $kx^2/2$ in a harmonic oscillator—or parts that have different time scales, for example, Zeeman precession and relaxation in a nuclear spin system at high magnetic field. For a time step t such that $\|\mathbf{H}t\|_2 \ll 1$, the following *Suzuki-Trotter approximations* [115] to the exact propagator $\mathbf{P}(t) = \exp(-i\mathbf{H}t)$ generated by a composite Hamiltonian $\mathbf{H} = \mathbf{H}_1 + \ldots + \mathbf{H}_K$ are popular:

$$\begin{aligned} \mathbf{P}_1(t) &= \mathrm{e}^{-i\mathbf{H}_1 t} \cdots \mathrm{e}^{-i\mathbf{H}_K t}, \quad \mathbf{P}_2(t) = \mathbf{P}_1^\dagger(-t/2)\mathbf{P}_1(t/2) \\ \mathbf{P}_4(t) &= \mathbf{P}_2(at)\mathbf{P}_2(at)\mathbf{P}_2((1-4a)t)\mathbf{P}_2(at)\mathbf{P}_2(at) \end{aligned} \tag{4.133}$$

where $a = 1/(4 - 4^{1/3})$. Errors are estimated using the CBH formula (Sect. 4.3.3), for example:

$$\mathrm{e}^{\mathbf{A}t}\mathrm{e}^{\mathbf{B}t} = \mathrm{e}^{(\mathbf{A}+\mathbf{B})t + [\mathbf{A},\mathbf{B}]t^2/2 + \ldots} \tag{4.134}$$

and likewise for the schemes of higher order [116] that are engineered to zero out as many deviation terms from $\mathbf{H}_1 + \ldots + \mathbf{H}_K$ as possible in the CBH merger of the approximant.

4.4 Perturbation Theories

Quantum theory often uses matrix eigenvalues and eigenvectors, but diagonalisation is numerically expensive. For this reason, much thought has been given to methods of casting operator matrices into a sum of something that is cheap to diagonalise (*reference problem*), and something that is relatively small in some norm (*perturbation*). This is the subject of *time-independent perturbation theory*, of which we consider two logistically different formulations in this section.

A related problem is the efficient evaluation of the effect of a time-dependent perturbation on a reference system for which an eigen system or a time-domain solution are known. A particular version of this problem is the evaluation of the transition probabilities per unit time that a perturbation is causing within the energy level structure of the reference system. This is *time-dependent perturbation theory*.

The discussion of both theories below is given from the point of view of numerical computing—it is assumed that the reader has the matrices as numerical arrays and no inclination for attacking the problem with a pencil. Thus, the formulations presented here are *numerically* friendly.

4.4.1 Rayleigh-Schrödinger Perturbation Theory

Given a diagonal matrix **A** with different real elements on the diagonal, we would like to obtain, as cheaply as possible and approximately if we must, the eigen system for a perturbed matrix **A** + **B**:

$$[\mathbf{A}+\mathbf{B}]\mathbf{V}=\mathbf{V}\mathbf{E} \tag{4.135}$$

where **B** is Hermitian, **V** is unitary and has the required eigenvectors in columns, and **E** is diagonal and contains the corresponding eigenvalues on the diagonal. Explicit diagonalisation of **A** + **B** is undesirable because it is expensive compared to multiplication, particularly when the matrices are sparse.

To find a cheaper approximation [117, 118], we use the uniqueness theorem for the Taylor series [91]: two series in the same parameter that are equal must also have equal coefficients. We insert a real parameter λ into the perturbed matrix to make it $\mathbf{A}+\lambda\mathbf{B}$ and look for a solution of the following form:

$$\mathbf{V}=\sum_{n=0}^{\infty}\lambda^n\mathbf{V}^{(n)},\qquad \mathbf{E}=\sum_{n=0}^{\infty}\lambda^n\mathbf{E}^{(n)} \tag{4.136}$$

where $\mathbf{V}^{(0)}=\mathbf{1}$ and $\mathbf{E}^{(0)}=\mathbf{A}$ because **A** is already diagonal. We also have orthogonality and normalisation condition $\mathbf{V}^{\dagger}\mathbf{V}=\mathbf{1}$ on the eigenvectors because $\mathbf{A}+\lambda\mathbf{B}$ is Hermitian.

We will skip the dreary slog through inserting Eq. (4.136) into Eq. (4.135), opening brackets, equating the coefficients in front of different powers of λ, and solving the resulting system of equations to obtain

$$\begin{aligned}&\mathbf{E}^{(1)}=\operatorname{diag}[\mathbf{B}],\quad \mathbf{V}^{(1)}=\mathbf{Q}\odot\mathbf{B}\\&\mathbf{E}^{(k)}=\operatorname{diag}\left[\mathbf{B}\mathbf{V}^{(k-1)}\right],\quad \mathbf{V}^{(k)}=\mathbf{Q}\odot\left[\mathbf{B}\mathbf{V}^{(k-1)}-\sum_{m=1}^{k-1}\mathbf{V}^{(k-m)}\mathbf{E}^{(m)}\right]\end{aligned} \tag{4.137}$$

where diag zeroes the off-diagonal elements of a matrix, $\odot$ denotes element-wise multiplication, and the energy denominator matrix **Q** is

$$Q_{nk}=\begin{cases}0 & n=k\\ \left(E_{kk}^{(0)}-E_{nn}^{(0)}\right)^{-1} & n\neq k\end{cases} \tag{4.138}$$

There is only one operation here that has cubic worst-case complexity scaling with the Hamiltonian dimension—the $\mathbf{BV}^{(k-1)}$ product. Therefore, the cost of a perturbation theory treatment to n^{th} order is approximately n matrix–matrix multiplications. Because spin operators are always sparse, this cost is in practice much smaller than the cost of diagonalisation.

A sufficient convergence condition is for $\|\mathbf{B}\|_2$ to be smaller than any eigenvalue gap in $\mathbf{A}$; a practical convergence test is to check the norm of $\mathbf{V}^{(k)}$. Because diagonal elements may be redistributed arbitrarily between $\mathbf{A}$ and $\mathbf{B}$, there is some scope for convergence acceleration strategies.

4.4.2 Van Vleck Perturbation Theory

Rayleigh-Schrödinger perturbation theory does not respect the group-theoretical structure (Sect. 1.6) of spin dynamics—Eq. (4.137) contains associative multiplication operations that take matrices outside of the Lie algebra spanned by nested commutators of $\mathbf{A}$ and $\mathbf{B}$. At the same time, the BCH equation (Sect. 4.3.2) requires any effective Hamiltonian to reside in that algebra. It is, therefore, advantageous to seek a perturbation theory that would stay within the Lie algebra of the problem [119].

The initial setting is the same as it was in RSPT (Sect. 4.4.1)—we need a cheap way towards the eigen system of a diagonal reference matrix $\mathbf{A}$ perturbed by a Hermitian matrix $\mathbf{B}$:

$$(\mathbf{A}+\mathbf{B})\mathbf{V}=\mathbf{V}\mathbf{E} \tag{4.139}$$

where $\mathbf{V}$ is unitary and has the required eigenvectors in columns, and $\mathbf{E}$ is diagonal and contains the corresponding eigenvalues on the diagonal. Unitary automorphisms of a Lie algebra are accomplished by its adjoint exponential action (Sect. 1.5) on itself. In particular, there is an automorphism that diagonalises $\mathbf{A}+\mathbf{B}$:

$$\mathbf{E}=\exp(-\mathbf{G})[\mathbf{A}+\mathbf{B}]\exp(\mathbf{G}) \tag{4.140}$$

where $\mathbf{G}$ is required to be skew-Hermitian because its exponential is unitary. After comparing this with Eq. (4.139), we conclude that $\mathbf{V}=\exp(\mathbf{G})$.

We will start by observing that $\mathbf{E}=\mathbf{A}+\mathbf{W}$, where $\mathbf{W}$ is a diagonal matrix giving the correction to the eigenvalues of $\mathbf{A}$ brought about by the perturbation:

$$\mathbf{W}=\exp(-\mathbf{G})[\mathbf{A}+\mathbf{B}]\exp(\mathbf{G})-\mathbf{A}=\sum_{n=0}^{\infty}\frac{1}{n!}\mathrm{ad}_{\mathbf{G}}^{(n)}(\mathbf{A}+\mathbf{B})-\mathbf{A} \tag{4.141}$$

and proceed by expressing $\mathbf{W}$ and $\mathbf{G}$ as power series in the same parameter λ:

$$\mathbf{W} = \sum_{k=1}^{\infty} \lambda^k \mathbf{W}_k, \qquad \mathbf{G} = \sum_{k=1}^{\infty} \lambda^k \mathbf{G}_k \tag{4.142}$$

Substituting these expressions into Eq. (4.141), and equating terms with equal powers of λ produces a system of commutator equations for $\mathbf{G}_k$ and $\mathbf{W}_k$. After the perturbation is split into the diagonal part $\mathbf{B}_\mathrm{D} = \mathrm{diag}(\mathbf{B})$ and the off-diagonal part $\mathbf{B}_\mathrm{X} = \mathbf{B} - \mathbf{B}_\mathrm{D}$, an eye-glazingly boring elaboration yields

$$\begin{aligned}
\mathbf{G}_1 &= \mathbf{Q} \odot \mathbf{B}_\mathrm{X} \\
\mathbf{G}_2 &= \mathbf{Q} \odot [\mathbf{B}_\mathrm{D}, \mathbf{G}_1] \\
\mathbf{G}_3 &= \mathbf{Q} \odot \left([\mathbf{B}_\mathrm{D}, \mathbf{G}_2] + \tfrac{1}{3}[[\mathbf{B}_\mathrm{X}, \mathbf{G}_1], \mathbf{G}_1]\right) \\
\mathbf{G}_4 &= \mathbf{Q} \odot \begin{pmatrix} [\mathbf{B}_\mathrm{D}, \mathbf{G}_3] + \frac{1}{3}[[\mathbf{B}_\mathrm{X}, \mathbf{G}_1], \mathbf{G}_2] \\ + \frac{1}{3}[[\mathbf{B}_\mathrm{X}, \mathbf{G}_2], \mathbf{G}_1] \end{pmatrix} \\
\mathbf{G}_5 &= \mathbf{Q} \odot \begin{pmatrix} [\mathbf{B}_\mathrm{D}, \mathbf{G}_4] + \frac{1}{3}[[\mathbf{B}_\mathrm{X}, \mathbf{G}_1], \mathbf{G}_3] \\ + \frac{1}{3}[[\mathbf{B}_\mathrm{X}, \mathbf{G}_2], \mathbf{G}_2] + \frac{1}{3}[[\mathbf{B}_\mathrm{X}, \mathbf{G}_3], \mathbf{G}_1] \\ - \frac{1}{45}[[[[\mathbf{B}_\mathrm{X}, \mathbf{G}_1], \mathbf{G}_1], \mathbf{G}_1], \mathbf{G}_1] \end{pmatrix}
\end{aligned}
\qquad
\begin{aligned}
\mathbf{W}_1 &= \mathbf{B}_\mathrm{D} \\
\mathbf{W}_2 &= \tfrac{1}{2}[\mathbf{B}_\mathrm{X}, \mathbf{G}_1] \\
\mathbf{W}_3 &= \tfrac{1}{2}[\mathbf{B}_\mathrm{X}, \mathbf{G}_2] \\
\mathbf{W}_4 &= \frac{1}{2}[\mathbf{B}_\mathrm{X}, \mathbf{G}_3] \\
&\quad - \frac{1}{24}[[[\mathbf{B}_\mathrm{X}, \mathbf{G}_1], \mathbf{G}_1], \mathbf{G}_1] \\
\mathbf{W}_5 &= \frac{1}{2}[\mathbf{B}_\mathrm{X}, \mathbf{G}_4] \\
&\quad - \frac{1}{24}[[[\mathbf{B}_\mathrm{X}, \mathbf{G}_1], \mathbf{G}_1], \mathbf{G}_2] \\
&\quad - \frac{1}{24}[[[\mathbf{B}_\mathrm{X}, \mathbf{G}_1], \mathbf{G}_2], \mathbf{G}_1] \\
&\quad - \frac{1}{24}[[[\mathbf{B}_\mathrm{X}, \mathbf{G}_2], \mathbf{G}_1], \mathbf{G}_1]
\end{aligned} \tag{4.143}$$

with the same definitions of $\mathbf{Q}$ as in Eq. (4.137). No simple recurrence relations to arbitrary order appear to exist. Energy corrections resulting from this procedure are order-by-order identical to those obtained by RSPT, and the procedure is more expensive, but the advantage is that the eigenvector matrix generator $\mathbf{G}$ is a collection of commutators, many of which may be analytical or even zero. Preconditioning opportunities and the sufficient convergence condition are the same as those of RSPT.

4.4.3 Dyson Perturbation Theory

Consider the equation of motion for the propagator $\mathbf{P}(t)$, such that $|\psi(t)\rangle = \mathbf{P}(t)|\psi(0)\rangle$, obtained by substituting this definition into Schrödinger's equation:

$$\frac{\partial \mathbf{P}(t)}{\partial t} = -i\mathbf{H}(t)\mathbf{P}(t), \quad \mathbf{P}(0) = \mathbf{1} \tag{4.144}$$

Both sides of the equation may be integrated to yield

$$\mathbf{P}(t) = \mathbf{1} - i\int_0^t \mathbf{H}(t')\mathbf{P}(t')\mathrm{d}t' \tag{4.145}$$

Using $\mathbf{P}^{(0)}(t) = \mathbf{1}$ as an initial guess and applying this equation repeatedly yields the following iteration:

$$\mathbf{P}^{(k+1)}(t) = \mathbf{1} - i\int_0^t \mathbf{H}(t')\mathbf{P}^{(k)}(t')\mathrm{d}t' \tag{4.146}$$

When this loop is unrolled, we obtain the already familiar (Sect. 4.1) *Dyson series* [92]:

$$\mathbf{P}(t) = \mathbf{1} - i\int_0^t dt'\mathbf{H}(t') - \int_0^t \mathrm{d}t' \int_0^{t'} \mathrm{d}t''\mathbf{H}(t')\mathbf{H}(t'') + \dots \tag{4.147}$$

It converges rapidly when $\|\mathbf{H}(t)t\|_2 \ll 1$, and therefore finds much use in time-dependent perturbation theories, where the norm in question is assumed to be small.

Consider now a setting where $\mathbf{H}(t) = \mathbf{H}_0 + \mathbf{V}(t)$, with $\mathbf{H}_0$ a diagonal matrix (meaning that the dynamics under it is numerically cheap and analytically simple) and a Hermitian perturbation term $\mathbf{V}(t)$. In the interaction representation (denoted by upper R index, Sect. 4.3.1) with respect to $\mathbf{H}_0$:

$$\begin{gathered} \mathbf{P}^{\mathrm{R}}(t) = \exp[+i\mathbf{H}_0 t]\mathbf{P}(t), \quad \mathbf{V}^{\mathrm{R}}(t) = \exp[+i\mathbf{H}_0 t]\mathbf{V}(t)\exp[-i\mathbf{H}_0 t] \\ \dot{\mathbf{P}}^{\mathrm{R}}(t) = -i\mathbf{V}^{\mathrm{R}}(t)\mathbf{P}^{\mathrm{R}}(t), \quad \mathbf{P}^{\mathrm{R}}(0) = \mathbf{1} \end{gathered} \tag{4.148}$$

For a system that was in the nth eigenstate of $\mathbf{H}_0$ at time zero, the exact expression for the probability of being found in the kth eigenstate at time t is

$$\begin{aligned} P_{n\to k}(t) &= |\langle k|\mathbf{P}(t)|n\rangle|^2 = \left|\langle k|\exp[-i\mathbf{H}_0 t]\mathbf{P}^{\mathrm{R}}(t)|n\rangle\right|^2 \\ &= \left|\mathrm{e}^{-i\omega_k t}\langle k|\mathbf{P}^{\mathrm{R}}(t)|n\rangle\right|^2 = \left|\langle k|\mathbf{P}^{\mathrm{R}}(t)|n\rangle\right|^2 \end{aligned} \tag{4.149}$$

where the choice of the representation predictably does not influence the result. The computational complexity of this expression is the same as that of solving the TDSE. However, Eq. (4.147) now provides less expensive approximate expressions in the interaction representation

$$
\begin{aligned}
P_{n\to k}(t) &= \left|c^{(0)}_{n\to k}(t) + c^{(1)}_{n\to k}(t) + c^{(2)}_{n\to k}(t) + \ldots\right|^2 \\
c^{(0)}_{n\to k}(t) &= \delta_{nk},\ c^{(1)}_{n\to k}(t) = -i\int_0^t \mathrm{d}t' \langle k|\mathbf{V}^{\mathrm{R}}(t')|n\rangle \\
c^{(2)}_{n\to k}(t) &= -\int_0^t \mathrm{d}t' \int_0^{t'} \mathrm{d}t'' \langle k|\mathbf{V}^{\mathrm{R}}(t')\mathbf{V}^{\mathrm{R}}(t'')|n\rangle
\end{aligned}
\tag{4.150}
$$

and so on for higher order corrections. To express elements of matrix products via the elements of the individual matrices, we insert the unit operator $\mathbf{1} = \sum_m |m\rangle\langle m|$ in between, for example:

$$
c^{(2)}_{n\to k}(t) = -\sum_m \int_0^t \mathrm{d}t' \int_0^{t'} \mathrm{d}t'' \langle k|\mathbf{V}^{\mathrm{R}}(t')|m\rangle\langle m|\mathbf{V}^{\mathrm{R}}(t'')|n\rangle
\tag{4.151}
$$

When these expressions are returned into the lab frame using the definition of $\mathbf{V}^{\mathrm{R}}(t)$ in Eq. (4.148), and the exponentials of $\mathbf{H}_0$ are applied to its eigenstates, we get

$$
\begin{aligned}
c^{(0)}_{n\to k}(t) &= \delta_{nk}, \quad c^{(1)}_{n\to k}(t) = -i\int_0^t \mathrm{d}t' V_{kn}(t')\mathrm{e}^{+i\omega_{kn}t'} \\
c^{(2)}_{n\to k}(t) &= -\sum_m \int_0^t \mathrm{d}t' \int_0^{t'} \mathrm{d}t'' V_{km}(t')\mathrm{e}^{+i\omega_{km}t'} V_{mn}(t'')\mathrm{e}^{+i\omega_{mn}t''} \\
V_{kn}(t) &= \langle k|\mathbf{V}(t)|n\rangle, \quad \omega_{kn} = \omega_k - \omega_n
\end{aligned}
\tag{4.152}
$$

and so on for higher orders. These expressions are the foundation of time-domain perturbation and relaxation theories. In the limit of $t \to \infty$, the first-order coefficient $c^{(1)}_{n\to k}(t)$ becomes a Fourier transform and its physical meaning emerges: the transition probability is proportional to the square amplitude that the perturbation has at the transition frequency. The second-order coefficient then depends on the product of the amplitudes that the perturbation has across every possible pair of transitions that connects the source and the destination energy level through some intermediate energy level m.

4.4.4 Fermi's Golden Rule

Time-dependent perturbation theory has important analytical solutions. Consider a monochromatic perturbation with an operator $\mathbf{V}$ and a frequency ω:

$$\mathbf{H}(t) = \mathbf{H}_0 + \mathbf{V}\cos(\omega t), \quad T = 2\pi/\omega \tag{4.153}$$

To first order in Dyson's perturbation theory, the transition probability is

$$P^{(1)}_{n\to k}(t) = \left|c^{(1)}_{n\to k}(t)\right|^2 = \ldots = 4|V_{kn}|^2(\omega_{kn}-\omega)^{-2}\sin^2\left[\tfrac{1}{2}(\omega_{kn}-\omega)t\right] \tag{4.154}$$

In the limit of matching perturbation and transition frequencies, the expression becomes

$$\lim_{\omega\to\omega_{kn}} P^{(1)}_{n\to k}(t) = |V_{kn}|^2 t^2 \tag{4.155}$$

The matrix element $V_{kn} = \langle k|\mathbf{V}|n\rangle$ that determines the transition rate is called *transition moment*; under weak perturbations with $\|\mathbf{V}t\|_2 \ll 1$, the transition probability is proportional to its absolute square.

When transitions happen to a continuum of states (never for spin systems, but we will encounter baths in Chap. 6 where that is the case), we must extend Eq. (4.154) to state densities:

$$P^{(1)}_{n\to k}(t) = \int \frac{\sin^2\left[\frac{1}{2}(\omega_{kn}-\omega)t\right]}{\left[\frac{1}{2}(\omega_{kn}-\omega)\right]^2}|V_{kn}|^2\rho(\omega_{kn})d\omega_{kn} \tag{4.156}$$

where integration is performed over the density of the destination state. If we assume the matrix element V_{kn} to be the same across that density, we are left with a convolution between $\rho(\omega_{kn})$ and the square of a sinc function, which is a fair approximation to the delta function, therefore:

$$P^{(1)}_{n\to k}(t) \approx |V_{kn}|^2 \int \frac{\sin^2\left[\frac{1}{2}(\omega_{kn}-\omega)t\right]}{\left[\frac{1}{2}(\omega_{kn}-\omega)\right]^2}\rho(\omega_{kn})d\omega_{kn} \approx 2\pi|V_{kn}|^2\rho(\omega_{kn})t \tag{4.157}$$

The derivative of this probability, called *transition rate*, is time-independent:

$$W^{(1)}_{n\to k} = \frac{dP^{(1)}_{n\to k}}{dt} \approx 2\pi|V_{kn}|^2\rho(\omega_{kn}) \tag{4.158}$$

This is *Fermi's golden rule* (actually Dirac's [120]), it will be used later to compute the intensity of spectroscopic transitions. The same treatment to second order in perturbation theory yields

$$W_{n\to k}^{(2)} = 2\pi \left| \sum_{m\neq n} \frac{\langle k|\mathbf{V}|m\rangle\langle m|\mathbf{V}|n\rangle}{\omega_n - \omega_m} \right|^2 \rho(\omega_{kn}) \tag{4.159}$$

An important special case (*Rabi's formula*) is the probability of finding a two-state system in eigenstate 2 of $\mathbf{H}_0$ when the initial condition is pure eigenstate 1 and the perturbation is constant. The problem dimension is small enough for an exact analytical solution [93]:

$$\mathbf{H} = \mathbf{H}_0 + \mathbf{V} = \begin{pmatrix} \omega_1 & V_{12} \\ V_{12}^* & \omega_2 \end{pmatrix}, \quad P_1(0) = 1$$

$$P_{1\to2}(t) = \left|\langle 2|e^{-i\mathbf{H}t}|1\rangle\right|^2 = \ldots = \frac{4|V_{12}|^2}{\omega_{12}^2 + 4|V_{12}|^2} \sin^2\left(\frac{1}{2}t\sqrt{\omega_{12}^2 + 4|V_{12}|^2}\right) \tag{4.160}$$

in the limit of a weak perturbation $4|V_{12}|^2 \ll \omega_{12}^2$:

$$P_{1\to2}(t) \approx \frac{4|V_{12}|^2}{\omega_{12}^2} \sin^2\left(\frac{1}{2}\omega_{12}t\right) \tag{4.161}$$

where the average over the evolution period under $\mathbf{H}_0$ is again approximately proportional to the absolute square of the transition moment.

4.5 Resonance Fields

When the perturbation is electromagnetic, the transition moment in Eq. (4.158) determines the number of photons absorbed and/or emitted by the system per unit time, and therefore the intensity of the spectral line. It does however require: (a) the perturbation frequency ω to match the gap between energy levels $\{n, k\}$; (b) the transition moment V_{nk} to be non-zero. When the perturbation frequency is fixed and some instrumental parameter is swept, this brings us to the problem of finding *resonance fields*—such values of the coefficient α in the Hamiltonian $\mathbf{H} = \alpha\mathbf{H}_A + \mathbf{H}_B$ as would make the difference between any two eigenvalues of $\mathbf{H}$ equal to the specified frequency ω. The name *resonance field* comes from magnetic resonance spectroscopy, where $\alpha\mathbf{H}_A$ is the Zeeman Hamiltonian that is varied by sweeping the magnet field, and $\mathbf{H}_B$ contains all other interactions. Many electron spin resonance experiments involve weak microwave irradiation at a fixed frequency, and a magnet with a variable field. An important question is finding the values of that field at which the difference between any two energy levels is equal to the microwave frequency, and calculating the corresponding transition moments.

4.5.1 Eigenfields Method

Consider the eigen system $\{|n\rangle, \omega_n\}$ that satisfies

$$[\alpha \mathbf{H}_\mathrm{A} + \mathbf{H}_\mathrm{B}]|n\rangle = \omega_n|n\rangle \tag{4.162}$$

The resonance fields problem consists in finding such values of α at which the difference between any two eigenvalues is equal to the perturbation frequency. Consider therefore two eigenvectors separated by ω:

$$\begin{cases} (\alpha \mathbf{H}_\mathrm{A} + \mathbf{H}_\mathrm{B})|u\rangle = \lambda|u\rangle \\ (\alpha \mathbf{H}_\mathrm{A} + \mathbf{H}_\mathrm{B})|v\rangle = (\lambda - \omega)|v\rangle \end{cases} \tag{4.163}$$

Multiplying the first one by $\langle v|$ and the Hermitian conjugate of the second one by $|u\rangle$, we get

$$\begin{cases} (\alpha \mathbf{H}_\mathrm{A} + \mathbf{H}_\mathrm{B})|u\rangle\langle v| = \lambda|u\rangle\langle v| \\ |u\rangle\langle v|(\alpha \mathbf{H}_\mathrm{A} + \mathbf{H}_\mathrm{B}) = |u\rangle\langle v|(\lambda - \omega) \end{cases} \tag{4.164}$$

After subtracting the first equation from the second one, we obtain

$$|u\rangle\langle v|(\alpha \mathbf{H}_\mathrm{A} + \mathbf{H}_\mathrm{B}) - (\alpha \mathbf{H}_\mathrm{A} + \mathbf{H}_\mathrm{B})|u\rangle\langle v| = -\omega|u\rangle\langle v| \tag{4.165}$$

After some re-grouping, we get

$$(|u\rangle\langle v|\mathbf{H}_\mathrm{B} - \mathbf{H}_\mathrm{B}|u\rangle\langle v|) + \omega|u\rangle\langle v| = \alpha(\mathbf{H}_\mathrm{A}|u\rangle\langle v| - |u\rangle\langle v|\mathbf{H}_\mathrm{A}) \tag{4.166}$$

The use of commutation superoperators (Sect. 4.2.4) simplifies the notation:

$$(\omega \mathbf{1} - \mathcal{H}_\mathrm{B})\boldsymbol{\rho} = \alpha \mathcal{H}_\mathrm{A}\boldsymbol{\rho} \tag{4.167}$$

where $\boldsymbol{\rho}=|u\rangle\langle v|$, $\mathcal{H}_\mathrm{B} = \mathbf{1} \otimes \mathbf{H}_\mathrm{B} - \mathbf{H}_\mathrm{B}^\mathrm{T} \otimes \mathbf{1}$, and $\mathcal{H}_\mathrm{A} = \mathbf{1} \otimes \mathbf{H}_\mathrm{A} - \mathbf{H}_\mathrm{A}^\mathrm{T} \otimes \mathbf{1}$. This is an instance of the generalised eigensystem problem which may be solved numerically. Transition moments with respect to a Hermitian perturbation operator $\mathbf{V}$ are calculated as

$$V_{vu} = \langle v|\mathbf{V}|u\rangle = \langle \mathrm{vec}(\mathbf{V})|\boldsymbol{\rho}\rangle \tag{4.168}$$

This method is elegant and exact [121], but computationally expensive. Some efficiency savings are possible: (a) the problem lives in Liouville space that may be restricted (Sect. 7.3); (b) $\boldsymbol{\rho}$ is a dyadic tensor (Sect. 9.1) that has fewer independent elements than a vector of the same dimension; (c) when the Pauli basis (Sect. 1.6.3.2) is used to build spin Hamiltonians, $\mathbf{H}_\mathrm{A}$ and $\mathbf{H}_\mathrm{B}$ have known Kronecker structures, are Hermitian and very sparse (Sect. 9.1.2); (d) in practical magnetic resonance, $\mathbf{H}_\mathrm{A}$ is a sum of single-spin operators (Zeeman Hamiltonian) and $\mathbf{H}_\mathrm{B}$ is a sum of two-spin operators (coupling Hamiltonian).

4.5.2 Adaptive Trisection Method

When state-space restriction (Sect. 7.3) is not applicable, a more efficient way of finding the resonance fields is to stay in Hilbert space and search the specified interval $[\alpha_\mathrm{L}, \alpha_\mathrm{R}]$ for points where any two Hamiltonian eigenvalues have the specified frequency difference ω. A reliable numerical method, inspired by the Stoll-Schweiger bisection method [122], is

1. Trisect the interval to obtain a four-point grid $[\alpha_\mathrm{L}, \alpha_\mathrm{M1}, \alpha_\mathrm{M2}, \alpha_\mathrm{R}]$ where the two midpoints are located at 1/3 and 2/3 of the interval. At each point of the grid, calculate the eigensystem:

$$[\alpha_k \mathbf{H}_\mathrm{A} + \mathbf{H}_\mathrm{B}]\mathbf{V}_k = \mathbf{V}_k \boldsymbol{\Omega}_k, \quad k \in \{\mathrm{L},\ \mathrm{M1},\ \mathrm{M2},\ \mathrm{R}\} \tag{4.169}$$

   where $\mathbf{V}_k$ has eigenvectors as columns, and $\boldsymbol{\Omega}_k$ is a diagonal matrix with the vector of eigenvalues $\boldsymbol{\omega}_k$ on the diagonal. Sort the eigen system by energy—numerical diagonalisation routines do not order their output. Tracing eigenvector correspondence between grid points is not necessary and will not work—exact degeneracies are common in spin Hamiltonians.

2. Find eigenvalue derivatives with respect to α using the Hellmann–Feynman theorem (Sect. 2.6.8):

$$\boldsymbol{\omega}'_k = \langle \mathbf{V}_k \mid \mathbf{H}_\mathrm{A} \mathbf{V}_k \rangle^\mathrm{T} \tag{4.170}$$

   where the scalar product is to be applied column by column of $\mathbf{V}_k$—therefore $\boldsymbol{\omega}'_k$ is a column vector. Do not keep the eigenvectors, but do record which pairs of energy levels have significant transition moments under the perturbation Hamiltonian $\mathbf{H}_\mathrm{T}$—this will later be used for efficiency screening:

$$\mathbf{T}_k = \left(\left| \mathbf{V}_k^\dagger \mathbf{H}_\mathrm{T} \mathbf{V}_k \right|^{\odot 2} > \varepsilon_\mathrm{T} \right) \tag{4.171}$$

where the modulus and the square are applied element-wise, ε_{T} is a transition drop tolerance, and the elements of $\mathbf{T}_k$ are Boolean "true" or "false"; using a sparse matrix is recommended.

3. Use the eigenvalue array $\boldsymbol{\omega}$ and its derivatives $\boldsymbol{\omega}'$ at α_{L} and α_{R} to interpolate the eigenvalues with a cubic Hermite spline over the $[\alpha_{\mathrm{L}}, \alpha_{\mathrm{R}}]$ interval. Hermite spline approximations to the eigenvalue arrays at the two midpoints are

$$\begin{aligned} \boldsymbol{\omega}_{\mathrm{M1}} &\approx \frac{1}{27}[20\boldsymbol{\omega}(\alpha_{\mathrm{L}}) + 4\boldsymbol{\omega}'(\alpha_{\mathrm{L}})\Delta\alpha - 2\boldsymbol{\omega}'(\alpha_{\mathrm{R}})\Delta\alpha + 7\boldsymbol{\omega}(\alpha_{\mathrm{R}})] \\ \boldsymbol{\omega}_{\mathrm{M2}} &\approx \frac{1}{27}[7\boldsymbol{\omega}(\alpha_{\mathrm{L}}) + 2\boldsymbol{\omega}'(\alpha_{\mathrm{L}})\Delta\alpha - 4\boldsymbol{\omega}'(\alpha_{\mathrm{R}})\Delta\alpha + 20\boldsymbol{\omega}(\alpha_{\mathrm{R}})] \end{aligned} \tag{4.172}$$

where $\Delta\alpha = \alpha_{\mathrm{R}} - \alpha_{\mathrm{L}}$. Re-sort the arrays and compare them to the machine precision accuracy eigenvalues obtained from the diagonalisations performed in Step 1.

4. If the differences between spline-interpolated and machine-precision eigenvalues are larger than the specified energy accuracy tolerance ε_{E}, subdivide the grid into smaller intervals by calling Step 1 recursively for $[\alpha_{\mathrm{L}}, \alpha_{\mathrm{M1}}]$, $[\alpha_{\mathrm{M1}}, \alpha_{\mathrm{M2}}]$, $[\alpha_{\mathrm{M2}}, \alpha_{\mathrm{R}}]$. Alternatively, if all differences are smaller than the tolerance, send each of the three intervals to Step 5.

5. There is now a controlled-accuracy spline approximation to the eigenvector array in some interval $[\beta_{\mathrm{L}}, \beta_{\mathrm{R}}]$ that is one of the sub-intervals of $[\alpha_{\mathrm{L}}, \alpha_{\mathrm{R}}]$, with eigenvalue arrays and their derivatives already computed at the edges. Points within this interval such that a pair of splines differs by exactly ω can now be found analytically. For each eigenvalue $\omega^{(k)}$:

   (a) Calculate cubic Hermite spline interpolation coefficients:

$$\begin{bmatrix} c_3^{(k)} \\ c_2^{(k)} \\ c_1^{(k)} \\ c_0^{(k)} \end{bmatrix} = \begin{bmatrix} 1 & 2 & 1 & -2 \\ -2 & -3 & -1 & 3 \\ 1 & 0 & 0 & 0 \\ 0 & 1 & 0 & 0 \end{bmatrix} \begin{bmatrix} \omega'^{(k)}(\beta_L)\Delta\beta \\ \omega^{(k)}(\beta_L) \\ \omega'^{(k)}(\beta_R)\Delta\beta \\ \omega^{(k)}(\beta_R) \end{bmatrix} \tag{4.173}$$

   (b) Calculate minimum and maximum value of the spline interpolant on $[\beta_{\mathrm{L}}, \beta_{\mathrm{R}}]$ interval:

$$\begin{aligned} \{\beta_m\} &= \left\{\beta_L,\ \mathrm{roots}\left[3c_3^{(k)}\beta^2 + 2c_2^{(k)}\beta + c_1^{(k)} = 0,\ \beta_L \le \beta \le \beta_R\right],\ \beta_R\right\} \\ \omega_{\min}^{(k)} &= \min\left\{c_3^{(k)}\beta_m^3 + c_2^{(k)}\beta_m^2 + c_1^{(k)}\beta_m + c_0^{(k)}\right\} \\ \omega_{\max}^{(k)} &= \max\left\{c_3^{(k)}\beta_m^3 + c_2^{(k)}\beta_m^2 + c_1^{(k)}\beta_m + c_0^{(k)}\right\} \end{aligned} \tag{4.174}$$

   where k enumerates the energy levels.

(c) Identify all eigenvalue index pairs $\{n, k\}$, such that

$$\begin{gathered}\omega_{\min}^{(n)} - \omega_{\max}^{(k)} < \omega < \omega_{\max}^{(n)} - \omega_{\min}^{(k)} \\ [\mathbf{T}_{\mathrm{L}}]_{nk} = [\mathbf{T}_{\mathrm{R}}]_{nk} = 1\end{gathered} \tag{4.175}$$

This screening stage is necessary because the number of eigenvalue pairs is quadratic in the Hilbert space dimension; an unscreened calculation would be expensive.

(d) For each index pair that has survived the screening, solve the cubic equation for the points where the corresponding spline interpolants are exactly ω apart:

$$\{\beta_m\} = \mathrm{roots}\left\{\left(c_3^{(n)} - c_3^{(k)}\right)\beta^3 + \left(c_2^{(n)} - c_2^{(k)}\right)\beta^3 + \left(c_1^{(n)} - c_1^{(k)}\right)\beta + \left(c_0^{(n)} - c_0^{(k)}\right) = \omega\right\} \tag{4.176}$$

Eliminate complex roots, as well as the roots that fall outside the $[\beta_{\mathrm{L}}, \beta_{\mathrm{R}}]$ interval. Collect all surviving roots across all grid intervals into a global resonance field list $\{\alpha_m\}$.

6. At every resonance field α_m, calculate the eigensystem:

$$(\alpha_m \mathbf{H}_{\mathrm{A}} + \mathbf{H}_{\mathrm{B}})\mathbf{V}_m = \mathbf{V}_m \mathbf{\Omega}_m, \tag{4.177}$$

find the eigenvalue index pairs $\{n, k\}$ such that

$$\left|\omega_m^{(n)} - \omega_m^{(k)} - \omega\right| < \varepsilon_{\mathrm{E}}, \tag{4.178}$$

and compute the transition moments:

$$t_{n,k}^{(m)} = \left\langle \mathbf{v}_m^{(n)} \middle| \mathbf{H}_{\mathrm{T}} \middle| \mathbf{v}_m^{(k)} \right\rangle. \tag{4.179}$$

Return resonance fields, transition moments, and eigenpairs to the user.

As it often happens in numerical simulations, *Matlab* code is shorter and more readable than its human language description—the code is available in the **eigenfields.m** function of *Spinach* [87].

4.6 Symmetry Factorisation

A *symmetry* is a transformation of a physical system that leaves its Hamiltonian unchanged. The set of such operations is closed under inversion and superposition, and it has a unit element—it is, therefore, a group (Sect. 1.5). Elements of this group commute with the Hamiltonian, and therefore share eigenfunctions or eigenvectors with it. A good example is label permutations on the three identical proton spins (denoted S) of a ^{13}C (denoted L) methyl group in liquid-state NMR spectroscopy:

$$\mathbf{H} = \omega_{\mathrm{L}} \mathbf{L}_{\mathrm{Z}} + \omega_{\mathrm{S}} \left(\mathbf{S}_{\mathrm{Z}}^{(1)} + \mathbf{S}_{\mathrm{Z}}^{(2)} + \mathbf{S}_{\mathrm{Z}}^{(3)} \right) + 2\pi J_{\mathrm{CH}} \left(\mathbf{S}_{\mathrm{Z}}^{(1)} + \mathbf{S}_{\mathrm{Z}}^{(2)} + \mathbf{S}_{\mathrm{Z}}^{(3)} \right) \mathbf{L}_{\mathrm{Z}} \tag{4.180}$$

Another common case is liquid-state ESR spectroscopy, where all nuclei with similar isotropic hyperfine coupling constants to the same electron can be declared equivalent under the permutation group.

Matrix representations of symmetry operators can be reduced efficiently—the reduction process is analytical. Because it partitions the basis into symmetry-invariant subspaces, it also block-diagonalises the Hamiltonian with computational benefits (individual blocks may be simulated separately) and additional physical insights (the blocks often behave differently). This is the reason for the popularity of group theory in quantum mechanics [123]; this section describes the subject in the context of spin.

4.6.1 Symmetry-Adapted Linear Combinations

Consider basis sets of wavefunctions $\{|\psi_k\rangle\}$ and operators $\{\mathbf{O}_k\}$ of a multi-spin system. Both may be generated as direct products of the basis sets pertaining to individual spins:

$$\begin{gathered} |\psi_k\rangle = \left|\psi_k^{(1)}\right\rangle \otimes \left|\psi_k^{(2)}\right\rangle \otimes \left|\psi_k^{(3)}\right\rangle \cdots \\ \mathbf{O}_k = \mathbf{O}_k^{(1)} \otimes \mathbf{O}_k^{(2)} \otimes \mathbf{O}_k^{(3)} \cdots \end{gathered} \tag{4.181}$$

where $\left|\psi_k^{(n)}\right\rangle$ is the state that the nth spin has in the kth element of the composite system basis. Likewise, $\mathbf{O}_k^{(n)}$ is the spin operator acting on the nth spin in the kth element of the operator basis.

When this construction is used, the action by a symmetry operation amounts to a permutation of the direct product components. *Symmetry-adapted linear combinations* (SALCs) spanning irreducible representations of the symmetry group G are then obtained by multiplying the character string (Sect. 1.5.5) of each irrep by the actions of the corresponding group elements (Sect. 1.5.3) on the basis set:

$$\mathbf{O}_k^{(\Gamma)} = \frac{1}{N}\sum_{g\in G}\chi_g^{(\Gamma)} g(\mathbf{O}_k), \quad \left|\psi_k^{(\Gamma)}\right\rangle = \frac{1}{N}\sum_{g\in G}\chi_g^{(\Gamma)} g(|\psi_k\rangle) \tag{4.182}$$

where N is the normalisation constant, the summation runs over the elements g of the group, $\chi_g^{(\Gamma)}$ is the character of irreducible representation Γ of the group element g, $g(\mathbf{O}_k)$ is the result of the action by that group element on the basis operator $\mathbf{O}_k$, and $g(|\psi_k\rangle)$ is the result of the action by that group element on the basis wavefunction $|\psi_k\rangle$. If the irrep is not one-dimensional, SALCs will come out non-orthogonal, but can always be orthogonalised efficiently because the dimension of the relevant subspace is tiny—equal to the dimension of the irrep.

If multiple sets of symmetry-related spins are present, the total system symmetry group is a product of the individual groups, and the total character table is, therefore, the direct product of character tables of the individual groups. For irreducible representations Γ and Λ of groups $\mathcal{G}$ and $\mathcal{H}$, respectively:

$$\mathcal{G}\times\mathcal{H} = \{gh | g\in\mathcal{G}, h\in\mathcal{H}\}, \quad \chi_{gh}^{(\Gamma\times\Lambda)} = \chi_g^{(\Gamma)}\chi_h^{(\Lambda)} \tag{4.183}$$

The SALC procedure is agnostic to the spin quantum number—Eqs. (4.181) and (4.182) make no reference to the nature of the direct product basis, which may, therefore, contain spin operators of any rank. It is logistically efficient because Eq. (4.182) amounts simply to permuting columns in the basis set descriptor table (Sect. 7.1), and Eq. (4.183) is just a Kronecker product of character tables (Sect. 1.5.5). The cost of generating SALCs is quadratic in the dimension of the basis set, but the benefit from the resulting block-diagonalisation of the Hamiltonian (Fig. 4.2) or Liouvillian (Fig. 4.3) is cubic because the matrix is broken up into smaller blocks corresponding to non-interacting subspaces.

4.6.2 Liouville Space Symmetry Treatment

In Hilbert space simulations, all irreducible representations are usually populated. However, unless the symmetry had been broken at some point in the system's past, only the fully symmetric irrep is active in Liouville space [124]. This is best illustrated using the singlet state of a two-spin system. In Hilbert space, the singlet *wavefunction* changes sign under spin permutation operation $\hat{P}_{12}$:

$$\hat{P}_{12}(|\alpha\beta\rangle - |\beta\alpha\rangle) = |\beta\alpha\rangle - |\alpha\beta\rangle = -(|\alpha\beta\rangle - |\beta\alpha\rangle) \tag{4.184}$$

It, therefore, belongs to the antisymmetric irrep of the permutation group. The three triplet states belong to the symmetric irrep. However, in Liouville space, the singlet *projector* does not change sign:

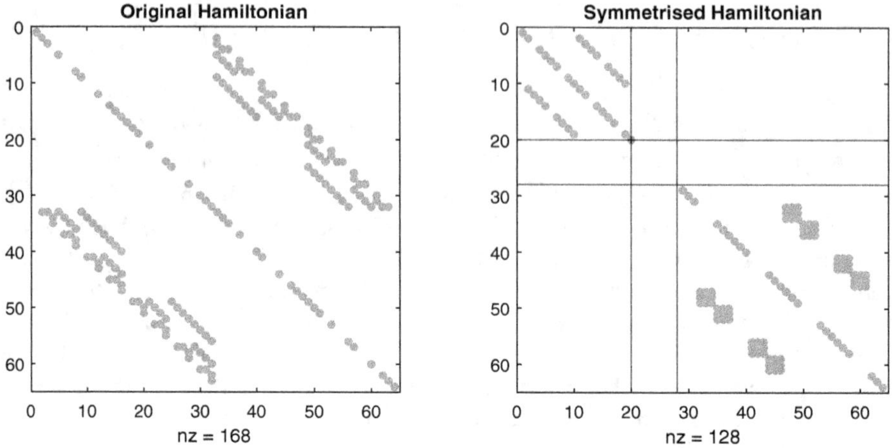

Fig. 4.2 Block structure emerging in the spin Hamiltonian operator (Hilbert space) of a radical pair with four equivalent spin-1/2 nuclei after symmetry factorisation under the S_4 permutation group. Grey dots indicate non-zero elements. The Hamiltonian includes isotropic Zeeman interactions for all particles and equal isotropic hyperfine couplings between one of the electrons and the four nuclei

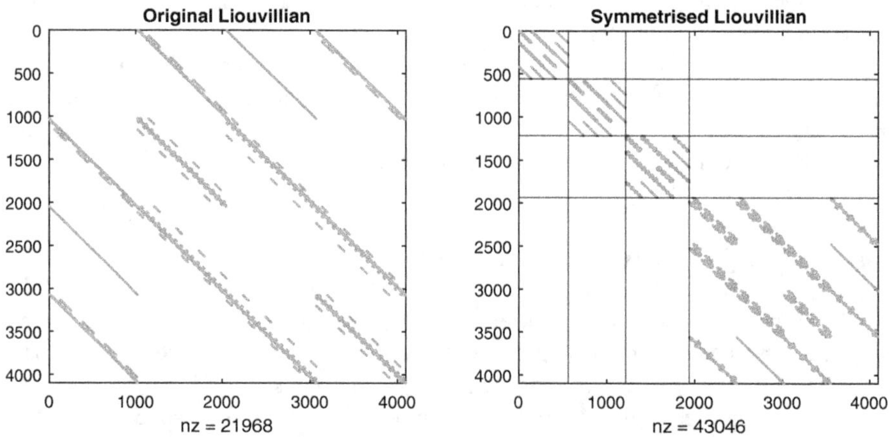

Fig. 4.3 Block structure emerging in the spin Hamiltonian commutation superoperator (Liouville space) of a radical pair with four equivalent spin-1/2 nuclei after symmetry factorisation under the S_4 permutation group. Grey dots indicate non-zero elements. The Liouvillian includes isotropic Zeeman interactions for all particles and equal isotropic hyperfine couplings between one of the electrons and the four nuclei. Only the fully symmetric irreducible representation block (top left) is populated unless the symmetry is broken and subsequently restored

$$\hat{P}_{12}(|\alpha\beta\rangle - |\beta\alpha\rangle)(\langle\alpha\beta| - \langle\beta\alpha|)\hat{P}_{12}^{\dagger} = (|\alpha\beta\rangle - |\beta\alpha\rangle)(\langle\alpha\beta| - \langle\beta\alpha|) \tag{4.185}$$

More generally, any wavefunction belonging to an irrep other than the fully symmetric one

$$g(|\psi\rangle) = e^{i\varphi}|\psi\rangle, \quad \varphi \in \mathbb{R}, \quad g \in \mathcal{G} \tag{4.186}$$

yields a projector that lives in the fully symmetric irrep because the phase multiplier cancels:

$$g(|\psi\rangle\langle\psi|) = e^{i\varphi}|\psi\rangle\langle\psi|e^{-i\varphi} = |\psi\rangle\langle\psi| \tag{4.187}$$

Unless the symmetry had been broken in the system's past, coherences connecting different irreps have no way or arising: the equation of motion obeys the symmetry by definition, the thermal equilibrium state inherits particle permutation symmetry of the Hamiltonian, and any user-supplied initial condition is symmetric by definition with respect to the spins that the user had declared equivalent.

Thus, in Liouville space, SALCs of basis operators not belonging to the fully symmetric irrep of the system symmetry group may be dropped from the basis because they do not get populated. The construction of fully symmetric SALCs is efficient because all characters of the fully symmetric irrep are equal to 1:

$$\mathbf{O}_k^{(\mathrm{A_{1g}})} = \frac{1}{N}\sum_{g \in G} g(\mathbf{O}_k) \tag{4.188}$$

The resulting Liouville space dimension reduction factor is equal to the order of the group. In the rare situations where other irreps are populated (e.g. by the user's decision in the initial condition) and therefore must be tracked in Liouville space, there is still a significant efficiency gain because the Liouvillian is block-diagonal (Fig. 4.3) in the symmetry-adapted basis.

4.6.3 Total Spin Representation

Another approach to symmetry factorisation stems from the total spin representation. For a given set of identical spins, the direct product representation of their algebra may be reduced by diagonalising the Casimir operator (Sect. 1.5.12) and one of the three generators, conventionally $\mathbf{S}_\mathrm{Z}$. The procedures are described in Sect. 2.5; they are only efficient when the factorisation is applied to each subset of identical spins *before* the direct product representation for the entire spin system is constructed. Factorisation by the total spin achieves the same final result of block-diagonalising the Hamiltonian operator (Hilbert space) and commutation superoperator (Liouville space).

4.7 Product Operator Formalism

Magnetic resonance spectroscopy has a powerful semi-analytical formalism that provides physical insight into time-domain spin dynamics and enables straightforward analysis of common experiments [125]. It uses the fact that density matrix can be eliminated from the equation of motion, and the time dynamics problem reformulated entirely in terms of observables. For a specific observable $O(t)$:

$$\begin{aligned} \frac{\partial}{\partial t}\langle \mathbf{O}\rangle &= \mathrm{Tr}\left(\mathbf{O}\left(\frac{\partial}{\partial t}\boldsymbol{\rho}\right)\right) = -i\mathrm{Tr}(\mathbf{O}[\mathbf{H},\boldsymbol{\rho}]) \\ &= i\mathrm{Tr}([\mathbf{H},\mathbf{O}]\boldsymbol{\rho}) = i\langle[\mathbf{H},\mathbf{O}]\rangle \end{aligned} \tag{4.189}$$

where $\mathbf{O}$ is the operator of the observable. Because the interactions in the Hamiltonian are at most two-spin operators, the commutators on the right-hand side are straightforward. When some relevant subset of observables is chosen, the result is a system of equations not unlike those seen in chemical kinetics—the subject with which chemists have much experience and intuition.

In particular, when the operator basis is chosen to be direct products of single-spin operators and the interactions are considered one at a time, the dynamics prescribed by Eq. (4.189) may be represented by simple rotation diagrams. On the education side, these diagrams are the foundation of magnetic resonance spectroscopy and imaging.

4.7.1 Evolution Under Zeeman Interactions

Consider the evolution of a single isotropically shielded spin L in a strong and uniform magnetic field directed along the Z-axis. The Hamiltonian is just the Zeeman interaction $\omega\mathbf{L}_\mathrm{Z}$, and a convenient basis set are Cartesian spin operators $\{\mathbf{L}_\mathrm{X},\mathbf{L}_\mathrm{Y},\mathbf{L}_\mathrm{Z}\}$. The commutators are the structure relations of the $\mathfrak{su}(2)$ algebra in Eq. (1.125); placing those into Eq. (4.189) yields

$$\begin{cases} \dfrac{\partial}{\partial t}\langle \mathbf{L}_\mathrm{X}\rangle = -\omega\langle \mathbf{L}_\mathrm{Y}\rangle \\ \dfrac{\partial}{\partial t}\langle \mathbf{L}_\mathrm{Y}\rangle = +\omega\langle \mathbf{L}_\mathrm{X}\rangle \\ \dfrac{\partial}{\partial t}\langle \mathbf{L}_\mathrm{Z}\rangle = 0 \end{cases} \tag{4.190}$$

where $\langle \mathbf{L}_{\{\mathrm{X,Y,Z}\}}\rangle = L_{\{\mathrm{X,Y,Z}\}}$ notation will be used for observables from now on. Equation (4.190) is a special case of Bloch equations [94] describing circular precession of the $[L_\mathrm{X} \quad L_\mathrm{Y} \quad L_\mathrm{Z}]$ vector around the magnetic field vector. This may be seen from the solution produced by $L_\mathrm{X} = 1/2$, $L_\mathrm{Y} = L_\mathrm{Z} = 0$ initial condition:

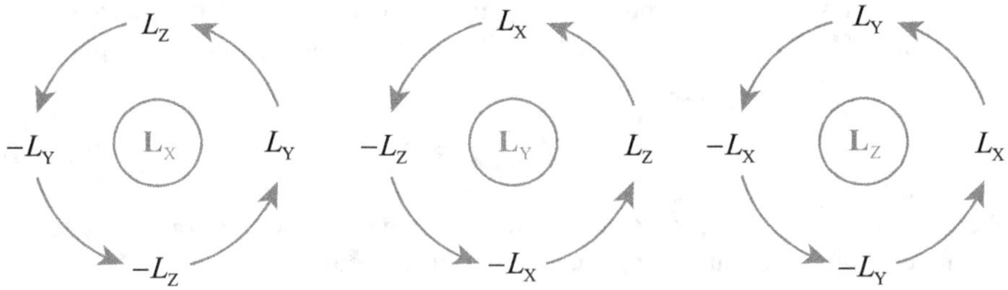

Fig. 4.4 Exponential action diagrams by SU(2) generators, indicated in the circles, on the $\mathfrak{su}(2)$ Lie algebra. Physically, the generators of SU(2) correspond to observable magnetisation operators, and this picture may therefore be interpreted as a magnetic moment precessing around the external magnetic field

$$L_X(t) = \frac{1}{2}\cos(\omega t), \quad L_Y(t) = \frac{1}{2}\sin(\omega t), \quad L_Z(t) = 0 \tag{4.191}$$

After a similar treatment for the magnetic field directed along the X- and Y-axes of the laboratory frame (corresponding to $\mathbf{H} = \omega\mathbf{L}_X$ and $\mathbf{H} = \omega\mathbf{L}_Y$ respectively), the following diagrams summarise the dynamics:

From the Lie algebraic point of view, these are $\mathfrak{su}(2)$ group action diagrams by the generators indicated in the central circles on the $\mathfrak{su}(2)$ algebra. From the physical point of view, these diagrams are interpreted using spin state classification discussed in Sect. 4.2.3—as rotations in the subspaces spanned by the two observables on the outside of the circle, generated by the operator that appears on the inside. Mathematically, the diagrams in Fig. 4.4 describe the following propagator group orbits:

$$\begin{aligned} e^{-i\omega\mathbf{L}_Z t}\mathbf{L}_X e^{+i\omega\mathbf{L}_Z t} &= \mathbf{L}_X\cos(\omega t) + \mathbf{L}_Y\sin(\omega t) \\ e^{-i\omega\mathbf{L}_Z t}\mathbf{L}_Y e^{+i\omega\mathbf{L}_Z t} &= \mathbf{L}_Y\cos(\omega t) - \mathbf{L}_X\sin(\omega t) \\ e^{-i\omega\mathbf{L}_Z t}\mathbf{L}_Z e^{+i\omega\mathbf{L}_Z t} &= \mathbf{L}_Z \end{aligned} \tag{4.192}$$

but they may also be viewed (using the correspondence between operators and states discussed in Sect. 4.2.3) as rotations of the Cartesian components of the magnetic moment vector. This latter picture dominates chemistry literature, where only the observables are considered:

$$\begin{aligned} L_X &\xrightarrow{\omega\mathbf{L}_Z} L_X\cos(\omega t) + L_Y\sin(\omega t) \\ L_Y &\xrightarrow{\omega\mathbf{L}_Z} L_Y\cos(\omega t) - L_X\sin(\omega t) \\ L_Z &\xrightarrow{\omega\mathbf{L}_Z} L_Z \end{aligned} \tag{4.193}$$

The same rules apply to product states that involve other spins—because the Hamiltonian $\mathbf{H} = \omega \mathbf{L}_{\{X,Y,Z\}}$ commutes with operators acting on other spins, the same rotation diagrams apply to

$$\cdots \otimes \mathbf{A} \otimes \mathbf{L}_{\{X,Y,Z\}} \otimes \mathbf{C} \otimes \cdots \tag{4.194}$$

where the operators **A**, **C**, etc. on other spins are arbitrary. Because operators acting on different spins commute, the general case of the Zeeman interaction Hamiltonian being a linear combination of all single-spin operators in the system (Sects. 3.1.6 and 3.2.1):

$$\mathbf{H} = \sum_k \mathbf{L}^{(k)} \cdot \mathbf{Z}_k \cdot \mathbf{B} \tag{4.195}$$

reduces to the case considered above. Any exponential action by **H** splits into a product of actions by single-spin operators; those actions may be considered one at a time.

4.7.2 Evolution Under Spin–Spin Couplings

In the general case of an arbitrary interaction tensor, the product operator formalism offers no cognitive or logistical advantages over Liouville—von Neumann equation; brute-force numerics is the best way forward. However, in the common case of "weak" spin–spin coupling (Sect. 4.3.8.4), simple evolution diagrams do exist [125]. The weak interaction Hamiltonian is

$$\mathbf{H} = \omega_{\mathrm{C}} \mathbf{L}_{\mathrm{Z}} \mathbf{S}_{\mathrm{Z}} \tag{4.196}$$

where $\omega_{\mathrm{C}} = 2\pi J$ in the case of heteronuclear J-coupling. In the case of heteronuclear dipolar coupling, $\omega_{\mathrm{C}} = -(\mu_0/4\pi)\left(\gamma_{\mathrm{L}}\gamma_{\mathrm{S}}\hbar/r_{\mathrm{LS}}^5\right)\left[3(z_{\mathrm{L}} - z_{\mathrm{S}})^2 - r_{\mathrm{LS}}^2\right]$, and so on for other interactions (Sect. 3.2.1).

The longitudinal magnetisation of both spins commutes with the Hamiltonian in Eq. (4.196) and therefore remans invariant. The commutation relations for the transverse magnetisation are

$$\begin{aligned} [\mathbf{L}_{\mathrm{Z}}\mathbf{S}_{\mathrm{Z}}, \mathbf{L}_{\mathrm{X}}] &= [\mathbf{L}_{\mathrm{Z}}, \mathbf{L}_{\mathrm{X}}]\mathbf{S}_{\mathrm{Z}} = +i\mathbf{L}_{\mathrm{Y}}\mathbf{S}_{\mathrm{Z}}, \\ [\mathbf{L}_{\mathrm{Z}}\mathbf{S}_{\mathrm{Z}}, \mathbf{L}_{\mathrm{Y}}\mathbf{S}_{\mathrm{Z}}] &= [\mathbf{L}_{\mathrm{Z}}, \mathbf{L}_{\mathrm{Y}}]\mathbf{S}_{\mathrm{Z}}^2 = -(i/4)\mathbf{L}_{\mathrm{X}} \\ [\mathbf{L}_{\mathrm{Z}}\mathbf{S}_{\mathrm{Z}}, \mathbf{L}_{\mathrm{Y}}] &= [\mathbf{L}_{\mathrm{Z}}, \mathbf{L}_{\mathrm{Y}}]\mathbf{S}_{\mathrm{Z}} = -i\mathbf{L}_{\mathrm{X}}\mathbf{S}_{\mathrm{Z}}, \\ [\mathbf{L}_{\mathrm{Z}}\mathbf{S}_{\mathrm{Z}}, \mathbf{L}_{\mathrm{X}}\mathbf{S}_{\mathrm{Z}}] &= [\mathbf{L}_{\mathrm{Z}}, \mathbf{L}_{\mathrm{X}}]\mathbf{S}_{\mathrm{Z}}^2 = +(i/4)\mathbf{L}_{\mathrm{Y}} \end{aligned} \tag{4.197}$$

where the factor of 4 in the denominators comes from $\mathbf{S}_{\mathrm{Z}}^2 = \mathbf{1}/4$ for spin 1/2. For higher spin quantum numbers, the commutator would be different because $\mathbf{S}_{\mathrm{Z}}^2$

would not be a multiple of the unit matrix; those cases are onerous, and numerical treatments are the best way forward.

With the commutators in place, Eq. (4.189) yields the following equations of motion for the observables when both spins are 1/2:

$$\begin{cases} \frac{\partial}{\partial t} L_X = -\omega_C L_Y S_Z \\ \frac{\partial}{\partial t} L_Y S_Z = +\frac{\omega_C}{4} L_X \end{cases}, \quad \begin{cases} \frac{\partial}{\partial t} L_Y = +\omega_C L_X S_Z \\ \frac{\partial}{\partial t} L_X S_Z = -\frac{\omega_C}{4} L_Y \end{cases} \tag{4.198}$$

where angular brackets are now dropped, and the composite symbols like $L_Y S_Z$ are to be viewed as single variables, not products. The dynamics is again rotational, but the trajectories appear to be elliptical. For example, when the initial condition is L_X (left) or L_Y(right):

$$\begin{cases} L_X(t) = \cos(\omega_C t/2) \\ L_Y S_Z(t) = \frac{1}{2}\sin(\omega_C t/2) \end{cases}, \quad \begin{cases} L_Y(t) = \cos(\omega_C t/2) \\ L_X S_Z(t) = -\frac{1}{2}\sin(\omega_C t/2) \end{cases} \tag{4.199}$$

Moving the factor of 2 into the definition of the two-spin order turns ellipses into circles:

$$\begin{cases} L_X(t) = \cos(\omega_C t/2) \\ 2L_Y S_Z(t) = \sin(\omega_C t/2) \end{cases}, \quad \begin{cases} L_Y(t) = \cos(\omega_C t/2) \\ 2L_X S_Z(t) = -\sin(\omega_C t/2) \end{cases} \tag{4.200}$$

This normalisation transformation works fine in this system of two spin-1/2 particles, but becomes problematic in more general cases because the multipliers misbehave (Sect. 1.6.3.3)—a silly accounting trick of adding a non-interacting ghost spin at the other end of the Universe makes them change. Note also that the frequency is half of what occurs in the Hamiltonian in Eq. (4.196). This may be fixed using the same normalisation trick—for example, for the weak J-coupling (Sect. 3.2.7) in NMR spectroscopy:

$$\mathbf{H} = \pi J (2\mathbf{L}_Z \mathbf{S}_Z) \tag{4.201}$$

and thus the rotation frequency is πJ. The corresponding rotation diagrams are

Product operator formalism is well developed; instructions on dealing with more complicated systems and interactions may be found in [125]. There are *Mathematica* extensions that automate it [126].

The nature of direct product basis sets, and the following property of Kronecker products

$$(\mathbf{A} \otimes \mathbf{B} \otimes \cdots)(\mathbf{C} \otimes \mathbf{D} \otimes \cdots) = (\mathbf{AC}) \otimes (\mathbf{BD}) \otimes \cdots \tag{4.202}$$

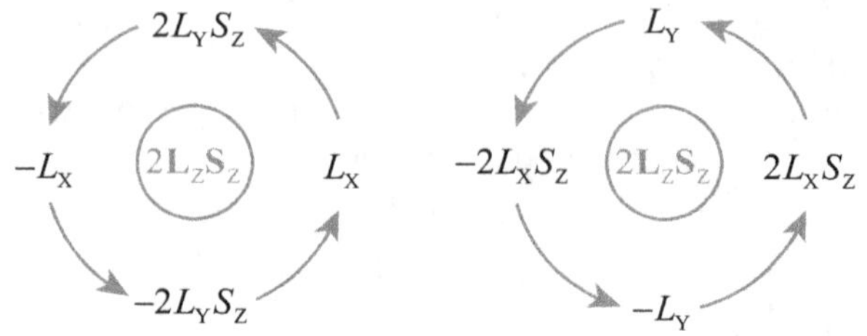

Fig. 4.5 Exponential action diagrams by SU(4) generator indicated in the circles on the $\mathfrak{su}(4)$ Lie algebra. Physically, the indicated generator of SU(4) corresponds to the weak spin–spin coupling (Sect. 4.3.8.4), and this picture may therefore be interpreted as rotational dynamics between a transverse magnetisation direction of spin L and its perpendicular direction where the sign of the magnetisation depends on the longitudinal projection state of the partner spin S

means that the evolution diagrams in Figs. 4.4 and 4.5 apply to systems with an arbitrary number of spectator spins. Another extension appears when we notice that the exponential propagation relations in Eq. (4.192) are the consequence of the commutation relations between the generators of $\mathfrak{su}(2)$—any other system of operators that follows the same commutation rules as $\{\mathbf{S}_\text{X}, \mathbf{S}_\text{Y}, \mathbf{S}_\text{Z}\}$ would follow the same diagram as Fig. 4.4. One example is Fig. 4.5, and many more may be found in specialist literature.

4.7.3 Example: Ideal Pulse

Product operator formalism has pedagogic value—rigorous descriptions of magnetic resonance experiments may be smuggled into chemistry and biology departments because the use of quantum mechanics is not overt. It also helps make intuitive sense of complicated spin processes because dynamics on Lie algebras is mapped into what looks like rotational motion. A good example is evolution under strong radiofrequency or microwave pulses, where an adjoint exponential action:

$$e^{-i\omega\mathbf{L}_\text{Y}t}\mathbf{L}_\text{Z}e^{+i\omega\mathbf{L}_\text{Y}t} = \mathbf{L}_\text{Z}\cos(\omega t) + \mathbf{L}_\text{X}\sin(\omega t) \tag{4.203}$$

requiring detailed knowledge of Chap. 1 material is represented by a three-dimensional rotation:

$$L_\text{Z} \overset{\omega\mathbf{L}_\text{Y}}{\longrightarrow} L_\text{Z}\cos(\omega t) + L_\text{X}\sin(\omega t) \overset{\omega t=\pi/2}{\longrightarrow} L_\text{X} \tag{4.204}$$

which is accessible to a liberal arts major. Here, the product of pulse frequency $\omega = -\gamma B_1$ and duration t is the flip angle. In magnetic resonance pulse sequence diagrams, it is common to specify ωt product in radians, and to leave the choice of B_1 and t to the user because equipment settings differ.

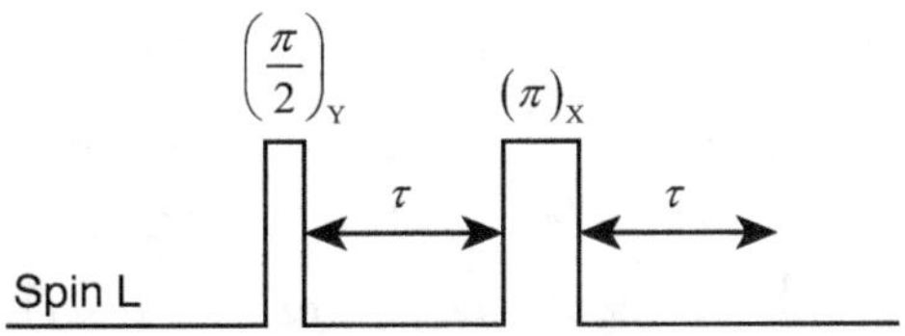

Fig. 4.6 One of the many possible spin echo experiments

4.7.4 Example: Spin Echo

Product operator formalism provides a simple but rigorous description of the *spin echo* [127]—an important element of magnetic resonance experiments on heterogeneous samples, such as powders in solid-state NMR and tissues in MRI, where the Larmor frequency ω_k may be different for each spin k in the ensemble. The pulse sequence contains two pulses and two delays (Fig. 4.6):

In the figure, the where pulses are specified by their generator (X and Y subscripts correspond to $\mathbf{L}_X$ and $\mathbf{L}_Y$ evolution generators) and the effective flip angle $\varphi = -\gamma B_1 t$, where t is the duration of the pulse. If the initial condition is Z magnetisation, then Fig. 4.4 indicates that at the end of the first pulse:

$$L_Z^{(k)} \xrightarrow{(\pi/2)_Y} L_Z^{(k)} \cos\left(\frac{\pi}{2}\right) + L_X^{(k)} \sin\left(\frac{\pi}{2}\right) = L_X^{(k)} \tag{4.205}$$

where $(\pi/2)_Y$ is a shorthand for the evolution under $\omega \mathbf{L}_Y$ for a time t such that $\omega t = \pi/2$. The system is now in the state $\sum_k \mathbf{L}_X^{(k)}$—all spins have the same phase. However, because their precession frequencies are different, they would go out of phase during the evolution period τ:

$$L_X^{(k)} \xrightarrow{\omega_k \mathbf{L}_Z^{(k)}} L_X^{(k)} \cos(\omega_k \tau) + L_Y^{(k)} \sin(\omega_k \tau) \tag{4.206}$$

This can be undesirable, for example in MRI, because the total transverse magnetisation is reduced when there is a distribution in the transverse precession phases. However, this dephasing is reversed after the π_X pulse is applied, which flips the signs of $L_Y^{(k)}$:

$$L_X^{(k)} \cos(\omega_k \tau) + L_Y^{(k)} \sin(\omega_k \tau) \xrightarrow{\pi_X} L_X^{(k)} \cos(\omega_k \tau) - L_Y^{(k)} \sin(\omega_k \tau) \tag{4.207}$$

Then, at the end of the second evolution period τ:

$$\begin{aligned} & L_{\mathrm{X}}^{(k)}\cos(\omega_k\tau) - L_{\mathrm{Y}}^{(k)}\sin(\omega_k\tau) \xrightarrow{\omega \mathbf{L}_{\mathrm{Z}}^{(k)}} \\ & \quad \left[L_{\mathrm{X}}^{(k)}\cos(\omega_k\tau) + L_{\mathrm{Y}}^{(k)}\sin(\omega_k\tau)\right]\cos(\omega_k\tau) \\ & \quad -\left[L_{\mathrm{Y}}^{(k)}\cos(\omega_k\tau) - L_{\mathrm{X}}^{(k)}\sin(\omega_k\tau)\right]\sin(\omega_k\tau) \\ & \quad = L_{\mathrm{X}}^{(k)}\left[\cos^2(\omega_k\tau) + \sin^2(\omega_k\tau)\right] = L_{\mathrm{X}}^{(k)} \end{aligned} \tag{4.208}$$

where the magnetisation is again frequency-independent—all spins have the same phase. The π pulse of the spin-echo experiment is called refocusing pulse because its effect is to bring the ensemble magnetisation back into a coherent state.

4.7.5 Example: Magnetisation Transfer

A popular building block of magnetic resonance experiments is magnetisation transfer through weak scalar coupling [128]. Consider a liquid state sample containing ^{1}H–^{15}N spin pairs in a high-field magnet, and an experiment in which two independent RF transmitters are tuned exactly to ^{1}H and ^{15}N Zeeman frequencies. These transmitters are assumed to be powerful enough that millisecond-scale J-coupling evolution may be ignored during microsecond-scale RF pulses. In the rotating frame, the Hamiltonian is

$$\mathbf{H} = \begin{cases} \omega_{1\mathrm{L}}(\mathbf{L}_{\mathrm{X}}\cos\varphi_{\mathrm{L}} + \mathbf{L}_{\mathrm{Y}}\sin\varphi_{\mathrm{L}}) + \omega_{1\mathrm{S}}(\mathbf{S}_{\mathrm{X}}\cos\varphi_{\mathrm{S}} + \mathbf{S}_{\mathrm{Y}}\sin\varphi_{\mathrm{S}}) & \text{during hard pulses} \\ \omega_{\mathrm{C}}\mathbf{L}_{\mathrm{Z}}\mathbf{S}_{\mathrm{Z}},\ \omega_{\mathrm{C}} = 2\pi J & \text{during free evolution} \end{cases} \tag{4.209}$$

where $\omega_{1\mathrm{L},1\mathrm{S}}$ are nutation frequencies under the pulses applied to the indicated spins, $\varphi_{\mathrm{L,S}}$ are phases of those pulses in the rotating frame, and ω_{C} is the angular frequency of the scalar coupling. Full quantum mechanical treatment of the pulse sequence in Fig. 4.7 either analytically or numerically would be a considerable undertaking, but product operator formalism makes the analysis straightforward.

Either analytically or numerically would be a considerable undertaking, but product operator formalism makes the analysis straightforward.

Consider the initial condition where protons (L spin) are magnetised on the Z-axis, but the initial magnetisation of ^{15}N (S spin) is negligible; thus $\boldsymbol{\rho}_0 \propto \mathbf{L}_{\mathrm{Z}}$ in the high-temperature limit (Sect. 6.9). The first pulse makes proton magnetisation transverse (middle diagram in Fig. 4.4):

$$L_{\mathrm{Z}} \xrightarrow{(\pi/2)_{\mathrm{Y}}} L_{\mathrm{X}} \tag{4.210}$$

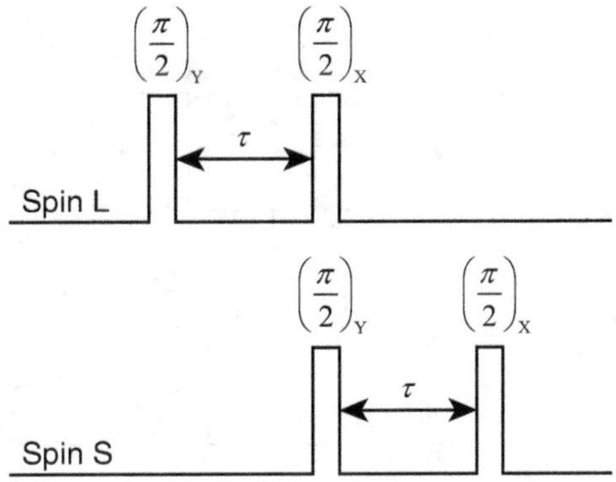

Fig. 4.7 A magnetisation transfer pulse sequence that converts longitudinal magnetisation of spin L into longitudinal magnetisation on spin S in a system where both spins are exactly on resonance with their corresponding control channel transmitters and the spin–spin coupling is "weak" in the sense of only having the ZZ term (Sect. 4.3.8.4)

Subsequent evolution under $\omega_C \mathbf{L}_Z \mathbf{S}_Z$ rotates L_X towards the two-spin order (left diagram in Fig. 4.5). Choosing a delay τ such that $\omega_C \tau = \pi$ brings the magnetisation completely into the two-spin order:

$$L_X \xrightarrow{\omega_C \mathbf{L}_Z \mathbf{S}_Z} L_X \cos\left(\frac{\omega_C}{2}\tau\right) + 2L_Y S_Z \sin\left(\frac{\omega_C}{2}\tau\right) \xrightarrow{a\tau=\pi} 2L_Y S_Z \tag{4.211}$$

The next pair of pulses moves the state where the proton is transverse and nitrogen longitudinal into the state where the proton is longitudinal, and the nitrogen is transverse:

$$2L_Y S_Z \xrightarrow{(\pi/2)_X \text{ on } L} 2L_Z S_Z \xrightarrow{(\pi/2)_Y \text{ on } S} 2L_Z S_X \tag{4.212}$$

The next evolution period (right diagram in Fig. 4.5) rotates the resulting two-spin order towards transverse nitrogen magnetisation. Choosing again the evolution delay such that $\omega_C \tau = \pi$ yields

$$2L_Z S_X \xrightarrow{\omega_C \mathbf{L}_Z \mathbf{S}_Z} 2L_Z S_X \cos\left(\frac{\omega_C}{2}\tau\right) + S_Y \sin\left(\frac{\omega_C}{2}\tau\right) \xrightarrow{\omega_C \tau=\pi} S_Y \tag{4.213}$$

The optional last pulse (left diagram in Fig. 4.4) makes nitrogen magnetisation longitudinal:

$$S_Y \xrightarrow{(\pi/2)_X \text{ on } S} S_Z \tag{4.214}$$

If the two nuclei are not exactly on resonance with the corresponding RF transmitters, or there exists a distribution of precession frequencies, spin-echo stages—in the form of π pulses—are inserted into the evolution periods to make sure that the offsets are refocused (Fig. 4.8).

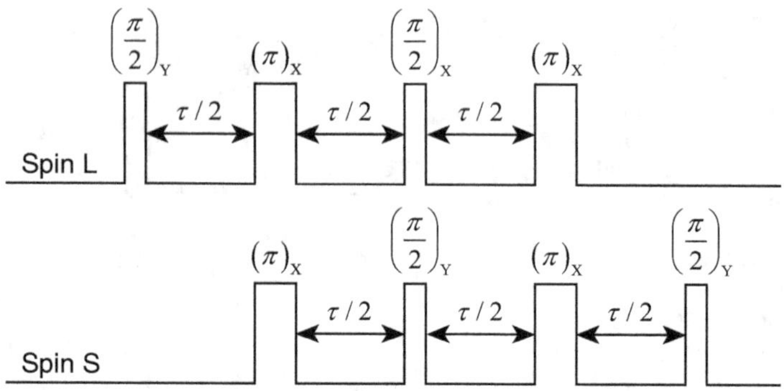

Fig. 4.8 A magnetisation transfer pulse sequence that converts longitudinal magnetisation of spin L into longitudinal magnetisation on spin S in a system where spins might not be on resonance with their corresponding control channel transmitters and the spin–spin coupling is "weak" in the sense of only having the ZZ term (Sect. 4.3.8.4)

A similar analysis shows that this sequence accomplishes the same magnetisation transfer, but it is also resilient to Larmor frequency offsets on both spins.

4.8 Floquet Theory

Spin Hamiltonians are often time-periodic: external fields used to manipulate the spins are electromagnetic, anisotropic interactions are attenuated in some experiments by spinning the sample, and sequences used for decoupling contain repeating patterns of pulses. This suggests that the Fourier series may be a useful way of looking at the dynamics. Indeed, Gaston Floquet demonstrated in 1883 [129] that solutions of linear ODEs with periodic coefficients have periodic factors. In particular, solutions of the LvN equation

$$\frac{\partial}{\partial t}\boldsymbol{\rho}(t) = -i\mathcal{L}(t)\boldsymbol{\rho}(t), \quad \mathcal{L}(t) = \mathcal{H}(t) + i\mathcal{R} \tag{4.215}$$

with a periodic Hamiltonian commutation superoperator

$$\mathcal{H}(t+T) = \mathcal{H}(t) \quad \Rightarrow \quad \mathcal{H}(t) = \sum_n \mathcal{H}_n e^{in\omega t}, \quad \omega = 2\pi/T, \quad n \in \mathbb{Z} \tag{4.216}$$

must have the following form:

$$\boldsymbol{\rho}(t) = \sum_n \boldsymbol{\rho}_n(t) e^{in\omega t}, \quad \omega = 2\pi/T, \quad n \in \mathbb{Z} \tag{4.217}$$

This is useful because problems with time-dependent periodic generators (magic angle spinning, off-resonance RF and MW fields, etc.) may be converted into problems with a time-independent generator in a space of higher dimension. This section describes the procedure.

4.8.1 Single-Mode Floquet Theory

We start by writing a Fourier expansion of the Hamiltonian commutation superoperator:

$$\mathcal{H}(t) = \sum_n \mathcal{H}_n e^{in\omega t}, \quad \omega = 2\pi/T, \quad n \in \mathbb{Z} \tag{4.218}$$

and substituting the harmonic series from Eqs (4.217) and (4.218) into the LvN equation:

$$\frac{\partial}{\partial t}\left[\sum_n \boldsymbol{\rho}_n(t) e^{in\omega t}\right] = -i \sum_{m,k} \mathcal{H}_m \boldsymbol{\rho}_k(t) e^{i(m+k)\omega t} \tag{4.219}$$

The double sum on the right is inconveniently indexed; it helps to replace $m + k \to n$, and therefore also $m \to n - k$. The derivative on the left may be opened up:

$$\sum_n \left(\frac{\partial}{\partial t}\boldsymbol{\rho}_n(t) + in\omega\boldsymbol{\rho}_n(t)\right)\mathrm{e}^{in\omega t} = \sum_n \left(-i\sum_k \mathcal{H}_{n-k}\boldsymbol{\rho}_k(t)\right)\mathrm{e}^{in\omega t} \tag{4.220}$$

Because ω and t are arbitrary, the uniqueness theorem for Fourier expansions requires these sums to be equal element-by-element:

$$\left\{\frac{\partial}{\partial t}\boldsymbol{\rho}_n(t) = -i\sum_k \mathcal{H}_{n-k}\boldsymbol{\rho}_k(t) - in\omega\boldsymbol{\rho}_n(t)\right._n \tag{4.221}$$

This is now a system of linear first-order ODEs for the harmonic components of the state vector. Truncating the expansion in Eq. (4.217) at some finite multiple k of the Hamiltonian frequency (user-selectable, increase until convergence) and stacking $\boldsymbol{\rho}_n(t)$ vertically in the order of descending indices to make a longer vector $\vec{\boldsymbol{\rho}}(t)$ yields the following block matrix equation:

$$\begin{aligned} \frac{\partial}{\partial t}\vec{\boldsymbol{\rho}}(t) &= -i\mathcal{F}\vec{\boldsymbol{\rho}}(t) \\ \mathcal{F} &= \mathbf{T}\{\mathcal{H}_{-k}, \cdots, \mathcal{H}_{+k}\} + \boldsymbol{\Omega}^{(k)} \otimes \mathbf{1} \\ \boldsymbol{\Omega}^{(k)} &= \mathrm{diag}[+k\omega, \cdots, -k\omega] \end{aligned} \tag{4.222}$$

where $\mathbf{1}$ is a unit matrix of the same dimension as $\mathcal{H}$, and the block-Toeplitz matrix $\mathbf{T}\{\ldots\}$ is obtained by placing the arguments $\{\mathcal{H}_{-k},\cdots,\mathcal{H}_{+k}\}$ along the corresponding diagonals of a block matrix [130].

Relative to Eq. (4.215), the dimension of the problem has increased. However, the problem also became easier: the evolution generator is now time-independent. Equation (4.222) may be solved using any of the numerical methods developed for the time-independent Schrödinger equation (Sect. 4.9 and Chapter 9).

4.8.2 Effective Hamiltonian in Floquet Theory

The equation of motion obtained in the previous section lives in a direct product of a Hilbert space populated by density matrices $\boldsymbol{\rho}_n(t)$ and a Fourier space spanned by the complex harmonics $e^{in\omega t}$. The evolution generator is not diagonal in either Hilbert or Fourier space:

$$\mathcal{F}=\begin{pmatrix} & & \vdots & & \\ & \mathcal{H}_0+\omega\mathbf{1} & \mathcal{H}_{+1} & \mathcal{H}_{+2} & \\ \cdots & \mathcal{H}_{-1} & \mathcal{H}_0 & \mathcal{H}_{+1} & \cdots \\ & \mathcal{H}_{-2} & \mathcal{H}_{-1} & \mathcal{H}_0-\omega\mathbf{1} & \\ & & \vdots & & \end{pmatrix} \tag{4.223}$$

Diagonalising this matrix with respect to the Fourier part of the product space (*i.e.* reshuffling the blocks to make it block-diagonal) is not an obvious necessity. What that would do is partition the dynamics into non-interacting and non-intersecting orbits (Sect. 1.5.3) under $\exp(-i\mathcal{F}t)$ propagator group action. The advantage of such a step is that most initial conditions only populate the central block of the resulting block-diagonal structure. Other blocks are unimportant because they never become active—the idea of the effective Hamiltonian in Floquet theory is to only keep the central block. This approach is fundamentally different from the average Hamiltonian theory (Sect. 4.3.8) because there is no averaging.

Analytically, the block-diagonalization is best accomplished using Van Vleck perturbation theory, where the $\mathbf{Q}$ matrix (Sect. 4.4.2) is simple because the frequencies in the Fourier basis are equally spaced. We set the scene by splitting the matrix into block-diagonal and block-off-diagonal part:

$$\mathcal{F}=\mathbf{1}\otimes\mathcal{H}_0+\boldsymbol{\Omega}^{(k)}\otimes\mathbf{1}+\mathbf{T}\{\mathcal{H}_{-k},\cdots,0,\cdots\mathcal{H}_{+k}\} \tag{4.224}$$

and treating the Toeplitz term as the perturbation. Tedious application of Eq. (4.143) yields [131]

$$
\begin{aligned}
&\mathcal{H}_{\text{eff}} = \mathcal{H}_{\text{eff}}^{(0)} + \mathcal{H}_{\text{eff}}^{(1)} + \mathcal{H}_{\text{eff}}^{(2)} + \dots \\
&\mathcal{H}_{\text{eff}}^{(0)} = \mathcal{H}_0, \ \mathcal{H}_{\text{eff}}^{(1)} = -\frac{1}{2}\sum_{n\neq 0}\frac{[\mathcal{H}_{-n}, \mathcal{H}_n]}{\omega n} \\
&\mathcal{H}_{\text{eff}}^{(2)} = \frac{1}{2}\sum_{n\neq 0}\frac{[\mathcal{H}_{-n}, [\mathcal{H}_0, \mathcal{H}_n]]}{\omega^2 n^2} - \frac{1}{3}\sum_{\substack{n\neq 0, k\neq 0,\\ n\neq k}}\frac{[\mathcal{H}_{-n}, [\mathcal{H}_k, \mathcal{H}_{n-k}]]}{\omega^2 nk}
\end{aligned}
\tag{4.225}
$$

where the sums are short because the highest harmonic rank in the Hamiltonian in Eq. (4.218) is rarely bigger than 2. It is clear from the appearance of the denominators that this theory works well when ω is much greater than the 2-norm of the spin Hamiltonian commutation superoperator.

4.8.3 Multi-mode Floquet Theory

When multiple non-commensurate frequencies, here collected into a vector $\boldsymbol{\omega}$, are present in the Hamiltonian, the summation index n in the harmonic expansion also becomes a vector $\mathbf{n}$, and the sum runs over all components of that vector:

$$
\boldsymbol{\rho}(t) = \sum_{\mathbf{n}} \boldsymbol{\rho}_{\mathbf{n}}(t) e^{i(\mathbf{n}\cdot\boldsymbol{\omega})t}, \qquad \mathcal{H}(t) = \sum_{\mathbf{n}} \mathcal{H}_{\mathbf{n}} e^{i(\mathbf{n}\cdot\boldsymbol{\omega})t}
\tag{4.226}
$$

Performing the same substitutions into the LvN equation as in Sect. 4.8.1 yields

$$
\frac{\partial}{\partial t}\left[\sum_{\mathbf{n}} \boldsymbol{\rho}_{\mathbf{n}}(t) e^{i(\mathbf{n}\cdot\boldsymbol{\omega})t}\right] = -i\sum_{\mathbf{m},\mathbf{k}} \mathcal{H}_{\mathbf{m}} \boldsymbol{\rho}_{\mathbf{k}}(t) e^{i[(\mathbf{m}+\mathbf{k})\cdot\boldsymbol{\omega}]t}
\tag{4.227}
$$

After the same index replacement ($\mathbf{m}+\mathbf{k} \to \mathbf{n}$, $\mathbf{m} \to \mathbf{n}-\mathbf{k}$) and differentiation:

$$
\sum_{\mathbf{n}}\left(\frac{\partial}{\partial t}\boldsymbol{\rho}_{\mathbf{n}}(t) + i(\mathbf{n}\cdot\boldsymbol{\omega})\boldsymbol{\rho}_{\mathbf{n}}(t)\right) e^{i(\mathbf{n}\cdot\boldsymbol{\omega})t} = \sum_{\mathbf{n}}\left(-i\sum_{\mathbf{k}} \mathcal{H}_{\mathbf{n}-\mathbf{k}}\boldsymbol{\rho}_{\mathbf{k}}(t)\right) e^{i(\mathbf{n}\cdot\boldsymbol{\omega})t}
\tag{4.228}
$$

Because $\boldsymbol{\omega}$ and t are arbitrary, these sums must be equal element-by-element:

$$
\left\{\frac{\partial}{\partial t}\boldsymbol{\rho}_{\mathbf{n}}(t) = -i\sum_{\mathbf{k}} \mathcal{H}_{\mathbf{n}-\mathbf{k}}\boldsymbol{\rho}_{\mathbf{k}}(t) - i(\mathbf{n}\cdot\boldsymbol{\omega})\boldsymbol{\rho}_{\mathbf{n}}(t)\right.\Bigg._{\mathbf{n}}
\tag{4.229}
$$

which has the same structure as Eq. (4.221), except **n** and **k** are now vector indices [131]. Numerically, the best way to proceed is to unroll the vector index into a linear index, and then use the same block Toeplitz construction as in Eq. (4.222) to obtain a flat matrix representation.

4.8.4 Floquet-Magnus Expansion

The most general form of the Floquet theorem [132] states that when the matrix $\mathbf{A}(t)$ in

$$\frac{\mathrm{d}\mathbf{Y}}{\mathrm{d}t} = \mathbf{A}(t)\mathbf{Y} \tag{4.230}$$

is periodic with a period T, the solution has the form

$$\mathbf{Y}(t) = \mathbf{P}(t)\exp(t\mathbf{F}) \tag{4.231}$$

where $\mathbf{P}(t)$ is periodic with the same period, and $\mathbf{F}$ is constant. When we substitute Eq. (4.231) back into Eq. (4.230) and simplify, we obtain

$$\mathbf{P}'(t) = \mathbf{A}(t)\mathbf{P}(t) - \mathbf{P}(t)\mathbf{F}, \quad \mathbf{P}(0) = 0 \tag{4.232}$$

Using an exponential ansatz $\mathbf{P}(t) = \exp(\mathbf{\Lambda}(t))$ yields an equation of motion for $\mathbf{\Lambda}(t)$ in the same way as this happens in Magnus theory (Sect. 4.3.5):

$$\frac{\mathrm{d}}{\mathrm{d}t}\mathbf{\Lambda} = \sum_{k=0}^{\infty}\frac{B_k}{k!}\mathrm{ad}_{\mathbf{\Lambda}}^{k}\left(\mathbf{A} + (-1)^{k+1}\mathbf{F}\right) \tag{4.233}$$

where B_k are Bernoulli's numbers. This leads to the following form for Eq. (4.231):

$$\begin{gathered}\mathbf{\Lambda}(t) = \sum_{n=1}^{\infty}\mathbf{\Lambda}_n(t), \quad \mathbf{F} = \sum_{n=1}^{\infty}\mathbf{F}_n \\ \mathbf{Y}(t) = \exp\left[\sum_{n=1}^{\infty}\mathbf{\Lambda}_n(t)\right]\exp\left[t\sum_{n=1}^{\infty}\mathbf{F}_n\right]\end{gathered} \tag{4.234}$$

that is called *Floquet-Magnus expansion* [132]; it is free of time-ordered exponentials. Plugging the series from Eq. (4.234) into Eq. (4.233) and equating the terms of the same order yields recursive expressions:

$$\begin{gathered}\frac{\mathrm{d}}{\mathrm{d}t}\mathbf{\Lambda}_n = \sum_{k=0}^{n-1}\frac{B_k}{k!}\left(\mathbf{W}_n^{(k)} + (-1)^{k+1}\mathbf{T}_n^{(k)}\right) \\ \mathbf{W}_n^{(k)} = \sum_{m=1}^{n-k}\left[\mathbf{\Lambda}_m, \mathbf{W}_{n-m}^{(k-1)}\right], \quad \mathbf{W}_1^{(0)} = \mathbf{A}, \quad \mathbf{W}_n^{(0)} = \mathbf{0} \\ \mathbf{T}_n^{(k)} = \sum_{m=1}^{n-k}\left[\mathbf{\Lambda}_m, \mathbf{T}_{n-m}^{(k-1)}\right], \quad \mathbf{T}_n^{(0)} = \mathbf{F}_n\end{gathered} \tag{4.235}$$

that are easy to evaluate because there is no $\mathbf{\Lambda}_n(t)$ on the right-hand side. The first two terms are

$$\mathbf{F}_1 = \frac{1}{T}\int_0^T \mathbf{A}(x)dx,\ \mathbf{\Lambda}_1(t) = \int_0^t \mathbf{A}(x)\mathrm{dx} - t\mathbf{F}_1$$

$$\mathbf{F}_2 = \frac{1}{2T}\int_0^T [\mathbf{A}(x) + \mathbf{F}_1, \mathbf{\Lambda}_1(x)]\mathrm{dx} \tag{4.236}$$

$$\mathbf{\Lambda}_2(t) = \frac{1}{2}\int_0^t [\mathbf{A}(x) + \mathbf{F}_1, \mathbf{\Lambda}_1(x)]\mathrm{dx} - t\mathbf{F}_2$$

where the similarity with the average Hamiltonian theory expressions is obvious, but the recursive expression is less complex. An empirically sufficient condition reported by Casas [132] for the absolute convergence of this expansion everywhere in $t \in [0, T]$ is $\int_0^T \|\mathbf{A}(t)\|\mathrm{dt} < 1/5$.

4.9 Numerical Time Propagation

Most time-domain simulation problems in spin dynamics reduce to solving Lie equation [7]

$$\frac{\mathrm{d}}{\mathrm{d}t}\mathbf{x}(t) = \mathbf{A}(\mathbf{x}(t), t)\mathbf{x}(t) \tag{4.237}$$

for the evolution of a state vector $\mathbf{x}(t)$ under the influence of a matrix $\mathbf{A}(\mathbf{x}(t), t)$, which may include unitary dynamics, dissipation, chemical kinetics, and spatial transport. When the equation of motion is non-uniform, it may be returned into the form given in Eq. (4.237) by the following transformation:

$$\begin{gathered}\frac{\mathrm{d}}{\mathrm{d}t}\mathbf{x}(t) = \mathbf{A}(\mathbf{x}(t), t)\mathbf{x}(t) + \mathbf{y}(t) \\ \Updownarrow \\ \frac{\mathrm{d}}{\mathrm{d}t}\begin{pmatrix} 1 \\ \mathbf{x}(t) \end{pmatrix} = \begin{pmatrix} 0 & 0 \\ \mathbf{y}(t) & \mathbf{A}(\mathbf{x}(t), t) \end{pmatrix}\begin{pmatrix} 1 \\ \mathbf{x}(t) \end{pmatrix}\end{gathered} \tag{4.238}$$

In spin dynamics, numerical accuracy requirements for the solutions of Eq. (4.237) can be extreme: for example, cryogenic DNP simulations must accurately track nuclear relaxation (order of 0.01 Hz) in the presence of electron Zeeman interaction (order of 100 GHz) for minutes of physical time. This eliminates common ODE solvers, such as Runge–Kutta [133,134], from the list of possibilities. In double-precision arithmetic, only one class of algorithms works consistently here—geometric integrators [135] that respect the group-theoretical structure of the problem and always act by a matrix exponential.

4.9.1 Product Integrals

Consider first the situation where the evolution generator in Eq. (4.237) does not depend on the state. The solution, for an infinitesimal time increment $\mathrm{d}t$, then is

$$\mathbf{x}(t+\mathrm{d}t) = \exp[\mathbf{A}(t)\mathrm{d}t]\mathbf{x}(t) \tag{4.239}$$

This may be verified by expanding the exponential into a Taylor series and keeping the leading order in $\mathrm{d}t$. The solution over a finite period $t \in [a, b]$ is then obtained by dividing it into a set of intervals

$$\{[t_0, t_1], [t_1, t_2], \dots, [t_{n-1}, t_n]\}, \quad a = t_0 < t_1 < t_2 < \dots < t_n = b \tag{4.240}$$

picking points $\tau_k \in [t_k, t_{k+1}]$ arbitrarily within those intervals, and taking the limit in which the number of intervals goes to infinity and the size of each interval goes to zero:

$$\mathbf{x}(b) = \lim_{\Delta t_k \to 0} \left[\overleftarrow{\prod_k} \exp[\mathbf{A}(\tau_k)\Delta t_k] \right] \mathbf{x}(a) \tag{4.241}$$

where the arrow indicates the direction of a time-ordered product—individual matrix exponentials need not commute. This procedure is reminiscent of the construction of Riemann's integral (which is the limit of a sum [136]), hence the term *product integral.*

4.9.2 Example: Time-Domain NMR

Consider a high-field liquid-state NMR experiment on an isolated CH_2 group with the spinless ^{12}C nucleus. In the interaction representation with respect to the Zeeman Hamiltonian (Sect. 4.3.8.1), the two protons A and B would in general have different offset frequencies $\omega_{A,B}$ and a scalar coupling $\omega_C = 2\pi J$:

$$\mathbf{H} = \omega_A \mathbf{S}_Z^{(A)} + \omega_B \mathbf{S}_Z^{(B)} + \omega_C \left(\mathbf{S}_X^{(A)} \mathbf{S}_X^{(B)} + \mathbf{S}_Y^{(A)} \mathbf{S}_Y^{(B)} + \mathbf{S}_Z^{(A)} \mathbf{S}_Z^{(B)} \right) \tag{4.242}$$

In simulations of this type, *Matlab* is convenient. The first stage is to define single-spin operators:

```
% Define spin operators
Sx=[0, 1/2; 1/2, 0];
Sy=[0, -1i/2; 1i/2, 0];
Sz=[1/2, 0; 0, -1/2];
```

Two-spin operators are generated using Eq. (2.60), by taking Kronecker products with unit matrices:

```
% Build two-spin operators
SAx=kron(Sx,eye(2)); SBx=kron(eye(2),Sx);
SAy=kron(Sy,eye(2)); SBy=kron(eye(2),Sy);
SAz=kron(Sz,eye(2)); SBz=kron(eye(2),Sz);
```

These operators are used to build the Hamiltonian. Typical parameters would be Zeeman offsets of 200 Hz for spin A and 400 Hz for spin B, and the scalar coupling of 40 Hz:

```
% Build the Hamiltonian
omega_A=2*pi*200; omega_B=2*pi*400; omega_C=2*pi*40;
H=omega_A*SAz+omega_B*SBz+omega_C*(SAx*SBx+SAy*SBy+SAz*SBz);
```

The initial condition $\boldsymbol{\rho}_0$ will be transverse magnetisation on both spins:

$$\boldsymbol{\rho}_0 = \mathbf{S}_{\mathrm{X}}^{(\mathrm{A})} + \mathbf{S}_{\mathrm{X}}^{(\mathrm{B})} \tag{4.243}$$

Modern NMR instruments use heterodyne detection, with the in-phase and the out-of-phase signal in the rotating frame combined into a complex number. Because all relations are linear, this is mathematically equivalent to having a non-Hermitian *coil state*:

$$\mathbf{O} = \left(\mathbf{S}_{\mathrm{X}}^{(\mathrm{A})} + \mathbf{S}_{\mathrm{X}}^{(\mathrm{B})}\right) + i\left(\mathbf{S}_{\mathrm{Y}}^{(\mathrm{A})} + \mathbf{S}_{\mathrm{Y}}^{(\mathrm{B})}\right) \tag{4.244}$$

In MATLAB syntax, these relations become

```
% Initial and detection state
rho=SAx+SBx; coil=(SAx+SBx)+1i*(SAy+SBy);
```

We will use Eq. (4.40) for time propagation, but we must first decide the time step Δt. To sample an oscillatory signal without loss of information, we need at least two points per period of the fastest oscillation (*Nyquist-Shannon sampling theorem* [137]). The highest frequency in the system is the largest eigenvalue of the Hamiltonian. For Hermitian matrices, that is the definition of the 2-norm (Sect. 1.4.5), which we can here afford to calculate because our matrices are small:

```
% Decide the time step
time_step=(1/2)*pi/norm(H,2);
% Build the propagator
P=expm(-1i*H*time_step);
```

Time propagation and detection are done in a loop, where the observable is recorded into an array (called free induction decay) using Eq. (4.35), and then a time step is taken using Eq. (4.33):

```
% Time evolution, 2048 steps
for n=1:2048
   fid(n)=trace(coil*rho);
   rho=P*rho*P';
end
```

The number of steps is decided experimentally, it is dictated by the transverse relaxation rate and the target frequency resolution; our calculation here does not include a relaxation model, and we therefore simply pick some suitably large number.

We now have the trajectory for X and Y projections of the spin, and what remains is cosmetic data processing—magnetic resonance data is usually presented in the frequency domain. We have no relaxation in this model, and therefore the signal does not decay. To avoid unphysical wobbles in the Fourier transform, we will manually push it towards zero; this is called apodisation [138]:

```
% Apply signal apodization
decay=exp(-5*linspace(0,1,2048));
fid=fid.*decay;
```

We will instruct *MATLAB* to pad the right edge of the time domain signal with zeroes prior to running the Fourier transform (this zero-filling [139] procedure improves the sampling of the frequency spectrum) and to shift the zero frequency to the centre (*MATLAB*'s default is to have it at the edge) (Fig. 4.9):

```
% Fourier transform with zerofill
spectrum=fftshift(fft(fid,8196));
```

Plotting then yields a pair of doublets that is often seen in NMR spectroscopy (Fig. 4.9). Computer science technicalities of making such calculations more efficient are discussed in Chap. 9.

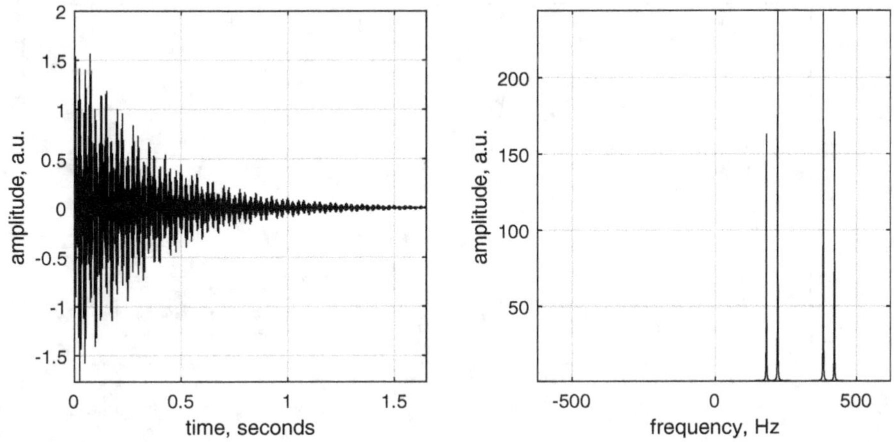

Fig. 4.9 Time-domain simulation of the proton spin dynamics in a CH_2 group with the Hamiltonian in Eq. **Left:** real part of the apodised free induction decay. **Right:** real part of the spectrum

4.9.3 Lie-Group Methods, State-Independent Generator

A prominent feature of Eq. (4.241) is that time evolution is generated by exponential actions (Sect. 1.5.3), the set of which is at least a semigroup (the generator may be dissipative). Group theory being central to spin dynamics, the most accurate numerical solvers for Eq. (4.237) are those that respect it. Accordingly, the strategy must be to solve approximately the equation of motion for the generator $\mathbf{\Omega}(t)$

$$\boldsymbol{\rho}(t)=\exp\{\mathbf{\Omega}(t)\}\boldsymbol{\rho}(0) \tag{4.245}$$

and then apply its exact exponential to the state. The equation for $\mathbf{\Omega}(t)$ is [100]:

$$\frac{\mathrm{d}\mathbf{\Omega}}{\mathrm{d}t}=-i\sum_{m=0}^{\infty}\frac{B_m}{m!}\underbrace{[\mathbf{\Omega},[\mathbf{\Omega},...[\mathbf{\Omega},\mathbf{H}]]]}_{m},\qquad \mathbf{\Omega}(0)=0 \tag{4.246}$$

where B_m are Bernoulli's numbers. Methods wherein this equation is solved approximately using standard numerical methods (*e.g.* Runge–Kutta [133,134]) and the exponential of $\mathbf{\Omega}(t)$ is then applied to the state vector are called *Lie-group methods* [135].

Within this framework, the popular piecewise-constant Hamiltonian approximation corresponds to the midpoint rule:

$$\boldsymbol{\rho}_{k+1}\leftarrow\exp\{-i\mathcal{H}_{\mathrm{M}}\Delta t\}\boldsymbol{\rho}_k \tag{4.247}$$

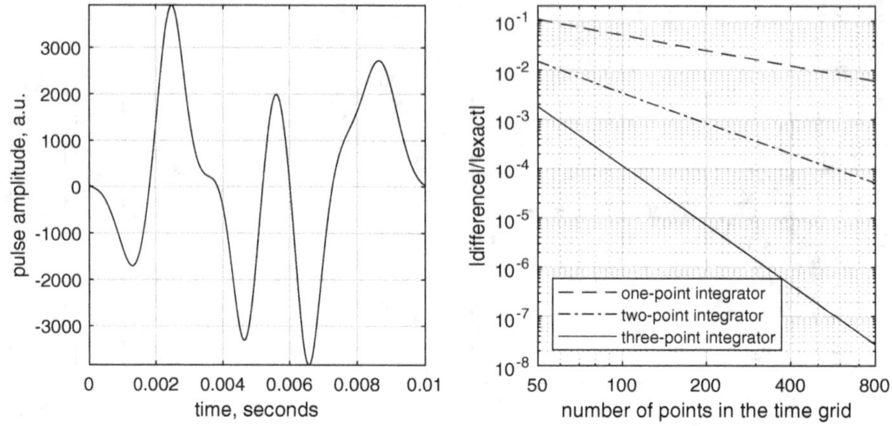

Fig. 4.10 **(Left)** Veshtort-Griffin E1000B band-selective pulse [140]. **(Right)** Final state accuracy as a function of discretisation point count using: (dashed) piecewise-constant approximation; (dash-dot) two-point second-order integrator in Eq. (4.248); (solid) three-point fourth-order integrator in Eq. (4.249)

The best two- and three-point rules are

$$\boldsymbol{\rho}_{k+1} \leftarrow \exp\left\{-i\left(\frac{\mathcal{H}_{\mathrm{L}}+\mathcal{H}_{\mathrm{R}}}{2}+\frac{i\sqrt{3}}{12}[\mathcal{H}_{\mathrm{L}},\mathcal{H}_{\mathrm{R}}]\Delta t\right)\Delta t\right\}\boldsymbol{\rho}_k \tag{4.248}$$

$$\boldsymbol{\rho}_{k+1} \leftarrow \exp\left\{-i\left(\frac{\mathcal{H}_{\mathrm{L}}+4\mathcal{H}_{\mathrm{M}}+\mathcal{H}_{\mathrm{R}}}{6}+\frac{i[\mathcal{H}_{\mathrm{L}},\mathcal{H}_{\mathrm{R}}]\Delta t}{12}\right)\Delta t\right\}\boldsymbol{\rho}_k \tag{4.249}$$

where L, R, and M subscripts indicate the left edge, the right edge, and the midpoint of the Δt interval. The performance is illustrated in Fig. 4.10.

In Eqs. (4.247)–(4.249), it is not necessary to compute matrix exponentials or even matrix products explicitly—those expressions may be evaluated (Sect. 4.9.6) using only matrix–vector products. The same expressions for the propagators apply to double-sided propagation in Hilbert space.

4.9.4 Lie-Group Methods, State-Dependent Generator

There are situations (radiation damping, second-order chemical kinetics, low-temperature relaxation theories, etc.) when the evolution generator depends on both the time and the state vector:

$$\frac{\partial}{\partial t}\boldsymbol{\rho} = -i\mathcal{H}(t,\boldsymbol{\rho})\boldsymbol{\rho} \tag{4.250}$$

In this case, the simplest second-order product quadrature algorithm estimates the generator at the midpoint of each interval and then uses the estimate to propagate the state through the interval:

$$\begin{aligned} \mathcal{H}_{\mathrm{L}} &\leftarrow \mathcal{H}(t_{\mathrm{L}}, \boldsymbol{\rho}_{\mathrm{L}}), \\ \boldsymbol{\rho}_{\mathrm{M}} &\leftarrow \exp\left(-\tfrac{i}{2}(t_{\mathrm{R}} - t_{\mathrm{L}})\mathcal{H}_{\mathrm{L}}\right)\boldsymbol{\rho}_{\mathrm{L}} \\ \mathcal{H}_{\mathrm{M}} &\leftarrow \mathcal{H}(t_{\mathrm{M}}, \boldsymbol{\rho}_{\mathrm{M}}), \\ \boldsymbol{\rho}_{\mathrm{R}} &\leftarrow \exp(-i(t_{\mathrm{R}} - t_{\mathrm{L}})\mathcal{H}_{\mathrm{M}})\boldsymbol{\rho}_{\mathrm{L}} \end{aligned} \tag{4.251}$$

The simplest fourth-order method continues by estimating the generator at the right edge of the interval from the second-order approximation of the state, and then using Eq. (4.249) to build an effective generator over the propagation step:

$$\begin{aligned} \mathcal{H}_{\mathrm{R}} &\leftarrow \mathcal{H}(t_{\mathrm{R}}, \boldsymbol{\rho}_{\mathrm{R}}) \\ \boldsymbol{\rho}_{\mathrm{R}} &\leftarrow \exp\left\{-i\left(\frac{\mathcal{H}_{\mathrm{L}} + 4\mathcal{H}_{\mathrm{M}} + \mathcal{H}_{\mathrm{R}}}{6} + \frac{i[\mathcal{H}_{\mathrm{L}}, \mathcal{H}_{\mathrm{R}}]\Delta t}{12}\right)\Delta t\right\}\boldsymbol{\rho}_{\mathrm{L}} \end{aligned} \tag{4.252}$$

Many variations and refinements exist for these methods [135]. Here too, expensive matrix–matrix multiplication and matrix exponentiation operations may be avoided, and the procedure reformulated (Sect. 4.9.6) entirely in terms of sparse matrix–vector products.

4.9.5 Matrix Exponential and Logarithm

Much has been written about computing matrix exponentials [141]. In the context of spin dynamics—large and sparse matrices—the comical truth is that the Taylor series [91] with scaling and squaring works best: it is compatible with dissipative dynamics (Chebyshev polynomial series [142] can diverge with non-Hermitian matrices), only involves matrix multiplications (Padé approximation [143] needs a costly and perilous matrix inverse), uses minimal memory resources (only one monomial is stored), and only needs approximate scaling (Newton [144] and Chebyshev [142] polynomial series are less forgiving). Taylor series also makes it easy to control the accumulation of insignificant non-zeroes when both the matrix and its exponential are sparse—a decisive consideration in spin dynamics where Hamiltonians are guaranteed to be sparse in the Pauli basis set [145]. The following algorithm for $\exp(\mathbf{A})$, implemented in *propagator.m* function of *Spinach* [87], can deal with spin Hamiltonian dimensions in the millions; it has survived 20 years of abuse by project students:

1. Obtain an upper bound $\|\mathbf{A}\|_{2,\mathrm{ub}}$ on the (expensive) 2-norm of $\mathbf{A}$ by computing the cheaper 1-norm (Matlab, Fortran, and other languages that store matrices column-wise) or infinity-norm (C++, Java, and other languages that store

matrices row-wise). If $\mathbf{A}$ is implicit (*e.g.* a polyadic object, Sect. 9.1), use Hager's norm estimation algorithm [146].

2. Obtain the number of squaring operations n_{sq} and scale the matrix:

$$n_{sq} = \max\left\{0, \mathrm{ceil}\left(\ln\|\mathbf{A}\|_{2,\mathrm{ub}}\right)\right\}, \quad \mathbf{B} \leftarrow \mathrm{chop}[2^{-n_{sq}}\mathbf{A}, \varepsilon] \tag{4.253}$$

   This scaling guarantees monotonic convergence of the Taylor series for the exponential and therefore minimises round-off errors in finite precision arithmetic. The chop function drops the elements with the absolute value smaller than the user-specified tolerance ε, usually machine precision. This is necessary because $i\mathbf{H}\Delta t$ matrices can contain inconsequentially small non-zeroes whose presence in the sparse array would reduce the efficiency unless they are dropped.

3. Converge the Taylor series for the exponential of $\mathbf{B}$:

$$\begin{gathered}\mathbf{P} = \mathbf{P}_0 + \mathbf{P}_1 + \mathbf{P}_2 + \ldots, \ \mathbf{P}_0 = \mathbf{1} \\ \mathbf{P}_n = \begin{cases} \mathrm{chop}[(1/n)\mathbf{B}\mathbf{P}_{n-1}, \varepsilon] \\ \text{or} \\ \mathrm{chop}[(1/n)\mathbf{P}_{n-1}\mathbf{B}, \varepsilon] \end{cases}\end{gathered} \tag{4.254}$$

   where left and right multiplications by $\mathbf{B}$ are algebraically equivalent, but may take different wall clock times depending on matrix storage format (row- or column-major, sparse or full) and location (CPU or GPU). The chop function improves efficiency because $\mathbf{P}_n$ either is sparse, or becomes sparse after a few iterations. It also creates a cheap convergence condition: iterations are stopped when there are no non-zeroes left in $\mathbf{P}_n$.

4. Square the propagator $\mathbf{P}$ up to the original time step by applying $\mathbf{P} \leftarrow \mathrm{chop}\left[\mathbf{P}^2, \varepsilon\right]$ operation n_{sq} times. The chop function again prevents the accumulation of insignificant non-zeroes and improves the efficiency in the spin dynamics context.

For large calculations involving repeated evaluations of expensive matrix functions, the hashing and caching wrapper described in Sect. 9.3.2 is recommended. When the structure of the Hamiltonian permits, Suzuki-Trotter expansions (Sect. 4.3.9) and other matrix exponential manipulation tools (Sect. 4.3) may be beneficial because they either reduce the number of matrix exponentials that need to be (re)computed, or improve the sparsity of the Hamiltonian and the propagator.

For large and sparse matrices appearing in dissipative spin dynamics simulations, the best way to compute matrix logarithm is the inverse of the procedure described above [97]: repeated square roots of the propagator are taken until it acquires the form $\mathbf{P} = \mathbf{1} + \mathbf{Q}$ where $\|\mathbf{Q}\|_{2,\mathrm{ub}} \ll 1$. This brings the problem into the rapid monotonic convergence region of the Taylor series for the logarithm

$$\ln(\mathbf{1}+\mathbf{Q}) = \sum_{n=1}^{\infty} (-1)^{n+1} n^{-1} \mathbf{Q}^n \tag{4.255}$$

and the series is computed with the same care about matrix storage format and insignificant non-zeroes. In common with the matrix exponential algorithm above, only Taylor series is compatible with large sparse matrices and dissipative dynamics at this stage; rational approximations may run out of memory.

4.9.6 Matrix Exponential-Times-Vector

An efficient family of algorithms exists for the calculation of $\exp(\mathbf{A})\mathbf{v}$ products that bypasses the calculation of matrix exponential. The opportunity is visible in the Taylor expansion:

$$\exp(\mathbf{A})\mathbf{v} = \left[\sum_{n=0}^{\infty} \mathbf{A}^n/n!\right]\mathbf{v} = \sum_{n=0}^{\infty} \frac{1}{n!}\mathbf{A}(\mathbf{A}\cdots(\mathbf{A}\mathbf{v})) \tag{4.256}$$

where operations may be reordered to use only matrix–vector products. This is particularly advantageous in situations (*e.g.* Section 9.1) where only an implicit representation is available for the matrix $\mathbf{A}$.

Recent literature advocates sophisticated methods where the problem is projected into the Krylov subspace of $\mathbf{A}$ and $\mathbf{v}$ spanned by the set of products $\{\mathbf{v}, \mathbf{A}\mathbf{v}, \mathbf{A}^2\mathbf{v}, \ldots\}$, the product $\exp(\mathbf{A})\mathbf{v}$ is computed inside the Krylov subspace, and then projected back [147]. However, it may be argued that the expansion coefficients that the Krylov procedure seeks to obtain are already known from Eq. (4.256), and there is little to be gained by going through the expensive orthogonalisation-projection process. In the context of spin dynamics—sparse matrices and vectors—practical benchmarks confirm that a scaled implementation of Eq. (4.256) is faster than the Krylov subspace method, particularly when $\mathbf{v}$ has multiple columns. The algorithm implemented in *step.m* function of *Spinach* [87] is as follows:

1. Obtain an upper bound $\|\mathbf{A}\|_{2,\mathrm{ub}}$ on the (expensive) 2-norm of $\mathbf{A}$ by computing (cheaper) 1-norm (MATLAB, Fortran, and other languages that store matrices column-wise) or infinity-norm (C++, Java, and other languages that store matrices row-wise). If $\mathbf{A}$ is implicit (*e.g.* a polyadic object, Sect. 9.1), use Hager's norm estimation algorithm [146].

2. Subdivide the time step into $k = \mathrm{ceil}\left(\|\mathbf{A}\|_{2,\mathrm{ub}}\right)$ smaller steps—this ensures that the convergence of the series in Eq. (4.256) for $\exp(\mathbf{A}/k)\mathbf{v}$ is monotonic. Scale the vector by dividing out its 1-norm—this improves numerical accuracy in finite-precision arithmetic.

3. For each of the k sub-steps, converge the Taylor series in Eq. (4.256) with respect to the cheapest possible convergence criterion: the number of elements with the absolute value exceeding machine precision. The same clean-up of the sparse array index as in Sect. 4.9.5 is to be applied at every iteration of the Taylor series summation process.

If the number of subdivision steps exceeds the dimension of $\mathbf{A}$, the computational complexity advantage relative to exponentiating $\mathbf{A}$ disappears, but memory advantage remains—$\exp(\mathbf{A})$ may be less sparse than $\mathbf{A}$, and therefore impossible to store.

4.9.7 Bidirectional Propagation

Many magnetic resonance experiments have indirectly incremented evolution times. A typical case is HNCO [148], where the state vector at the detection stage has the form:

$$\begin{aligned} &\boldsymbol{\rho}(t_1, t_2, t_3) = \mathrm{e}^{-i\mathcal{L}t_3}\mathcal{P}_3\mathrm{e}^{-i\mathcal{L}t_2/2}\mathcal{M}_2\mathrm{e}^{-i\mathcal{L}t_2/2}\mathcal{P}_2\mathrm{e}^{-i\mathcal{L}t_1/2}\mathcal{M}_1\mathrm{e}^{-i\mathcal{L}t_1/2}\mathcal{P}_1\boldsymbol{\rho}_0, \\ &\mathcal{L} = \mathcal{H} + i\mathcal{R} \end{aligned} \tag{4.257}$$

where $\boldsymbol{\rho}_0$ is the initial density matrix, $\mathcal{H}$ is the Hamiltonian commutation superoperator, $\mathcal{R}$ is the relaxation superoperator, $\mathcal{P}_n$ are preparation pulse and delay propagators, and $\mathcal{M}_n$ are propagators of refocusing pulses in the middle of evolution periods. In protein spin systems, the dimension of $\boldsymbol{\rho}_0$ can be in the millions, and the number of time discretisation points in the thousands in each of the three dimensions—it is clear that explicit storage of $\boldsymbol{\rho}(t_1, t_2, t_3)$ is out of the question.

This problem may be circumvented by noting that semigroups are associative (Sect. 1.5)—simulations can be partially run "backwards", even in the presence of relaxation because Dirac brackets in the projection onto the detection state $\boldsymbol{\sigma}$ may be shifted, for example:

$$\langle \boldsymbol{\sigma} \mid \boldsymbol{\rho}(t_1, t_2, t_3)\rangle = \left\langle \boldsymbol{\sigma}\mathrm{e}^{-i\mathcal{L}t_3}\mathcal{P}_3\mathrm{e}^{-i\mathcal{L}t_2/2}\mathcal{M}_2 \;\middle|\; \mathrm{e}^{-i\mathcal{L}t_2/2}\mathcal{P}_2\mathrm{e}^{-i\mathcal{L}t_1/2}\mathcal{M}_1\mathrm{e}^{-i\mathcal{L}t_1/2}\mathcal{P}_1\boldsymbol{\rho}_0\right\rangle \tag{4.258}$$

In this case, a 3D HNCO simulation splits into one forward 2D simulation from the initial state, one backward 2D simulation from the detection state and one inner product in the middle (Fig. 4.11). The reduction in storage requirements is considerable—instead of a $512 \times 512 \times 512 \times 10^6$ (or thereabouts) array $\boldsymbol{\rho}(t_1, t_2, t_3)$ at the end of the t_3 period in Eq. (4.257), the two arrays on either side of the inner product Eq. (4.258) have dimensions of $512 \times 512 \times 10^6$ and better sparsity [150].

Bidirectional propagation approach retains parallelisation opportunities (Sect. 9.2): different t_1 increments may be evolved independently in t_2 forward, and different t_3 increments may be evolved independently in t_2 backward. The final inner product can also be computed in parallel.

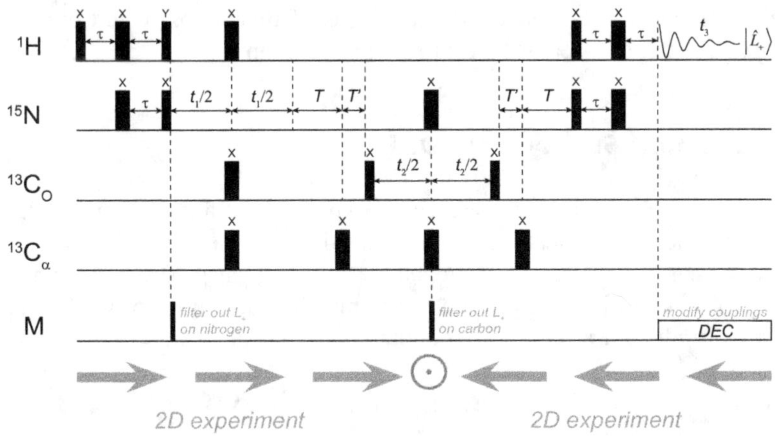

Fig. 4.11 Bidirectional propagation method schematic for the simulation of 3D HNCO NMR experiment [149]. Time is run forward from the initial condition to the middle of the t_2 period and backward from the detection state to the middle of the t_2 period. Both halves have the computational complexity of a 2D simulation, and their scalar product generates the required 3D free induction decay. The channel labelled M ("magic") represents analytical coherence selection and decoupling that are achieved by directly modifying the system state vector or Hamiltonian—the simulation need not implement phase cycles and pulsed field gradients literally. Reproduced with permission from [150]

4.9.8 Steady States of Dissipative Systems

In dissipative systems (Chap. 6) with time-independent Hamiltonians, an elegant solution to the steady-state problem uses the fact that the time derivative vanishes at the steady state. For example, in the high-temperature limit of magnetic resonance simulations:

$$\begin{gathered} \frac{\partial \boldsymbol{\rho}}{\partial t} = -i\mathcal{H}\boldsymbol{\rho} + \mathcal{R}\left(\boldsymbol{\rho} - \boldsymbol{\rho}_{\mathrm{eq}}\right) \\ \Downarrow \\ \mathbf{0} = -i\mathcal{H}\boldsymbol{\rho}_{\infty} + \mathcal{R}\left(\boldsymbol{\rho}_{\infty} - \boldsymbol{\rho}_{\mathrm{eq}}\right) \\ \Downarrow \\ \boldsymbol{\rho}_{\infty} = (-i\mathcal{H} + \mathcal{R})^{-1}\mathcal{R}\boldsymbol{\rho}_{\mathrm{eq}} \end{gathered} \tag{4.259}$$

where $\boldsymbol{\rho}_{\infty}$ is the steady state and $\boldsymbol{\rho}_{\mathrm{eq}}$ is the thermal equilibrium state. In Liouville space representation, there are only sparse matrix–vector products here, and it is not necessary to actually invert the matrix in the brackets—efficient algorithms exist (practical experience favours iLU preconditioned GMRES [151]) that only need matrix–vector products; they can also handle cases where the matrix is defined implicitly as a tensor structure (Sect. 9.1). This approach is useful in the simulation of dynamic nuclear polarisation where billions of time steps would otherwise be necessary to reach the steady state in the time domain.

4.9.9 Example: Steady-State DNP

Consider the dynamic nuclear polarisation experiment [152,153], in which microwave irradiation of electron spin transitions produces (through various electron-nuclear interactions and relaxation mechanisms) strongly non-thermal nuclear spin polarisation. The laboratory frame spin Hamiltonian, with continuous microwave irradiation at a fixed frequency, is time dependent:

$$\begin{aligned}\mathbf{H} = &\sum_k \mathbf{E}^{(k)}\cdot\mathbf{Z}^{(k)}\cdot(\mathbf{B}_0+\mathbf{B}_1\cos(\omega_{\mathrm{MW}}t)) + \sum_m \mathbf{N}^{(m)}\cdot\mathbf{Z}^{(m)}\cdot\mathbf{B}_0\\ &+\sum_{j,k}\mathbf{E}^{(j)}\cdot\mathbf{D}^{(j,k)}\cdot\mathbf{E}^{(k)} + \sum_{k,m}\mathbf{E}^{(k)}\cdot\mathbf{A}^{(k,m)}\cdot\mathbf{N}^{(m)}\\ &+\sum_{m,n}\mathbf{N}^{(m)}\cdot\mathbf{Q}^{(m,n)}\cdot\mathbf{N}^{(n)}+\ldots\end{aligned} \tag{4.260}$$

where $\mathbf{E}^{(k)}$ and $\mathbf{N}^{(m)}$ are electron and nuclear Cartesian spin operator vectors, $\mathbf{B}_0 = \begin{bmatrix}0 & 0 & B_0\end{bmatrix}^{\mathrm{T}}$ is the static magnetic field, $\mathbf{B}_1$ is the magnetic field associated with the microwave irradiation (the frequency ω_{MW} is assumed to be far enough away from nuclear Zeeman transitions to ignore $\mathbf{B}_1$ in the nuclear part), $\mathbf{Z}^{(k)}$ are Zeeman tensors of the indicated particles, $\mathbf{D}^{(j,k)}$ are zero-field splittings and inter-electron couplings, $\mathbf{Q}^{(m,n)}$ are quadrupolar interactions and inter-nuclear couplings, and $\mathbf{A}^{(k,m)}$ are hyperfine interaction tensors. Explicit expressions for all these tensors are given in Chap. 3.

As discussed in Sect. 3.5.7, it is here advantageous to perform an interaction representation transformation with respect to the electron Zeeman Hamiltonian matched to the microwave frequency:

$$\mathbf{H}_0 = \omega_{\mathrm{MW}}\sum_k \mathbf{E}_{\mathrm{Z}}^{(k)} \tag{4.261}$$

This improves the condition number—the 2-norm is reduced by three orders of magnitude because electron Zeeman frequencies become frequency offsets from ω_{MW}—and therefore permits the use of the average Hamiltonian theory because its validity condition (Sect. 4.3.8) becomes satisfied. To first order in AHT, the microwave irradiation terms become time-independent:

$$\begin{aligned}\mathbf{H}_{\mathrm{MW}}^{\mathrm{R}}(t) &= a_{\mathrm{MW}}\cos(\omega_{\mathrm{MW}}t)\mathrm{e}^{+i\omega_{\mathrm{MW}}\mathbf{E}_{\mathrm{Z}}t}\mathbf{E}_{\mathrm{X}}\mathrm{e}^{-i\omega_{\mathrm{MW}}\mathbf{E}_{\mathrm{Z}}t} \\ &= a_{\mathrm{MW}}\cos(\omega_{\mathrm{MW}}t)[\mathbf{E}_{\mathrm{X}}\cos(\omega_{\mathrm{MW}}t)+\mathbf{E}_{\mathrm{Y}}\sin(\omega_{\mathrm{MW}}t)] \\ &= \frac{1}{2}a_{\mathrm{MW}}\mathbf{E}_{\mathrm{X}}+\frac{1}{2}a_{\mathrm{MW}}[\mathbf{E}_{\mathrm{X}}\cos(2\omega_{\mathrm{MW}}t)+\mathbf{E}_{\mathrm{Y}}\sin(2\omega_{\mathrm{MW}}t)] \\ \frac{1}{T}\int_0^T &\mathbf{H}_{\mathrm{MW}}^{\mathrm{R}}(t)\mathrm{d}t = \frac{1}{2}a_{\mathrm{MW}}\mathbf{E}_{\mathrm{X}}, \quad T=\frac{2\pi}{\omega_{\mathrm{MW}}}\end{aligned} \tag{4.262}$$

Similar transformations (Sect. 4.3.7) retain secular parts of the inter-electron couplings, as well as secular and pseudosecular parts of the electron-nuclear couplings, in the interaction part. The same happens to the relaxation superoperator (Chap. 6). In Liouville space, the equation of motion becomes

$$\frac{\partial \boldsymbol{\rho}}{\partial t} = -i\mathcal{H}\boldsymbol{\rho}+\mathcal{R}(\boldsymbol{\rho}-\boldsymbol{\rho}_{\mathrm{eq}}) \tag{4.263}$$

At the steady state ($t=\infty$), the time derivative is zero, and the steady-state density matrix $\boldsymbol{\rho}_\infty$, at the point where microwave irradiation is balanced out by relaxation, is obtained as

$$-i\mathcal{H}\boldsymbol{\rho}_\infty+\mathcal{R}(\boldsymbol{\rho}_\infty-\boldsymbol{\rho}_{\mathrm{eq}})=0 \quad \Rightarrow \quad \boldsymbol{\rho}_\infty=(-i\mathcal{H}+\mathcal{R})^{-1}\mathcal{R}\boldsymbol{\rho}_{\mathrm{eq}} \tag{4.264}$$

As discussed in Sect. 4.9.8, this may be computed efficiently. The steady-state nuclear magnetisation $\langle \mathbf{N}_{\mathrm{Z}}\rangle_\infty$ is then obtained from $\boldsymbol{\rho}_\infty$:

$$\langle \mathbf{N}_{\mathrm{Z}}\rangle_\infty = \mathrm{Tr}(\mathbf{N}_{\mathrm{Z}}\boldsymbol{\rho}_\infty) \tag{4.265}$$

In this situation, a time-domain calculation would not have been practical, even in the rotating frame: DNP systems can take many seconds of physical time to reach their steady states.

Other Degrees of Freedom

5

Some parameters of spin Hamiltonians (inter-particle distances, amplitudes of electromagnetic fields, chemical bond angles, *etc.*) are classical variables in all but the most exotic cases. Accordingly, the interactions discussed in this chapter are less rigorously derived than those we had considered in Chap. 3—we rely on physical intuition and classical analogies to produce empirical Hamiltonians whose justification is that they explain experimental observations and have predictive power.

5.1 Static Parameter Distributions

Ensembles of spin systems are rarely uniform: members can differ in orientations and amplitudes of internal interactions, they can be positioned in locations with different external fields, and there may be a distribution of concentrations across a macroscopic sample. A typical instrument only sees the overall magnetisation of the sample—at some point in the simulation process, ensemble averaging must be performed. This section deals with the cases where the distributions are static.

Strictly speaking, ensemble integration is unrelated to spin dynamics; it belongs instead to the domain of numerical linear algebra. It is discussed here because ensemble averaging problems in spin dynamics are arguably the hardest anywhere in physics, and the methods which must be deployed are consequently brutal in their requirements on computing resources and sophistication. Some of the best performing methods are semi-analytical; there, the algebra becomes entangled with spin physics in subtle and practically important ways—it is not just a numerical quadrature.

I. KUPROV, *Spin*,
https://doi.org/10.1007/978-3-031-05607-9_5

5.1.1 General Framework

In a system where the Hamiltonian depends on a classical parameter vector $\mathbf{x}$ and time, the Hamiltonian commutation superoperator and the density matrix may be written as

$$\mathcal{H}(\mathbf{x},t)=\sum_m g_m(\mathbf{x})\mathcal{H}_m(t),\quad \boldsymbol{\rho}(\mathbf{x},t)=\sum_k g_k(\mathbf{x})\boldsymbol{\rho}_k(t) \tag{5.1}$$

where $\{g_m\}$ is a set of orthonormal functions of $\mathbf{x}$. This is a consequence of the fact that any well-behaved function of two variables can be expanded in products of functions of the individual variables. We will require the set $\{g_m(\mathbf{x})\}$ to be closed under multiplication, and therefore to span an associative algebra (Sect. 1.5.6) with the following structure relations:

$$\begin{aligned} g_n g_k &= \sum_m c_{nkm} g_m \\ c_{nkm} &= \langle g_m \mid g_n g_k\rangle = \int g_m^*(\mathbf{x}) g_n(\mathbf{x}) g_k(\mathbf{x}) \mathrm{dV} \end{aligned} \tag{5.2}$$

where the integration is carried over the domain of $\mathbf{x}$, and $\mathrm{d}V$ is the volume element of the corresponding vector space. Placing Eq. (5.1) into the Liouville-von Neumann equation yields

$$\frac{\partial}{\partial t}\boldsymbol{\rho}(\mathbf{x},t)=-i\mathcal{H}(\mathbf{x},t)\boldsymbol{\rho}(\mathbf{x},t)\quad\Rightarrow\quad \sum_k g_k\frac{\partial\boldsymbol{\rho}_k}{\partial t}=-i\sum_{nkm} c_{nkm} g_m \mathcal{H}_n \boldsymbol{\rho}_k \tag{5.3}$$

Taking an inner product on both sides with each g_m yields the following system:

$$\left\{\frac{\partial\boldsymbol{\rho}_m}{\partial t}=-i\sum_k \mathcal{H}_{km}\boldsymbol{\rho}_k,\quad \mathcal{H}_{km}=\sum_n c_{nkm}\mathcal{H}_n\right._m \tag{5.4}$$

After the state vectors are stacked vertically, the equation acquires the following block structure:

$$\frac{\partial}{\partial t}\begin{bmatrix}\boldsymbol{\rho}_0\\ \boldsymbol{\rho}_1\\ \vdots\end{bmatrix}=\begin{bmatrix}\mathcal{H}_{00} & \mathcal{H}_{01} & \cdots\\ \mathcal{H}_{10} & \mathcal{H}_{11} & \cdots\\ \vdots & \vdots & \ddots\end{bmatrix}\begin{bmatrix}\boldsymbol{\rho}_0\\ \boldsymbol{\rho}_1\\ \vdots\end{bmatrix} \tag{5.5}$$

When the functions in $\{g_m(\mathbf{x})\}$ do not overlap (meaning that $c_{nkm}=\delta_{nk}\delta_{km}$, for example when a set of discrete locations in the parameter space is considered), Eq. (5.5) is block-diagonal—the task is reduced to solving a set of independent simulation problems. This set may be continuous (for example, with orientation or location distributions), and may need to be integrated over.

5.1.2 Gaussian Quadratures

One way to integrate an observable $f(\mathbf{x})$ over a smooth distribution in the parameter vector $\mathbf{x}$ is to pick a basis set $\{g_k(\mathbf{x})\}$ of the space of all integrable functions of $\mathbf{x}$, preferably such that the expansion

$$f(\mathbf{x}) = \sum_k \alpha_k g_k(\mathbf{x}) \tag{5.6}$$

converges rapidly. If the integrals of the basis functions are known, they can be combined with the same coefficients α_k to obtain the integral of $f(\mathbf{x})$. Common choices for the basis functions are orthogonal polynomials (Chebyshev [142], Legendre [34], Laguerre [153], Hermite [154], etc.) and eigenfunctions of system-specific operators, for example spherical harmonics [33].

In practice, the function $f(\mathbf{x})$ is only known at specific points $\{\mathbf{x}_n\}$—the complexity of its evaluation may be such that a budget exists on how many points are available. Thus, when a numerical quadrature is used to approximate an integral of $f(\mathbf{x})$ over some domain $\mathbf{\Omega}$:

$$\int_{\Omega} f(\mathbf{x}) dV_{\Omega} \approx \sum_n w_n f(\mathbf{x}_n), \quad \mathbf{x}_n \in \mathbf{\Omega} \tag{5.7}$$

a good method is the one that uses few points, yet integrates (to a specified precision) a large number of basis functions $g_k(\mathbf{x})$. This is the essence of a *Gaussian quadrature* [155]—we pick the basis, pick the largest rank $k_{\max}$ that is to be integrated to the desired precision, and find the smallest set $\{w_n, \mathbf{x}_n\}$ of weights and locations that suffice.

Gaussian quadratures are well researched, with plug-and-play implementations available in all major programming languages. Given a function handle for $f(\mathbf{x})$, they generate $\{w_n, \mathbf{x}_n\}$ and compute the sum in Eq. (5.7), with adaptive domain subdivision to meet the accuracy target. The minutiae of their implementation are outside the scope of this book. There is only one specific case that we must consider explicitly due to its importance in spin physics—powder averages.

5.1.3 Gaussian Spherical Quadratures

A thorny problem in spin dynamics of disordered systems (powders, glasses, etc.) is the need to average the simulation over all possible orientations. This corresponds to integration, over SO(3) or its two-angle quotient $\mathrm{SO}(3)/\mathrm{SO}(2) \simeq \mathrm{S}^2$ (surface of a sphere), of a density matrix or an observable.

Because we are dealing with the rotation group, an appropriate basis is Wigner D functions (Sect. 1.6.2.4) that we have previously used (Sect. 3.2.4) for Hamiltonian rotations:

$$\mathbf{H}(\Omega, t) = \mathbf{H}_{\text{iso}} + \sum_{l=2,4,6} \sum_{km} \mathfrak{D}_{km}^{(l)}(\Omega) \mathbf{Q}_{km}^{(l)}(t) \tag{5.8}$$

$$\boldsymbol{\rho}(\Omega, t) = \sum_{l=0}^{\infty} \sum_{km} \mathfrak{D}_{km}^{(l)}(\Omega) \boldsymbol{\rho}_{km}^{(l)}(t), \quad -l \leq k, m \leq l \tag{5.9}$$

where $\mathbf{H}_{\text{iso}}$ is the orientation-independent (isotropic) part of the Hamiltonian, $\mathbf{Q}_{km}^{(l)}(t)$ are irreducible components of its anisotropic part, and Ω is a parametrisation of the rotation group (Sect. 1.6.2). Mathematically, the sum in Eq. 5.8 could go to infinity, but experimentally encountered interactions (Chap. 3) have low spherical ranks. However, the sum in Eq. 5.9 does—and that is our problem—go to infinity.

The algebra of Wigner D ("darstellung" [123]) functions is closed under the product operation, they are orthogonal, and only $\mathfrak{D}_{00}^{(0)}$ has a non-zero spherical integral:

$$\mathfrak{D}_{k_1 m_1}^{(l_1)}(\Omega) \mathfrak{D}_{k_2 m_2}^{(l_2)}(\Omega) = \sum_{L=|l_1 - l_2|}^{l_1 + l_2} \sum_{MN} C_{l_1, k_1, l_2, k_2}^{L,M} C_{l_1, m_1, l_2, m_2}^{L,N} \mathfrak{D}_{MN}^{(L)}(\Omega) \tag{5.10}$$

$$\left\langle \mathfrak{D}_{k_1 m_1}^{(l_1)}(\Omega) \,\middle|\, \mathfrak{D}_{k_2 m_2}^{(l_2)}(\Omega) \right\rangle = \frac{1}{8\pi^2} \int \mathfrak{D}_{k_1 m_1}^{(l_1)*}(\Omega) \mathfrak{D}_{k_2 m_2}^{(l_2)}(\Omega) \mathrm{d}\Omega = \frac{\delta_{l_1 l_2} \delta_{k_1 k_2} \delta_{m_1 m_2}}{2l_1 + 1} \tag{5.11}$$

$$\frac{1}{8\pi^2} \int \mathfrak{D}_{km}^{(l)}(\Omega) d\Omega = \delta_{0l} \delta_{0m} \delta_{0k} \tag{5.12}$$

and thus the powder average density matrix is contained in $\boldsymbol{\rho}_{00}^{(0)}(t)$—the objective is to extract it with the highest possible accuracy using the smallest number of simulations at specific orientations. We do that by computing $\boldsymbol{\rho}(\Omega, t)$ at specific orientations Ω_n and averaging them with some weights.

That is hard because of *rank explosion*—although the Hamiltonian in Eq. 5.8 is low rank, the rank of the density matrix in Eq. 5.9 is unbounded. This is because the propagator (Sect. 4.9.5) is a series in the powers of the Hamiltonian, and therefore powers of Wigner D functions. Equation 5.10 indicates that multiplying Wigner D functions increases their rank. This can generate sharp orientation dependences in the observables (Fig. 5.1): the rank can go into thousands—a difficult situation compared to, for example, electronic structure theory where ranks exceeding a dozen are uncommon.

At the time of writing, mathematically optimal Gaussian spherical quadratures for SO(3) and S^2 remain unknown, but close shots do exist [157]. A practical accuracy benchmark is the difference between the analytical integral of each basis

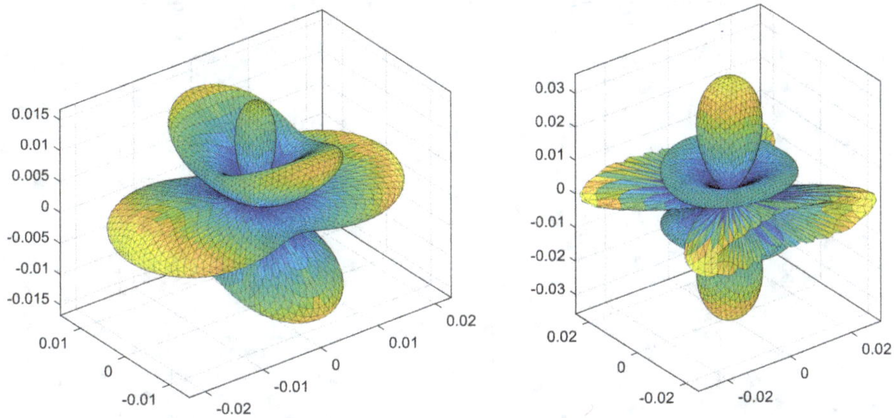

Fig. 5.1 Singlet yield anisotropy in a magnetosensitive radical pair reaction for **(left)** the case of a rapid recombination reaction where the evolution under the anisotropic hyperfine coupling has little time to take place, and **(right)** when the recombination reaction is ten times slower. Note the sharp ridges on the right diagram, poorly sampled even on this rank 131 Lebedev grid [156]

function and the numerical approximation. In the case of orientation averaging, the basis functions are $\mathfrak{D}_{km}^{(l)}$ (three-angle averaging), $Y_m^{(l)}$ (two-angle averaging), and $Y_0^{(l)}$ (one-angle averaging). Thus, the performance metrics as functions of the spherical rank l for one-, two-, and three-angle quadratures respectively are

$$\begin{aligned} S_1(l) &= \left\| \sum_k w_k Y_0^{(l)}(\theta_k) \right\| - \delta_{0,l} \\ S_2(l) &= \left\| \sum_k w_k \mathbf{Y}^{(l)}(\theta_k, \varphi_k) \right\| - \delta_{0,l} \\ S_3(l) &= \left\| \sum_k w_k \mathfrak{D}^{(l)}(\alpha_k, \beta_k, \gamma_k) \right\| - \delta_{0,l} \end{aligned} \tag{5.13}$$

where $Y_0^{(l)}(\theta_k)$ is the zero-projection spherical harmonic of rank l that does not depend on φ (Sect. 2.3), $\mathbf{Y}^{(l)}(\theta_k, \varphi_k)$ is a vector of $Y_m^{(l)}(\theta_k, \varphi_k)$ spherical harmonics with $-l \le m \le l$, $\mathfrak{D}^{(l)}$ is a Wigner D matrix of rank l (Sect. 1.6.2.4), and all norms are 2-norms. These metrics are used in the discussion below.

For S^2, nearly optimal grids are available—from group theory, the minimum number of points required to build a rotationally invariant grid [158] that would integrate spherical harmonics up to rank L exactly is $(L+1)^2/3$; close shots were published by Lebedev, whose rank 131 grid [156] has 5810 points against the optimal count of 5808. The construction of *Lebedev grids* involves large and badly conditioned systems of non-linear equations for point locations and weights,

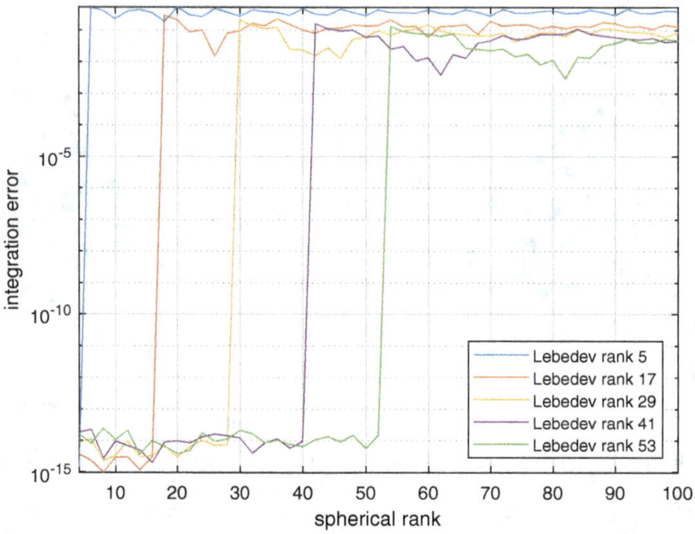

Fig. 5.2 Integration error of some Lebedev grids in 64-bit IEEE754 arithmetic as a function of spherical rank. Only even ranks are shown—due to the inversion symmetry of the grid, odd ranks are not a good indicator

requiring arbitrary precision arithmetic. At the time of writing, it remains an arduous manual process, but the grids are tabulated.

Lebedev grids are only recommended when guarantees exist on the maximum spherical rank present in the integrand, for example in electronic structure theory. Using a Lebedev grid for an integrand that has a higher spherical rank than the grid will result in catastrophic loss of accuracy (Fig. 5.2). This is the Achilles heel of Gaussian quadratures in general, and this is where the hillbilly grids from the next section come useful—they are less accurate than Gaussian quadratures, but they fail gracefully.

5.1.4 Heuristic Spherical Quadratures

Dozens of integration grids were proposed by people trying various geometric patterns and intuitions about what a good one might look like (Fig. 5.3). Compared to Lebedev quadratures, these hillbilly grids are as bad as each other (Fig. 5.4). They are not rotationally invariant; the intuition behind them—that points should be spread evenly with weights obtained by tessellation—is incorrect. Advantages of heuristic grids are at best logistical: some can be subdivided efficiently, and some have convenient topology or symmetry for specific domains, such as magnetic resonance or computer graphics.

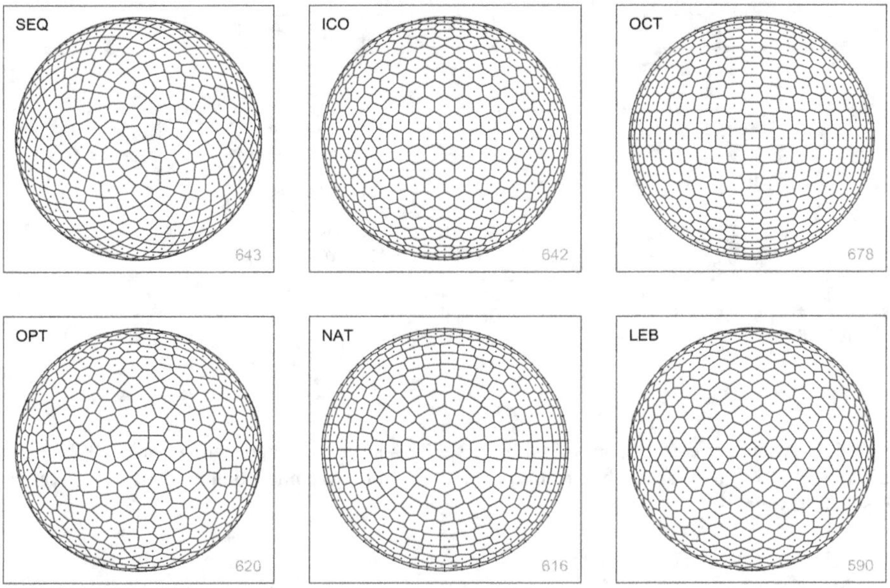

Fig. 5.3 Common types of heuristic spherical quadrature grids, obtained by: **(SEQ)** sampling some functions or sequences (Fibonacci grid [159] shown) to obtain a nearly uniform point distribution on a sphere; **(ICO, OCT)** subdividing facets of some polyhedron and performing further transformations (symmetrisation [160], maximising the volume of the convex hull [161], etc.) to even out the point spacing (icosahedral [161] and octahedral [162] cases shown); **(OPT)** optimising some heuristic score function (in this case, repulsion energy [163]); **(NAT)** borrowing uniform-looking patterns from the natural word (Igloo gird [164]). In all cases, point weights are chosen to be proportional to the solid angles of the Voronoi cells [165, 166]. As a reference point, **(LEB)**: exact, rotationally invariant, and very nearly optimal Lebedev grid—neither uniform, nor uniformly weighted [157]

Still, the history of spin dynamics compels me to catalogue some heuristic grids because they appear in significant publications. A few of them may be used as starting points for the adaptive quadratures discussed in Sect. 5.1.6; random grids will be skipped—those are inefficient and hardly ever used.

5.1.4.1 Geometric Pattern Girds

These quadratures are obtained by sampling functions and sequences that generate uniform-looking distributions of points on the unit sphere. Weights are usually obtained by Voronoi tessellation followed by the calculation of solid angles of Voronoi cells [165, 166]. Popular ones include:

1. *Igloo grids* are obtained by uniform discretisation of latitude and choosing the number of longitudinal points to be proportional to the length of the circle of latitude [164]:

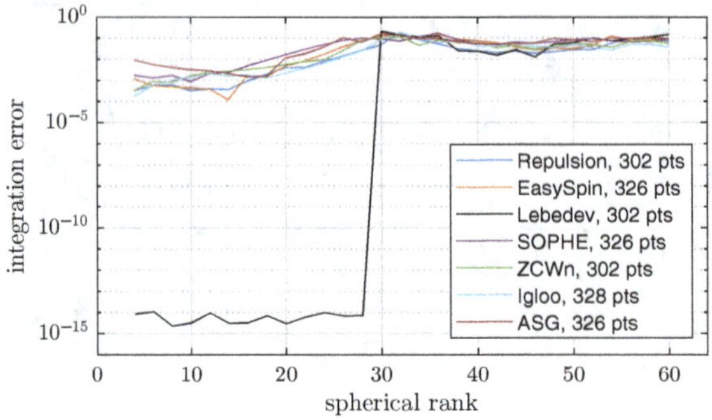

Fig. 5.4 Accuracy comparison in 64-bit IEEE754 arithmetic between a rank 29 Lebedev quadrature [157] and some heuristic quadratures with a similar point count and Voronoi tessellation weights

$$\begin{aligned}&\theta_{jk}=\pi k/(n-1),\quad \varphi_{jk}=2\pi j/m_k\\&0\le k\le n-1,\quad 0\le j\le m_k-1\\&m_k=\text{floor}[2(n-1)\sin\theta_k+1/2]\end{aligned} \tag{5.14}$$

where n is the number of latitudes in the gird. The total number of points grows quadratically with n. This grid has inversion symmetry.

2. *Fibonacci grids* are derived from Fibonacci lattices. Three common variants are spherical Fibonacci (FIB [159]), Zaremba-Conroy-Wolfsberg (ZCW [167–169]), and an approximate limit of ZCW permitting arbitrary point count (ZCWn [170]):

$$\begin{cases}k=-n,\ldots,n\\ \theta_k^{\text{FIB}}=\arccos\left(\frac{2k}{2n+1}\right)\\ \varphi_k^{\text{FIB}}=2\pi k/\Phi\end{cases}\quad \begin{cases}k=0,\ldots,F_{n+2}-1\\ \theta_k^{\text{ZCW}}=\arccos\left(\frac{2k}{F_{n+2}}-1\right)\\ \varphi_k^{\text{ZCW}}=2\pi kF_n/F_{n+2}\end{cases}\quad \begin{cases}k=0,\ldots,n-1\\ \theta_k^{\text{ZCWn}}=\arccos\left(\frac{2k}{n}-1\right)\\ \varphi_k^{\text{ZCWn}}=2\pi k/\Phi^2\end{cases} \tag{5.15}$$

where $\Phi=\lim_{n\to\infty}(F_{n+1}/F_n)=\left(1+\sqrt{5}\right)/2$ is the golden ratio. For a given n, these grids contain $2n+1$, F_{n+2}, and n points respectively, where F_n are Fibonacci's numbers [338].

3. *Triangular grids* are obtained by subdividing the faces of an octahedron and either projecting the result on the unit sphere (Alderman-Solum-Grant grid [162] , O_h symmetry):

$$\begin{aligned}\theta_{kj}^{\mathrm{ASG}} &= \arccos\left((n-k-j)/r_{kj}\right)\\ \varphi_{kj}^{\mathrm{ASG}} &= \arctan(j,k)\\ r_{kj} &= \sqrt{k^2+j^2+(n-k-j)^2}\\ k &= 0,\ldots,n; \quad j=0,\ldots,n-k\end{aligned} \tag{5.16}$$

or distributing points using an Igloo-like scheme (SOPHE grid [171], D_{4h} symmetry) :

$$\begin{aligned}\theta_{kj}^{\mathrm{SOPHE}} &= (\pi/2)(k/n), \quad k=0,\ldots,n\\ \varphi_{kj}^{\mathrm{SOPHE}} &= (\pi/2)(j/k), \quad j=0,\ldots,k\end{aligned} \tag{5.17}$$

with optional symmetrisation back into the octahedral group (EasySpin grid [160] , O_h symmetry):

$$\begin{aligned}&\theta_{kj}^{\mathrm{A}} = (\pi/2)(k/n), \quad \theta_{kj}^{\mathrm{B}} = (\pi/2)((n-j)/n), \quad \theta_{kj}^{\mathrm{C}} = (\pi/2)((n-k+j)/n)\\ &\varphi_{kj}^{\mathrm{A}} = (\pi/2)(j/k), \quad \varphi_{kj}^{\mathrm{B}} = (\pi/2)((k-j)/(n-j)),\\ &\varphi_{kj}^{\mathrm{C}} = (\pi/2)((n-k)/(n-k+j))x_{kj}^{\mathrm{ES}} = x_{kj}^{\mathrm{A}} + y_{kj}^{\mathrm{B}} + z_{kj}^{\mathrm{C}},\\ &\quad y_{kj}^{\mathrm{ES}} = y_{kj}^{\mathrm{A}} + z_{kj}^{\mathrm{B}} + x_{kj}^{\mathrm{C}}, \quad z_{kj}^{\mathrm{ES}} = z_{kj}^{\mathrm{A}} + x_{kj}^{\mathrm{B}} + y_{kj}^{\mathrm{C}}\end{aligned} \tag{5.18}$$

In Eqs. 5.16–5.20, only the first octant is specified; other points are related by symmetry.

Other geometric pattern quadratures (see the reviews by Crăciun [170, 172], Ponti [173], and Eden [174]) may be obtained by subdividing the faces of an icosahedron, by discretising one or more spirals on a sphere, through various ways of mapping rectangular domains onto a sphere, etc. Hemisphere and octant sub-grids offer logistical advantages in magnetic resonance because many of those systems follow subgroups of O_h.

5.1.4.2 Optimisation Grids

A number of spherical quadratures may be obtained from a heuristic assumption that certain functions of point locations should be maximised or minimised. There are grids (primary literature cited by Crăciun [172]) that maximise minimal point distance, minimise maximal point distance, maximise the volume of the convex hull, minimise repulsion energy with respect to various potentials, maximise the determinant of the interpolation matrix, or make all point weights maximally similar. Compared to Lebedev grids, the outcomes are just as bad (Fig. 5.5) as those of pattern grids (Fig. 5.4). Unless symmetry is constrained during optimisation, the result is neither symmetric nor a subset of a larger grid in the same class.

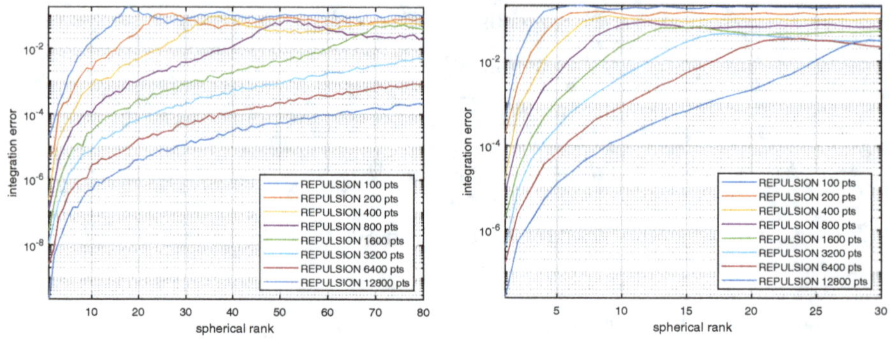

Fig. 5.5 Integration error of two-angle (left) and three-angle (right) REPULSION grids [163] as a function of the rank of spherical harmonics and Wigner D functions respectively

In the context of spin dynamics, a popular choice is REPULSION [163] grids (Fig. 5.3, OPT panel) that are obtained by placing a user-specified number of points randomly on a sphere and minimising their "repulsion energy" with respect to a monotonic distance-dependent potential.

5.1.4.3 Heuristic Weight Selection

Many heuristic spherical quadrature proposals assume—incorrectly—that the weight of each point should be proportional to the area of its Voronoi cell on the unit sphere. For a finite set $\{\mathbf{r}_n\}$ of distinct points on a metric manifold, the Voronoi cell of $\mathbf{r}_k$ is defined as the set of all points whose distance to $\mathbf{r}_k$ is smaller than their distance to any other point in $\{\mathbf{r}_n\}$. On a unit sphere, an appropriate measure of distance is the arc length σ between two points defined by their radius vectors $\mathbf{n}_1$ and $\mathbf{n}_2$:

$$\sigma(\mathbf{n}_1, \mathbf{n}_2) = \arctan[|\mathbf{n}_1 \times \mathbf{n}_2|, (\mathbf{n}_1 \cdot \mathbf{n}_2)] \tag{5.19}$$

Details of Voronoi tessellation algorithms are outside the scope of this book. They are efficient—the current standard has $O(N \log N)$ complexity with the number of points in the set [175]. The cells are convex spherical polygons, each defined by an ordered set of points (Fig. 5.3). Their area is the sum of the areas of all curvilinear triangles defined by adjacent polygon vertices and the parent grid point. The oriented area (sign is determined by the direction of the surface normal vector) of the curvilinear triangle defined on a unit sphere by radius vectors $\mathbf{n}_{1,2,3}$ of the three corners is

$$S(\mathbf{n}_1, \mathbf{n}_2, \mathbf{n}_3) = 2\arctan[\det[\mathbf{n}_1 \quad \mathbf{n}_2 \quad \mathbf{n}_3], (\mathbf{n}_1 \cdot \mathbf{n}_2) + (\mathbf{n}_2 \cdot \mathbf{n}_3) + (\mathbf{n}_3 \cdot \mathbf{n}_1) + 1] \tag{5.20}$$

where the numerically stable two-argument arctangent is used. A counter-example is Lebedev grids [156, 157], where some weights deviate from the areas of their Voronoi cells, and an attempt to use tessellation weights leads to a deterioration of the accuracy.

A more defensible weight assignment procedure [176] makes use of the integration residual metrics in Eq. (5.13) and attempts to find such weights as would drive those metrics to zero. Consider a grid with N points that is tasked with accurate integration of all spherical ranks up to and including $L_{\max}$:

$$\left\{\sum_{k=1}^{N} w_k Y_m^{(l)}(\theta_k,\varphi_k)=\delta_{l,0}\right.\Bigg._{l,m} \tag{5.21}$$

This is a system of $(L_{\max}+1)^2$ linear equations for the weights w_k; it may be reduced to a system of $(L_{\max}+1)(L_{\max}+2)/2$ equations if a zero constraint is imposed on the imaginary part. For a given set of grid points, the system is solved using regularised pseudoinverse when it is underdetermined, in the least squares sense when it is overdetermined, and exactly when the number of grid points matches the number of equations. The accuracy of the resulting quadratures is still inferior to that of similarly sized Lebedev grids [156, 157] by many orders of magnitude.

5.1.5 Direct Product Quadratures

From the group-theoretical point of view, an SO(3) quadrature grid may be represented at the level of the Lie algebra by a discrete set of points in the parameter space, but also at the level of the Lie group as a discrete subset of group operations. These two points of view yield two methods for generating larger quadrature grids from smaller ones:

(a) **Lie algebra picture**. In this picture, a grid is a discretisation of the parameter space, which for SO(3) is isomorphic to $\mathbb{R}^3$ (Sect. 1.5.7), which may in turn be viewed as a direct product of three instances of $\mathbb{R}$. Thus, two- and three-angle grids may be assembled as set direct products of one-dimensional grids along the individual angles, for example:

$$\{\theta_k, v_k\}\otimes\{\varphi_n, w_n\}=\{(\theta_k,\varphi_n), v_k w_n\} \tag{5.22}$$

where a grid of latitude points θ_k with weights v_k is combined by a set direct product with a grid of longitude points φ_n with weights w_n.

Direct product grids inherit the accuracy of their parents, but they are inefficient —the number of points in a direct product of Gaussian quadratures that matches the accuracy of the corresponding Lebedev grids is much greater. However, when the optimal grid is not known (for example, for three-angle quadratures), this is the only practical option.

(b) **Lie group picture**. In this picture, the quadrature grid is viewed as a discrete subset of SO(3) group operations. Larger grids may be obtained from smaller ones by taking set direct products with respect to the group operation, for example:

$$\{\mathbf{R}_k^{\mathrm{A}}, w_k^{\mathrm{A}}\} \otimes \{\mathbf{R}_n^{\mathrm{B}}, w_n^{\mathrm{B}}\} = \{\mathbf{R}_k^{\mathrm{A}}\mathbf{R}_n^{\mathrm{B}}, w_k^{\mathrm{A}}w_n^{\mathrm{B}}\} \tag{5.23}$$

Essentially, one grid is tiled using the operations and weights of the other. When $\mathbf{R}_k^{\mathrm{A}}$ and $\mathbf{R}_n^{\mathrm{B}}$ have orthogonal generators, the result is identical to the Lie algebra picture. Because SO(3) is closed under the superposition of rotations, but its two-angle quotient is not, a product of two-angle grids is in general a three-angle grid.

5.1.6 Adaptive Spherical Quadratures

The quadratures described above are uniform: they integrate *any* spherical function up to the rank specified with a guaranteed accuracy. However, they are inefficient when the function is itself non-uniform. In that case, it makes sense to start with some minimal grid and subdivide it locally as necessary. The methods described in this section make use of the algebraic structure of spin dynamics where spherical ranks of *propagators* are unbounded, but the spherical ranks of their *generators* are guaranteed to be small: as we saw in Sect. 3.1, spin Hamiltonians with spherical rank above six are uncommon.

5.1.6.1 Adaptive Spherical Grid Subdivision

Consider a spherical grid $\{\mathbf{r}_k, \Delta_{nkm}\}$ specified by unit Cartesian vectors $\mathbf{r}_k$, a list of Delaunay triangles Δ_{nkm}, and some quadrature rule $\Im[\mathbf{f}(\mathbf{r}), \Delta]$ that approximates the integral of an array-valued function $\mathbf{f}(\mathbf{r})$ over a sufficiently small triangle Δ. The following is the integration algorithm used by *Spinach*; it locally subdivides the grid to ensure that the overall quadrature meets the specified accuracy target:

Loop over the Delaunay triangles Δ_{nkm} of the initial grid.

Subdivide the triangle $\Delta = \{\mathbf{r}_1, \mathbf{r}_2, \mathbf{r}_3\}$ by finding the three edge midpoints:

$$\mathbf{r}_{12} = \frac{\mathbf{r}_1 + \mathbf{r}_2}{\|\mathbf{r}_1 + \mathbf{r}_2\|_2}, \quad \mathbf{r}_{23} = \frac{\mathbf{r}_2 + \mathbf{r}_3}{\|\mathbf{r}_2 + \mathbf{r}_3\|_2}, \quad \mathbf{r}_{31} = \frac{\mathbf{r}_3 + \mathbf{r}_1}{\|\mathbf{r}_3 + \mathbf{r}_1\|_2} \tag{5.24}$$

Apply the quadrature rule $\Im[\mathbf{f}(\mathbf{r}), \Delta]$ to the four resulting sub-triangles

$$\begin{aligned} \Delta_{\mathrm{A}} &= \{\mathbf{r}_1, \mathbf{r}_{12}, \mathbf{r}_{31}\}, \quad \Delta_{\mathrm{B}} = \{\mathbf{r}_{12}, \mathbf{r}_2, \mathbf{r}_{23}\} \\ \Delta_{\mathrm{C}} &= \{\mathbf{r}_{31}, \mathbf{r}_{23}, \mathbf{r}_3\}, \quad \Delta_{\mathrm{D}} = \{\mathbf{r}_{12}, \mathbf{r}_{23}, \mathbf{r}_{31}\} \end{aligned} \tag{5.25}$$

Judge the approximation quality using some norm of the difference:

$$\|\Im[\mathbf{f}(\mathbf{r}),\Delta_{\mathrm{A}}]+\Im[\mathbf{f}(\mathbf{r}),\Delta_{\mathrm{B}}]+\Im[\mathbf{f}(\mathbf{r}),\Delta_{\mathrm{C}}]+\Im[\mathbf{f}(\mathbf{r}),\Delta_{\mathrm{D}}]-\Im[\mathbf{f}(\mathbf{r}),\Delta]\| \quad (5.26)$$

between the sum over the four sub-triangles and the value computed for the original triangle. If this measure is small enough, return the sum. Otherwise, call the procedure recursively for each of the four sub-triangles, and return the sum of the outputs.

End loop.

5.1.6.2 Adaptive Transition Moment Integral

The procedure described in the previous section needs the quadrature rule $\Im[\mathbf{f}(\mathbf{r}),\Delta]$ that returns an inexpensive approximation of the integral of $\mathbf{f}(\mathbf{r})$ over the spherical triangle Δ. This rule is entirely at the user's discretion, and some are better than others. We will discuss here one particular method, proposed by Voitländer [177], which improved field-swept EPR simulation capabilities when it was implemented in *EasySpin* [178]. An example is given in Fig. (5.6). Consider the situation where we seek to integrate, with respect to the system orientation, the intensity of a spectroscopic transition at a particular magnetic field B:

$$s(B)=\oiint_{\Delta} f(p(\mathbf{r}),a(\mathbf{r}),w(\mathbf{r}),B)\mathrm{d}S \quad (5.27)$$

where Δ is a small triangle and $f(\ldots)$ is the probability per unit time of a spectroscopic transition with a field position $p(\mathbf{r})$, amplitude $a(\mathbf{r})$, and width $w(\mathbf{r})$, all obtained from Fermi's golden rule (Sect. 4.4.4). From the algebraic form of Fermi's golden rule, the following observations can be made about the function under the integral:

- The intensity $a(\mathbf{r})$ is multiplicative:

$$f[p(\mathbf{r}),a(\mathbf{r}),w(\mathbf{r}),B] \rightarrow a(\mathbf{r})f[p(\mathbf{r}),w(\mathbf{r}),B] \quad (5.28)$$

- The position $p(\mathbf{r})$ is an offset:

$$a(\mathbf{r})f[p(\mathbf{r}),w(\mathbf{r}),B] \rightarrow a(\mathbf{r})f[w(\mathbf{r}),B-p(\mathbf{r})] \quad (5.29)$$

- The shape is a convolution:

$$a(\mathbf{r})f[w(\mathbf{r}),B-p(\mathbf{r})] \rightarrow a(\mathbf{r})\delta[B-p(\mathbf{r})]*f[w(\mathbf{r}),B] \quad (5.30)$$

where $\delta(x)$ is Dirac's delta function. Angular dependences of intensity and width are slow; the chief problem here that generates sharp transients and makes the integration difficult is the angular dependence in the argument of the delta function.

From the adaptive grid subdivision discussed in the previous section, we only have the values of the intensity, offset, and width at the three corners of the triangle. The following combination of numerical integrators works well in practice:

- Midpoint rule for the intensity and the width:

$$\begin{aligned} s(B) &= \oiint_{\Delta} a(\mathbf{r})\delta[B - p(\mathbf{r})] * f[w(\mathbf{r}), B]\mathrm{d}S \\ &\approx a_{\mathrm{M}} f[w_{\mathrm{M}}, B] * \oiint_{\Delta} \delta[B - p(\mathbf{r})]\mathrm{d}S \\ a_{\mathrm{M}} &= [a(\mathbf{r}_1) + a(\mathbf{r}_2) + a(\mathbf{r}_3)]/3 \\ w_{\mathrm{M}} &= [w(\mathbf{r}_1) + w(\mathbf{r}_2) + w(\mathbf{r}_3)]/3 \end{aligned} \tag{5.31}$$

- Trilinear interpolation followed by analytical evaluation for the remaining position integral because that is the simplest method that can handle the delta function. The interpolant is obtained from the known values at the three vertices of the triangle; the integral is analytical:

$$s(B) = a_{\mathrm{M}} S_{\Delta} f(w_{\mathrm{M}}, B) * \begin{cases} (B - p_1)(p_2 - p_1)^{-1}(p_3 - p_1)^{-1}, & p_1 < B < p_2 \\ (p_3 - B)(p_3 - p_2)^{-1}(p_3 - p_1)^{-1}, & p_2 < B < p_3 \\ 0 & \text{otherwise} \end{cases} \tag{5.32}$$

  The content of the curly bracket is a normalised triangle with corners at $p_1 < p_2 < p_3$. Its convolutions with Lorentzian and Gaussian line shapes are analytical [179], although the expressions are unpleasant—a *MATLAB* implementation is available in *Spinach*.

For field swept electron spin resonance spectra, this adaptive quadrature outperforms Lebedev grids (Fig. 5.6). For time-domain simulations, an equivalent method does not at present exist.

5.1.7 Example: DEER Kernel with Exchange

A rare example of an entirely analytical solution to the orientational averaging problem is pulsed dipolar spectroscopy—a popular technique for distance measurement between unpaired electrons [180]. Consider the laboratory frame Hamiltonian for a two-electron system with a Zeeman, dipole–dipole, and exchange coupling:

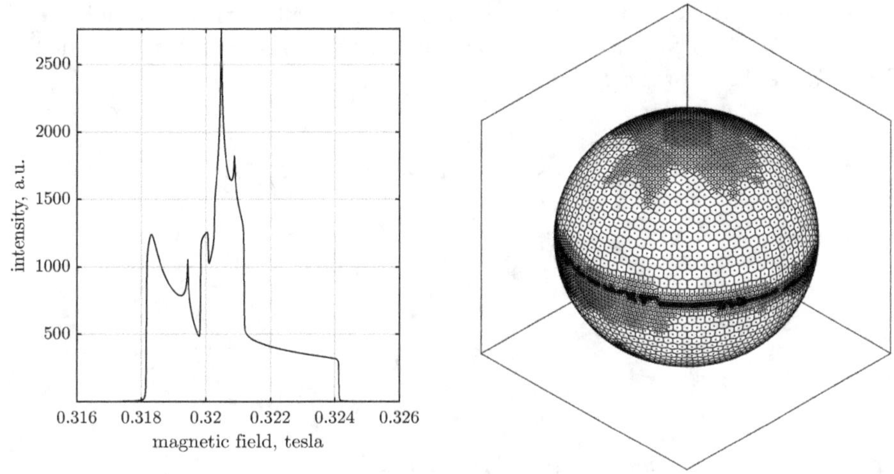

Fig. 5.6 **(Left)** A simulation of a field-swept ESR spectrum of a nitroxide radical using the eigenfields method with an adaptive spherical quadrature starting from a rank 6 Stoll grid. **(Right)** Voronoi cells of the adaptive grid at the convergence point

$$\begin{aligned}\mathbf{H} = \omega_1 \mathbf{S}_Z^{(1)} + \omega_2 \mathbf{S}_Z^{(2)} - \omega_{DD}\sqrt{6}\sum_{m=-2}^{2}\mathbf{T}_{2,m}\mathfrak{D}_{m,0}^{(2)}(\alpha,\beta,\gamma) \\ + \omega_{EX}\left[\mathbf{S}_X^{(1)}\mathbf{S}_X^{(2)} + \mathbf{S}_Y^{(1)}\mathbf{S}_Y^{(2)} + \mathbf{S}_Z^{(1)}\mathbf{S}_Z^{(2)}\right], \quad \omega_{DD} = \frac{\mu_0}{4\pi}\frac{\gamma_1\gamma_2\hbar}{r^3}\end{aligned} \tag{5.33}$$

where (α, β, γ) are Euler angles (Sect. 1.6.2) and $\mathfrak{D}_{m,0}^{(2)}$ are second-rank Wigner D-functions (Sect. 3.3.4). After applying the rotating frame (Sect. 4.3.7) with respect to the Zeeman Hamiltonian

$$\mathbf{H}_0 = \omega_1 \mathbf{S}_Z^{(1)} + \omega_2 \mathbf{S}_Z^{(2)} \tag{5.34}$$

and using the fact that DEER pulse sequence refocuses Zeeman frequency offsets, we obtain the following rotating frame Hamiltonian in the weak coupling limit (Sect. 4.3.8.4):

$$\mathbf{H}^R = \left(\omega_{DD}\left[1 - 3\cos^2(\beta)\right] + \omega_{EX}\right)\mathbf{S}_Z^{(1)}\mathbf{S}_Z^{(2)} \tag{5.35}$$

This approximation is applicable when $|\omega_1 - \omega_2| \gg |\omega_{DD}|, |\omega_{EX}|$; this holds when the difference between the pump and observe frequencies strongly exceeds both couplings. When acting on a transverse magnetization state (Sect. 4.7.2), this Hamiltonian generates the following oscillation in the observable dynamics:

$$\langle \mathbf{S}_X \rangle \propto \cos\left[\left(\omega_{DD}\left[1 - 3\cos^2(\beta)\right] + \omega_{EX}\right)t\right] \tag{5.36}$$

What is detected experimentally is an average over all orientations and over a distribution in the isotropic exchange couplings. The orientational average (called *DEER kernel* [181]) is analytical:

$$\begin{aligned}\gamma(r,t) &= \int_0^\pi \cos\left[\left(\omega_{\mathrm{DD}}\left[1-3\cos^2(\theta)\right]+\omega_{\mathrm{EX}}\right)t\right]\sin\theta\mathrm{d}\theta \\ &= \sqrt{\frac{\pi}{6\omega_{\mathrm{DD}}t}}\left[\cos[(\omega_{\mathrm{DD}}+\omega_{\mathrm{EX}})t]\mathrm{FrC}\left[\sqrt{\frac{6\omega_{\mathrm{DD}}t}{\pi}}\right]\right. \\ &\qquad \left. +\sin[(\omega_{\mathrm{DD}}+\omega_{\mathrm{EX}})t]\mathrm{FrS}\left[\sqrt{\frac{6\omega_{\mathrm{DD}}t}{\pi}}\right]\right]\end{aligned} \tag{5.37}$$

where *normalised Fresnel integrals* [182] are defined as

$$\mathrm{FrC}(x)=\int_0^x \cos\left(\pi t^2/2\right)\mathrm{d}t, \quad \mathrm{FrS}(x)=\int_0^x \sin\left(\pi t^2/2\right)\mathrm{d}t \tag{5.38}$$

Because $\gamma(r,t)$ Eq. 5.37 refers to a specific distance, the case where a distribution of distances exists within the sample may be treated by taking the convolution of $\gamma(r,t)$ with the distance distribution.

The kernel in Eq. 5.37 stays simple when a Gaussian distribution with the mean μ_{EX} and standard deviation σ_{EX} in the exchange coupling is assumed:

$$p(\omega_{\mathrm{EX}})=\frac{1}{\sqrt{2\pi}\sigma_{\mathrm{EX}}}\exp\left[-(\omega_{\mathrm{EX}}-\mu_{\mathrm{EX}})^2/2\sigma_{\mathrm{EX}}^2\right] \tag{5.39}$$

because ω_{EX} only occurs under simple trigonometric functions, for which:

$$\begin{aligned}\cos[(\omega_{\mathrm{DD}}+\omega_{\mathrm{EX}})t]*p(\omega_{\mathrm{EX}}) &= \mathrm{e}^{-\frac{t^2\sigma_{\mathrm{EX}}^2}{2}}\cos[(\omega_{\mathrm{DD}}+\mu_{\mathrm{EX}})t] \\ \sin[(\omega_{\mathrm{DD}}+\omega_{\mathrm{EX}})t]*p(\omega_{\mathrm{EX}}) &= \mathrm{e}^{-\frac{t^2\sigma_{\mathrm{EX}}^2}{2}}\sin[(\omega_{\mathrm{DD}}+\mu_{\mathrm{EX}})t]\end{aligned} \tag{5.40}$$

The updated expression for the DEER kernel is

$$\gamma(r,\mu_{\mathrm{EX}},\sigma_{\mathrm{EX}},t)=\sqrt{\frac{\pi\exp(-t^2\sigma_{\mathrm{EX}}^2)}{6\omega_{\mathrm{DD}}t}}\begin{pmatrix}\cos[(\omega_{\mathrm{DD}}+\mu_{\mathrm{EX}})t]\mathrm{FrC}\left[\sqrt{\frac{6\omega_{\mathrm{DD}}t}{\pi}}\right] \\ +\sin[(\omega_{\mathrm{DD}}+\mu_{\mathrm{EX}})t]\mathrm{FrS}\left[\sqrt{\frac{6\omega_{\mathrm{DD}}t}{\pi}}\right]\end{pmatrix} \tag{5.41}$$

where μ_{EX} is the average exchange coupling and σ_{EX} is the standard deviation of its distribution.

5.2 Dynamics in Classical Degrees of Freedom

In room temperature chemical systems, a popular approximation is that the dynamics in the classical degrees of freedom does affect quantum spin dynamics (via the time dependencies it creates in the spin Hamiltonian), but the spins do not talk back because the energies of spin interactions are tiny—the spin state of the system does not influence its spatial distribution or classical dynamics. Examples are diffusion and flow in magnetic resonance imaging [183], and magic angle spinning in solid-state NMR [184].

In this approximation, classical ensemble dynamics is described by a probability density $p(\mathbf{x}, t)$ of finding a system in a point $\mathbf{x}$ of the classical parameter manifold at time t. The flux $\mathbf{j}(\mathbf{x}, t)$ of this probability density is determined by the local velocity $\mathbf{v}(\mathbf{x}, t)$

$$\mathbf{j}(\mathbf{x}, t) = \mathbf{v}(\mathbf{x}, t)p(\mathbf{x}, t) \tag{5.42}$$

and the velocity depends on the equation of motion. The total probability must be conserved—the local reduction per unit time must, therefore, be equal to the divergence of the flux [185]:

$$\frac{\partial}{\partial t}p(\ldots) = -\mathrm{div}[\mathrm{flux}[p(\ldots)]] \quad \Leftrightarrow \quad \frac{\partial p(\mathbf{x}, t)}{\partial t} = -\nabla_{\mathbf{x}} \cdot \mathbf{j}(\mathbf{x}, t) \tag{5.43}$$

where $\mathbf{x}$ can be Cartesian coordinates, various angles, amplitudes and phases of time- and position-dependent external fields, phases of various rotors, etc. When classical and quantum dynamics coexist, the scalar probability in this equation becomes a density matrix. The derivation is exceedingly laborious (Sect. 6.5.1), here we will only present the final result:

$$\frac{\partial \boldsymbol{\rho}(\mathbf{x}, t)}{\partial t} = -i\mathcal{L}(\mathbf{x}, t)\boldsymbol{\rho}(\mathbf{x}, t) + \mathcal{M}(\mathbf{x}, t)\boldsymbol{\rho}(\mathbf{x}, t) \tag{5.44}$$

in which $\mathcal{L}(\mathbf{x}, t) = \mathcal{H}(\mathbf{x}, t) + i\mathcal{R} + i\mathcal{K}$ is the Liouvillian that is responsible for spin dynamics and $\mathcal{M}(\mathbf{x}, t)$ is the spatial dynamics generator that controls diffusion, flow, sample spinning, and other types of classical dynamics in the laboratory space. Equation (5.44) resembles Fokker-Planck equation of motion for the scalar probability density [186, 187], this approach may, therefore, be called *Fokker–Planck formalism* [188].

5.2.1 Spatial Dynamics Generators

Expressions for $\mathcal{M}(\mathbf{x}, t)$ come from classical mechanics. A few relevant equations of motion for the probability density are given below. The spatial dynamics generators are in square brackets.

1. Flow along a coordinate x and counterclockwise circular motion with a phase φ

$$\frac{\partial p(x,t)}{\partial t} = \left[-v(t)\frac{\partial}{\partial x}\right]p(x,t), \qquad \frac{\partial p(\varphi,t)}{\partial t} = \left[-\omega(t)\frac{\partial}{\partial \varphi}\right]p(\varphi,t), \tag{5.45}$$

   where v is linear velocity and ω is angular velocity.
2. Circular motion around a specific axis $\mathbf{n} = \begin{bmatrix} n_X & n_Y & n_Z \end{bmatrix}$ in three dimensions

$$\frac{\partial p(\mathbf{r},t)}{\partial t} = \left[-\omega(t)\left(n_X\hat{L}_X + n_Y\hat{L}_Y + n_Z\hat{L}_Z\right)\right]p(\mathbf{r},t), \tag{5.46}$$

   where $\{\hat{L}_X, \hat{L}_Y, \hat{L}_Z\}$ are angular momentum operators from Eq. (2.21), and ω is angular velocity.
3. Translational diffusion in a potential $U(\mathbf{r})$ in an isotropic medium

$$\frac{\partial p(\mathbf{r},t)}{\partial t} = \left[\frac{\sigma^2\nabla^2}{2\gamma^2} - \frac{\nabla\cdot\mathbf{f}(\mathbf{r})}{\gamma}\right]p(\mathbf{r},t), \qquad \mathbf{f}(\mathbf{r}) = -\nabla U(\mathbf{r}), \tag{5.47}$$

   where σ is the amplitude of the fluctuating force that generates the diffusion and γ is the friction constant of the medium.
4. Isotropic rotational diffusion in three dimensions

$$\frac{\partial p(\Omega,t)}{\partial t} = \left[D_R\left(\hat{L}_X^2 + \hat{L}_Y^2 + \hat{L}_Z^2\right)\right]p(\Omega,t), \tag{5.48}$$

   where Ω is a shorthand for Euler angles or other orientation parameters, $\{\hat{L}_X, \hat{L}_Y, \hat{L}_Z\}$ are angular momentum operators from Eq. (2.21) and D_R is the rotational diffusion coefficient.

For example, the stochastic Liouville equation (Sect. 6.5) used in electron spin resonance [189] is obtained by using the rotational diffusion term:

$$\frac{\partial \boldsymbol{\rho}(\Omega,t)}{\partial t} = -i\mathcal{L}(\Omega,t)\boldsymbol{\rho}(\Omega,t) + D_R\left(\hat{L}_X^2 + \hat{L}_Y^2 + \hat{L}_Z^2\right)\boldsymbol{\rho}(\Omega,t) \tag{5.49}$$

and the Bloch-Torrey equations [183], used in magnetic resonance imaging, by adding a translational diffusion term in combination with the Bloch sphere representation of single-spin dynamics:

$$\frac{\partial \boldsymbol{\rho}(\mathbf{r},t)}{\partial t} = -i\mathcal{L}(\mathbf{r},t)\boldsymbol{\rho}(\mathbf{r},t) + D_T\left(\frac{\partial^2}{\partial x^2} + \frac{\partial^2}{\partial y^2} + \frac{\partial^2}{\partial y^2}\right)\boldsymbol{\rho}(\mathbf{r},t) \tag{5.50}$$

Expressions for other types of motion may be found in the literature [190, 191]. Although spatial dynamics generators can depend on time, they are commonly static in practice.

5.2.2 Algebraic Structure of the Problem

From the algebraic perspective, the Fokker–Planck equation operates in a direct product of spatial and spin coordinates. For logistical reasons (to do with data layouts in *MATLAB*), it is preferable to have spatial coordinates as the first term in the Kronecker product and spin coordinates as the second term. Thus, the general form of a purely spatial operator is $\mathbf{A}\otimes\mathbf{1}$, the general form of a pure spin operator is $\mathbf{1}\otimes\mathbf{B}$ and the form of an operator that correlates spatial and spin degrees of freedom (for example, a pulsed field gradient in MRI) is $\mathbf{A}\otimes\mathbf{B}$. The spin subspace has the usual direct product structure (Sect. 4.2.5) and for the spatial coordinates it is reasonable to put all periodic coordinates ahead of all non-periodic ones. All of this results in the following product algebra for the evolution generators

$$[\mathfrak{so}_2(\mathbb{R})]^{\otimes m}\otimes[\mathfrak{gl}_k(\mathbb{R})]\otimes[\mathfrak{su}_n(\mathbb{R})], \tag{5.51}$$

where the first term in the direct product accounts for the (possibly multiple) radiofrequency irradiations and uniaxial sample rotations, all generated by $\mathfrak{so}_2(\mathbb{R})$ Lie algebras, the second term is responsible for translational and diffusional dynamics generated by the general linear algebra $\mathfrak{gl}_k(\mathbb{R})$ acting on a k-dimensional real space, and the last term is the special unitary algebra $\mathfrak{su}_n(\mathbb{R})$ generating the dynamics of an n-state quantum system (Sect. 2.5.5). Schematically:

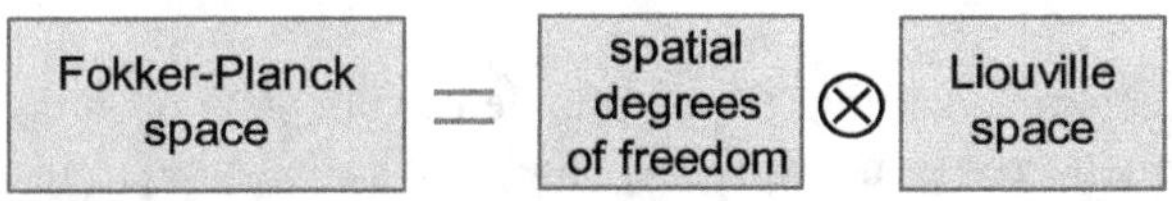

where the Liouville space (Sect. 4.2.4) is the vectorisation of the density operator space acted upon by the adjoint representation of $\mathfrak{su}(n)$

Projection back (into either the laboratory space or the Liouville space) is accomplished by re-indexing the state vector to have spatial degrees of freedom along one dimension and spin degrees of freedom along the other. Projection into the lab space is then done by taking a scalar product with the desired spin state; projection into the spin space is done by integrating over the dimension corresponding to the spatial degrees of freedom. An implementation of this formalism that avoids opening the Kronecker products in Eq. 5.51 is described in Chap. 9.

5.2.3 Matrix Representations

Spin operators are matrices by nature; their representation is determined by the basis set, for which the best option depends on the size of the spin system. With fewer than ten spins, the standard Pauli matrix product basis (Sect. 4.2.5) works well:

$$\mathbf{S}^{(k)}_{\{X,Y,Z\}} = \mathbf{1} \otimes \mathbf{1} \otimes \cdots \otimes \mathbf{S}_{\{X,Y,Z\}} \otimes \cdots \otimes \mathbf{1} \tag{5.52}$$

where $\mathbf{1}$ is a unit matrix of the dimension equal to the multiplicity of the corresponding spin and $\mathbf{S}_{\{X,Y,Z\}}$ are single-spin operators that occur in the kth position in the direct product. Larger spin systems are best handled with the restricted state-space approximation (Chap. 7), wherein unimportant and unpopulated states are dropped to keep the problem dimension manageable.

The situation is more delicate with spatial distributions and differential operators generating processes like diffusion and flow. When spatial coordinates are discretised, those operators are only approximately representable by matrices; the choice of the approximation depends on the context—either finite difference matrices or spectral differentiation matrices depending on the accuracy requirements and boundary conditions of each individual case. This is discussed in detail below.

5.2.3.1 Distributed Initial Conditions and Detection States

Consider a discretisation of spatial coordinates into a number of non-overlapping spatial locations, called voxels. Any location-dependent spin state would have the following expansion:

$$\boldsymbol{\rho} = \sum_k \varphi_k \otimes \boldsymbol{\rho}_k \tag{5.53}$$

where φ_k are vectors specifying the amplitude of spin state $\boldsymbol{\rho}_k$ at each voxel of the sample; in magnetic resonance imaging, they are called phantoms. For the initial condition, the amplitude variation may come, for example, from chemical concentration distributions.

The same algebraic structure pertains to detection states. In the case of inductive detection used in magnetic resonance imaging:

$$\begin{aligned} \boldsymbol{\sigma} = {} & \varphi_X \otimes \sum_n (1+\delta_n)\gamma_n \mathbf{S}^{(n)}_X + \varphi_Y \otimes \sum_n (1+\delta_n)\gamma_n \mathbf{S}^{(n)}_Y \\ & + \varphi_Z \otimes \sum_n (1+\delta_n)\gamma_n \mathbf{S}^{(n)}_Z \end{aligned} \tag{5.54}$$

where the spatial distribution vectors $\varphi_{X,Y,Z}$ are called *coil profiles*, they determine how sensitive the coil is to each Cartesian component of nuclear induction at each voxel. The sums contain Zeeman detection state vectors weighted by the corresponding magnetogyric ratios and chemical shifts.

5.2.3.2 Distributed Evolution Generators

In the Fokker-Planck space, the general form of a location-dependent spin evolution generator is

$$\mathfrak{F} = \sum_k \mathbf{\Phi}_k \otimes \mathfrak{L}_k, \quad \mathbf{\Phi}_k = \mathrm{diag}(\varphi_k) \tag{5.55}$$

where $\mathbf{\Phi}_k$ are diagonal matrices with voxel amplitude vectors φ_k on the diagonal; their role is to specify the amplitudes of Liouville space spin evolution generators $\mathfrak{L}_k$ in each voxel.

Evolution generators may be dissipative (e.g., relaxation superoperators responsible for tissue contrast in MRI) or coherent—for example, for evolution under B_Z magnetic field gradient $[g_X \quad g_Y \quad g_Z]$:

$$\mathfrak{F}_{\mathrm{PFG}}(t) = -\begin{bmatrix} g_X(t)\mathbf{X} \otimes \mathbf{1}_Y \otimes \mathbf{1}_Z + \\ g_Y(t)\mathbf{1}_X \otimes \mathbf{Y} \otimes \mathbf{1}_Z + \\ g_Z(t)\mathbf{1}_X \otimes \mathbf{1}_Y \otimes \mathbf{Z} \end{bmatrix} \otimes \sum_n (1+\delta_n)\gamma_n \mathcal{S}_Z^{(n)} \tag{5.56}$$

where $\mathcal{S}_Z^{(n)}\boldsymbol{\rho} = \left[\mathbf{S}_Z^{(n)}, \boldsymbol{\rho}\right]$, $\{\mathbf{X}, \mathbf{Y}, \mathbf{Z}\}$ are diagonal matrices containing Cartesian voxel coordinates on the diagonal, and $\{\mathbf{1}_X, \mathbf{1}_Y, \mathbf{1}_Z\}$ are unit matrices of appropriate dimensions. Here, the square bracket contains the time-dependent spatial distribution of the amplitude of the Zeeman interaction commutation superoperator $\sum_k (1+\delta_n)\gamma_n \mathcal{S}_Z^{(n)}$. The expression in the square brackets may be extended to non-linear field profiles produced by realistic hardware by adding higher order polynomial terms like $\mathbf{X}^2 \otimes \mathbf{Y} \otimes \mathbf{1}_Z$.

Another common distributed evolution generator appears when magnetic fields of radiofrequency coils and microwave resonators are not uniform. The generator has the same structure as Eq. (5.56), except the time-dependent amplitude coefficients are differently located and all three Zeeman interaction operators are involved:

$$\begin{aligned} \mathfrak{F}_{\mathrm{TR}}(t) = a(t)\Bigg[&\mathbf{\Phi}_X \otimes \sum_n (1+\delta_n)\gamma_n \mathcal{S}_X^{(n)} \\ &+ \mathbf{\Phi}_Y \otimes \sum_n (1+\delta_n)\gamma_n \mathcal{S}_Y^{(n)} + \mathbf{\Phi}_Z \otimes \sum_n (1+\delta_n)\gamma_n \mathcal{S}_Z^{(n)}\Bigg] \end{aligned} \tag{5.57}$$

Here, $\mathbf{\Phi}_{X,Y,Z}$ are matrices with coil profiles $\varphi_{X,Y,Z}$ on the diagonal, and $a(t)$ is proportional, in the case of nuclear magnetic resonance, to the current passing through the coil.

5.2.3.3 Diffusion and Flow Generators

Generators of spin-independent dynamics, such as diffusion and flow, act on spin by a unit matrix—therefore, a matrix representation only needs to be constructed for the spatial part. Given a concentration profile $c(\mathbf{r},t)$, the diffusion flux is given by Fick's first law [185], and the hydrodynamic flux is the product of concentration and flow velocity. The total flux therefore is

$$\mathbf{j}(\mathbf{r},t) = \mathbf{v}(\mathbf{r},t)c(\mathbf{r},t) - \mathbf{D}(\mathbf{r},t)\nabla c(\mathbf{r},t) \tag{5.58}$$

where $\nabla = [\partial/\partial x \quad \partial/\partial y \quad \partial/\partial z]^{\mathrm{T}}$ is the gradient operator, $\mathbf{v}(\mathbf{r},t)$ is the flow velocity field, and $\mathbf{D}(\mathbf{r},t)$ is the translational diffusion tensor field. We take both fields as given—the assumption is that spin processes in question are too weak to influence diffusion or hydrodynamics. Any established solver [192, 193] may therefore be used to obtain spatial dynamics before one starts the spin dynamics simulations covered here.

Conservation of matter requires a local decrease in concentration to be equal to the divergence of its flux:

$$\frac{\partial}{\partial t}c(\mathbf{r},t) = -\mathrm{div}[\mathbf{j}(\mathbf{r},t)] = \left[-\nabla^{\mathrm{T}}\cdot\mathbf{v}(\mathbf{r},t) - \mathbf{v}^{\mathrm{T}}(\mathbf{r},t)\cdot\nabla + \nabla^{\mathrm{T}}\cdot\mathbf{D}(\mathbf{r},t)\cdot\nabla\right]c(\mathbf{r},t) \tag{5.59}$$

For the gradient operator acting on the vectorised array of concentrations in every voxel:

$$\nabla = \begin{bmatrix} \partial/\partial x \\ \partial/\partial y \\ \partial/\partial z \end{bmatrix} \quad \Rightarrow \quad [\![\nabla]\!] = \begin{bmatrix} [\![\partial/\partial x]\!]\otimes\mathbf{1}_{\mathrm{Y}}\otimes\mathbf{1}_{\mathrm{Z}} \\ \mathbf{1}_{\mathrm{X}}\otimes[\![\partial/\partial y]\!]\otimes\mathbf{1}_{\mathrm{Z}} \\ \mathbf{1}_{\mathrm{X}}\otimes\mathbf{1}_{\mathrm{Y}}\otimes[\![\partial/\partial z]\!] \end{bmatrix} \tag{5.60}$$

where $[\![\partial/\partial x]\!]$ denotes a matrix representation of $\partial/\partial x$ on a finite grid, and $\mathbf{1}_{\{\mathrm{X,Y,Z}\}}$ are unit matrices of appropriate dimensions. For diffusion and flow, finite difference matrices [194] suffice.

5.2.3.4 Phase Turning Generators

Matrix representations of phase turning generators should not be obtained using finite differences because they need to be very precise—a magic angle spinning simulation in NMR can go through tens of thousands of rotor cycles. For uniform grids with periodic boundary conditions, Fourier differentiation matrices [195] are recommended:

$$\left[\!\!\left[\frac{\partial}{\partial\varphi}\right]\!\!\right]_{nk} = \begin{cases} \frac{(-1)^{n+k}}{2}\cot\left[\frac{(n-k)\pi}{N}\right] & n \neq k \\ 0 & n = k \end{cases}. \tag{5.61}$$

Operators for higher derivatives are obtained by taking powers of this matrix.

5.2.4 Periodic Motion, Diffusion, and Flow

For counterclockwise periodic flow with a frequency ω and phase φ, the generator is $\partial/\partial\varphi$:

$$\exp\left[-\omega t\frac{\partial}{\partial\varphi}\right]f(\varphi)=f(\varphi-\omega t) \tag{5.62}$$

Selecting a uniform periodic phase grid makes this generator a time-independent matrix, a good choice is Eq. (5.61). When the spin evolution generator $\mathfrak{L}$ depends on this phase (for example, in the Zeeman interaction term under an oscillating magnetic field), the equation of motion is

$$\frac{\partial}{\partial t}\mathbf{\rho}(\varphi,t)=\left[-i\mathfrak{L}(\varphi,t)-\omega(t)\frac{\partial}{\partial\varphi}\right]\mathbf{\rho}(\varphi,t) \tag{5.63}$$

with an obvious extension to situations when there are two or more periodic processes:

$$\begin{aligned}&\frac{\partial}{\partial t}\mathbf{\rho}(\varphi_1,\varphi_2,\ldots,t)\\&=\left[-i\mathfrak{L}(\varphi_1,\varphi_2,\ldots,t)-\omega_1(t)\frac{\partial}{\partial\varphi_1}\right.\\&\left.-\omega_2(t)\frac{\partial}{\partial\varphi_2}-\ldots\right]\mathbf{\rho}(\varphi_1,\varphi_2,\ldots,t)\end{aligned} \tag{5.64}$$

At the numerical implementation level, the evolution generator is structured as a sparse block-diagonal matrix with each block corresponding to a point on the phase grid. The job of the phase increment generator is to move block populations around during evolution. The matrix form of Eq. 5.63 is

$$\begin{aligned}\frac{\partial}{\partial t}\begin{pmatrix}\mathbf{\rho}(\varphi_1,t)\\\mathbf{\rho}(\varphi_2,t)\\\vdots\\\mathbf{\rho}(\varphi_N,t)\end{pmatrix}=&-i\begin{pmatrix}\mathfrak{L}(\varphi_1)&&&\\&\mathfrak{L}(\varphi_2)&&\\&&\cdots&\\&&&\mathfrak{L}(\varphi_N)\end{pmatrix}\begin{pmatrix}\mathbf{\rho}(\varphi_1,t)\\\mathbf{\rho}(\varphi_2,t)\\\vdots\\\mathbf{\rho}(\varphi_N,t)\end{pmatrix}\\&-\omega\left(\left[\!\left[\frac{\partial}{\partial\varphi}\right]\!\right]\otimes\mathbf{1}\right)\begin{pmatrix}\mathbf{\rho}(\varphi_1,t)\\\mathbf{\rho}(\varphi_2,t)\\\vdots\\\mathbf{\rho}(\varphi_N,t)\end{pmatrix}\end{aligned} \tag{5.65}$$

where N is the number of points in the phase grid, the rotor turning generator matrix $[\![\partial/\partial\varphi]\!]$ comes from Eq. (5.61), and $\mathbf{1}$ is a unit matrix of the same dimension as the spin evolution generator. This treatment of periodic time dependencies in the Hamiltonian is convenient because the phase turning generator is time-independent for fixed frequencies, and trivially time-dependent when the frequencies change. When the Liouvillian is dissipative, this equation may be solved for the steady-state orbit of the system in the same way Eq. (4.259) was solved for the steady state; this is particularly useful in DNP simulations.

To accommodate translational diffusion and flow, we add generators from Eqs. (5.45) and (5.47). In the simplest case of uniform diffusion and flow in one dimension:

$$\frac{\partial}{\partial t}\boldsymbol{\rho}(z,t)=\left[-i\mathcal{L}(z,t)+D_{\mathrm{T}}\frac{\partial^2}{\partial z^2}-v\frac{\partial}{\partial z}\right]\boldsymbol{\rho}(z,t), \tag{5.66}$$

where D_{T} is the translational diffusion coefficient, v is the flow velocity, z is the sample coordinate, and finite difference matrix representations of the spatial derivative operators are sufficient. Extension to multiple spatial coordinates is accomplished by adding further spatial dynamics generators.

5.2.5 Connection to Floquet Theory

Floquet theory (Sect. 4.8) is a variable separation solution of Eq. The relationship is best illustrated for uniparametric non-dissipative periodic Hamiltonian:

$$\frac{\partial}{\partial t}\boldsymbol{\rho}(\varphi,t)=-i\mathcal{H}(\varphi)\boldsymbol{\rho}(\varphi,t)-\omega\frac{\partial}{\partial\varphi}\boldsymbol{\rho}(\varphi,t) \tag{5.67}$$

Because the dynamics is periodic, the following expansions are possible:

$$\boldsymbol{\rho}(\varphi,t)=\sum_n\boldsymbol{\rho}_n(t)\exp(in\varphi),\quad \mathcal{H}(\varphi)=\sum_k\mathcal{H}_k\exp(ik\varphi) \tag{5.68}$$

where Fourier indices n and k run over all integers. After substitution, Eq. 5.67 becomes

$$\begin{aligned}\sum_n\exp(in\varphi)\left(\frac{\partial}{\partial t}\boldsymbol{\rho}_n(t)\right)&=\sum_{kn}\exp(i(k+n)\varphi)(-i\mathcal{H}_k\boldsymbol{\rho}_n(t))\\&\quad-i\omega_{\mathrm{S}}\sum_n n\exp(in\varphi)\boldsymbol{\rho}_n(t)\end{aligned} \tag{5.69}$$

After rotating $(m=n+k)$ the summation indices of the double sum and equating the coefficients of the same Fourier factors, we obtain

$$\left\{\frac{\partial}{\partial t}\boldsymbol{\rho}_m(t)=-i\sum_n\mathcal{H}_{m-n}\boldsymbol{\rho}_n(t)-im\omega_{\mathrm{S}}\boldsymbol{\rho}_m(t)\right._m \tag{5.70}$$

which is identical to Eq. (4.221) in the section dealing with unimodal Floquet theory. The latter may therefore be viewed as a variable separation solution of Eq. 5.63.

5.3 Chemical Reactions

Time-domain electronic structure theory treatment of most chemical reactions is computationally intractable, and we are compelled to work at the level of the spin Hamiltonian which hides electronic structure theory inside effective parameters such as chemical shielding (Chapter 3). At that level, mathematical models of spin dynamics during chemical processes may be classified as follows:

1. *Spin-independent closed chemical reactions*—those where the chemical process changes amplitudes of interaction terms in the spin Hamiltonian (for example conformational mobility), but does not change the number of spins or perform non-unitary operations on individual spin systems. The spin state of the system does not influence stoichiometric rates or yields of the chemical reaction. Ensembles can dephase due to the noise chemistry creates in the spin Hamiltonian.

2. *Spin-independent open chemical reactions*—those where the chemical process connects the system to the outside world (for example, proton exchange with the solvent) and can perform non-unitary operations: either change the number of spins or reset the state of a spin: a proton departing into the solvent is not necessarily the proton that would come back. Ensembles can also dephase due to the noise chemistry creates in the spin Hamiltonian. However, the spin state of the system still does not influence stoichiometric rates or yields of the chemical reaction.

3. *Spin-selective chemical reactions*—those where different spin states have different chemical behaviours, for example, spin-correlated radical pairs where the singlet state is reactive (because electrons can make a stable chemical bond) but the triplet state is not.

In all three cases, the equation of motion may become non-linear because differential equations describing chemical kinetics may contain products of concentrations. It can also become distributed over classical coordinates—because concentrations need not be the same in all points of the sample—with the attendant complications caused by diffusion and flow.

5.3.1 Networks of First-Order Spin-Independent Reactions

A simple example that illustrates the general idea is an ensemble of molecules that alternate between two conformations A and B. The two conformations have different geometries, chemical shifts, and J-couplings; their spin Hamiltonians and relaxation superoperators are, therefore, different linear combinations, but of the same basis operators. The equation of motion is obtained by taking a Kronecker product between the spin state space (of whatever dimension) and the two-dimensional chemical space:

$$\left[\frac{\mathrm{d}}{\mathrm{dt}}\boldsymbol{\rho}=-i\mathcal{H}\boldsymbol{\rho}+\mathcal{R}\boldsymbol{\rho}\right]\otimes\left[\frac{\mathrm{d}}{\mathrm{dt}}\begin{pmatrix}A\\B\end{pmatrix}=\begin{pmatrix}-k_{+} & +k_{-}\\+k_{+} & -k_{-}\end{pmatrix}\begin{pmatrix}A\\B\end{pmatrix}\right]$$
$$\Downarrow$$
$$\frac{\mathrm{d}}{\mathrm{dt}}\begin{bmatrix}\boldsymbol{\rho}_{\mathrm{A}}\\\boldsymbol{\rho}_{\mathrm{B}}\end{bmatrix}=\left[-i\begin{pmatrix}\mathcal{H}_{\mathrm{A}} & \mathbf{0}\\\mathbf{0} & \mathcal{H}_{\mathrm{B}}\end{pmatrix}+\begin{pmatrix}\mathcal{R}_{\mathrm{A}} & \mathbf{0}\\\mathbf{0} & \mathcal{R}_{\mathrm{B}}\end{pmatrix}+\begin{pmatrix}-k_{+}\mathbf{1} & +k_{-}\mathbf{1}\\+k_{+}\mathbf{1} & -k_{-}\mathbf{1}\end{pmatrix}\right]\begin{bmatrix}\boldsymbol{\rho}_{\mathrm{A}}\\\boldsymbol{\rho}_{\mathrm{B}}\end{bmatrix} \tag{5.71}$$

Here, the Hamiltonian and the relaxation superoperator act as they normally do in each of the two conformers, and the role of the kinetics superoperator is to shuffle the populations of the spin states between the two conformers at the rate of the corresponding chemical reactions. In the lower part of Eq. 5.71, density matrices are re-defined to include concentrations which are then also interpreted as probability densities. Conceptually, this is arguable—we could keep concentrations separate and density matrices normalised. This divarication does a fine job of sorting theorists into those who use what they publish, and those who do not: an attempt to keep concentrations separate puts them into denominators [196]; the resulting numerical stability problems in finite precision arithmetic make that option impractical.

The simple form of the kinetics superoperator in Eq. 5.71 requires the basis set of the spin state space to be the same, and sorted in the same way, on either side of the reaction arrow. When it is not, unit matrices must be replaced with appropriate permutation or projection matrices. Extending Eq. 5.71 to larger networks of first-order reactions (for example, Markov state models [197], where molecular dynamics is approximated by a network of exchanging conformations) is a simple matter of having more blocks on the diagonals of the spin operator arrays, and a bigger kinetic matrix.

5.3.2 Networks of Arbitrary Spin-Independent Reactions

When second- and higher-order chemical reactions are present, the linear Liouville-von Neumann equation must be combined with non-linear equations describing chemical kinetics. We must use the assumption made at the start of this section—that chemistry is spin-independent—and proceed as follows:

1. Solve the chemical kinetics equations using standard methods (for example, Runge–Kutta [133, 134]) and obtain concentrations of all substances $c_k(t)$ as functions of time.

2. For the spin system of every molecule in the reaction network, build the state space basis set $\{\mathbf{v}_n^{(k)}\}$, Hamiltonian commutation superoperator $\mathcal{H}_k$, relaxation superoperator $\mathcal{R}_k$, and the concentration-weighted initial spin state $\boldsymbol{\rho}_k(0)$ where the index k runs over substances. Basis sets composed of direct products of individual spin operators (Sect. 7.1) are recommended here.

3. For every reaction, build a matching table of basis spin states on either side of the reaction arrow: which state goes where with what coefficient, which product states stay intact and which get broken up as the reaction takes place in either direction.

4. Convert the matching tables into reaction generators $\mathcal{G}_r$ that take the concatenated spin state vector of all molecules and apply the rth reaction by permuting populations of relevant states of relevant spins as its reaction prescribes. This makes $\mathcal{G}_r$ a generator of the time evolution semigroup (Sect. 1.5); such generators act by matrix exponentials (Sect. 1.5.3)—the blocks of $\mathcal{G}_r$ corresponding to molecules that do not participate in the rth reaction must, therefore, be zero. Populations of multi-spin product states that end up becoming inter-molecular must be taken from the source but not sent to any destination.

5. From the concentrations obtained in Item (1), calculate the time-dependent rate $k_r(t)$ of each magnetisation transport process. The time evolution generator acting on the collection of spin systems as a result of the rth reaction taking place is then $k_r(t)\mathcal{G}_r$.

6. Assemble the equation of motion with the now time-dependent generator:

$$\frac{\mathrm{d}}{\mathrm{d}t}\begin{bmatrix}\boldsymbol{\rho}_\mathrm{A}\\ \boldsymbol{\rho}_\mathrm{B}\\ \vdots\end{bmatrix} = \left[-i\begin{pmatrix}\mathcal{H}_\mathrm{A} & \mathbf{0} & \cdots\\ \mathbf{0} & \mathcal{H}_\mathrm{B} & \cdots\\ \vdots & \vdots & \end{pmatrix} + \begin{pmatrix}\mathcal{R}_\mathrm{A} & \mathbf{0} & \cdots\\ \mathbf{0} & \mathcal{R}_\mathrm{B} & \cdots\\ \vdots & \vdots & \end{pmatrix} + \sum_r k_r(t)\mathcal{G}_r\right]\begin{bmatrix}\boldsymbol{\rho}_\mathrm{A}\\ \boldsymbol{\rho}_\mathrm{B}\\ \vdots\end{bmatrix} \tag{5.72}$$

and solve it using geometric integrators (Section 4.9.4) for time-dependent generators.

5.3.3 Chemical Transport of Multi-spin Orders

Let us look at the process of building transport generators $\mathcal{G}_r$ in more detail, using nuclear magnetic resonance as an illustration. When a nucleus moves from position n in a reactant to position m in a product, it takes its spin state with it—the coefficients of all krons involving the operators of that spin in the reactant spin state basis are moved to the corresponding krons in the product spin state basis:

$$\cdots\underbrace{\otimes\mathbf{S}\otimes}_{n}\cdots \quad\rightarrow\quad \cdots\underbrace{\otimes\mathbf{S}\otimes}_{m}\cdots \tag{5.73}$$

where $\alpha \in \{z, +, -, ...\}$ is an index running over the elements of the local state space of that spin. This is always the case for single-spin operators, but the transport of multi-spin orders is more subtle.

When multiple nuclei move together from a reactant to a product, their correlated spin states are preserved. For example, when two nuclei in a correlated state $\mathbf{S}_\alpha \otimes \mathbf{L}_\beta$ with $\alpha, \beta \in \{z, +, -, ...\}$ move from positions $\{k, m\}$ in a reactant to positions $\{p, q\}$ in a product, the coefficient in front of the corresponding basis states:

$$\cdots\underbrace{\otimes\mathbf{S}\otimes}_{k}\cdots\underbrace{\otimes\mathbf{L}\otimes}_{m}\cdots \quad\rightarrow\quad \cdots\underbrace{\otimes\mathbf{S}\otimes}_{p}\cdots\underbrace{\otimes\mathbf{L}\otimes}_{q}\cdots \tag{5.74}$$

must be forwarded. The corresponding element of $\mathcal{G}_r$ is then a combination of projectors that subtracts the coefficient from the source state, and adds it to the destination:

$$\mathcal{G}_r = \ldots - |\text{source}\rangle\langle\text{source}| + |\text{destin}\rangle\langle\text{source}| + \ldots \tag{5.75}$$

However, when the nuclei participating in a correlated state end up in different molecules, we must account for the fact that those molecules are unlikely to meet again. This assumption is implicit in the law of mass action [198]; here it effectively creates a relaxation mechanism—the coefficients of such states must be counted as lost. This is done by not forwarding them to any destination:

$$\mathcal{G}_r = \ldots - |\text{source}\rangle\langle\text{source}| + \ldots \tag{5.76}$$

in which case the exponential action by $k_r(t)\mathcal{G}_r$ damps the state at the reaction rate.

5.3.4 Flow, Diffusion, and Noise Limits

Networks of unidirectional chemical processes may be used to approximate continuous flow processes, including cyclic flows such as magic angle spinning. As a demonstration, consider a set of uniformly spaced bins $\{1, 2, \ldots, N\}$ and a cyclic chain of first-order transport processes between them:

$$\begin{array}{ccc} A_1 & \rightarrow & A_2 \\ \uparrow & & \downarrow \\ \cdots & \leftarrow & A_3 \end{array} \qquad \frac{\mathrm{d}}{\mathrm{d}t}\begin{pmatrix} A_1 \\ A_2 \\ A_3 \\ \vdots \end{pmatrix} \propto \begin{pmatrix} -1 & 0 & 0 & \cdots \\ 1 & -1 & 0 & \cdots \\ 0 & 1 & -1 & \cdots \\ \vdots & \vdots & \vdots & \end{pmatrix}\begin{pmatrix} A_1 \\ A_2 \\ A_3 \\ \vdots \end{pmatrix} \tag{5.77}$$

It is clear from direct inspection that the matrix here is a finite difference approximation of the first derivative that would converge to $\partial/\partial x$ in the limit of infinitely many bins. Exponential action by a derivative does indeed produce a flow of probability density along the corresponding coordinate:

$$\exp\left(-vt\frac{\partial}{\partial x}\right)p(x) = p(x - vt) \tag{5.78}$$

This connection is appealing, but using the actual matrix in Eq. (5.77) to approximate $\partial/\partial x$ is numerically inefficient. In the context of periodic flows on finite grids, the most accurate approximation for $\partial/\partial x$ is the Fourier differentiation matrix in Eq. (5.61). For non-periodic flows, high-accuracy finite-difference approximations [194] are recommended.

Networks of bidirectional transport processes converge to continuous diffusion, for example:

$$\begin{array}{ccc} A_1 & \rightleftarrows & A_2 \\ \downarrow\uparrow & & \downarrow\uparrow \\ \cdots & \rightleftarrows & A_3 \end{array} \qquad \frac{d}{dt}\begin{pmatrix} A_1 \\ A_2 \\ A_3 \\ \vdots \end{pmatrix} \propto \begin{pmatrix} -2 & 1 & 0 & \cdots \\ 1 & -2 & 1 & \cdots \\ 0 & 1 & -2 & \cdots \\ \vdots & \vdots & \vdots & \end{pmatrix}\begin{pmatrix} A_1 \\ A_2 \\ A_3 \\ \vdots \end{pmatrix} \tag{5.79}$$

where the matrix is clearly a finite difference approximation to the diffusion generator $\partial^2/\partial x^2$ whose exponential does indeed act by convolution with the Green's function of the diffusion equation:

$$\exp\left(Dt\frac{\partial^2}{\partial x^2}\right)p(x) = \left[\frac{1}{\sqrt{4\pi Dt}}\exp\left(-\frac{x^2}{4Dt}\right)\right] * p(x) \tag{5.80}$$

where the star denotes a convolution operation. Diffusion is easier to handle numerically than flow; low-accuracy finite-difference approximations generally suffice. This makes finite networks of exchanging locations (produced, for example, by Markov state models [197]) an appealing alternative to logistically complicated continuum descriptions of local dynamics such as the stochastic Liouville equation (Sect. 6.5).

In some situations, it may be expedient to ignore entirely the details of chemical dynamics and treat it simply as a source of noise in the spin Hamiltonian: every time a particular spin is transported from one chemical environment to another, its various interaction tensors switch to new values. This limit is considered in detail in Sect. 6.3, where it fits as a special case into Redfield's relaxation theory.

5.3.5 Spin-Selective Chemical Elimination

Consider a situation where a particular spin wavefunction $|\psi\rangle$ permits the electronic structure to react, for example, by making it possible to form a chemical bond, with the result that the molecule is taken out of the ensemble. A first-order reaction of this kind must be draining the entire cross (row and column) of the density matrix: not only the probability of $|\psi\rangle$, but also its ensemble correlations.

At the time of writing, there are unsettled debates about this process, but one reasonable treatment is presented here. Wavefunction coefficients are square roots of probability: it would not be outrageous to assume that the quantum state dependent first-order kinetics would, therefore, drain off-diagonal ensemble correlation terms $|\psi\rangle\langle\varphi|$ of the density matrix at half the rate of the ensemble probability term $|\psi\rangle\langle\psi|$. The corresponding superoperator will not be Lindbladian (Sect. 6.8) because the process is not trace-preserving: we are removing systems from the ensemble.

To obtain the generator of this quantum state selective elimination process, consider its action on an otherwise static ensemble of spin systems. If $\mathbf{P} = |\psi\rangle\langle\psi|$, the relevant row in the density matrix is $\mathbf{P}\boldsymbol{\rho}$, the relevant column is $\boldsymbol{\rho}\mathbf{P}$, and the diagonal element is $\mathbf{P}\boldsymbol{\rho}\mathbf{P}$. We expect our reaction to fade off-diagonal elements $\mathbf{P}\boldsymbol{\rho} + \boldsymbol{\rho}\mathbf{P} - 2\mathbf{P}\boldsymbol{\rho}\mathbf{P}$ exponentially at the rate $k/2$ and the diagonal element at the rate k:

$$\exp(\mathcal{K}t)\boldsymbol{\rho}_0 = \boldsymbol{\rho}_0 - \left(1 - \mathrm{e}^{-kt/2}\right)(\mathbf{P}\boldsymbol{\rho}_0 + \boldsymbol{\rho}_0\mathbf{P} - 2\mathbf{P}\boldsymbol{\rho}_0\mathbf{P}) - \left(1 - \mathrm{e}^{-kt}\right)\mathbf{P}\boldsymbol{\rho}_0\mathbf{P} \quad (5.81)$$

This action is constructed to follow first-order kinetics: at $t = 0$ we have the initial condition $\boldsymbol{\rho}_0$, and at $t = \infty$ the cross has faded: $\boldsymbol{\rho}_0 - (\mathbf{P}\boldsymbol{\rho}_0 + \boldsymbol{\rho}_0\mathbf{P} - \mathbf{P}\boldsymbol{\rho}_0\mathbf{P})$. Equation (5.81) defines a semigroup orbit (Sect. 1.5.3) parametrised by time. Using Eq. (5.1) to extract the generator from a finite semigroup action (Sect. 1.5.7), we obtain the *Johnson-Merrifield kinetics superoperator* [199]:

$$\mathcal{K}\boldsymbol{\rho} = \frac{\partial}{\partial t}\left[\exp(\mathcal{K}t)\boldsymbol{\rho}\right]_{t=0} = -\frac{k}{2}(\mathbf{P}\boldsymbol{\rho} + \boldsymbol{\rho}\mathbf{P}) \quad (5.82)$$

Semigroup generators add algebraically, and therefore, for multiple first-order chemical reactions out of states with different orthogonal spin wavefunctions $|\psi_n\rangle$ with projectors $\mathbf{P}_n = |\psi_n\rangle\langle\psi_n|$:

$$\mathcal{K}\boldsymbol{\rho} = -\sum_n \frac{k_n}{2}(\mathbf{P}_n\boldsymbol{\rho} + \boldsymbol{\rho}\mathbf{P}_n) \quad (5.83)$$

This dubious model is about the only thing that is agreed upon; a vigorous and ongoing discussion of this subject may be found in the primary literature.

5.4 Spin-Rotation Coupling

Consider a rigid molecule undergoing rotational motion in its translational rest frame. In classical physics, two types of magnetic interactions proceed from circular currents created by the orbital motion of electric charges: with the external magnetic field and with the magnetic dipoles associated with nuclear and electron spin. When the molecule is viewed as a collection of point charges, the magnetic induction at the origin from each charge q moving at location $\mathbf{r}$ with velocity $\mathbf{v}$ is (*Biot-Savart law* [200]):

$$\mathbf{B} = \frac{\mu_0}{4\pi} q \frac{\mathbf{v} \times \mathbf{r}}{r^3} \tag{5.84}$$

This may be rewritten using linear momentum $\mathbf{p} = m\mathbf{v}$, and then angular momentum $\mathbf{L} = \mathbf{r} \times \mathbf{p}$:

$$\mathbf{B} = \frac{\mu_0}{4\pi} q \frac{\mathbf{p} \times \mathbf{r}}{mr^3} = -\frac{\mu_0}{4\pi} \frac{q}{mr^3} \mathbf{L} \tag{5.85}$$

where m is the mass of the charge. The induction is here linear in angular momentum; Zeeman interaction is bilinear in spin and magnetic induction (Sect. 3.2)—therefore, the general form of the spin-rotation interaction Hamiltonian must be

$$\hat{\mathbf{H}}_{\mathrm{SR}} = -\mathbf{S} \cdot \mathbf{M} \cdot \hat{L} \tag{5.86}$$

where $\mathbf{S} = [\mathbf{S}_\mathrm{X}, \mathbf{S}_\mathrm{Y}, \mathbf{S}_\mathrm{Z}]^\mathrm{T}$ is a vector of nuclear spin operator matrices, $\hat{L} = [\hat{L}_\mathrm{X}, \hat{L}_\mathrm{Y}, \hat{L}_\mathrm{Z}]^\mathrm{T}$ is a vector of angular momentum operators of the rigid molecule, $\mathbf{M}$ is a real 3×3 matrix called *spin-rotation coupling tensor*, and the minus is there for historical reasons. When the current loops (created by rotating charges) are viewed as magnetic moments, a similar treatment leads to the rotational Zeeman interaction [201]:

$$\hat{H}_{\mathrm{RZ}} = -\frac{\mu_\mathrm{N}}{\hbar} \mathbf{B} \cdot \mathbf{g}_\mathrm{R} \cdot \hat{L} \tag{5.87}$$

where $\mathbf{g}_\mathrm{R}$ is the *rotational g-tensor*. This interaction enters the background Hamiltonian when the simultaneous dynamics in rotational and spin degrees of freedom is considered. Both tensors may be obtained using the energy derivative formalism (Sect. 3.1.8); evaluation of the derivatives is a technical matter [201], dealt with in electronic structure theory (Chap. 3 and [201, 202]).

5.4.1 Molecules as Rigid Rotors

The rigid rotor Hamiltonian follows from the classical expression for the energy of a rotating body:

$$E = \frac{1}{2}\boldsymbol{\omega}^{\mathrm{T}} \cdot \mathbf{I}_{\mathrm{CM}} \cdot \boldsymbol{\omega}, \qquad \mathbf{I}_{\mathrm{CM}} = \sum_k m_k (\mathbf{r}_k - \mathbf{r}_{\mathrm{CM}})(\mathbf{r}_k - \mathbf{r}_{\mathrm{CM}})^{\mathrm{T}} \tag{5.88}$$

where m_k is the mass of the kth point particle, $\mathbf{r}_k$ is its Cartesian coordinate vector, $\mathbf{r}_{\mathrm{CM}}$ are the coordinates of the centre of mass, $\boldsymbol{\omega}$ is the angular velocity vector, and $\mathbf{I}_{\mathrm{CM}}$ is the tensor of inertia (a symmetric 3×3 matrix) relative to the centre of mass. This expression is quantised using the classical mechanics relation $\mathbf{L} = \mathbf{I}\omega$ connecting angular velocity and angular momentum:

$$\hat{H}_{\mathrm{RR}} = \frac{\hat{L} \cdot \mathbf{I}^{-1} \cdot \hat{L}}{2} \tag{5.89}$$

When the inertia tensor is isotropic, this Hamiltonian is the multiple of the total angular momentum operator $\hat{L}^2$. When the tensor is axial, it may be rearranged into a linear combination of $\hat{L}^2$ and $\hat{L}_{\mathrm{Z}}^2$:

$$\hat{H}_{\mathrm{RR}} = \frac{\hat{L}_{\mathrm{X}}^2 + \hat{L}_{\mathrm{Y}}^2}{2I_{\perp}} + \frac{\hat{L}_{\mathrm{Z}}^2}{2I_{\parallel}} = \frac{\hat{L}_{\mathrm{X}}^2 + \hat{L}_{\mathrm{Y}}^2 + \hat{L}_{\mathrm{Z}}^2}{2I_{\perp}} + \left(\frac{1}{2I_{\perp}} - \frac{1}{2I_{\perp}}\right)\hat{L}_{\mathrm{Z}}^2 \tag{5.90}$$

In both cases (Sect. 6.3.6.2) the eigenfunctions are elements of Wigner D matrices $\mathfrak{D}_{km}^{(l)}$ indexed by the integer orbital quantum number $l \geq 0$ and two integer projection indices $k, m \in [-l, l]$. The rhombic inertia tensor case is similar to the equation solved in Sect. 6.4.4 for rhombic rotational diffusion.

To account for centrifugal distortions, the Hamiltonian is augmented by higher powers of angular momentum operators—essentially, a Taylor series [203]. In the principal axis frame of the inertia tensor:

$$\hat{H} = \hat{H}_{\mathrm{RR}} + \frac{1}{4}\sum_{\mu\nu\xi\zeta} \tau_{\mu\nu\xi\zeta} \hat{L}_{\mu}\hat{L}_{\nu}\hat{L}_{\xi}\hat{L}_{\zeta} + \ldots \tag{5.91}$$

where $\tau_{\mu\nu\xi\zeta}$ are system-specific coefficients without a clear physical meaning.

5.4.2 Internal Rotations and Haupt Effect

A good example of rotational degrees of freedom influencing spin is the low-temperature spin dynamics in the freely rotating methyl group of crystalline γ-picoline [204]. Assuming that the electronic structure theory is in the ground state, the dominant part of the Hamiltonian is

$$\hat{\mathbf{H}} = \hat{H}_{\mathrm{RR}} \otimes \mathbf{1}_{\mathrm{S}} + \mathbf{1}_{\mathrm{RR}} \otimes \mathbf{H}_{\mathrm{NZ}} + \mathbf{1}_{\mathrm{RR}} \otimes \mathbf{H}_{\mathrm{DD}} \tag{5.92}$$

where $\hat{H}_{\mathrm{RR}}$ is the rigid rotor Hamiltonian, $\mathbf{H}_{\mathrm{Z}}$ is the nuclear Zeeman interaction, and $\mathbf{H}_{\mathrm{DD}}$ contains inter-nuclear dipolar couplings that depend on the methyl group turning angle φ. For uniaxial rotation, the perpendicular moment of inertia is infinite, and the Hamiltonian in Eq. (5.90) simplifies

$$\hat{H}_{\mathrm{RR}} = -\frac{\hat{L}_{\mathrm{Z}}^2}{2I_{\|}} = -\frac{\hbar^2}{2I_{\|}}\frac{\partial^2}{\partial\varphi^2} \tag{5.93}$$

where $I_{\|}$ is the moment of inertia of the methyl group around its axis of rotation. The Zeeman part of the Hamiltonian is $\mathbf{H}_{\mathrm{NZ}} = -\gamma_{\mathrm{H}} B_0 \left(\mathbf{S}_{\mathrm{Z}}^{(1)} + \mathbf{S}_{\mathrm{Z}}^{(2)} + \mathbf{S}_{\mathrm{Z}}^{(3)}\right)$, and the three differently rotated inter-nuclear terms present in the dipole–dipole interaction Hamiltonian

$$\begin{aligned} \mathbf{H}_{\mathrm{DD}}(\varphi) &= -\sqrt{6}\frac{\mu_0}{4\pi}\frac{\gamma_1\gamma_2\hbar}{r^3}\left(\hat{R}(\varphi)\mathbf{T}_{2,0}^{(1,2)} + \hat{R}\left(\varphi + \frac{2\pi}{3}\right)\mathbf{T}_{2,0}^{(2,3)} + \hat{R}\left(\varphi + \frac{4\pi}{3}\right)\mathbf{T}_{2,0}^{(3,1)}\right) \\ \mathbf{T}_{2,0}^{(n,k)} &= +\sqrt{\frac{2}{3}}\left(\mathbf{S}_{\mathrm{Z}}^{(n)}\mathbf{S}_{\mathrm{Z}}^{(k)} - \frac{1}{4}\left(\mathbf{S}_{+}^{(n)}\mathbf{S}_{-}^{(k)} + \mathbf{S}_{-}^{(n)}\mathbf{S}_{+}^{(k)}\right)\right) \end{aligned} \tag{5.94}$$

contain two contributions each—the positioning rotations that take inter-nuclear vectors from the Z-axis into the molecular frame of reference, and the turning rotation with the phase φ that enters the Hamiltonian in Eq. The expressions for the action by the rotation superoperator on an irreducible spherical tensor are given in Sect. 3.2.2; for our purposes here, it suffices to note that the dipole–dipole interaction Hamiltonian depends on φ. This means that $\hat{H}_{\mathrm{RR}}$ and $\mathbf{H}_{\mathrm{DD}}$ do not commute; this results, via coherent (Liouville-von Neumann equation contains a commutator) and relaxation driven mechanisms, in the transfer of polarisation from the rotor degrees of freedom to the nuclear dipolar order [205], and from there to Zeeman order. This process is called the *Haupt effect*.

5.5 Spin-Phonon Coupling

Moving an atom in a crystal lattice perturbs the spin Hamiltonian, both directly (e.g. inter-nuclear dipolar coupling) and through the associated perturbation of the electronic structure. The general case of simultaneous quantum dynamics in nuclear position, electronic structure, and spin is intractable. Common approximations (Born—Oppenheimer, harmonic lattice vibrations, separation of timescales, etc.) lead in practice to predictions that can be orders of magnitude away from experimental data [206]. The appearance of a rigorous treatment below is just that—at the time of writing, there is no quantitative theory.

The root of the problem is in the relationship between frequencies probed in spin dynamics (MHz to THz) and the crystal size in a chemical sample. The speed of sound in a typical crystalline solid is $\sim 5 \times 10^3$ m/s, meaning that the wavelength of a standing wave is in the millimetres for MHz frequencies and nanometres for THz frequencies. The slow end of that interval is influenced by crystal shape, chemical and isotopic purity, and defect density. At the fast end of the interval, the harmonic approximation and the timescale separations break down. Still, it is possible to collect irreproducible and incomputable unknowns into physically motivated coefficients that can be fitted to experimental data.

5.5.1 Harmonic Oscillator

A one-dimensional harmonic oscillator of mass m with a spring force constant k and displacement x has the following Hamiltonian, which we translate into canonical variables:

$$\hat{H} = \frac{\hat{p}^2}{2m} + \frac{k\hat{x}^2}{2} = \hbar\omega\left(\hat{\eta}^2 + \hat{\xi}^2\right), \quad \hat{\xi} = \sqrt{\frac{m\omega}{2\hbar}}\hat{x}, \quad \hat{\eta} = \frac{\hat{p}}{\sqrt{2\hbar m\omega}} \tag{5.95}$$

where $\omega = (k/m)^{1/2}$ is the classical angular frequency of the oscillator, and $[\hat{\xi}, \hat{\eta}] = i/2$ as befits canonical coordinate and momentum. A lengthy exercise in calculus [207] yields the eigensystem in which we will index the eigenfunctions $|n\rangle$ by their energy; their explicit expressions will not be needed:

$$\hat{H}|n\rangle = E_n|n\rangle, \quad E_n = \left(n + \frac{1}{2}\right)\hbar\omega, \quad n \in \{0, 1, 2, \ldots\} \tag{5.96}$$

It is convenient to define *creation and annihilation operators* [120]

$$\hat{a}^\dagger = \hat{\xi} - i\hat{\eta}, \quad \hat{a} = \hat{\xi} + i\hat{\eta}, \quad \left[\hat{a}, \hat{a}^\dagger\right] = 1 \tag{5.97}$$

that increase and decrease the eigenfunction index:

$$\hat{a}^\dagger|n\rangle = \sqrt{n+1}|n+1\rangle, \quad \hat{a}|n\rangle = \sqrt{n}|n-1\rangle \tag{5.98}$$

In terms of these operators, the Hamiltonian and the *displacement operator* are

$$\hat{H} = \frac{\hbar\omega}{2}\left(\hat{a}\hat{a}^\dagger + \hat{a}^\dagger\hat{a}\right), \quad \hat{x} = \sqrt{\frac{2\hbar}{m\omega}}\frac{\hat{a} + \hat{a}^\dagger}{2} \tag{5.99}$$

Matrix representations of raising and lowering operators (and therefore of $\hat{H}$ and $\hat{x}$) are obtained from Eq. 5.98; they are infinite-dimensional but may be truncated at some suitable energy level. Another useful operator returns the energy level number:

$$\hat{n} = \hat{a}^{\dagger}\hat{a}, \quad \hat{n}|n\rangle = n|n\rangle \tag{5.100}$$

When written in terms of this operator, the Hamiltonian acquires a particularly simple form:

$$\hat{H} = \hbar\omega(\hat{n} + 1/2) \tag{5.101}$$

that may be used to derive the unobvious Eq. 5.96. Creation and annihilation operators are eigenoperators of the Hamiltonian commutation:

$$\left[\hat{H}, \hat{a}\right] = -\hbar\omega\hat{a}, \quad \left[\hat{H}, \hat{a}^{\dagger}\right] = +\hbar\omega\hat{a}^{\dagger} \tag{5.102}$$

and therefore the exponential propagation relations are (angular frequency units):

$$e^{-i\hat{H}t}\hat{a}e^{+i\hat{H}t} = \hat{a}e^{-i\omega t}, \qquad e^{-i\hat{H}t}\hat{a}^{\dagger}e^{+i\hat{H}t} = \hat{a}^{\dagger}e^{+i\omega t} \tag{5.103}$$

These correspond to Heisenberg picture evolution rules; when combined with the thermodynamic expectation values (i.e. traces with the thermal equilibrium state) for the population number operator $\hat{n} = \hat{a}^{\dagger}\hat{a}$ and the related operator $\hat{a}\hat{a}^{\dagger}$:

$$\left\langle \hat{a}^{\dagger}\hat{a} \right\rangle = \left(e^{\hbar\omega/kT} - 1\right)^{-1}, \quad \left\langle \hat{a}\hat{a}^{\dagger} \right\rangle = \left(1 - e^{-\hbar\omega/kT}\right)^{-1} \tag{5.104}$$

they yield the position autocorrelation function that will be useful in spin-phonon relaxation theories:

$$\begin{aligned} \langle x(t)x(0) \rangle &= \frac{\hbar}{2m\omega}\left(\left\langle \hat{a}^{\dagger}\hat{a} \right\rangle e^{+i\omega t} + \left\langle \hat{a}\hat{a}^{\dagger} \right\rangle e^{-i\omega t}\right) \\ &= \frac{\hbar}{2m\omega}\left[\left(e^{\hbar\omega/kT} - 1\right)^{-1} e^{+i\omega t} + \left(1 - e^{-\hbar\omega/kT}\right)^{-1} e^{-i\omega t}\right] \end{aligned} \tag{5.105}$$

A counter-intuitive observation is that this function has a non-zero imaginary part.

5.5.2 Harmonic Crystal Lattice

For a harmonic crystal lattice—multiple atoms assumed to be linked by harmonic springs—we have

$$\hat{H} = \sum_{i} \frac{\hat{p}_i^2}{2m_i} + \frac{1}{2}\sum_{i,j} k_{ij}\hat{r}_i\hat{r}_j \tag{5.106}$$

where r_i is the displacement from the energy minimum position along the ith Cartesian coordinate and m_i is the mass of the corresponding atom. When the real and symmetric force constant matrix $\mathbf{K}$ is diagonalised, the Hamiltonian acquires the following form:

$$\hat{H} = \sum_i \hat{H}_i, \quad \hat{H}_i = \frac{\hat{\pi}_i^2}{2\mu_i} + \frac{k_i \hat{\rho}_i^2}{2} \tag{5.107}$$

where μ_i is the effective mass associated with the displacement along the ith eigenvector (called *normal mode*) of $\mathbf{K}$, and k_i is the corresponding effective force constant. The reason for this transformation is simplicity: Cartesian displacements in Eq. 5.106 interacted with one another, but normal mode displacements in Eq. 5.107 do not. The mathematical connection between the two descriptions follows from the definition of eigenvector and the fact that eigenvectors of $\mathbf{K}$ are orthonormal:

$$\mathbf{K}\vec{\kappa}^{(i)} = k_i \vec{\kappa}^{(i)}, \quad \hat{\rho}_i = \sum_j \kappa_j^{(i)} \hat{r}_j, \qquad \hat{r}_i = \sum_j \kappa_i^{(j)} \hat{\rho}_j \tag{5.108}$$

We now apply a canonical variable transformation similar to the one in Eq. 5.95:

$$\begin{gathered} \hat{H}_i = \frac{\hat{\pi}_i^2}{2\mu_i} + \frac{k_i \hat{\rho}_i^2}{2} = \hbar\omega_i \left(\hat{\eta}_i^2 + \hat{\xi}_i^2\right), \quad \omega_i = \sqrt{\frac{k_i}{\mu_i}} \\ \hat{\xi}_i = \sqrt{\frac{\mu_i \omega_i}{2\hbar}} \hat{\rho}_i, \quad \hat{\eta}_i = \frac{\hat{\pi}_i}{\sqrt{2\hbar\mu_i\omega_i}} \end{gathered} \tag{5.109}$$

and introduce similar creation and annihilation operators for each normal mode:

$$\begin{gathered} \hat{a}_i^\dagger = \hat{\xi}_i - i\hat{\eta}_i, \quad \hat{a}_i = \hat{\xi}_i + i\hat{\eta}_i \\ \hat{a}_i^\dagger |n_i\rangle = \sqrt{n_i + 1}|n_i + 1\rangle, \quad \hat{a}_i|n_i\rangle = \sqrt{n_i}|n_i - 1\rangle \end{gathered} \tag{5.110}$$

with the result that the Hamiltonian and the mode displacement operator become

$$\hat{H} = \sum_i \frac{\hbar\omega_i}{2} \left(\hat{a}_i \hat{a}_i^\dagger + \hat{a}_i^\dagger \hat{a}_i\right), \quad \hat{\rho}_i = \sqrt{\frac{2\hbar}{\mu_i\omega_i}} \frac{\hat{a}_i + \hat{a}_i^\dagger}{2} \tag{5.111}$$

with matrix representations computed from the actions of raising and lowering operators on the eigenfunctions specified in Eq. (5.110). The energy level number operator for each mode, and the corresponding form of the Hamiltonian, are the same as Eqs. 5.100 and 5.101 for each normal mode:

$$\hat{n}_i = \hat{a}_i^{\dagger}\hat{a}_i, \quad \hat{H} = \sum_i \hbar\omega_i(\hat{n}_i + 1/2) \tag{5.112}$$

5.5.3 Spin-Displacement Coupling

Within the Born–Oppenheimer approximation, the position of an atom is defined by the coordinates of the nucleus. For small classical adiabatic nuclear displacements $\mathbf{r}$ from the energy minimum of the electronic structure, the perturbation in the spin Hamiltonian $\mathbf{H}$ may be approximated by a Taylor series:

$$\mathbf{H}(\mathbf{r}) = \mathbf{H}_0 + \left(\frac{\partial\mathbf{H}}{\partial\mathbf{r}}\right)_{\mathbf{r}=\mathbf{0}}\mathbf{r} + \frac{1}{2}\mathbf{r}^{\mathrm{T}}\left(\frac{\partial^2\mathbf{H}}{\partial\mathbf{r}^{\mathrm{T}}\partial\mathbf{r}}\right)_{\mathbf{r}=\mathbf{0}}\mathbf{r} + O\left(|\mathbf{r}|^3\right) \tag{5.113}$$

Here, $\partial\mathbf{H}/\partial r_k$ is the linear response and $\partial^2\mathbf{H}/\partial r_n\partial r_k$ is the quadratic response of the spin Hamiltonian to displacements r_n and r_k of the nuclear coordinates. At the time of writing, both sets of derivatives may be computed accurately—either analytically (Sect. 3.2.1) or numerically, by calling an electronic structure theory package repeatedly from a finite difference scheme.

In the current theories of spin-lattice coupling, four draconian approximations are made. In order to connect Eq. 5.113 to a quantum mechanical description of vibrations in a crystal lattice, we shall: (a) assume that nuclear position dynamics is adiabatic—that the electronic structure stays in the ground state; (b) truncate the Taylor expansion in Eq. (5.113) at the quadratic term; (c) assume that crystal lattice is perfect: infinite, with no impurities, defects, or isotope distributions; (d) assume that crystal lattice vibrations are harmonic. The latter two assumptions are almost never true—the resulting theories are qualitative at best. Still, we now have a Hamiltonian that couples spin operators to lattice displacement operators:

$$\begin{gathered}
\hat{\mathbf{H}} = \mathbf{H}_0 + \sum_{\alpha}\mathbf{H}'_{\alpha}\hat{r}_{\alpha} + \frac{1}{2}\sum_{\alpha\beta}\mathbf{H}''_{\alpha,\beta}\hat{r}_{\alpha}\hat{r}_{\beta} \\
\hat{r}_{\alpha} = \sum_j\sqrt{\frac{\hbar}{2\mu_j\omega_j}}\kappa_{\alpha}^{(j)}\left(\hat{a}_j + \hat{a}_j^{\dagger}\right), \\
\mathbf{H}'_{\alpha} = \partial\mathbf{H}/\partial r_{\alpha}, \mathbf{H}''_{\alpha,\beta} = \partial^2\mathbf{H}/\partial r_{\alpha}\partial r_{\beta}
\end{gathered} \tag{5.114}$$

Performing the substitutions and replacing individual atom displacement operators with normal mode displacement operators yields

$$\hat{\mathbf{H}} = \mathbf{H}_0 + \sum_{\alpha j} f_{\alpha j} \mathbf{H}'_{\alpha} \left(\hat{a}_j + \hat{a}_j^{\dagger} \right) + \sum_{\alpha \beta i j} s_{\alpha \beta i j} \mathbf{H}''_{\alpha,\beta} \left(\hat{a}_j + \hat{a}_j^{\dagger} \right) \left(\hat{a}_i + \hat{a}_i^{\dagger} \right)$$
$$f_{\alpha j} = \sqrt{\frac{\hbar}{2\mu_j \omega_j}} \kappa_{\alpha}^{(j)}, \quad s_{\alpha \beta i j} = \frac{\hbar}{4} \frac{\kappa_{\alpha}^{(j)} \kappa_{\beta}^{(i)}}{\sqrt{\mu_j \mu_i \omega_j \omega_i}} \tag{5.115}$$

In this Hamiltonian, spin operators $\mathbf{H}'_{\alpha}$ in the first-order atom displacement response are coupled to single (de-)excitations (by $\hat{a}_i^{\dagger}$ and $\hat{a}_i$) of the phonon modes in the crystal lattice; spin operators $\mathbf{H}''_{\alpha,\beta}$ in the second-order atom displacement response are coupled to simultaneous (de-)excitations (by $\hat{a}_j^{\dagger}\hat{a}_i^{\dagger}$ and $\hat{a}_j\hat{a}_i$) of phonon mode pairs, and also to Raman transitions performed by $\hat{a}_j^{\dagger}\hat{a}_i$ and $\hat{a}_j\hat{a}_i^{\dagger}$.

Because all coefficients in Eq (5.115) are computable and all matrix representations are known, the problem is now reduced to numerical linear algebra: standard matrix–vector solvers (Sect. 4.9) and perturbation theories (Sect. 4.4) for Schrödinger and Liouville-von Neumann equation may be used. Spin-phonon relaxation theories are discussed in Sect. 6.10.7.

5.6 Coupling to Quantised Electromagnetic Fields

So far, electromagnetic fields have appeared in our Hamiltonians, starting with Dirac's equation in Sect. 2.5.1, as Maxwell's scalar and vector potentials—relativistic, but not quantum mechanical in their own equations of motion. For large ensembles of spin systems at room temperature, controlled and detected by room temperature electronics, that is a good approximation. The situation changes when we move to consider the dynamics of a single spin interacting with a comparably sized electromagnetic cavity at temperatures below $\hbar\omega/k_{\mathrm{B}}$: we must treat the cavity quantum mechanically and must find an equation of motion describing the collective dynamics that includes the coupling between the cavity and the spin.

5.6.1 LC Circuit Quantisation

Consider a simple LC circuit with capacitance C and inductance L. The canonical variables are the flux φ through the inductor and the charge q on the capacitor [208]. After the same mathematical treatment as in Eq. (5.95–5.99), the Hamiltonian, written in terms of creation and annihilation operators, is

$$\begin{aligned} \hat{H} &= \frac{\hat{q}^2}{2C} + \frac{\hat{\varphi}^2}{2L} = \cdots = \hbar\omega_{\mathrm{R}}\left(\hat{a}^{\dagger}\hat{a} + \tfrac{1}{2}\right) \\ \hat{q} &= i\sqrt{\frac{\hbar}{2Z}}\left(\hat{a}^{\dagger} - \hat{a}\right), \quad \hat{\varphi} = \sqrt{\frac{\hbar Z}{2}}\left(\hat{a}^{\dagger} + \hat{a}\right) \end{aligned} \tag{5.116}$$

where $\omega_{\mathrm{R}} = 1/\sqrt{LC}$ is resonator frequency and $Z = \sqrt{L/C}$ is resonator impedance. Elaborate derivations in the literature are an illusion—the canonical quantisation procedure itself is a fudge [209].

5.6.2 Spin-Cavity Coupling

Maxwell equations of motion for electric and magnetic fields inside an electromagnetic cavity in vacuum with perfectly conductive walls and in the absence of charges are

$$\begin{aligned} &\nabla \cdot \mathbf{E} = 0, \quad \nabla \cdot \mathbf{B} = 0 \\ &\nabla \times \mathbf{E} = -\frac{\partial \mathbf{B}}{\partial t}, \quad \nabla \times \mathbf{B} = \frac{1}{c^2}\frac{\partial \mathbf{E}}{\partial t} \end{aligned} \tag{5.117}$$

Conductive walls create $\mathbf{n} \times \mathbf{E} = 0$ boundary condition at the walls, where $\mathbf{n}$ is the normal vector of the wall surface. Taking the curl of the leftmost equation and rearranging $\nabla \times \nabla \times \mathbf{E} = \nabla(\nabla \cdot \mathbf{E}) - \nabla^2\mathbf{E}$ yields a pair of wave equations for the electric and the magnetic fields:

$$\frac{1}{c^2}\frac{\partial^2 \mathbf{E}}{\partial t^2} = \nabla^2\mathbf{E}, \quad \frac{1}{c^2}\frac{\partial^2 \mathbf{B}}{\partial t^2} = \nabla^2\mathbf{B} \tag{5.118}$$

Because the expression for the energy is isomorphic to Eqs. 5.116 and 5.95:

$$E = \frac{1}{2}\int \left(\varepsilon_0\mathbf{E}^2 + \mu_0\mathbf{B}^2\right)dV \tag{5.119}$$

another instance of the same arithmetic [208] yields the following expressions for the electric field (notionally directed along X) and magnetic field (along Y) operators of one-dimensional (along Z, length L) cavity modes in terms of creation and annihilation operators of what looks like a harmonic oscillator:

$$\begin{aligned} \hat{E}_{\mathrm{X}}^{(k)}(z,t) &= \sqrt{\frac{\hbar\omega_{\mathrm{C}}^{(k)}}{\varepsilon_0 V}}\left(\hat{a}^{\dagger} + \hat{a}\right)\sin\left(\omega_{\mathrm{C}}^{(k)} z/c\right) \\ \hat{B}_{\mathrm{Y}}^{(k)}(z,t) &= i\sqrt{\frac{\mu_0\hbar\omega_{\mathrm{C}}^{(k)}}{V}}\left(\hat{a}^{\dagger} - \hat{a}\right)\cos\left(\omega_{\mathrm{C}}^{(k)} z/c\right) \end{aligned} \tag{5.120}$$

where V is the effective volume of the cavity, the mode frequency is $\omega_C^{(k)} = \pi n/L$, and n is a positive integer. For an isotropically shielded spin with a magnetogyric ratio γ, Zeeman interaction will couple the magnetic field of the cavity to the corresponding spin operator:

$$\hat{\mathbf{H}}_{\mathrm{INT}} = -\gamma\hat{B}_{\mathrm{Y}} \otimes \mathbf{S}_{\mathrm{Y}} = -g_0\left(\hat{a}^\dagger - \hat{a}\right) \otimes (\mathbf{S}_+ - \mathbf{S}_-)$$
$$g_0 = \frac{\gamma}{2}\sqrt{\frac{\mu_0\hbar\omega_C^{(k)}}{V}}\cos\left(\omega_C^{(k)}z/c\right) \tag{5.121}$$

where $\mathbf{S}_\pm$ are spin operators defined in Sect. 1.6.3. When we add the Hamiltonian of the cavity itself and the Zeeman interaction due to the external magnetic field, we get

$$\hat{\mathbf{H}} = \mathbf{1}_{\mathrm{C}} \otimes \mathbf{H}_{\mathrm{Z}} + \hat{H}_{\mathrm{C}} \otimes \mathbf{1}_{\mathrm{S}} + \hat{\mathbf{H}}_{\mathrm{INT}}$$
$$\mathbf{H}_{\mathrm{Z}} = \omega_{\mathrm{S}}\mathbf{S}_{\mathrm{Z}}, \qquad \hat{H}_{\mathrm{C}} = \omega_{\mathrm{C}}\left(\hat{a}^\dagger\hat{a} + \tfrac{1}{2}\right) \tag{5.122}$$

where $\omega_{\mathrm{S}} = -\gamma B_0$ is the Zeeman frequency that the spin has due to the presence of the static field directed along the Z-axis of the cavity.

When the cavity is tuned to have its resonance frequency ω_{C} close to the spin Zeeman frequency ω_{S}, an average Hamiltonian theory treatment (Sect. 4.3.8) yields a further simplification. In the interaction representation with respect to $\hat{\mathbf{H}}_0 = \mathbf{1}_{\mathrm{C}} \otimes \mathbf{H}_{\mathrm{Z}} + \hat{H}_{\mathrm{C}} \otimes \mathbf{1}_{\mathrm{S}}$, the spin-cavity coupling term becomes

$$\begin{aligned}
\hat{\mathbf{H}}_{\mathrm{INT}}^{\mathrm{R}}(t) &= \mathrm{e}^{+i\hat{\mathbf{H}}_0 t}\hat{\mathbf{H}}_{\mathrm{INT}}\mathrm{e}^{-i\hat{\mathbf{H}}_0 t} \\
&= -g_0\left(\hat{a}^\dagger \mathrm{e}^{+i\omega_{\mathrm{C}}t} - \hat{a}\mathrm{e}^{-i\omega_{\mathrm{C}}t}\right) \otimes \left(\mathbf{S}_+ e^{+i\omega_{\mathrm{S}}t} - \mathbf{S}_-\mathrm{e}^{-i\omega_{\mathrm{S}}t}\right) \\
&= -g_0\Big(\hat{a}^\dagger \otimes \mathbf{S}_+\mathrm{e}^{+i(\omega_{\mathrm{C}}+\omega_{\mathrm{S}})t} - \hat{a} \otimes \mathbf{S}_+\mathrm{e}^{-i(\omega_{\mathrm{C}}-\omega_{\mathrm{S}})t} \\
&\quad - \hat{a}^\dagger \otimes \mathbf{S}_-\mathrm{e}^{+i(\omega_{\mathrm{C}}-\omega_{\mathrm{S}})t} + \hat{a} \otimes \mathbf{S}_-\mathrm{e}^{-i(\omega_{\mathrm{C}}+\omega_{\mathrm{S}})t}\Big)
\end{aligned} \tag{5.123}$$

When $|\omega_{\mathrm{C}} - \omega_{\mathrm{S}}| \ll |\omega_{\mathrm{C}} + \omega_{\mathrm{S}}|$, the average of this expression over the period of $\omega_{\mathrm{C}} + \omega_{\mathrm{S}}$ is

$$\bar{\hat{\mathbf{H}}}_{\mathrm{INT}}^{\mathrm{R}}(t) \approx g_0\left(\hat{a} \otimes \mathbf{S}_+\mathrm{e}^{-i(\omega_{\mathrm{C}}-\omega_{\mathrm{S}})t} + \hat{a}^\dagger \otimes \mathbf{S}_-\mathrm{e}^{+i(\omega_{\mathrm{C}}-\omega_{\mathrm{S}})t}\right) \tag{5.124}$$

When this is returned to the laboratory frame, we obtain the *Jaynes-Cummings Hamiltonian* [210]:

$$\begin{aligned}
\hat{\mathbf{H}} &= \mathbf{1}_{\mathrm{C}} \otimes \mathbf{H}_{\mathrm{Z}} + \hat{H}_{\mathrm{C}} \otimes \mathbf{1}_{\mathrm{S}} + \hat{\hat{\mathbf{H}}}_{\mathrm{INT}} \\
\mathbf{H}_{\mathrm{Z}} &= \omega_{\mathrm{S}} \mathbf{S}_{\mathrm{Z}}, \quad \hat{H}_{\mathrm{C}} = \omega_{\mathrm{C}} \left(\hat{a}^{\dagger} \hat{a} + \tfrac{1}{2} \right) \\
\hat{\hat{\mathbf{H}}}_{\mathrm{INT}} &= g_0 \left(\hat{a} \otimes \mathbf{S}_{+} + \hat{a}^{\dagger} \otimes \mathbf{S}_{-} \right)
\end{aligned} \tag{5.125}$$

In physical terms, spin excitation grabs a photon from the cavity and a spin de-excitation pops a photon back. Matrix representations of creation and annihilation operators are, strictly speaking (Sect. 5.5.1), infinite dimensional, but they can be truncated at finite temperatures. Therefore, the problem is now directly computable—we have a Kronecker product and an exercise in numerical linear algebra.

Beyond this simple outline treatment, things get messy—Hamiltonians of real-life cavities depend on the details of their construction. Dissipative terms can arise due to the presence of transmission lines; those matters [211] are outside the scope of this book.

Dissipative Spin Dynamics

6

In the order of increasing sophistication, three strategies are used to account for spin relaxation. It may be treated empirically, by introducing phenomenological damping and population exchange terms into the equations of motion. At a higher level, we may observe that relaxation is caused by the presence of noise in the spin Hamiltonian, and use perturbation theories. At the highest level, we may consider explicitly the environment which either creates the noise, or makes the state space big enough that nothing returns into the subspace of interest on the time scale of the experiment. In this chapter we walk this path in reverse order—from the sophisticated theories down to the simple ones.

The fundamental symmetries discussed in Chap. 2 require the evolution of an isolated quantum system to be unitary. Any relaxation theory is, therefore, necessarily an approximation. A popular starting point is to consider an ensemble of identical and isolated systems, focus on a small subsystem, and account for the presence of the rest of the system in some effective way. For historical reasons (going back to flasks immersed in heated water baths by early thermodynamics researchers), the small and interesting part is called *the system* (S), and the remaining part is called *the bath* (B).

It bears notice that a system without dissipative processes, undergoing perfectly symplectic (classical mechanics) or unitary (quantum mechanics) evolution, is called *perpetual motion machine of the third kind.* In the context of spin dynamics, a tragic and cautionary example is the notorious "scalable spin quantum computer"—much written about, never to be built.

The magnetic resonance community has always viewed relaxation not as a nuisance, but as a source of information on inter-atomic distances [212] and molecular dynamics [213]. In magnetic resonance imaging, relaxation creates tissue contrast [214]. On the simulation side, modern polynomial complexity scaling methods (Chap. 7) derive their efficiency from the assumption that relaxation makes much of the state space unreachable [215].

I. KUPROV, *Spin*,
https://doi.org/10.1007/978-3-031-05607-9_6

6.1 Small Quantum Bath: Adiabatic Elimination

When a quantum mechanical treatment of the bath is affordable and there is a clear timescale separation between the system and the bath, an elegant and numerically friendly approach exists, based on partitioning the state space into the "slow" subspace (pure system degrees of freedom, state vector $\boldsymbol{\rho}_0$) and the "fast" subspace (bath degrees of freedom and system-bath correlations, state vector $\boldsymbol{\rho}_1$). With that separation in place, the equation of motion acquires the following structure:

$$\frac{\mathrm{d}}{\mathrm{d}t}\begin{pmatrix}\boldsymbol{\rho}_0\\ \boldsymbol{\rho}_1\end{pmatrix} = -i\begin{pmatrix}\mathcal{L}_{00} & \mathcal{L}_{01}\\ \mathcal{L}_{10} & \mathcal{L}_{11}\end{pmatrix}\begin{pmatrix}\boldsymbol{\rho}_0\\ \boldsymbol{\rho}_1\end{pmatrix} \quad \Leftrightarrow \quad \begin{cases} d\boldsymbol{\rho}_0/dt = -i\mathcal{L}_{00}\boldsymbol{\rho}_0 - i\mathcal{L}_{01}\boldsymbol{\rho}_1\\ d\boldsymbol{\rho}_1/dt = -i\mathcal{L}_{10}\boldsymbol{\rho}_0 - i\mathcal{L}_{11}\boldsymbol{\rho}_1\end{cases} \tag{6.1}$$

where $\mathcal{L}_{nk}$ are the corresponding blocks of the Liouvillian (Hamiltonian commutation superoperator plus any pre-existing dissipative superoperators). If we assume that the bath remains in the thermal equilibrium, this implies that $d\boldsymbol{\rho}_1/dt = 0$, and therefore (from the second equation) that $\boldsymbol{\rho}_1 = -\mathcal{L}_{11}^{-1}\mathcal{L}_{10}\boldsymbol{\rho}_0$. After placing this into the first equation, we obtain an effective equation of motion for the system:

$$\frac{\mathrm{d}\boldsymbol{\rho}_0}{\mathrm{d}t} = -i\mathcal{L}_{00}\boldsymbol{\rho}_0 + i\mathcal{L}_{01}\mathcal{L}_{11}^{-1}\mathcal{L}_{10}\boldsymbol{\rho}_0 \tag{6.2}$$

where the second term on the right-hand side is dissipative. This equation captures the long-term evolution of $\boldsymbol{\rho}_0$, and incorporates the rapid evolution and decoherence of the bath in an effective way.

Generalisations to less restrictive assumptions about the bath are straightforward —we can, for example, allow it to have $d\boldsymbol{\rho}_1/dt \neq 0$ but require $d^2\boldsymbol{\rho}_1/dt^2 = 0$, in which case

$$\frac{\mathrm{d}\boldsymbol{\rho}_0}{\mathrm{d}t} = -i\left(1 + \mathcal{L}_{01}\mathcal{L}_{11}^{-2}\mathcal{L}_{10}\right)^{-1}\left(\mathcal{L}_{00} - \mathcal{L}_{01}\mathcal{L}_{11}^{-1}\mathcal{L}_{10}\right)\boldsymbol{\rho}_0 \tag{6.3}$$

for time-independent Liouvillians. When its assumptions are valid, this *adiabatic elimination* framework [216] is appealing because its numerical implementation is simple. Its weakness is that some model of dissipation for the bath must be assumed a priori, otherwise, the inverse of $\mathcal{L}_{11}$ would not exist. Adiabatic elimination is only practical for the smallest of systems, a good example being nuclear spin relaxation in the presence of a rapidly relaxing lanthanide ion (Sect. 6.9.7).

6.2 Large Quantum Bath: Hubbard Theory

The long-winded and precarious mathematics you will now see is unavoidable—the *Hubbard theory* [217] equations derived in this section are central to dissipative spin dynamics. Consider a time-independent Hamiltonian for a system (S) and a bath (B) connected by a bilinear interaction:

$$\mathbf{H} = \mathbf{H}_0 + \mathbf{H}_1, \qquad \mathbf{H}_0 = \mathbf{H}_0^{\mathrm{S}} \otimes \mathbf{1}^{\mathrm{B}} + \mathbf{1}^{\mathrm{S}} \otimes \mathbf{H}_0^{\mathrm{B}}, \qquad \mathbf{H}_1 = \sum_{nk} a_{nk} \mathbf{S}_n \otimes \mathbf{B}_k \tag{6.4}$$

where $\mathbf{H}_0^{\mathrm{S}}$ is the local system Hamiltonian, $\mathbf{H}_0^{\mathrm{B}}$ is the local bath Hamiltonian, $\{\mathbf{S}_n\}$ is a basis set of operators acting on the system, and $\{\mathbf{B}_k\}$ is a basis set of operators acting on the bath. The sum is necessary because the dynamics of different bath modes may be correlated. We are at liberty to choose both basis sets to be commutation eigenoperators of the corresponding local Hamiltonians:

$$\left[\mathbf{H}_0^{\mathrm{S}}, \mathbf{S}_n\right] = \omega_n^{\mathrm{S}} \mathbf{S}_n \qquad \left[\mathbf{H}_0^{\mathrm{B}}, \mathbf{B}_k\right] = \omega_k^{\mathrm{B}} \mathbf{B}_k \tag{6.5}$$

where ω_n^{S} and ω_k^{B} are frequencies of the corresponding dynamical modes of the system and the bath; these basis operators need not be Hermitian. It follows from Eqs. (6.5) and (4.99) that

$$\begin{cases} \mathrm{e}^{+i\mathbf{H}_0^{\mathrm{S}}t} \mathbf{S}_n \mathrm{e}^{-i\mathbf{H}_0^{\mathrm{S}}t} = \mathrm{e}^{+i\omega_n^{\mathrm{S}}t} \mathbf{S}_n \\ \mathrm{e}^{+i\mathbf{H}_0^{\mathrm{S}}t} \mathbf{S}_n^{\dagger} \mathrm{e}^{-i\mathbf{H}_0^{\mathrm{S}}t} = \mathrm{e}^{-i\omega_n^{\mathrm{S}}t} \mathbf{S}_n^{\dagger} \end{cases} \qquad \begin{cases} \mathrm{e}^{+i\mathbf{H}_0^{\mathrm{B}}t} \mathbf{B}_k \mathrm{e}^{-i\mathbf{H}_0^{\mathrm{B}}t} = \mathrm{e}^{+i\omega_k^{\mathrm{B}}t} \mathbf{B}_k \\ \mathrm{e}^{+i\mathbf{H}_0^{\mathrm{B}}t} \mathbf{B}_k^{\dagger} \mathrm{e}^{-i\mathbf{H}_0^{\mathrm{B}}t} = \mathrm{e}^{-i\omega_k^{\mathrm{B}}t} \mathbf{B}_k^{\dagger} \end{cases} \tag{6.6}$$

This will be useful later. We are also at liberty to balance the overall Hamiltonian in Eq (6.4) so that the Boltzmann average of the interaction term over the bath degrees of freedom is zero:

$$\sum_{nk} a_{nk} \mathbf{S}_n \mathrm{Tr}\left[\boldsymbol{\rho}_{\mathrm{eq}}^{\mathrm{B}} \mathbf{B}_k\right] = 0, \qquad \boldsymbol{\rho}_{\mathrm{eq}}^{\mathrm{B}} = \frac{\exp\left(-\hbar \mathbf{H}_0^{\mathrm{B}} / k_{\mathrm{B}} T\right)}{\mathrm{Tr}\left[\exp\left(-\hbar \mathbf{H}_0^{\mathrm{B}} / k_{\mathrm{B}} T\right)\right]} \tag{6.7}$$

This is always possible because the sum in Eq. (6.7) is a pure system operator—if this sum is not zero, it can be subtracted out and placed into $\mathbf{H}_0^{\mathrm{S}}$.

We now go into the interaction representation (Sect. 4.3.1) with respect to the local Hamiltonian:

$$\boldsymbol{\rho}(t) = \mathrm{e}^{-i\mathbf{H}_0 t} \boldsymbol{\rho}^{\mathrm{R}}(t) \mathrm{e}^{+i\mathbf{H}_0 t}, \qquad \mathbf{H}_1 = \mathrm{e}^{-i\mathbf{H}_0 t} \mathbf{H}_1^{\mathrm{R}}(t) \mathrm{e}^{+i\mathbf{H}_0 t} \tag{6.8}$$

After placing this into the Liouville–von Neumann equation (Sect. 4.2.1) and simplifying, we get

$$\frac{\partial}{\partial t}\boldsymbol{\rho}^{\mathrm{R}}(t) = -i\left[\mathbf{H}_1^{\mathrm{R}}(t), \boldsymbol{\rho}^{\mathrm{R}}(t)\right] \tag{6.9}$$

Because $\mathbf{H}_0^{\mathrm{S}} \otimes \mathbf{1}^{\mathrm{B}}$ and $\mathbf{1}^{\mathrm{S}} \otimes \mathbf{H}_0^{\mathrm{B}}$ commute, the exponential of $\mathbf{H}_0$ simplifies (Sect. 4.3.7):

$$\exp\left[\pm i\left(\mathbf{H}_0^{\mathrm{S}} \otimes \mathbf{1}^{\mathrm{B}} + \mathbf{1}^{\mathrm{S}} \otimes \mathbf{H}_0^{\mathrm{B}}\right)t\right] = \exp\left[\pm i\mathbf{H}_0^{\mathrm{S}}t\right] \otimes \exp\left[\pm i\mathbf{H}_0^{\mathrm{B}}t\right] \tag{6.10}$$

Therefore, the transformations in Eq. (6.8) are local to the system and the bath:

$$\begin{aligned} \mathbf{S}_n^{\mathrm{R}}(t) &= \mathrm{e}^{+i\mathbf{H}_0^{\mathrm{S}}t}\mathbf{S}_n\mathrm{e}^{-i\mathbf{H}_0^{\mathrm{S}}t} = \mathrm{e}^{+i\omega_n{}^{\mathrm{S}}t}\mathbf{S}_n \\ \mathbf{B}_k^{\mathrm{R}}(t) &= \mathrm{e}^{+i\mathbf{H}_0^{\mathrm{B}}t}\mathbf{B}_k\mathrm{e}^{-i\mathbf{H}_0^{\mathrm{B}}t} = \mathrm{e}^{+i\omega_k{}^{\mathrm{B}}t}\mathbf{B}_k \end{aligned} \tag{6.11}$$

and correspond to Heisenberg representation evolution of system and bath operators under their own non-interacting Hamiltonians $\mathbf{H}_0^{\mathrm{S}}$ and $\mathbf{H}_0^{\mathrm{B}}$ in the laboratory frame. The interaction term becomes

$$\mathbf{H}_1^{\mathrm{R}}(t) = \sum_{nk} a_{nk}\mathbf{S}_n^{\mathrm{R}}(t) \otimes \mathbf{B}_k^{\mathrm{R}}(t) \tag{6.12}$$

We complete the stage setting by using the Dyson series (Sect. 4.4.3) solution for Eq. (6.9):

$$\frac{\partial}{\partial t}\boldsymbol{\rho}^{\mathrm{R}}(t) = -i\left[\mathbf{H}_1^{\mathrm{R}}(t), \boldsymbol{\rho}^{\mathrm{R}}(0)\right] - \int_0^t \left[\mathbf{H}_1^{\mathrm{R}}(t), \left[\mathbf{H}_1^{\mathrm{R}\dagger}(t'), \boldsymbol{\rho}^{\mathrm{R}}(0)\right]\right]dt' + \ldots \tag{6.13}$$

where one copy of the Hermitian interaction Hamiltonian was pasted in the conjugated form to facilitate subsequent mathematics. Up to here, everything is general and exact.

Our first approximation is truncating the Dyson series in Eq. (6.13) at the second term. As per Sect. 4.4.3, this is permissible when $\left\|\mathbf{H}_1^{\mathrm{R}}(t)t\right\|_2 \ll 1$, which is our first validity condition. We next assume that the system-bath coupling is sufficiently weak for the bath to remain in the thermal equilibrium:

$$\boldsymbol{\rho}(t) = \boldsymbol{\rho}^{\mathrm{S}}(t) \otimes \boldsymbol{\rho}_{\mathrm{eq}}^{\mathrm{B}} \quad \Rightarrow \quad \boldsymbol{\rho}^{\mathrm{R}}(t) = \boldsymbol{\rho}^{\mathrm{S,R}}(t) \otimes \boldsymbol{\rho}_{\mathrm{eq}}^{\mathrm{B}} \tag{6.14}$$

where $\boldsymbol{\rho}_{\mathrm{eq}}^{\mathrm{B}}$ did not get the R index because the bath equilibrium density matrix in Eq. (6.7) commutes with the bath interaction representation transformation—both the density matrix and the transformation are exponentials of the same Hamiltonian. Inserting Eq. (6.14) into Eq. (6.13) yields

$$\frac{\partial}{\partial t}\left[\boldsymbol{\rho}^{\mathrm{S,R}}(t)\otimes\boldsymbol{\rho}_{\mathrm{eq}}^{\mathrm{B}}\right]=-i\left[\mathbf{H}^{\mathrm{R}}(t),\boldsymbol{\rho}^{\mathrm{S,R}}(0)\otimes\boldsymbol{\rho}_{\mathrm{eq}}^{\mathrm{B}}\right]-\int_0^t\left[\mathbf{H}^{\mathrm{R}}(t),\left[\mathbf{H}^{\mathrm{R}\dagger}(t'),\boldsymbol{\rho}^{\mathrm{S,R}}(0)\otimes\boldsymbol{\rho}_{\mathrm{eq}}^{\mathrm{B}}\right]\right]\mathrm{d}t' \tag{6.15}$$

We now average this equation over the bath degrees of freedom by taking a trace over the bath parts of the direct products. The derivative on the left-hand side loses the bath part:

$$\mathrm{Tr}_{\mathrm{B}}\left[\frac{\partial}{\partial t}\left[\boldsymbol{\rho}^{\mathrm{S,R}}(t)\otimes\boldsymbol{\rho}_{\mathrm{eq}}^{\mathrm{B}}\right]\right]=\frac{\partial}{\partial t}\left[\boldsymbol{\rho}^{\mathrm{S,R}}(t)\mathrm{Tr}_{\mathrm{B}}\left[\boldsymbol{\rho}_{\mathrm{eq}}^{\mathrm{B}}\right]\right]=\frac{\partial}{\partial t}\boldsymbol{\rho}^{\mathrm{S,R}}(t) \tag{6.16}$$

because $\mathrm{Tr}_{\mathrm{B}}\left[\boldsymbol{\rho}_{\mathrm{eq}}^{\mathrm{B}}\right]=1$. A tedious substitution and simplification demonstrate that the first term on the right-hand side of Eq. (6.15) goes to zero:

$$\mathrm{Tr}_{\mathrm{B}}\left[\mathbf{H}^{\mathrm{R}}(t),\boldsymbol{\rho}^{\mathrm{S,R}}(0)\otimes\boldsymbol{\rho}_{\mathrm{eq}}^{\mathrm{B}}\right]=\cdots=0 \tag{6.17}$$

because of the choice we had made in Eq. (6.7). We are therefore left with the following:

$$\frac{\partial}{\partial t}\boldsymbol{\rho}^{\mathrm{S,R}}(t)=-\int_0^t\mathrm{Tr}_{\mathrm{B}}\left[\mathbf{H}^{\mathrm{R}}(t),\left[\mathbf{H}^{\mathrm{R}\dagger}(t'),\boldsymbol{\rho}^{\mathrm{S,R}}(0)\otimes\boldsymbol{\rho}_{\mathrm{eq}}^{\mathrm{B}}\right]\right]\mathrm{d}t' \tag{6.18}$$

where we can replace $\boldsymbol{\rho}^{\mathrm{S,R}}(0)$ with $\boldsymbol{\rho}^{\mathrm{S,R}}(t)$ because the difference is insignificant:

$$\boldsymbol{\rho}^{\mathrm{S,R}}(t)=\boldsymbol{\rho}^{\mathrm{S,R}}(0)+\left[\frac{\partial}{\partial t}\boldsymbol{\rho}^{\mathrm{S,R}}(t)\right]_{t=0}t+\cdots \tag{6.19}$$

Direct inspection of Eq. (6.18) shows that the second term in this sum is zero; later terms are insignificant when $\left\|\mathbf{H}_1^{\mathrm{R}}(t)t\right\|\ll 1$ because they correspond to the Dyson terms we had ignored. Thus,

$$\frac{\partial}{\partial t}\boldsymbol{\rho}^{\mathrm{S,R}}(t)=-\int_0^t\mathrm{Tr}_{\mathrm{B}}\left[\mathbf{H}^{\mathrm{R}}(t),\left[\mathbf{H}^{\mathrm{R}\dagger}(t'),\boldsymbol{\rho}^{\mathrm{S,R}}(t)\otimes\boldsymbol{\rho}_{\mathrm{eq}}^{\mathrm{B}}\right]\right]\mathrm{d}t' \tag{6.20}$$

We will now paste in the explicit form of the interaction Eq. (6.12). After a tedious rearrangement of commutators and cyclic permutations of matrix products under traces, we obtain

$$\frac{\partial}{\partial t}\boldsymbol{\rho}^{\mathrm{S,R}}(t)=-\sum_{\alpha\beta\mu\nu}a_{\alpha\beta}a_{\mu\nu}^{*}\int_{0}^{t}\left[\begin{array}{l}\left[\mathbf{S}_{\alpha}^{\mathrm{R}}(t),\mathbf{S}_{\mu}^{\mathrm{R}\dagger}(t')\boldsymbol{\rho}^{\mathrm{S,R}}(t)\right]\mathrm{Tr}\left(\mathbf{B}_{\nu}^{\mathrm{R}\dagger}(t')\boldsymbol{\rho}_{\mathrm{eq}}^{\mathrm{B}}\mathbf{B}_{\beta}^{\mathrm{R}}(t)\right)\\-\left[\mathbf{S}_{\alpha}^{\mathrm{R}}(t),\boldsymbol{\rho}^{\mathrm{S,R}}(t)\mathbf{S}_{\mu}^{\mathrm{R}\dagger}(t')\right]\mathrm{Tr}\left(\boldsymbol{\rho}_{\mathrm{eq}}^{\mathrm{B}}\mathbf{B}_{\nu}^{\mathrm{R}\dagger}(t')\mathbf{B}_{\beta}^{\mathrm{R}}(t)\right)\end{array}\right]\mathrm{d}t'. \tag{6.21}$$

Consider first the traces. The definition in Eq. (6.11) is identical to the definition of the Heisenberg representation dynamics. Thus, the traces in Eq. (6.21) are thermal equilibrium ensemble averages of products of differently timed observables—in other words, correlation functions. Statistical properties of the bath are the same today as they had been yesterday—correlation functions can, therefore, only depend on $\tau = t - t'$. In particular, we can shift the time axis origin backward by t':

$$\begin{aligned}\mathrm{Tr}_{\mathrm{B}}\left(\mathbf{B}_{\nu}^{\mathrm{R}\dagger}(t')\boldsymbol{\rho}_{\mathrm{eq}}^{\mathrm{B}}\mathbf{B}_{\beta}^{\mathrm{R}}(t)\right)&=\mathrm{Tr}_{\mathrm{B}}\left(\mathbf{B}_{\nu}^{\mathrm{R}\dagger}(0)\boldsymbol{\rho}_{\mathrm{eq}}^{\mathrm{B}}\mathbf{B}_{\beta}^{\mathrm{R}}(t-t')\right)=\mathrm{Tr}_{\mathrm{B}}\left(\mathbf{B}_{\nu}^{\dagger}\boldsymbol{\rho}_{\mathrm{eq}}^{\mathrm{B}}\mathbf{B}_{\beta}(\tau)\right)\\\mathrm{Tr}_{\mathrm{B}}\left(\boldsymbol{\rho}_{\mathrm{eq}}^{\mathrm{B}}\mathbf{B}_{\nu}^{\mathrm{R}\dagger}(t')\mathbf{B}_{\beta}^{\mathrm{R}}(t)\right)&=\mathrm{Tr}_{\mathrm{B}}\left(\boldsymbol{\rho}_{\mathrm{eq}}^{\mathrm{B}}\mathbf{B}_{\nu}^{\mathrm{R}\dagger}(0)\mathbf{B}_{\beta}^{\mathrm{R}}(t-t')\right)=\mathrm{Tr}_{\mathrm{B}}\left(\boldsymbol{\rho}_{\mathrm{eq}}^{\mathrm{B}}\mathbf{B}_{\nu}^{\dagger}\mathbf{B}_{\beta}(\tau)\right)\end{aligned} \tag{6.22}$$

where the rotating frame index R was dropped on the right-hand side for two reasons: (1) at time zero there is no difference between the laboratory frame and the rotating frame; (2) the dynamics in question is exactly Heisenberg representation dynamics, and we now start looking at it that way.

The relationship between the two traces may be obtained by: (1) using the definition of thermodynamic equilibrium in Eq. (6.7); (2) using Eq. (4.99) to castle $\mathbf{B}_{\nu}^{\dagger}$ and the exponential; (3) using Eq. (6.5) to get rid of the nested commutators; (4) summing the resulting Taylor series back into an exponential:

$$\mathbf{B}_{\nu}^{\dagger}\boldsymbol{\rho}_{\mathrm{eq}}^{\mathrm{B}}=\boldsymbol{\rho}_{\mathrm{eq}}^{\mathrm{B}}\mathbf{B}_{\nu}^{\dagger}\mathrm{e}^{-\hbar\omega_{\nu}\mathrm{B}/k_{\mathrm{B}}T} \tag{6.23}$$

we can now define bath correlation functions as follows:

$$b_{\nu\beta}(\tau)=\mathrm{Tr}_{\mathrm{B}}\left(\boldsymbol{\rho}_{\mathrm{eq}}^{\mathrm{B}}\mathbf{B}_{\nu}^{\dagger}\mathbf{B}_{\beta}(\tau)\right) \tag{6.24}$$

After we apply the same $\tau = t - t'$ substitution to the spin part of Eq. (6.21), it becomes

$$\frac{\partial}{\partial t}\boldsymbol{\rho}^{\mathrm{S,R}}(t)=-\sum_{\alpha\beta\mu\nu}a_{\alpha\beta}a_{\mu\nu}^{*}\int_{0}^{t}\left[\begin{array}{l}\left[\mathbf{S}_{\alpha}^{\mathrm{R}}(t),\mathbf{S}_{\mu}^{\mathrm{R}\dagger}(t-\tau)\boldsymbol{\rho}^{\mathrm{S,R}}(t)\right]\mathrm{e}^{-\hbar\omega_{\nu}\mathrm{B}/k_{\mathrm{B}}T}\\-\left[\mathbf{S}_{\alpha}^{\mathrm{R}}(t),\boldsymbol{\rho}^{\mathrm{S,R}}(t)\mathbf{S}_{\mu}^{\mathrm{R}\dagger}(t-\tau)\right]\end{array}\right]b_{\nu\beta}(\tau)\mathrm{d}\tau \tag{6.25}$$

Although the correlation functions $b_{\nu\beta}(\tau)$ clearly do not decay under unitary bath dynamics, a sleight of hand is commonly applied here, and the presence of a "decay" is assumed. This is the tragedy of Hubbard theory: it does not actually solve the dissipative dynamics problem—but only kicks it into the long grass by allowing the user to make assumptions about the bath rather than the system. Our third assumption shall be that the bath loses its memory (and therefore $b_{\nu\beta}(\tau)$ functions decay) so rapidly on the time scale of spin system evolution that the upper limit of the integral may be extended to infinity.

We now return into the into the laboratory frame by reversing the substitution in Eq. (6.8) and using the eigenoperator relation from Eq. (6.11). Another round of tedious transformations yields

$$\frac{\partial}{\partial t}\boldsymbol{\rho}^{\mathrm{S}}(t) = -i\left[\mathbf{H}_0^{\mathrm{S}}, \boldsymbol{\rho}^{\mathrm{S}}(t)\right] + \mathfrak{R}\boldsymbol{\rho}^{\mathrm{S}}(t)$$

$$\mathfrak{R}\boldsymbol{\rho}^{\mathrm{S}} = \sum_{\alpha\beta\mu\nu} a_{\alpha\beta} a_{\mu\nu}^{*}\left(\left[\mathbf{S}_\mu^{\dagger}\boldsymbol{\rho}^{\mathrm{S}}, \mathbf{S}_\alpha\right] \mathrm{e}^{-\hbar\omega_\nu{}^{\mathrm{B}}/k_{\mathrm{B}}T} + \left[\mathbf{S}_\alpha, \boldsymbol{\rho}^{\mathrm{S}}\mathbf{S}_\mu^{\dagger}\right]\right) \int_0^\infty b_{\nu\beta}(\tau)\mathrm{e}^{+i\omega_\mu^{\mathrm{S}}\tau} \mathrm{d}\tau \tag{6.26}$$

where the *relaxation superoperator* $\mathfrak{R}$ makes an appearance. It is a product of two parts: a collection of temperature-weighted spin operators in the brackets and the Fourier transform of the bath autocorrelation function evaluated at the system transition frequencies:

$$J_{\nu\beta}(\omega) = \int_0^\infty b_{\nu\beta}(\tau)\mathrm{e}^{+i\omega\tau}\mathrm{d}\tau \tag{6.27}$$

This integral, called *spectral density function*, has the physical meaning of the power spectrum of the bath: when the dynamics of the bath has components at the system transition frequencies, system relaxation takes place. Autocorrelation functions) may in practice be measured, simulated with methods like molecular dynamics, or assumed.

6.3 Implicit Classical Bath: Redfield Theory

When the dynamics of the bath is classical to a good approximation, the bath manifests itself as time dependence in the coefficients $a_k(t)$ in front of those terms in the spin Hamiltonian $\mathbf{H}_k^{\mathrm{S}}$ that depend on the coordinates of the bath:

$$\mathbf{H}^{\mathrm{S}} = \mathbf{H}_0^{\mathrm{S}} + \mathbf{H}_1^{\mathrm{S}}(t) = \mathbf{H}_0^{\mathrm{S}} + \sum_k a_k(t)\mathbf{H}_k^{\mathrm{S}} \tag{6.28}$$

As we have seen in the previous section, when correlation functions involving $a_k(t)$ have components at the system transition frequencies, relaxation takes place. This is the essence of *Redfield theory* [218], which connects spin relaxation rates to the statistical properties of the bath dynamics.

6.3.1 Redfield's Relaxation Superoperator

We must now go through another famously tangled derivation in which reasonable but unobvious assumptions must be applied in exactly the right order [339]; the central importance of the result in magnetic resonance compels me to present the process in full. We start with the Liouville–von Neumann equation for the density matrix $\boldsymbol{\rho}(t)$ in Liouville space, and split the Hamiltonian commutation superoperator:

$$\frac{\partial \boldsymbol{\rho}(t)}{\partial t} = -i[\mathcal{H}_0 + \mathcal{H}_1(t)]\boldsymbol{\rho}(t), \qquad \mathcal{H}\boldsymbol{\rho} = \mathbf{H}\boldsymbol{\rho} - \boldsymbol{\rho}\mathbf{H}, \qquad t \geq 0 \tag{6.29}$$

into the static part $\mathcal{H}_0$ (for example, chemical shifts and *J*-couplings) and the time-dependent part $\mathcal{H}_1$ that is modulated by the dynamics of the bath (for example, Brownian motion in solution).

We can demand, without loss of generality, that $\mathcal{H}_1(t)$ has a zero ensemble average:

$$\langle \mathcal{H}_1(t) \rangle = 0 \tag{6.30}$$

If this average is non-zero, it can be subtracted and put into $\mathcal{H}_0$. We then move into the interaction representation (*aka* "rotating frame", Sect. 4.3.1) with respect to $\mathcal{H}_0$:

$$\boldsymbol{\rho}(t) = \mathrm{e}^{-i\mathcal{H}_0 t}\boldsymbol{\rho}^{\mathrm{R}}(t), \qquad \mathcal{H}_1(t) = \mathrm{e}^{-i\mathcal{H}_0 t}\mathcal{H}_1^{\mathrm{R}}(t)\mathrm{e}^{+i\mathcal{H}_0 t} \tag{6.31}$$

After substitution and simplification, Eq. (6.29) becomes

$$\frac{\partial \boldsymbol{\rho}^{\mathrm{R}}(t)}{\partial t} = -i\mathcal{H}_1^{\mathrm{R}}(t)\boldsymbol{\rho}^{\mathrm{R}}(t) \tag{6.32}$$

We write the solution out using Dyson series (Sect. 4.4.3):

$$\boldsymbol{\rho}^{\mathrm{R}}(t)=\begin{bmatrix}\mathbf{1}-i\int\limits_0^t \mathcal{H}_1^{\mathrm{R}}(t_1)\mathrm{d}t_1-\int\limits_0^t\int\limits_0^{t_1}\mathcal{H}_1^{\mathrm{R}}(t_1)\mathcal{H}_1^{\mathrm{R}}(t_2)\mathrm{d}t_2\mathrm{d}t_1\\ +i\int\limits_0^t\int\limits_0^{t_1}\int\limits_0^{t_2}\mathcal{H}_1^{\mathrm{R}}(t_1)\mathcal{H}_1^{\mathrm{R}}(t_2)\mathcal{H}_1^{\mathrm{R}}(t_3)\mathrm{d}t_3\mathrm{d}t_2\mathrm{d}t_1+\cdots\end{bmatrix}\boldsymbol{\rho}^{\mathrm{R}}(0) \tag{6.33}$$

which is guaranteed to converge monotonically in the matrix 2-norm, and to be negligible beyond the double integral when

$$\|\mathcal{H}_1^{\mathrm{R}}(t)\|_2 t=\|\mathcal{H}_1(t)\|_2 t\ll 1 \qquad \forall t>0 \tag{6.34}$$

We will make this our *first assumption*, and only keep the single and the double integral in the Dyson series. We will also now take an ensemble average of both sides of the resulting equation. The initial condition $\boldsymbol{\rho}^{\mathrm{R}}(0)$ is the same in every member of the ensemble, and therefore Eq. (6.33) becomes

$$\langle\boldsymbol{\rho}^{\mathrm{R}}(t)\rangle=\langle[\text{Dyson Series}]\rangle\langle\boldsymbol{\rho}^{\mathrm{R}}(0)\rangle \tag{6.35}$$

The static Hamiltonian $\mathcal{H}_0$ is also the same across the ensemble, and therefore,

$$\langle\mathcal{H}_1^{\mathrm{R}}(t)\rangle=\mathrm{e}^{i\mathcal{H}_0 t}\langle\mathcal{H}_1(t)\rangle\mathrm{e}^{-i\mathcal{H}_0 t}=0 \tag{6.36}$$

because of the condition we had placed on $\mathcal{H}_1(t)$ in Eq. (6.30). We are now left with the following:

$$\langle\boldsymbol{\rho}^{\mathrm{R}}(t)\rangle=\left[\mathbf{1}-\int\limits_0^t\int\limits_0^{t_1}\langle\mathcal{H}_1^{\mathrm{R}}(t_1)\mathcal{H}_1^{\mathrm{R}}(t_2)\rangle\mathrm{d}t_2\mathrm{d}t_1\right]\langle\boldsymbol{\rho}^{\mathrm{R}}(0)\rangle \tag{6.37}$$

We will differentiate this and drop the angular brackets on the density matrix for convenience.

$$\frac{\partial}{\partial t}\boldsymbol{\rho}^{\mathrm{R}}(t)=-\left[\int\limits_0^t\langle\mathcal{H}_1^{\mathrm{R}}(t)\mathcal{H}_1^{\mathrm{R}}(t_1)\rangle\mathrm{d}t_1\right]\boldsymbol{\rho}^{\mathrm{R}}(0) \tag{6.38}$$

We again invoke the assumption made in Eq. (6.34) and note that it makes the difference between $\boldsymbol{\rho}^{\mathrm{R}}(0)$ and $\boldsymbol{\rho}^{\mathrm{R}}(t)$ on the right-hand side of Eq. (6.38) insignificant. This may be seen from the Taylor expansion:

$$\boldsymbol{\rho}^{\mathrm{R}}(t)=\boldsymbol{\rho}^{\mathrm{R}}(0)+\left[\frac{\partial}{\partial t}\boldsymbol{\rho}^{\mathrm{R}}(t)\right]_{t=0}t+\cdots \tag{6.39}$$

in which the second term on the right-hand side is zero by direct inspection of Eq. (6.38). This allows us to replace $\boldsymbol{\rho}^{\mathrm{R}}(0)$ with $\boldsymbol{\rho}^{\mathrm{R}}(t)$ on the right-hand side of Eq. (6.38):

$$\frac{\partial}{\partial t}\boldsymbol{\rho}^{\mathrm{R}}(t) = -\left[\int_0^t \left\langle \mathcal{H}_1^{\mathrm{R}}(t)\mathcal{H}_1^{\mathrm{R}}(t_1)\right\rangle \mathrm{d}t_1\right]\boldsymbol{\rho}^{\mathrm{R}}(t) \tag{6.40}$$

Any time-dependent Hermitian operator has the following expansion:

$$\mathcal{H}_1(t) = \sum_k q_k(t)\mathcal{Q}_k = \sum_m q_m^*(t)\mathcal{Q}_m^\dagger \tag{6.41}$$

where $q_k(t)$ are time-dependent scalar coefficients and $\mathcal{Q}_k = [\mathbf{Q}_k,]$ are time-independent orthonormal basis superoperators. In the case of rotationally modulated interactions, a convenient choice is Wigner D matrix elements and irreducible spherical tensor operators (Sect. 3.2.4). For later convenience, we will paste the second copy of the Hamiltonian in the conjugated form:

$$\frac{\partial}{\partial t}\boldsymbol{\rho}^{\mathrm{R}}(t) = -\sum_{km}\left[\int_0^t \left\langle q_k(t)q_m^*(t')\right\rangle \mathcal{Q}_k^{\mathrm{R}}(t)\mathcal{Q}_m^{\dagger\mathrm{R}}(t')\mathrm{d}t_1\right]\boldsymbol{\rho}^{\mathrm{R}}(t) \tag{6.42}$$

where the basis operators have been taken out of the ensemble averaging brackets because they are the same in every member of the ensemble.

We now make out *second assumption*: that $q_k(t)$ are stationary stochastic processes, and therefore their autocorrelation functions only depend on the absolute separation between t and t':

$$\left\langle q_k(t)q_m^*(t')\right\rangle = g_{km}(|t-t'|) \tag{6.43}$$

As we shall see later in this chapter, these are available analytically for many bath models. Performing the variable substitution $\tau = t - t'$ under the integral yields:

$$\frac{\partial}{\partial t}\boldsymbol{\rho}^{\mathrm{R}}(t) = -\sum_{km}\left[\int_0^t g_{km}(\tau)\mathcal{Q}_k^{\mathrm{R}}(t)\mathcal{Q}_m^{\dagger\mathrm{R}}(t-\tau)\mathrm{d}\tau\right]\boldsymbol{\rho}^{\mathrm{R}}(t) \tag{6.44}$$

Our *third assumption* is that the time t that satisfies Eq. (6.34) is nonetheless long enough that the autocorrelation function decays completely to zero by the time $\tau = t$. This is generally the case in magnetic resonance of small molecules in non-viscous solutions, where $||\mathcal{H}_1^{\mathrm{R}}(t)||_2^{-1}$ is in microseconds, and the autocorrelation function decays on a picosecond to nanosecond time scale. This allows us to extend the upper limit of the integration to infinity:

$$\frac{\partial}{\partial t}\boldsymbol{\rho}^{\mathrm{R}}(t) = -\sum_{km}\left[\int_0^{\infty} g_{km}(\tau)\mathcal{Q}_k^{\mathrm{R}}(t)\mathcal{Q}_m^{\dagger\mathrm{R}}(t-\tau)\mathrm{d}\tau\right]\boldsymbol{\rho}^{\mathrm{R}}(t) \tag{6.45}$$

After using Eqs. (6.31) to return to the laboratory frame, we get *Redfield relaxation superoperator*:

$$\begin{aligned}\frac{\partial\boldsymbol{\rho}(t)}{\partial t} &= -i\mathcal{H}_0\boldsymbol{\rho}(t) + \mathcal{R}\boldsymbol{\rho}(t)\\ \mathcal{R} &= -\sum_{km}\left[\int_0^{\infty} g_{km}(\tau)\mathcal{Q}_k e^{-i\mathcal{H}_0\tau}\mathcal{Q}_m^{\dagger}e^{+i\mathcal{H}_0\tau}\mathrm{d}\tau\right]\end{aligned} \tag{6.46}$$

in which the Hamiltonians may be reassembled into a form that is useful when they are available numerically, for example, from a molecular dynamics simulation [219]:

$$\mathcal{R} = -\left\langle \int_0^{\infty} \mathcal{H}_1(0)e^{-i\mathcal{H}_0\tau}\mathcal{H}_1(\tau)e^{+i\mathcal{H}_0\tau}\mathrm{d}\tau\right\rangle \tag{6.47}$$

Here, the ensemble average is taken over statistically independent molecular dynamics trajectories, and the integral is computed numerically over each trajectory. When this integral must be taken analytically, the best method is to use an auxiliary matrix relation (Sect. 8.2.4.5):

$$\exp\left[\begin{pmatrix}\mathbf{A} & \mathbf{B}\\ \mathbf{0} & \mathbf{C}\end{pmatrix}t\right] = \begin{pmatrix} e^{\mathbf{A}t} & e^{\mathbf{A}t}\int_0^t e^{-\mathbf{A}t_1}\mathbf{B}e^{\mathbf{C}t_1}\mathrm{d}t_1\\ \mathbf{0} & e^{\mathbf{C}t}\end{pmatrix} \tag{6.48}$$

Analytical Redfield theory can be too voluminous for manual derivations, but it is accessible to symbolic processing packages [220], particularly *Mathematica*.

6.3.2 Validity Range of Redfield Theory

In the derivation above, we have required the bath noise to be stationary (generally true and unproblematic) and also made the following—more significant—approximations:

1. Truncation of the Dyson series in Eq. (6.33) at the second order. This condition must be fulfilled until the correlation functions have decayed, meaning that

$$\|\mathcal{H}_1(t)\|_2\tau_{\max} \ll 1 \tag{6.49}$$

where $\tau_{\max}$ is the longest characteristic decay time found in $g_{km}(\tau)$. This is a useful prior condition for situations when physical insight into the bath dynamics is available.

2. Extension of the integration limit in Eq. (6.44) to infinity requires, after a tedious exercise with upper bounds on 2-norms of matrix integrals, that

$$\|\mathfrak{R}\|_2 \tau_{\max} \ll 1 \tag{6.50}$$

Remembering that the 2-norm is the largest singular value, we obtain an easily applicable posterior condition: *the use of Redfield theory is defensible when the slowest decay found in the correlation functions is much faster than the fastest relaxation process predicted by the theory.*

3. When the calculation of Redfield's integral is done numerically from spin Hamiltonian trajectories, additional convergence conditions are dictated by the quality of ensemble sampling and by the numerical accuracy of the integral. On the ensemble average side, the variance of the mean for the content of the angular bracket in Eq. (6.47)—abbreviated below as $\langle\cdots\rangle$—must be much smaller than the mean-squared:

$$\frac{1}{N}\left\|\left\langle(\cdots)(\cdots)^{\dagger}\right\rangle - \langle\cdots\rangle\langle\cdots\rangle^{\dagger}\right\| \ll \left\|\langle\cdots\rangle\langle\cdots\rangle^{\dagger}\right\| \tag{6.51}$$

where N is the number statistically independent trajectories. Accuracy conditions for numerical integration are standard textbook expressions discussed in [221].

The three conditions above rely on generous upper bounds. It is sometimes reported that Redfield theory matches experimental observations for correlation times up to an order of magnitude longer than these estimates suggest, but that is treacherous ground.

6.3.3 Spectral Density Functions

We are at liberty to require $\{\mathbf{Q}_k\}$ in Eq (6.41) to be eigenoperators of the static Hamiltonian:

$$[\mathbf{H}_0, \mathbf{Q}_k] = \omega_k \mathbf{Q}_k \quad \Rightarrow \quad \mathrm{e}^{-i\mathbf{H}_0\tau}\mathbf{Q}_k\mathrm{e}^{+i\mathbf{H}_0\tau} = \mathbf{Q}_k\mathrm{e}^{-i\omega_k\tau} \tag{6.52}$$

Such operators drive transitions between the energy levels of $\mathbf{H}_0$; the corresponding frequencies ω_k are transition frequencies. A good example is raising and lowering operators and Zeeman Hamiltonian:

$$[\omega \mathbf{S}_Z, \mathbf{S}_\pm] = \pm\omega \mathbf{S}_\pm \quad \Rightarrow \quad e^{-i\omega \mathbf{S}_Z \tau} \mathbf{S}_\pm e^{+i\omega \mathbf{S}_Z \tau} = \mathbf{S}_\pm e^{\pm i\omega\tau} \tag{6.53}$$

Because the adjoint representation of a Lie algebra is faithful (Sect. 1.5.9), we must then have

$$e^{-i\mathcal{H}_0\tau} \mathcal{Q}_k e^{+i\mathcal{H}_0\tau} = \mathcal{Q}_k e^{-i\omega_k\tau} \quad \Rightarrow \quad e^{-i\mathcal{H}_0\tau} \mathcal{Q}_m^\dagger e^{+i\mathcal{H}_0\tau} = \mathcal{Q}_m^\dagger e^{+i\omega_m\tau} \tag{6.54}$$

which reveals that the relaxation superoperator in Eq. (6.46) is a linear combination of Fourier transforms of bath correlation functions evaluated at the system transition frequencies:

$$\mathcal{R} = -\sum_{km} \mathcal{Q}_k \mathcal{Q}_m^\dagger \int_0^\infty g_{km}(\tau) e^{+i\omega_m\tau} d\tau \tag{6.55}$$

The Fourier transform of the correlation function is called *spectral density function*:

$$J(\omega) = \int_0^\infty g(\tau) e^{-i\omega\tau} d\tau \tag{6.56}$$

It has a physical meaning of energy density of the stochastic perturbation at the specified frequency, and relaxation therefore only happens when the noise created by the bath has components at the system transition frequencies. The function is complex-valued: the real part is responsible for relaxation proper, and the (usually much smaller) imaginary part contributes to the *dynamic frequency shift* [222].

6.3.4 A Simple Classical Example

This section contains a simple ab initio example of a Redfield type treatment for a physical system subject to a stochastic perturbation. Such elementary cases are few —for realistic spin systems, automated symbolic processing (*Mathematica* [220]) or numerical treatment (*Matlab* [83,102]) is normally needed.

Consider an ensemble of two-dimensional oscillators:

$$\begin{cases} x(t) = +\cos(\omega_0 t) \\ y(t) = -\sin(\omega_0 t) \end{cases} \quad \Leftrightarrow \quad \begin{cases} f(t) = x(t) + iy(t) \\ df(t)/dt = -i\omega_0 f(t) \end{cases} \tag{6.57}$$

where we modify the equation of motion to make the frequency slightly noisy, and make sure that the noise is different in each system of the ensemble, with a zero average:

$$\omega(t) = \omega_0 + \omega_1(t), \qquad |\omega_1(t)| \ll |\omega_0|, \qquad \langle \omega_1(t) \rangle = 0 \tag{6.58}$$

where angular brackets denote ensemble average. The initial condition will be the same across the ensemble. The equation of motion becomes

$$\frac{\mathrm{d}}{\mathrm{d}t} f(t) = -i[\omega_0 + \omega_1(t)] f(t) \tag{6.59}$$

Following Redfield's trail, we will assume that the noise in $\omega_1(t)$ is *stationary, i.e.,* that its statistical properties (mean, standard deviation, spectral power density, etc.) do not change with time—the objective is to express the ensemble-average solution via those time-independent statistical properties.

With the noise present, the systems will now be doing something else apart from just oscillating at the frequency ω_0. Let us call that something $\varphi(t)$ and look for a solution of the following form:

$$f(t) = \mathrm{e}^{-i\omega_0 t} \varphi(t) \tag{6.60}$$

This is a simple version of the interaction representation transformation—here literally a *rotating frame transformation* because multiplication by $\mathrm{e}^{-i\omega_0 t}$ is a rotation in the *XY* plane. Placing this into Eq. (6.59) and simplifying yields an equation that only involves $\omega_1(t)$ and $\varphi(t)$:

$$\frac{\mathrm{d}}{\mathrm{d}t} \varphi(t) = -i\omega_1(t) \varphi(t) \tag{6.61}$$

Following the path outlined in Sect. 6.3.1, we use the Dyson series—we repeatedly integrate Eq. (6.61) and substitute the resulting "solution" repeatedly back into Eq. (6.61) to obtain

$$\varphi(t) = \begin{bmatrix} 1 - i \int_0^t \omega_1(t_1) \mathrm{d}t_1 - \int_0^t \int_0^{t_1} \omega_1(t_1) \omega_1(t_2) \mathrm{d}t_2 \mathrm{d}t_1 \\ + i \int_0^t \int_0^{t_1} \int_0^{t_2} \omega_1(t_1) \omega_1(t_2) \omega_1(t_3) \mathrm{d}t_3 \mathrm{d}t_2 \mathrm{d}t_1 + \ldots \end{bmatrix} \varphi(0) \tag{6.62}$$

this expansion converges monotonically and the triple integral is negligible when $|\omega_1(t)| t \ll 1$. Truncating at the double integral and taking a derivative yields

$$\frac{\mathrm{d}}{\mathrm{d}t} \varphi(t) = \left[-i\omega_1(t) - \int_0^t \omega_1(t) \omega_1(t_1) \mathrm{d}t_1 \right] \varphi(0) \tag{6.63}$$

If we now average both sides over the ensemble, the assumption we had made in Eq. (6.58) makes the first term in the square brackets vanish

$$\left\langle \frac{\mathrm{d}}{\mathrm{d}t}\varphi(t)\right\rangle = \left[-\int_0^t \langle \omega_1(t)\omega_1(t_1)\rangle \mathrm{d}t_1\right]\langle \varphi(0)\rangle \tag{6.64}$$

In the first set of brackets, we can now recognise the autocorrelation function. For stationary noise with a root mean square amplitude a, the following applies

$$\langle \omega_1(t)\omega_1(t')\rangle = a^2 g(|t-t'|) \tag{6.65}$$

where $g(0) = 1$ and only the time difference appears because the statistical properties of stationary noise only depend on the relative timing of $\omega_1(t)$ and $\omega_1(t')$. Our objective of rewriting the problem via statistical properties of the noise is now accomplished.

By the same logic as in Eq. (6.39), the difference between $\langle \varphi(0)\rangle$ and $\langle \varphi(t)\rangle$ insignificant because the first-order term in the Taylor series is zero:

$$\langle \varphi(t)\rangle = \langle \varphi(0)\rangle + \langle \varphi'(0)\rangle t + \cdots \approx \langle \varphi(0)\rangle \tag{6.66}$$

After all of the above replacements are made, the equation of motion becomes

$$\frac{\mathrm{d}}{\mathrm{d}t}\varphi(t) = \left[-a^2\int_0^t g(t-t_1)\mathrm{d}t_1\right]\varphi(t) \tag{6.67}$$

where the modulus disappeared because $t' \le t$ and we have dropped the angular brackets on $\varphi(t)$ for convenience—it now refers to the ensemble average. Performing a $\tau = t - t_1$ variable substitution yields

$$\frac{\mathrm{d}}{\mathrm{d}t}\varphi(t) = \left[-a^2\int_0^t g(\tau)\mathrm{d}\tau\right]\varphi(t) \tag{6.68}$$

Assuming that the autocorrelation function decays rapidly on the time scale of rotating frame system evolution allows us to extend the integration limit to infinity. After using $\varphi(t) = \mathrm{e}^{i\omega_0 t}f(t)$ to return to the laboratory frame picture, we obtain:

$$\frac{\mathrm{d}}{\mathrm{d}t}f(t) = \left[-i\omega_0 - a^2 J(0)\right]f(t), \qquad J(0) = \int_0^\infty g(\tau)\mathrm{d}\tau \tag{6.69}$$

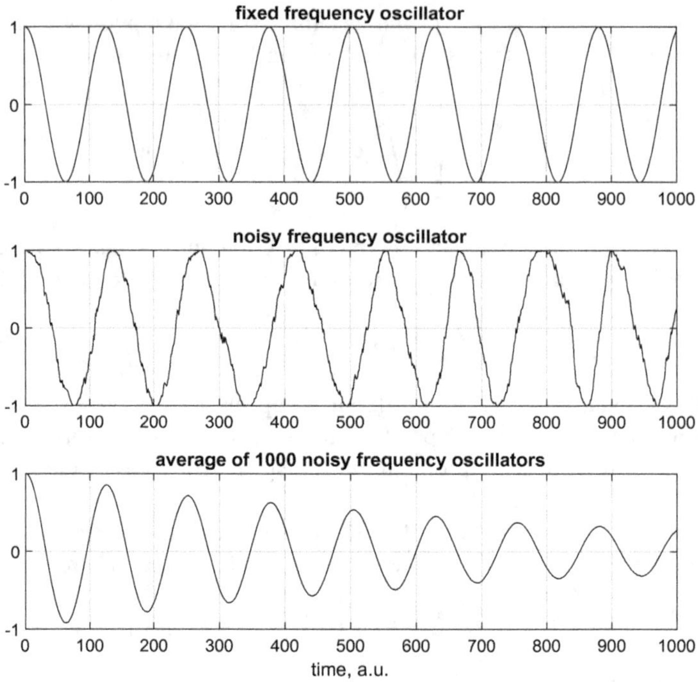

Fig. 6.1 An illustration of the fact that ensemble relaxation can proceed from the noise in the interaction parameters. The top trace is the coordinate of a fixed-frequency oscillator. The middle trace is the coordinate of an oscillator that has random noise in its frequency. The bottom trace is the average of coordinates of 1000 oscillators, each with its own random noise track

where we now see a rotation at the frequency ω_0 and a decay with the rate $a^2 J(0)$, in which the spectral density term $J(0)$ comes from the definition in Eq. (6.56). The negative quantity

$$R = -a^2 J(0) \tag{6.70}$$

is the *relaxation rate*—the presence of fast and weak noise in the frequency introduces a decay process into the ensemble average dynamics. The same conclusion may be obtained numerically (Fig. 6.1).

Important conclusions from the process described in this section are:

1. Each *individual* oscillator keeps going—the amplitude of its vector does not change, it is still a unit vector. What relaxes and goes to zero is the *ensemble average*.
2. Relaxation caused by stochastic noise is *irreversible*. The spin echo experiment would have been applicable if the frequency variation were static, but this is not the case here.

3. The relaxation rate is *quadratic* in the amplitude of the noise and depends on its frequency spectrum. Noise that does not hit the right frequencies would not cause relaxation.

6.3.5 Correlation Functions in General

Consider an ergodic classical bath with a state vector $\mathbf{x}$ and the following equation of motion for its probability density $p(\mathbf{x}, t)$ over the manifold of its possible states:

$$\frac{\partial}{\partial t}p(\mathbf{x}, t) = \hat{F}(\mathbf{x})p(\mathbf{x}, t) \qquad \Leftrightarrow \qquad \frac{\partial}{\partial t} = \hat{F}(\mathbf{x}) \tag{6.71}$$

where the bath dynamics generator $\hat{F}(\mathbf{x})$ is assumed to be time-independent. This equation has a simple connection to Redfield theory when the functions $q_n(\mathbf{x})$ in the spin Hamiltonian expansion

$$\mathbf{H}(\mathbf{x}) = \sum_n q_n(\mathbf{x})\mathbf{Q}_n \tag{6.72}$$

are chosen to be eigenfunctions of the bath dynamics generator:

$$\hat{F}(\mathbf{x})q_n(\mathbf{x}) = -\lambda_n q_n(\mathbf{x}) \tag{6.73}$$

where the eigenvalues λ_n need not be real and the minus reflects the assumption that the dynamics of the bath is dissipative. With this choice of basis, correlation functions and spectral densities can be evaluated analytically:

$$\begin{aligned} g_{nk}(\tau) &= \langle q_n^*(t)q_k(t+\tau)\rangle = \left\langle q_n^*(t)e^{[\partial/\partial t]\tau}q_k(st)\right\rangle \\ &= \langle q_n(\mathbf{x})|e^{\hat{F}\tau}|q_k(\mathbf{x})\rangle = \langle q_n(\mathbf{x})|q_k(\mathbf{x})\rangle \mathrm{e}^{-\lambda_k\tau} \end{aligned} \tag{6.74}$$

A corollary is that correlation functions are always multi-exponential for any bath with a linear equation of motion and a bilinear coupling to the system. Spectral densities are consequently always linear combinations of complex Lorentzian functions:

$$J_{nk}(\omega) = \langle q_n(\mathbf{x})|q_k(\mathbf{x})\rangle \int_0^\infty \mathrm{e}^{-\lambda_k\tau}\mathrm{e}^{-i\omega\tau}\mathrm{d}\tau = \frac{\langle q_n(\mathbf{x})|q_k(\mathbf{x})\rangle}{\lambda_k + i\omega} \tag{6.75}$$

From the numerical point of view, any linearly independent system of functions would work, so long as the evaluation of the matrix element $\langle q_n(\mathbf{x})|\mathrm{e}^{\hat{F}\tau}|q_k(\mathbf{x})\rangle$ can be performed. We will now use this approach to obtain commonly encountered correlation functions.

6.3.6 Rotational Diffusion Correlation Functions

As discussed in Sect. 3.2.4, a rotationally modulated Hamiltonian has the following form:

$$\mathbf{H}(\Omega) = \mathbf{H}_{\text{iso}} + \sum_{lkm} \mathfrak{D}_{km}^{(l)}(\Omega)\mathbf{Q}_{km}^{(l)} \tag{6.76}$$

where $\mathbf{H}_{\text{iso}}$ is the isotropic part, $\mathbf{Q}_{km}^{(l)}$ are rotational basis operators, $\mathfrak{D}_{km}^{(l)}(\Omega)$ are Wigner D functions (Sect. 1.6.2.4) of molecular orientation parametrised by some convention Ω (Sect. 1.6.2), l is a positive integer, and m, k are integer indices running from $-l$ to $+l$. The following *completeness and orthogonality relations* for Wigner D functions will be useful:

$$\delta(\Omega_2 - \Omega_1) = \sum_{lkm} \frac{2l+1}{8\pi^2} \mathfrak{D}_{km}^{(l)*}(\Omega_1)\mathfrak{D}_{km}^{(l)}(\Omega_2) \tag{6.77}$$

$$\left\langle \mathfrak{D}_{k_1m_1}^{(l_1)}(\Omega) \,\middle|\, \mathfrak{D}_{k_2m_2}^{(l_2)}(\Omega) \right\rangle = \frac{1}{8\pi^2} \int \mathfrak{D}_{k_1m_1}^{(l_1)*}(\Omega)\mathfrak{D}_{k_2m_2}^{(l_2)}(\Omega) d\Omega = \frac{\delta_{l_1l_2}\delta_{k_1k_2}\delta_{m_1m_2}}{2l_1+1} \tag{6.78}$$

These basis functions are enumerated by more than one index, but that is a cosmetic matter: matching Eq. (6.76) to Eq. (6.41) used in the derivation of Redfield theory is just a re-indexing operation.

6.3.6.1 Sphere

For an ensemble of spherical particles of radius r undergoing rotational diffusion in an isotropic solvent with dynamic viscosity η, the equation of motion for the probability density of orientations $p(\Omega, t)$ is

$$\frac{\partial}{\partial t} p(\Omega, t) = -D_{\text{R}}\left(\hat{L}_{\text{X}}^2 + \hat{L}_{\text{Y}}^2 + \hat{L}_{\text{Z}}^2\right)p(\Omega, t), \qquad D_{\text{R}} = \frac{k_{\text{B}}T}{8\pi\eta r^3} \tag{6.79}$$

where D_{R} is the rotational diffusion coefficient and $\hat{L}_{\text{XYZ}}$ are the differential operators of the three Cartesian components of angular momentum (Sect. 2.3), and Ω is a parametrisation of the rotation group (Sect. 1.6.2). The evolution propagator, therefore, is

$$\hat{P}(\tau) = \exp\left[-D_{\text{R}}\left(\hat{L}_{\text{X}}^2 + \hat{L}_{\text{Y}}^2 + \hat{L}_{\text{Z}}^2\right)\tau\right] \tag{6.80}$$

Wigner D functions are eigenfunctions of the generator:

$$\left[\hat{L}_{\text{X}}^2 + \hat{L}_{\text{Y}}^2 + \hat{L}_{\text{Z}}^2\right]\mathfrak{D}_{km}^{(l)}(\Omega) = l(l+1)\mathfrak{D}_{km}^{(l)}(\Omega) \tag{6.81}$$

which makes correlation functions easy to evaluate:

$$\begin{aligned}\left\langle \mathfrak{D}_{k_1m_1}^{(l_1)*}(t)\mathfrak{D}_{k_2m_2}^{(l_2)}(t+\tau)\right\rangle &= \left\langle \mathfrak{D}_{k_1m_1}^{(l_1)}\middle|\hat{P}(\tau)\middle|\mathfrak{D}_{k_2m_2}^{(l_2)}\right\rangle \\ &= \frac{\delta_{l_1l_2}\delta_{k_1k_2}\delta_{m_1m_2}}{2l_1+1}\mathrm{e}^{-Dl_1(l_1+1)\tau}\end{aligned} \tag{6.82}$$

Common spin interactions involve Wigner D functions of second spherical rank; the corresponding characteristic time is called *rotational correlation time* in magnetic resonance literature:

$$\tau_{\mathrm{C}} = \frac{1}{6D_{\mathrm{R}}} = \frac{4\pi\eta r^3}{3k_{\mathrm{B}}T} \tag{6.83}$$

When the elements of the direction cosine matrix (Sect. 1.6.2) are used instead of Wigner functions, the treatment tangles up significantly, but may be performed using *Mathematica*:

$$\langle r_{ab}(0)r_{cd}(\tau)\rangle = \frac{\delta_{ac}\delta_{bd}}{3}\mathrm{e}^{-2D\tau} \tag{6.84}$$

where $\{a,b,c,d\}$ indices run over $\{X,Y,Z\}$.

6.3.6.2 Symmetric Top

For a uniaxial ellipsoid with the long axis dimension $r_{\|}$ and perpendicular dimension $r_{\perp}$ in an isotropic solvent with dynamic viscosity η, the rotational diffusion equation is

$$\begin{aligned}\frac{\partial}{\partial t}p(\Omega,t) &= -\left[D_{\|}\hat{L}_{\mathrm{Z}}^2 + D_{\perp}\left(\hat{L}_{\mathrm{X}}^2+\hat{L}_{\mathrm{Y}}^2\right)\right]p(\Omega,t) \\ D_{\|} = \frac{k_{\mathrm{B}}T}{8\pi\eta r_{\|}^3}, &\quad D_{\perp} = \frac{k_{\mathrm{B}}T}{8\pi\eta r_{\perp}^3}\end{aligned} \tag{6.85}$$

where $D_{\|}$ and $D_{\perp}$ are the parallel and the perpendicular (to the long axis) rotational diffusion coefficients. The evolution propagator acquires the following form:

$$\hat{P}(\tau) = \exp\left\{-\left[D_{\|}\hat{L}_{\mathrm{Z}}^2 + D_{\perp}\left(\hat{L}_{\mathrm{X}}^2+\hat{L}_{\mathrm{Y}}^2\right)\right]\tau\right\} \tag{6.86}$$

Wigner D functions are still eigenfunctions of the diffusion operator and its exponential because it may be rearranged to only contain $\hat{L}_{\mathrm{X}}^2+\hat{L}_{\mathrm{Y}}^2+\hat{L}_{\mathrm{Z}}^2$ and $\hat{L}_{\mathrm{Z}}^2$, which commute

$$D_{\|}\hat{L}_{\mathrm{Z}}^2 + D_{\perp}\left(\hat{L}_{\mathrm{X}}^2+\hat{L}_{\mathrm{Y}}^2\right) = \left(D_{\|}-D_{\perp}\right)\hat{L}_{\mathrm{Z}}^2 + D_{\perp}\left(\hat{L}_{\mathrm{X}}^2+\hat{L}_{\mathrm{Y}}^2+\hat{L}_{\mathrm{Z}}^2\right) \tag{6.87}$$

The corresponding eigenvalues then make an appearance in the correlation function:

$$\left\langle \mathfrak{D}_{k_1 m_1}^{(l_1)*}(t)\mathfrak{D}_{k_2 m_2}^{(l_2)}(t+\tau)\right\rangle = \frac{\delta_{l_1 l_2}\delta_{k_1 k_2}\delta_{m_1 m_2}}{2l_1+1} e^{-\left[l_1(l_1+1)D_\perp + m_1^2\left(D_\parallel - D_\perp\right)\right]\tau} \quad (6.88)$$

Equation (6.85) assumes the molecular frame of reference to be the eigenframe of the rotational diffusion tensor—this condition must be observed during the spin system setup.

6.3.6.3 Asymmetric Top

For a general ellipsoid in an isotropic liquid, the rotational diffusion equation is

$$\begin{gathered} \frac{\partial}{\partial \tau} p(\Omega,t) = -\left[D_{XX}\hat{L}_X^2 + D_{YY}\hat{L}_Y^2 + D_{ZZ}\hat{L}_Z^2\right] p(\Omega,t) \\ D_{XX} = \frac{k_B T}{8\pi\eta r_X^3}, \qquad D_{YY} = \frac{k_B T}{8\pi\eta r_Y^3}, \qquad D_{ZZ} = \frac{k_B T}{8\pi\eta r_Z^3} \end{gathered} \quad (6.89)$$

in which r_{XYZ} are the three principal radii. The evolution propagator acquires the following form:

$$\hat{P}(\tau) = \exp\left\{-\left[D_{XX}\hat{L}_X^2 + D_{YY}\hat{L}_Y^2 + D_{ZZ}\hat{L}_Z^2\right]\tau\right\} \quad (6.90)$$

Wigner functions are no longer eigenfunctions of the bath dynamics generator. However, some linear combinations must then be eigenfunctions:

$$\begin{gathered} \Psi_{kq}^{(l)}(\Omega) = \frac{1}{N_q}\sum_m a_{qm}^{(l)} \mathfrak{D}_{km}^{(l)}(\Omega) \\ \left[D_{XX}\hat{L}_X^2 + D_{YY}\hat{L}_Y^2 + D_{ZZ}\hat{L}_Z^2\right]\Psi_{kq}^{(l)}(\Omega) = \lambda_q^{(l)}\Psi_{kq}^{(l)}(\Omega) \end{gathered} \quad (6.91)$$

where N_q is the normalisation coefficient, and the sum only needs to run over the second projection index because the generator does not mix ranks and does not act on the first projection index. Explicit expressions for the coefficients and eigenvalues may be obtained using machine algebra systems; for second rank Wigner

Table 6.1 Eigenvalues and eigenfunction expansion coefficients for the asymmetric top diffusion operator in Eq. (6.89)

$\mathbf{a}_{\mathbf{qm}}^{(2)}$	**m = −2**	**m = −1**	**m = 0**	**m = +1**	**m = +2**	$\lambda_{\mathbf{q}}^{(2)}$
q = −2	+1	0	Λ_D^-	0	+1	$2D_{XX} + 2D_{YY} + 2D_{ZZ} - 2\Delta_D$
q = −1	0	−1	0	+1	0	$D_{XX} + 4D_{YY} + D_{ZZ}$
q = 0	−1	0	0	0	+1	$D_{XX} + D_{YY} + 4D_{ZZ}$
q = +1	0	+1	0	+1	0	$4D_{XX} + D_{YY} + D_{ZZ}$
q = +2	+1	0	Λ_D^+	0	+1	$2D_{XX} + 2D_{YY} + 2D_{ZZ} + 2\Delta_D$

functions they are listed in Table 6.1, in which the ordering of q index values is arbitrary (it is not a projection index) and the shorthands are

$$\begin{aligned}\Delta_{\mathrm{D}} &= \sqrt{D_{\mathrm{XX}}^2 + D_{\mathrm{YY}}^2 + D_{\mathrm{ZZ}}^2 - D_{\mathrm{XX}}D_{\mathrm{YY}} - D_{\mathrm{XX}}D_{\mathrm{ZZ}} - D_{\mathrm{YY}}D_{\mathrm{ZZ}}} \\ \Lambda_{\mathrm{D}}^{\pm} &= \sqrt{\frac{2}{3}}\frac{D_{\mathrm{XX}} + D_{\mathrm{YY}} - 2D_{\mathrm{ZZ}} \pm 2\Delta_{\mathrm{D}}}{D_{\mathrm{XX}} - D_{\mathrm{YY}}}\end{aligned} \tag{6.92}$$

Placing a unit projector

$$\mathbf{1} = \sum_q \left|\Psi_{kq}^{(l)}(\Omega)\right\rangle\left\langle\Psi_{kq}^{(l)}(\Omega)\right| \tag{6.93}$$

into the correlation function integral and taking all scalar products yields

$$\left\langle \mathfrak{D}_{k_1m_1}^{(l_1)*}(t)\mathfrak{D}_{k_2m_2}^{(l_2)}(t+\tau)\right\rangle = \frac{\delta_{l_1l_2}\delta_{k_1k_2}}{2l_1+1}\sum_q a_{qm_1}^{(l)}a_{qm_2}^{(l)*}\exp\left[-\lambda_q^{(l)}\tau\right] \tag{6.94}$$

6.3.6.4 Lipari-Szabo Model

Rotational diffusion models discussed above assume that the spin system rotates as a rigid body. This is rarely the case; a more realistic picture must include the overall rotational motion, but also internal motion. A common situation in a well-folded protein in an aqueous solution is global tumbling with a rotational correlation time of a few nanoseconds, and restricted local motion of individual amino acids within the secondary and tertiary structure. An elegant analytical framework only exists for rotational motion, and only for situations when local motion is uncorrelated, in the local frame of reference, with the global motion [213].

Following the rotation group formalism described in Sect. 3.2.4, the orientation-dependent part of the spin Hamiltonian of each particular interaction with spherical rank l has the following form:

$$\mathbf{H}_1(t) = \sum_{kpm} \mathfrak{D}_{kp}^{(l)}(\Omega_{\mathrm{GM}})\mathfrak{D}_{pm}^{(l)}(\Omega_{\mathrm{LM}})\mathbf{Q}_{km}^{(l)} \tag{6.95}$$

where $\mathbf{Q}_{km}^{(l)}$ are irreducible spherical components of the spin Hamiltonian, Ω_{LM} are orientation parameters (for example Euler angles, Sect. 1.6.2) of the "local" motion (LM) relative to the molecular frame of reference, and Ω_{GM} are the orientation parameters of the "global" motion (GM) of the molecular frame relative to the laboratory frame.

When the ensemble average of Hamiltonian products is taken within Redfield theory (Sect. 6.3.1), the following correlation functions make an appearance:

$$g_{[\dots]}(\tau) = \left\langle \mathfrak{D}_{kp}^{(l)*}[\Omega_{\text{GM}}(t)]\mathfrak{D}_{pm}^{(l)*}[\Omega_{\text{LM}}(t)]\mathfrak{D}_{k'p'}^{(l)}[\Omega_{\text{GM}}(t+\tau)]\mathfrak{D}_{p'm'}^{(l)}[\Omega_{\text{LM}}(t+\tau)] \right\rangle \tag{6.96}$$

where the square bracket on the left-hand side contains all of the Wigner D function indices that occur on the right. After assuming that the internal motion is uncorrelated with the global motion, the correlation functions split into products of local and global correlation functions:

$$\begin{aligned} g_{[\dots]}(\tau) &= g_{[\dots]}^{\text{GM}}(\tau) g_{[\dots]}^{\text{LM}}(\tau) \\ g_{[\dots]}^{\text{GM}}(\tau) &= \left\langle \mathfrak{D}_{kp}^{(l)*}[\Omega_{\text{GM}}(t)]\mathfrak{D}_{k'p'}^{(l)}[\Omega_{\text{GM}}(t+\tau)] \right\rangle \\ g_{[\dots]}^{\text{LM}}(\tau) &= \left\langle \mathfrak{D}_{pm}^{(l)*}[\Omega_{\text{LM}}(t)]\mathfrak{D}_{p'm'}^{(l)}[\Omega_{\text{LM}}(t+\tau)] \right\rangle \end{aligned} \tag{6.97}$$

The global motion autocorrelation function decays to zero, but the local one does not. Given the assumption of a stationary stochastic process, the internal motion correlation function must decay and stabilise, but only at some fraction S^2 of the Wigner D function norm square:

$$\begin{aligned} g_{[\dots]}^{\text{LM}}(\infty) &= \left\langle \mathfrak{D}_{kp}^{(l)*}[\Omega_{\text{LM}}(0)]\mathfrak{D}_{k'p'}^{(l)}[\Omega_{\text{LM}}(\infty)] \right\rangle_t \\ &= S^2 \left\langle \mathfrak{D}_{kp}^{(l)*}[\Omega_{\text{LM}}]\mathfrak{D}_{k'p'}^{(l)}[\Omega_{\text{LM}}] \right\rangle_{\Omega_{\text{LM}}} = S^2 \frac{\delta_{kk'}\delta_{pp'}}{2l+1} \end{aligned} \tag{6.98}$$

This fraction is called the *order parameter*. It can be interpreted as a fraction of the full body angle that is spanned by the restriction cone of the internal motion. When the local motion is unrestricted, S^2 is equal to zero. As the motion gets increasingly restricted, S^2 approaches 1.

It is convenient to scale $g_{[\dots]}^{\text{GM}}(\tau)$ and $g_{[\dots]}^{\text{LM}}(\tau)$ so that, for non-orthogonal index combinations, they start at unity and decay to zero. With this scaling in place, the correlation function becomes

$$c_{[\dots]}(\tau) = c_{[\dots]}^{\text{GM}}(\tau)\left[S^2 + (1-S^2)c_{[\dots]}^{\text{LM}}(\tau)\right] \tag{6.99}$$

When both instances of rotational diffusion are isotropic and occur in flat potentials, the decays are monoexponential functions (Sect. 6.3.6.1); we will denote their characteristic times τ_{GM} and τ_{LM}. This leads to the following explicit expression for the correlation function and its Fourier transform:

$$\begin{aligned} c_{\text{LZ}}(\tau) &= \mathrm{e}^{-\tau/\tau_{\text{GM}}}\left[S^2 + (1-S^2)\mathrm{e}^{-\tau/\tau_{\text{LM}}}\right] \\ j_{\text{LZ}}(\omega) &= S^2 \frac{\tau_{\text{GM}}}{1+\omega^2\tau_{\text{GM}}^2} + (1-S^2)\frac{\tau_{\text{eff}}}{1+\omega^2\tau_{\text{eff}}^2}, \quad \frac{1}{\tau_{\text{eff}}} = \frac{1}{\tau_{\text{GM}}} + \frac{1}{\tau_{\text{LM}}} \end{aligned} \tag{6.100}$$

A convenient feature of this *Lipari-Szabo model* (the “model-free” moniker used in the paper title [213] is obviously too optimistic) is that this spectral density function may be used as a drop-in replacement of the rotational diffusion ones described above. In the limit of perfect local order, it reduces to rotational diffusion with the characteristic time τ_{GM}. In the limit of rapid local motion, it reduces to the appropriately scaled rotational diffusion with the characteristic time τ_{LM}. This latter feature makes the theory applicable to solid-state systems with rapid stochastic local motions.

6.4 Explicit Classical Bath: Stochastic Liouville Equation

Outside the simple cases and limits discussed above, analytical expressions for the correlation functions become unpleasant. Time scale separation assumptions of Hubbard and Redfield theories also bite—those methods are not applicable (Sect. 6.3.2) to slowly moving baths. In this section, we wipe the slate clean and consider a different approach that comes from the theory of stochastic differential equations.

For an individual molecule undergoing random spin-independent spatial motion, the LvN equation (Sect. 4.2.1) is a stochastic differential equation for the state vector $\boldsymbol{\rho}(t)$:

$$\frac{\partial}{\partial t}\boldsymbol{\rho}(t) = -i\mathcal{H}(\mathbf{x})\boldsymbol{\rho}(t) = -i(\mathcal{H}_0 + \mathcal{H}_1(\mathbf{x}))\boldsymbol{\rho}(t) \tag{6.101}$$

where $\mathcal{H}_0$ is an unchanging part of the Hamiltonian commutation superoperator and $\mathcal{H}_1$ is the part that depends on stochastic classical parameters $\mathbf{x}$, such as coordinates, angles, etc. This equation is exact but inconvenient—the spin system moves around and has to be followed. An equation of motion for the average density matrix $\boldsymbol{\rho}(\mathbf{x}, t)$ in the location $\mathbf{x}$ at time t would be easier to work with.

6.4.1 Derivation

Consider the probability density $p(\mathbf{x}, \boldsymbol{\rho}, t)$ of systems in a spatial location $\mathbf{x}$ in the spin state $\boldsymbol{\rho}$. We assume that systems undergo classical spatial motion with a linear evolution generator, and that quantum mechanical spin processes do not influence that motion. Thus, in the absence of spin dynamics:

$$\left.\frac{\partial p(\mathbf{x}, \boldsymbol{\rho}, t)}{\partial t}\right|_{\boldsymbol{\rho}} = \hat{G}(\mathbf{x})p(\mathbf{x}, \boldsymbol{\rho}, t) \tag{6.102}$$

where $\hat{G}(\mathbf{x})$ is the evolution generator of the classical motion. In the absence of spatial dynamics, the spin evolution is governed by Eq. (6.101) and therefore (after two applications of the chain rule):

$$p(\mathbf{x},\boldsymbol{\rho},t+\mathrm{d}t)|_{\mathbf{x}} = p\left(\mathbf{x},\mathrm{e}^{-i\mathcal{H}(\mathbf{x})\mathrm{d}t}\boldsymbol{\rho},t\right) \tag{6.103}$$

For an infinitesimal time increment, two more applications of the chain rule yield:

$$\begin{aligned} \ldots &= p(\mathbf{x},\boldsymbol{\rho}-i\mathcal{H}(\mathbf{x})\boldsymbol{\rho}\mathrm{d}t,t) \\ &= p(\mathbf{x},\boldsymbol{\rho},t) + \langle \nabla_{\boldsymbol{\rho}} p(\mathbf{x},\boldsymbol{\rho},t) \mid -i\mathcal{H}(\mathbf{x})\boldsymbol{\rho} \rangle \mathrm{d}t \end{aligned} \tag{6.104}$$

This gets us the remaining component of $\partial p/\partial t$:

$$\left.\frac{\partial p(\mathbf{x},\boldsymbol{\rho},t)}{\partial t}\right|_{\mathbf{x}} = \langle \nabla_{\boldsymbol{\rho}} p(\mathbf{x},\boldsymbol{\rho},t) \mid -i\mathcal{H}(\mathbf{x})\boldsymbol{\rho} \rangle \tag{6.105}$$

which can be rearranged using the product rule:

$$\ldots = -i\langle \nabla_{\boldsymbol{\rho}} | p(\mathbf{x},\boldsymbol{\rho},t)\mathcal{H}(\mathbf{x})\boldsymbol{\rho} \rangle + ip(\mathbf{x},\boldsymbol{\rho},t)\langle \nabla_{\boldsymbol{\rho}} | \mathcal{H}(\mathbf{x})\boldsymbol{\rho} \rangle \tag{6.106}$$

The second term in this expression is zero because $\mathcal{H}(\mathbf{x})$ is traceless

$$\langle \nabla_{\boldsymbol{\rho}} | \mathcal{H}\boldsymbol{\rho} \rangle = \sum_k \frac{\partial}{\partial \rho_k}[\mathcal{H}\boldsymbol{\rho}]_k = \sum_k \frac{\partial}{\partial \rho_k}\sum_m \mathcal{H}_{km}\rho_m = \sum_k \mathcal{H}_{kk} = 0 \tag{6.107}$$

After merging $(\partial p/\partial t)|_{\boldsymbol{\rho}}$ and $(\partial p/\partial t)|_{\mathbf{x}}$, we get the partial differential equation governing the dynamics of the probability density:

$$\frac{\partial}{\partial t} p(\mathbf{x},\boldsymbol{\rho},t) = -i\langle \nabla_{\boldsymbol{\rho}} | p(\mathbf{x},\boldsymbol{\rho},t)\mathcal{H}(\mathbf{x})\boldsymbol{\rho} \rangle + \hat{G}(\mathbf{x})p(\mathbf{x},\boldsymbol{\rho},t) \tag{6.108}$$

We can now obtain the equation of motion for the average density matrix $\boldsymbol{\rho}(\mathbf{x},t)$:

$$\boldsymbol{\rho}(\mathbf{x},t) = \int \boldsymbol{\rho}\, p(\mathbf{x},\boldsymbol{\rho},t)\mathrm{d}V_{\boldsymbol{\rho}} \tag{6.109}$$

by integrating the probability distribution over the volume $V_{\boldsymbol{\rho}}$ of the spin state space:

$$\begin{aligned} \frac{\partial}{\partial t}\boldsymbol{\rho}(\mathbf{x},t) &= \int \left(\frac{\partial}{\partial t}p(\mathbf{x},\boldsymbol{\rho},t)\right)\boldsymbol{\rho}\,\mathrm{d}V_{\boldsymbol{\rho}} \\ &= -i\int \langle \nabla_{\boldsymbol{\rho}} | p(\mathbf{x},\boldsymbol{\rho},t)\mathcal{H}(\mathbf{x})\boldsymbol{\rho} \rangle \boldsymbol{\rho}\,\mathrm{d}V_{\boldsymbol{\rho}} + \int \hat{G}(\mathbf{x})p(\mathbf{x},\boldsymbol{\rho},t)\boldsymbol{\rho}\,\mathrm{d}V_{\boldsymbol{\rho}} \end{aligned} \tag{6.110}$$

The second integral is easy:

$$\int \hat{G}(\mathbf{x})p(\mathbf{x},\boldsymbol{\rho},t)\boldsymbol{\rho}\,\mathrm{d}V_{\boldsymbol{\rho}} = \hat{G}(\mathbf{x})\int p(\mathbf{x},\boldsymbol{\rho},t)\boldsymbol{\rho}\,\mathrm{d}V_{\boldsymbol{\rho}} = \hat{G}(\mathbf{x})\boldsymbol{\rho}(\mathbf{x},t) \tag{6.111}$$

and the first integral can be taken using the multi-dimensional version of integration by parts:

$$\int_V f\langle\nabla|\mathbf{g}\rangle\mathrm{d}V = \oint_S f\langle\mathbf{g}|\mathrm{d}\mathbf{s}\rangle - \int_V \langle\mathbf{g}|\nabla f\rangle\mathrm{d}V \tag{6.112}$$

in which the surface integral would be zero because the integration surface can always be chosen to lie outside the unit ball containing the density matrix. Therefore,

$$\begin{aligned} -i\int \langle\nabla_{\boldsymbol{\rho}}\mid p(\mathbf{x},\boldsymbol{\rho},t)\mathcal{H}(\mathbf{x})\boldsymbol{\rho}\rangle\boldsymbol{\rho}\,\mathrm{d}V_{\boldsymbol{\rho}} &= -i\int p(\mathbf{x},\boldsymbol{\rho},t)\mathcal{H}(\mathbf{x})\boldsymbol{\rho}\,\mathrm{d}V_{\boldsymbol{\rho}} \\ = -i\mathcal{H}(\mathbf{x})\int \boldsymbol{\rho}\,p(\mathbf{x},\boldsymbol{\rho},t)\mathrm{d}V_{\boldsymbol{\rho}} &= -i\mathcal{H}(\mathbf{x})\boldsymbol{\rho}(\mathbf{x},t) \end{aligned} \tag{6.113}$$

and so we get a Fokker–Planck-type (because $\boldsymbol{\rho}$ is itself a probability density) equation of motion:

$$\frac{\partial}{\partial t}\boldsymbol{\rho}(\mathbf{x},t) = -i\mathcal{H}(\mathbf{x})\boldsymbol{\rho}(\mathbf{x},t) + \hat{G}(\mathbf{x})\boldsymbol{\rho}(\mathbf{x},t) \tag{6.114}$$

where $\mathcal{H}(\mathbf{x})$ is the Hamiltonian commutation superoperator and $\hat{G}(\mathbf{x})$ is the operator from the equation describing the classical spatial dynamics of the system.

This approach, although known as the *stochastic Liouville equation* [189] method, is only stochastic in the sense that the continuous Eq. (6.114) may be viewed as a probability density representation of certain stochastic differential equations. Its advantage over the relaxation theories discussed so far in this chapter is the absence of timescale separation approximations: it works for all timescales, from non-viscous liquids to solid powders. Two common styles of solving it differ in the treatment of the spatial part—it may either be solved on a discrete grid—in which case the spatial dynamics operator becomes a matrix approximating $\hat{G}(\mathbf{x})$ on that grid—or solved in a continuous representation using a spatial basis set (eigenfunctions of the spatial dynamics operator are generally a good choice). The numerical process has been described in Sect. 5.2; here we look at the analytical side.

6.4.2 General Solution

Consider an expansion of the Hamiltonian commutation superoperator in a complete orthonormal set of functions $q_k(\mathbf{x})$ of space variables and static superoperators $\mathcal{Q}_k$:

$$\mathcal{H}(\mathbf{x}) = \sum_n q_n(\mathbf{x})\mathcal{Q}_n, \qquad \hat{G}q_k(\mathbf{x}) = \sum_m g_{mk} q_m(\mathbf{x}), \qquad g_{mk} = \langle q_m|\hat{G}|q_k\rangle \tag{6.115}$$

The state vector would have a similar expansion, but with a time-dependent spin part:

$$\boldsymbol{\rho}(\mathbf{x},t) = \sum_k q_k(\mathbf{x})\boldsymbol{\rho}_k(t) \tag{6.116}$$

With this notation in place, Eq. (6.114) can be re-written as

$$\sum_k q_k \frac{\partial \boldsymbol{\rho}_k}{\partial t} = -i\sum_{nk} q_n q_k \mathcal{Q}_n \boldsymbol{\rho}_k(t) + \sum_{nk} g_{nk} q_n \boldsymbol{\rho}_k(t) \tag{6.117}$$

Taking the scalar product with each spatial basis function $q_m(\mathbf{x})$ in turn yields

$$\left\{ \frac{\partial \boldsymbol{\rho}_m(t)}{\partial t} = -i\sum_{nk} c_{nkm}\mathcal{Q}_n\boldsymbol{\rho}_k(t) + \sum_k g_{mk}\boldsymbol{\rho}_k(t) = \sum_k \left[-i\sum_n c_{nkm}\mathcal{Q}_n + g_{mk}\mathbf{1}\right]\boldsymbol{\rho}_k(t) \right. _m \tag{6.118}$$

where $\mathbf{1}$ is a unit matrix of the same dimension as $\mathcal{H}$, and the structure coefficients of the algebra of spatial functions are defined as

$$c_{nkm} = \int q_n(\mathbf{x}) q_k(\mathbf{x}) q_m^*(\mathbf{x}) \mathrm{d}V_{\mathbf{x}} \tag{6.119}$$

where $dV_{\mathbf{x}}$ is the volume element of the spatial coordinate space. After collecting a few terms:

$$\left\{ \frac{\partial \boldsymbol{\rho}_m(t)}{\partial t} = \sum_k \left[-i\mathcal{H}_{km} + g_{mk}\mathbf{1}\right]\boldsymbol{\rho}_k(t), \right. _m \qquad \mathcal{H}_{km} = \sum_n c_{nkm}\mathcal{Q}_n \tag{6.120}$$

This is a block matrix equation for the vertically concatenated vectors $\boldsymbol{\rho}_m(t)$; it may be solved using standard techniques. Although the sums over spatial functions are infinite, the fact that $\hat{G}(\mathbf{x})$ is negative definite for dissipative classical dynamics allows them to be truncated. Matrix dimensions can become infeasibly large, this problem is dealt with in Sect. 9.1 where we discuss polyadic objects.

6.4.3 Rotational Diffusion

For isotropic rotational diffusion, Wigner D functions are a natural spatial basis set because they are eigenfunctions of the rotational diffusion operator:

$$\begin{aligned} \mathcal{H}(\Omega) &= \sum_{lkm} \mathfrak{D}^{(l)}_{km}(\Omega)\mathcal{Q}_{km}, & \hat{G} &= -D_{\mathrm{R}}\left(\hat{L}_{\mathrm{X}}^2 + \hat{L}_{\mathrm{Y}}^2 + \hat{L}_{\mathrm{Z}}^2\right) \\ \boldsymbol{\rho}(\Omega, t) &= \sum_{lkm} \mathfrak{D}^{(l)}_{km}(\Omega)\boldsymbol{\rho}^{(l)}_{km}(t), & \hat{G}\mathfrak{D}^{(l)}_{km}(\Omega) &= -D_{\mathrm{R}}l(l+1)\mathfrak{D}^{(l)}_{km}(\Omega) \end{aligned} \tag{6.121}$$

where Ω is a parametrisation of SO(3), $\{\hat{L}_{\mathrm{X}}, \hat{L}_{\mathrm{Y}}, \hat{L}_{\mathrm{Z}}\}$ are its generators (Sect. 1.6.2), and D_{R} is the rotational diffusion coefficient. With this notation in place, the SLE becomes

$$\begin{aligned} \sum_{lkm} \mathfrak{D}^{(l)}_{km}(\Omega)\frac{\partial}{\partial t}\boldsymbol{\rho}^{(l)}_{km}(t) &= -i\sum_{jlkmpq} \mathfrak{D}^{(j)}_{km}(\Omega)\mathfrak{D}^{(l)}_{pq}(\Omega)\mathcal{Q}_{km}\boldsymbol{\rho}^{(l)}_{pq}(t) \\ &\quad - D_{\mathrm{R}}\sum_{lkm} l(l+1)\mathfrak{D}^{(l)}_{km}(\Omega)\boldsymbol{\rho}^{(l)}_{km}(t) \\ \mathfrak{D}^{(j)}_{km}(\Omega)\mathfrak{D}^{(l)}_{pq}(\Omega) &= \sum_{L=|l-j|}^{l+j}\sum_{MN} C^{L,M}_{j,k,l,p}C^{L,N}_{j,m,l,q}\mathfrak{D}^{(L)}_{MN}(\Omega) \end{aligned} \tag{6.122}$$

where $C^{L,M}_{l,m,l',m'}$ are Clebsch-Gordan coefficients (Sect. 2.5.4). After separating the ranks and projections of the Wigner D functions, we get

$$\left\{ \frac{\partial}{\partial t}\boldsymbol{\rho}^{(L)}_{MN}(t) = -i\sum_{jlkmpq} C^{L,M}_{j,k,l,p}C^{L,N}_{j,m,l,q}\mathcal{Q}_{km}\boldsymbol{\rho}^{(l)}_{pq}(t) - D_{\mathrm{R}}L(L+1)\boldsymbol{\rho}^{(L)}_{MN}(t) \right._{L,M,N} \tag{6.123}$$

If systems start off uniformly distributed and in the same spin state $\boldsymbol{\rho}_0$, the initial condition is

$$\boldsymbol{\rho}^{(l)}_{km}(0) = \begin{cases} \boldsymbol{\rho}_0 & l = k = m = 0 \\ \mathbf{0} & \text{otherwise} \end{cases} \tag{6.124}$$

This is now a special case of the block matrix Eq. (6.120) that is best solved numerically [340, 341]. Cases of axial and rhombic rotational diffusion tensors are treated identically with the spatial basis remaining the same (axial case) or modified as described in Sect. 6.4.1 (rhombic case). An important feature that guarantees convergence with respect to the rank L of Wigner D functions is the damping term in Eq. (6.123) that is quadratic with respect to that rank. For organic radicals in non-viscous liquids (correlation time of the order of nanoseconds) convergence is achieved around $L = 10$.

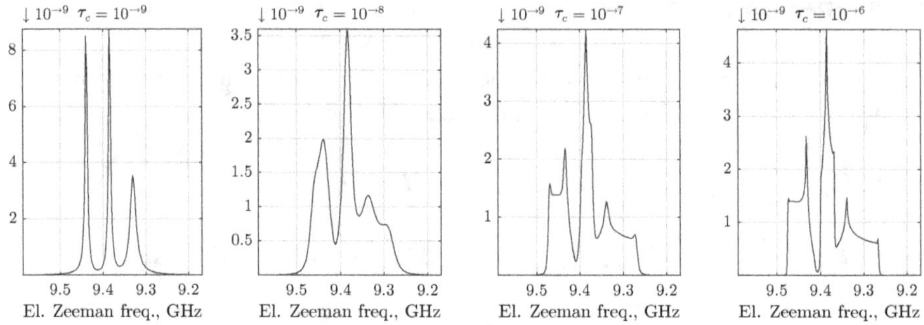

Fig. 6.2 Stochastic Liouville equation simulations of a frequency-swept electron spin resonance spectrum of a nitroxide radical with different isotropic rotational correlation times. Only the leftmost spectrum is within the applicability range of Redfield theory (Sect. 6.3.2); the rightmost spectrum closely resembles the solid powder pattern

6.4.4 Solid Limit of SLE

As classical descriptions of the bath go, Stochastic Liouville Equation formalism is superior to Redfield theory (Sect. 6.3.1) because it is non-perturbative with respect to the system-bath coupling, and therefore does not have to obey the associated validity conditions (Sect. 6.3.2). SLE remains applicable for all bath dynamics timescales all the way to the *solid limit* where the motion is so slow that, for example, the EPR spectrum in the rightmost panel of Fig. 6.2 is essentially a static powder average.

6.5 Generalised Cumulant Expansion

Because stochastic processes present in different terms of the equation of motion can become correlated, the only formally correct way to apply ensemble averaging is to the solution:

$$\langle \boldsymbol{\rho}(t) \rangle = \left\langle \overleftarrow{\exp}\left(-i \int_0^t \mathcal{H}(t')\mathrm{d}t' \right) \boldsymbol{\rho}(0) \right\rangle = \left\langle \overleftarrow{\exp}\left(-i \int_0^t \mathcal{H}(t')\mathrm{d}t' \right) \right\rangle \boldsymbol{\rho}(0) \tag{6.125}$$

Here, the exponential is time-ordered (Sect. 4.1), and the initial state can be taken out of the ensemble averaging bracket because is assumed to be the same for all members of the ensemble. If an effective generator $\mathcal{L}$ exists for this dynamics

$$\exp(-i\mathcal{L}t) = \left\langle \overleftarrow{\exp}\left(-i\int_0^t \mathcal{H}(t')\mathrm{d}t'\right)\right\rangle \tag{6.126}$$

it will be expressed (Sect. 4.3.2) through the logarithm of the right-hand side of this equation [223]. By far the easiest way to compute this is numerically (Sect. 4.9.5); our discussion here would focus on the alternative derivation this formalism offers for Redfield theory (Sect. 6.3).

6.5.1 Scalar Moments and Cumulants

Let $p(x)$ be the probability distribution of a stochastic variable x. The nth *moment* of this probability distribution is defined as the expectation value of x^n:

$$\langle x^n \rangle = \int_{-\infty}^{\infty} x^n p(x)\mathrm{d}x \tag{6.127}$$

and the *moment-generating function* (so-called because moments are its derivatives with respect to k at $k = 0$) as the expectation value of e^{kx}:

$$\left\langle \mathrm{e}^{kx} \right\rangle = \int_{-\infty}^{\infty} p(x)\mathrm{e}^{kx}\mathrm{d}x, \qquad \langle x^n \rangle = \left[\frac{\mathrm{d}^n}{\mathrm{d}k^n}\left\langle \mathrm{e}^{kx} \right\rangle\right]_{k=0} \tag{6.128}$$

Physically, a moment is a measure of the shape of the probability density: the first moment is the *mean value*, the second centred (mean is subtracted) moment is the *variance*, and the third standardised (divided by the appropriate power of the standard deviation to make it scale-invariant) moment is the *skewness*, and the fourth standardised moment is the *kurtosis*.

In statistical physics, an important moment-generating function is the *partition function*:

$$Z = \langle \exp(-\beta E) \rangle, \qquad \beta = 1/k_\mathrm{B}T \tag{6.129}$$

The fact that key properties (energy, entropy, etc.) are derivatives of its logarithm

$$\langle E \rangle = -\frac{\partial \ln Z}{\partial \beta}, \qquad \langle S \rangle = \frac{\partial}{\partial T}(k_\mathrm{B}T \ln Z) \tag{6.130}$$

suggests that the logarithm of the moment-generating function is important. It is called the *cumulant-generating function*; its derivatives $\langle\langle x^n \rangle\rangle$ are called *cumulants* of the stochastic variable x:

$$\langle\langle x^n \rangle\rangle = \left[\frac{\mathrm{d}^n}{\mathrm{d}k^n} \ln\langle \mathrm{e}^{kx} \rangle \right]_{k=0} \tag{6.131}$$

6.5.2 Joint Moments and Cumulants

For a set of stochastic variables $\{x_1, x_2, \ldots\}$ with the joint probability distribution $p(x_1, x_2, \ldots)$ the moment-generating function is the expectation value of the following exponential:

$$\langle \mathrm{e}^{k_1 x_1 + k_2 x_2 + \ldots} \rangle = \int p(x_1, x_2, \ldots) \mathrm{e}^{k_1 x_1 + k_2 x_2 + \ldots} \mathrm{d}V \tag{6.132}$$

where the integral is taken over the volume of the space containing $\{x_1, x_2, \ldots\}$. *Joint moments* are defined via partial derivatives of this function:

$$\langle x_1^{\alpha_1} x_2^{\alpha_2} \ldots \rangle = \left[\frac{\partial^{\alpha_1 + \alpha_2 + \cdots}}{\partial k_1^{\alpha_1} \partial k_2^{\alpha_2} \ldots} \langle \mathrm{e}^{k_1 x_1 + k_2 x_2 + \ldots} \rangle \right]_{k_i = 0} \tag{6.133}$$

and *joint cumulants* via partial derivatives of its logarithm:

$$\langle\langle x_1^{\alpha_1} x_2^{\alpha_2} \ldots \rangle\rangle = \left[\frac{\partial^{\alpha_1 + \alpha_2 + \ldots}}{\partial k_1^{\alpha_1} \partial k_2^{\alpha_2} \ldots} \ln\langle \mathrm{e}^{k_1 x_1 + k_2 x_2 + \ldots} \rangle \right]_{k_i = 0} \tag{6.134}$$

For a multivariate function $f(x_1, \ldots, x_1)$ the Taylor series around the origin is:

$$f(x_1, \ldots, x_N) = \sum_{\alpha_1 = 0}^{\infty} \cdots \sum_{\alpha_N = 0}^{\infty} \frac{1}{\alpha_1! \ldots \alpha_N!} \left[\frac{\partial^{\alpha_1 + \cdots + \alpha_N} f}{\partial x_1^{\alpha_1} \ldots \partial x_N^{\alpha_N}} \right]_{x_i = 0} x_1^{\alpha_1} \ldots x_N^{\alpha_N} \tag{6.135}$$

Therefore, the for the joint cumulant-generating function:

$$\ln\langle \exp(k_1 x_1 + \cdots + k_N x_N) \rangle = \sum_{\alpha_1 = 0}^{\infty} \cdots \sum_{\alpha_N = 0}^{\infty} \frac{\langle\langle x_1^{\alpha_1} \ldots x_N^{\alpha_N} \rangle\rangle}{\alpha_1! \ldots \alpha_N!} k_1^{\alpha_1} \ldots k_N^{\alpha_N} \tag{6.136}$$

A continuous limit of the sum on the left-hand side of this equation may be obtained by replacing $k_n x_n$ with $x(t_n)\Delta t_n$ and taking the Riemann integration limit of $\Delta t_n \to 0$:

$$\ln\left\langle \exp\left(\int_a^b x(t) \mathrm{d}t \right) \right\rangle = \sum_{n=1}^{\infty} \frac{1}{n!} \int_a^b \mathrm{d}t_1 \int_a^b \mathrm{d}t_2 \ldots \int_a^b \mathrm{d}t_n \langle\langle x(t_1) x(t_2) \ldots x(t_n) \rangle\rangle \tag{6.137}$$

After applying the same combinatorial trick as the one used in Eq. (4.9), we obtain

$$\ln\left\langle \exp\left(\int_a^b x(t)dt\right)\right\rangle = \sum_{n=1}^{\infty}\int_a^b \mathrm{d}t_1 \int_a^{t_1} \mathrm{d}t_2 \ldots \int_a^{t_{n-1}} \mathrm{d}t_n \langle\langle x(t_1)x(t_2)\ldots x(t_n)\rangle\rangle \tag{6.138}$$

In this case, no time ordering is necessary because all scalars commute.

6.5.3 Connection to Redfield Theory

To relate cumulant expansions to relaxation theory, we start again with the Dyson series solution of the Liouville–von Neumann equation (Sect. 4.2.1) in the interaction representation (Sect. 4.3.1) and observe that it must be equal to the time-ordered exponential solution:

$$\begin{aligned} &\overleftarrow{\exp}\left(-i\int_0^t \mathcal{H}_1^{\mathrm{R}}(t')\mathrm{d}t'\right) \\ &= 1 + \sum_{n=1}^{\infty}(-i)^n \int_0^t \mathrm{d}t_1 \int_0^{t_1} \mathrm{d}t_2 \ldots \int_0^{t_{n-1}} \mathrm{d}t_n \mathcal{H}_1^{\mathrm{R}}(t_1)\mathcal{H}_1^{\mathrm{R}}(t_2)\ldots\mathcal{H}_1^{\mathrm{R}}(t_n) \end{aligned} \tag{6.139}$$

Time ordering is now necessary because $\mathcal{H}_1^{\mathrm{R}}(t)$ need not commute with itself at different times. This brings us back to Eq. (6.126), where we can now take the logarithm of both sides:

$$\begin{aligned} &\ln\left\langle \overleftarrow{\exp}\left(-i\int_0^t \mathcal{H}^{\mathrm{R}}(t')\mathrm{d}t'\right)\right\rangle \\ &= \sum_{n=1}^{\infty}(-i)^n \int_0^t \mathrm{d}t_1 \int_0^{t_1} \mathrm{d}t_2 \ldots \int_0^{t_{n-1}} \mathrm{d}t_n \langle\langle \mathcal{H}^{\mathrm{R}}(t_1)\mathcal{H}^{\mathrm{R}}(t_2)\ldots\mathcal{H}^{\mathrm{R}}(t_n)\rangle\rangle \end{aligned} \tag{6.140}$$

When $\mathcal{H}_1^{\mathrm{R}}(t)$ is a centred stochastic process, the expressions for the first three cumulant terms via the corresponding moments (obtained by using the chain rule on the definition of the cumulant generating function) are identical to those obtained within Redfield theory (Sect. 6.3):

$$\begin{aligned} &\langle\langle \mathcal{H}_1^{\mathrm{R}}(t_1)\rangle\rangle = \langle \mathcal{H}_1^{\mathrm{R}}(t_1)\rangle = 0 \\ &\langle\langle \mathcal{H}_1^{\mathrm{R}}(t_1)\mathcal{H}_1^{\mathrm{R}}(t_2)\rangle\rangle = \langle \mathcal{H}_1^{\mathrm{R}}(t_1)\mathcal{H}_1^{\mathrm{R}}(t_2)\rangle \\ &\langle\langle \mathcal{H}_1^{\mathrm{R}}(t_1)\mathcal{H}_1^{\mathrm{R}}(t_2)\mathcal{H}_1^{\mathrm{R}}(t_3)\rangle\rangle = \langle \mathcal{H}_1^{\mathrm{R}}(t_1)\mathcal{H}_1^{\mathrm{R}}(t_2)\mathcal{H}_1^{\mathrm{R}}(t_3)\rangle \end{aligned} \tag{6.141}$$

The difference appears at the fourth cumulant, for which:

$$\begin{aligned}
&\langle\langle \mathcal{H}_1^{\mathrm{R}}(t_1)\mathcal{H}_1^{\mathrm{R}}(t_2)\mathcal{H}_1^{\mathrm{R}}(t_3)\mathcal{H}_1^{\mathrm{R}}(t_4)\rangle\rangle \\
&\quad = \langle \mathcal{H}_1^{\mathrm{R}}(t_1)\mathcal{H}_1^{\mathrm{R}}(t_2)\mathcal{H}_1^{\mathrm{R}}(t_3)\mathcal{H}_1^{\mathrm{R}}(t_4)\rangle \\
&\quad - \langle \mathcal{H}_1^{\mathrm{R}}(t_1)\mathcal{H}_1^{\mathrm{R}}(t_2)\rangle\langle \mathcal{H}_1^{\mathrm{R}}(t_3)\mathcal{H}_1^{\mathrm{R}}(t_4)\rangle \\
&\quad - \langle \mathcal{H}_1^{\mathrm{R}}(t_1)\mathcal{H}_1^{\mathrm{R}}(t_3)\rangle\langle \mathcal{H}_1^{\mathrm{R}}(t_2)\mathcal{H}_1^{\mathrm{R}}(t_4)\rangle \\
&\quad - \langle \mathcal{H}_1^{\mathrm{R}}(t_1)\mathcal{H}_1^{\mathrm{R}}(t_4)\rangle\langle \mathcal{H}_1^{\mathrm{R}}(t_2)\mathcal{H}_1^{\mathrm{R}}(t_3)\rangle
\end{aligned} \tag{6.142}$$

Once the Hamiltonian is expanded in products of spatial functions and spin operators, and the correlation functions are computed as described in Sect. 6.3.3, this formalism becomes an instance of Redfield theory with a different summation strategy for the Dyson series.

6.6 Secular and Diagonal Approximations

Consider a negative-definite relaxation superoperator $\mathcal{R}$ written in an orthogonal basis of spin states $\{\boldsymbol{\rho}_n\}$ that diagonalises the static part of the Hamiltonian commutation superoperator:

$$\mathcal{H}_0\boldsymbol{\rho}_n = [\mathbf{H}_0, \boldsymbol{\rho}_n] = \omega_n\boldsymbol{\rho}_n \tag{6.143}$$

In this picture, there are two classes of relaxation processes: (a) those corresponding to diagonal elements $r_{nn} = \langle\boldsymbol{\rho}_n|\mathcal{R}|\boldsymbol{\rho}_n\rangle/\langle\boldsymbol{\rho}_n|\boldsymbol{\rho}_n\rangle$ which damp the amplitudes of individual basis states (*self-relaxation*); (b) those corresponding to off-diagonal elements $r_{nk} = \langle\boldsymbol{\rho}_n|\mathcal{R}|\boldsymbol{\rho}_k\rangle/\sqrt{\langle\boldsymbol{\rho}_n|\boldsymbol{\rho}_n\rangle\langle\boldsymbol{\rho}_k|\boldsymbol{\rho}_k\rangle}$ that cause amplitudes to flow between different basis states (*cross-relaxation*).

Consider a cross-relaxation process with the rate r_{X} between two $\mathcal{H}_0$ eigenstates $\boldsymbol{\rho}_{\mathrm{A}}$ and $\boldsymbol{\rho}_{\mathrm{B}}$ whose self-relaxation rates are r_{S}, and whose frequencies ω_{A} and ω_{B} are different:

$$\frac{\mathrm{d}}{\mathrm{d}t}\begin{bmatrix}\langle A\rangle \\ \langle B\rangle\end{bmatrix} = \begin{bmatrix}-i\omega_{\mathrm{A}} - r_{\mathrm{S}} & -r_{\mathrm{X}} \\ -r_{\mathrm{X}} & -i\omega_B - r_{\mathrm{S}}\end{bmatrix}\begin{bmatrix}\langle A\rangle \\ \langle B\rangle\end{bmatrix}, \qquad \langle A\rangle = \mathrm{Tr}(\mathbf{A}\boldsymbol{\rho}) \tag{6.144}$$

The propagator is obtained by exponentiating the generator (Sect. 4.1):

$$\mathbf{U}(t) = \exp\left(\begin{bmatrix}-i\omega_{\mathrm{A}} - r_{\mathrm{S}} & -r_{\mathrm{X}} \\ -r_{\mathrm{X}} & -i\omega_{\mathrm{B}} - r_{\mathrm{S}}\end{bmatrix}t\right) \tag{6.145}$$

A tedious algebraic exercise yields the following expression and the following upper bound on the absolute value for the corner element $u_{\mathrm{X}}(t)$ of the propagator matrix $\mathbf{U}(t)$:

$$
\begin{aligned}
u_{\mathrm{X}}(t) &= -\frac{\sin\left[t\sqrt{(\omega_{\mathrm{B}}-\omega_{\mathrm{A}})^2-r_{\mathrm{X}}^2}\right] r_{\mathrm{X}}}{\sqrt{(\omega_{\mathrm{B}}-\omega_{\mathrm{A}})^2-r_{\mathrm{X}}^2}} e^{i(\omega_{\mathrm{B}}-\omega_{\mathrm{A}})t} e^{-r_{\mathrm{S}}t} \\
|u_{\mathrm{X}}| &\leq \frac{r_{\mathrm{X}}}{\sqrt{(\omega_{\mathrm{B}}-\omega_{\mathrm{A}})^2-r_{\mathrm{X}}^2}}
\end{aligned}
\tag{6.146}
$$

Thus, when the frequency separation $|\omega_{\mathrm{B}}-\omega_{\mathrm{A}}|$ is much greater than the cross-relaxation rate r_{X}, cross-relaxation effectively does not occur; the corresponding elements may be dropped from the matrix so long as some user-specified accuracy tolerance is observed.

This is unproblematic when calculations are performed in the laboratory frame—the above mathematics takes care of itself. However, interaction representations modify the frequency spectrum of $\mathcal{H}_0$ and must be treated with care. One way to proceed is to inspect the relaxation superoperator in the laboratory frame, zero out the ineffectual cross-relaxation terms, and only then to apply the interaction representation. This is called *secular cross-relaxation approximation*; it can be precarious in systems where small frequency differences are present. In situations where cross-relaxation processes are unimportant, a radical measure is to ignore all cross-relaxation terms; this is called *diagonal approximation*.

6.7 Group-Theoretical Aspects of Dissipation

Significant features of spin relaxation are: (a) irreversible ensemble dephasing; (b) return to thermal equilibrium. Neither concept pertains to an individual quantum system—we must, therefore, consider ensemble observables and equations of motion. Applying ensemble averaging to an observable $O(t)$

$$
\overline{\langle\psi|\mathbf{O}|\psi\rangle} = \overline{\mathrm{Tr}(|\psi\rangle\langle\psi|\mathbf{O})} = \mathrm{Tr}\left(\overline{|\psi\rangle\langle\psi|}\mathbf{O}\right) = \mathrm{Tr}(\boldsymbol{\rho}\mathbf{O}) \tag{6.147}
$$

brings up the density matrix; the overbar is omitted in this section because we will always be dealing with an ensemble average here. Assuming that the initial condition $\boldsymbol{\rho}_0$ is the same across the ensemble and applying the evolution law derived in Sect. 4.2 yields

$$
\boldsymbol{\rho}(t) = \mathcal{P}(t)\boldsymbol{\rho}_0, \qquad \mathcal{P}(t) = \overline{\overleftarrow{\exp}\left(-i\int_0^t \mathcal{H}(t')dt'\right)} \tag{6.148}
$$

where the Hamiltonian need not be the same across the ensemble: different systems may be seeing different random local fields or have a different spatial orientation. We now have a problem: an average of multiple unitary propagators need not be

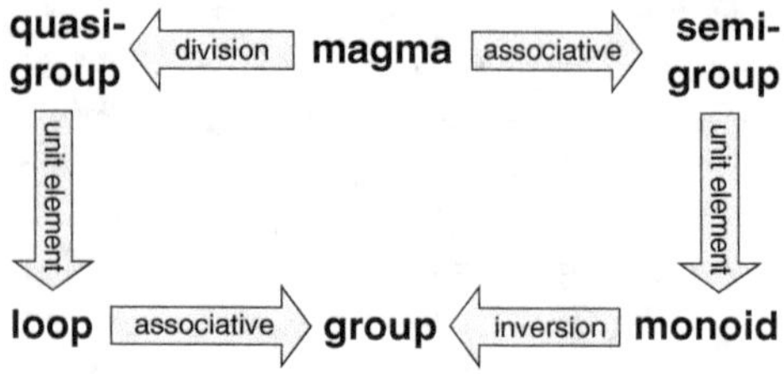

Fig. 6.3 A schematic of the relationship between a magma (set closed under a binary product), semigroup (add associativity), monoid (add unit element), and group (add inversion operation). An alternative path via quasigroup and loop is also shown

unitary, or an exponential of any matrix, or even invertible—the set of averaged propagators is not, in general, a group under superposition (Fig. 6.3).

Still, the evolution generator in Eq. (6.148) is a linear combination of exponentials of Hermitian matrices. For such objects we do have closure under superposition—by CBH formula (Sect. 4.3.3) a product of two such objects is itself a linear combination of exponentials of Hermitian matrices. We inherit associativity from matrix multiplication rules. A set with a binary operation that is closed under that operation is called a *magma*; an associative magma is called a *semigroup*. Strictly speaking, the obvious existence of a two-sided identity superoperator makes the set of all dissipative propagators a *monoid*, but the (still technically correct) semigroup designation is more common. Unlike a group (Sect. 1.5), a semigroup may have an *absorbing element*—something similar to a multiplicative zero.

Consider a finite-dimensional Hilbert state space with an algebra of bounded linear operators. Due to the physical nature of the density matrix (a table of probabilities and ensemble correlation coefficients, Sect. 4.2), any dissipative propagator must be

1. **Completely positive**. A linear map $\mathcal{P}: \mathbb{C}^{n\times n} \to \mathbb{C}^{n\times n}$ is called *positive* if $\mathcal{P}\mathbf{A}$ is Hermitian and positive semidefinite for all Hermitian and positive semidefinite $\mathbf{A} \in \mathbb{C}^{n\times n}$. $\mathcal{P}$ is called *completely positive* if all maps of the form $\mathcal{I} \otimes \mathcal{P}$, where $\mathcal{I}$ is a unit matrix of any dimension, are also positive.

2. **Trace-preserving**. A linear map $\mathcal{P}: \mathbb{C}^{n\times n} \to \mathbb{C}^{n\times n}$ is called *trace-preserving* if $\mathrm{Tr}(\mathcal{P}\mathbf{A}) = \mathrm{Tr}(\mathbf{A})$ for all Hermitian and positive semidefinite $\mathbf{A} \in \mathbb{C}^{n\times n}$.

The former requirement reflects the fact that probabilities must stay real and non-negative, and the latter is for the sum of all probabilities to remain equal to 1. Complete positivity is necessary because a Kronecker product with a unit matrix corresponds physically to having multiple independent copies of the same system; such a transformation must not influence the behaviour of physical theories. An example of a positive map that is not completely positive is the transpose map.

With some further technical conditions, continuous semigroups of propagators satisfying these properties are called *dynamical semigroups*. A lengthy algebraic exercise concludes that their generators must have the following *Lindblad form* [224]:

$$\mathcal{L}\boldsymbol{\rho} = -i[\mathbf{H},\boldsymbol{\rho}] + \sum_k \left(\left[\mathbf{V}_k\boldsymbol{\rho}, \mathbf{V}_k^\dagger \right] + \left[\mathbf{V}_k, \boldsymbol{\rho}\mathbf{V}_k^\dagger \right] \right) \tag{6.149}$$

where $\mathbf{H}, \mathbf{V}_k \in \mathbb{C}^{n\times n}$ are bounded linear operators and $\mathbf{H}$ is Hermitian. Without further approximations, none of the relaxation theories discussed above have this form; this is an unsolved problem.

6.8 Finite Temperature Effects

Redfield theory (Sect. 6.3) does not drive the system to the thermal equilibrium—the skew-Hermitian part of the relaxation superoperator in Eq. (6.47) is negative semidefinite, meaning that relaxation either has no effect, or drives the state amplitude to zero. That is the consequence of the classical description of the bath; no exact solutions for arbitrary temperatures are known at the time of writing.

Two approximate ways around this problem are suggested by the high-temperature limit of Hubbard theory (Sect. 6.2), which drives the quantity $(\boldsymbol{\rho}^{\mathrm{S}} - \boldsymbol{\rho}_{\mathrm{eq}}^{\mathrm{S}})$ to zero—the corresponding substitution may be attempted within the Redfield theory. Alternatively, Hubbard theory suggests that a thermal balancing multiplier—visible in the round brackets of Eq. (6.26)—may be retrofitted into Redfield's formalism.

6.8.1 Equilibrium Density Matrix

In an ensemble of identical systems at thermal equilibrium, there are no correlations and the probability of finding a system in the energy level n is given by Boltzmann's law [225]:

$$\mathbf{H}|\psi_n\rangle = E_n|\psi_n\rangle, \quad p_n = \frac{\exp(-E_n/k_{\mathrm{B}}T)}{\sum_m \exp(-E_m/k_{\mathrm{B}}T)} \tag{6.150}$$

These probabilities are the diagonal terms of the density matrix, therefore:

$$\boldsymbol{\rho}_{\mathrm{eq}} = \frac{\sum_n |\psi_n\rangle \exp(-E_n/k_{\mathrm{B}}T)\langle\psi_n|}{\sum_n \exp(-E_n/k_{\mathrm{B}}T)} = \frac{\exp(-\mathbf{H}/k_{\mathrm{B}}T)}{\mathrm{Tr}[\exp(-\mathbf{H}/k_{\mathrm{B}}T)]} \tag{6.151}$$

Numerical evaluation of $\boldsymbol{\rho}_{\mathrm{eq}}$ has caveats when $k_{\mathrm{B}}T \ll \|\mathbf{H}\|_2$ because exponentials of large numbers overflow finite precision arithmetic. The numerically stable approach implemented in *Spinach* [87] uses scaling and squaring: the equilibrium density matrix is first computed at the temperature $2^N T$, where

$N = \text{ceil}\left[\log_2 \|\mathbf{H}/k_{\text{B}}T\|_2\right]$, and then squared and divided by its own trace N times. The algorithm described in Sect. 4.9.5 is recommended for the matrix exponential.

The Liouville space expression involves the left side Hamiltonian product superoperator $\mathcal{H}^{(\text{L})}$:

$$\left|\boldsymbol{\rho}_{\text{eq}}\right\rangle = \frac{\exp\left(-\mathcal{H}^{(\text{L})}/k_{\text{B}}T\right)|\mathbf{1}\rangle}{\langle\mathbf{1}|\exp\left(-\mathcal{H}^{(\text{L})}/k_{\text{B}}T\right)|\mathbf{1}\rangle}, \qquad \mathcal{H}^{(\text{L})}|\boldsymbol{\rho}\rangle = |\mathbf{H}\boldsymbol{\rho}\rangle \tag{6.152}$$

where $|\mathbf{1}\rangle$ is a vectorisation of a unit matrix of the same dimension as $\boldsymbol{\rho}$. In practice, this unit state is propagated in imaginary time to $it = 1/k_{\text{B}}T$ and the result is divided by its scalar product with the unit state. Calculation of matrix exponentials is not necessary here—the algorithm described in Sect. 4.9.6 is recommended for the expm-times-vector operation.

From the computational point of view, Eqs. (6.150)–(6.152) are unsuitable for the absolute zero temperature because the limit is not numerically stable in finite-precision arithmetic. In Hilbert space, the lowest eigenvalue $E_{\min}$ and the corresponding eigenvector $|\psi_{\min}\rangle$ of the Hamiltonian should be obtained, and the density matrix constructed as $\boldsymbol{\rho}_{\text{eq}} = |\psi_{\min}\rangle\langle\psi_{\min}|$. In Liouville space, $\left|\boldsymbol{\rho}_{\text{eq}}\right\rangle$ is the lowest energy eigenvector of $\mathcal{H}^{(\text{L})}$, normalised so as to have a unit inner product with $|\mathbf{1}\rangle$. Krylov–Schur algorithm [226] that avoids full diagonalisation is recommended.

6.8.2 Inhomogeneous Thermalisation

The immediate appearance of the *inhomogeneous master equation* (IME)

$$\frac{\partial}{\partial t}|\boldsymbol{\rho}\rangle = -i\mathcal{H}|\boldsymbol{\rho}\rangle + \mathcal{R}\left|\boldsymbol{\rho} - \boldsymbol{\rho}_{\text{eq}}\right\rangle \tag{6.153}$$

makes it inconvenient to apply time propagation. There are two equivalent cosmetic transformations that eliminate the problem. Firstly, we could double the evolution generator dimension and observe that

$$\frac{\partial}{\partial t}\begin{bmatrix} \left|\boldsymbol{\rho}_{\text{eq}}\right\rangle \\ |\boldsymbol{\rho}\rangle \end{bmatrix} = \begin{bmatrix} \mathbf{0} & \mathbf{0} \\ -\mathcal{R} & -i\mathcal{H} + \mathcal{R} \end{bmatrix}\begin{bmatrix} \left|\boldsymbol{\rho}_{\text{eq}}\right\rangle \\ |\boldsymbol{\rho}\rangle \end{bmatrix} \tag{6.154}$$

which is again homogeneous, but the cost is a significant increase in computational complexity, even with sparse arithmetic. A more efficient and physically appealing way is to add just one row and one column [227]:

$$\frac{\partial}{\partial t}\begin{bmatrix} 1 \\ |\boldsymbol{\rho}\rangle \end{bmatrix} = \begin{bmatrix} 0 & \langle\mathbf{0}| \\ -\mathcal{R}\left|\boldsymbol{\rho}_{\text{eq}}\right\rangle & -i\mathcal{H} + \mathcal{R} \end{bmatrix}\begin{bmatrix} 1 \\ |\boldsymbol{\rho}\rangle \end{bmatrix} \tag{6.155}$$

This reveals that IME thermalisation is introducing a one-way coupling to the unit state. When the basis set is chosen to be Kronecker products of single-spin irreducible spherical tensors (Sect. 3.3.2), the first state in the basis descriptor is automatically the unit state (Sect. 7.1). That is the layout used by the IME thermalisation option in *Spinach*—in that case, the matrix dimension does not change at all, and the thermalisation amounts to writing $-\mathcal{R}|\boldsymbol{\rho}_{\mathrm{eq}}\rangle$ into the first column of the evolution generator.

6.8.3 Homogeneous Thermalisation

We will now use the tangent space transformation in Eq. (1.81) to obtain one of the many mathematically possible dynamical semigroup generators that drive the system to a finite temperature equilibrium state. For the infinitesimal $\beta = 1/k_{\mathrm{B}}T$ (high-temperature approximation), the expression for the equilibrium state vector comes from Eq. (6.152):

$$\begin{aligned} \boldsymbol{\rho}_{\mathrm{eq}} &\approx Z^{-1}(\mathbf{1}-\beta\mathbf{H}), \quad Z \approx \mathrm{Tr}(\mathbf{1}-\beta\mathbf{H}) = \mathrm{Tr}(\mathbf{1}) \\ |\mathbf{1}\rangle &= \mathrm{vec}(\mathbf{1}), \quad |\boldsymbol{\rho}_{\mathrm{eq}}\rangle \approx Z^{-1}\left(|\mathbf{1}\rangle - \beta\mathcal{H}^{(\mathrm{L})}|\mathbf{1}\rangle\right) \end{aligned} \tag{6.156}$$

where we have used the fact that the spin Hamiltonian is traceless. The inhomogeneous master equation may be rewritten as

$$\frac{\partial}{\partial t}|\boldsymbol{\rho}\rangle = -i\mathcal{H}|\boldsymbol{\rho}\rangle + \mathcal{R}\left[\mathbf{1} - |\boldsymbol{\rho}_{\mathrm{eq}}\rangle\langle\mathbf{1}|\right]|\boldsymbol{\rho}\rangle, \quad \langle\mathbf{1}\mid\boldsymbol{\rho}\rangle = \mathrm{Tr}(\boldsymbol{\rho}) = 1 \tag{6.157}$$

Inserting $|\boldsymbol{\rho}_{\mathrm{eq}}\rangle$ from Eq. (6.156) into Eq. (6.157) and observing that $\mathcal{R}|\mathbf{1}\rangle = |\mathbf{0}\rangle$ yields:

$$\frac{\partial}{\partial t}|\boldsymbol{\rho}\rangle = -i\mathcal{H}|\boldsymbol{\rho}\rangle + \mathcal{R}\left[\mathbf{1} + \beta Z^{-1}\mathcal{H}^{(\mathrm{L})}|\mathbf{1}\rangle\langle\mathbf{1}|\right]|\boldsymbol{\rho}\rangle \tag{6.158}$$

for a transformation with an infinitesimal β—meaning that the generator is $Z^{-1}\mathcal{H}^{(\mathrm{L})}|\mathbf{1}\rangle\langle\mathbf{1}|$. The finite transformation is therefore $\exp\left(\beta Z^{-1}\mathcal{H}^{(\mathrm{L})}|\mathbf{1}\rangle\langle\mathbf{1}|\right)$, and the equation of motion becomes [228]

$$\frac{\partial}{\partial t}|\boldsymbol{\rho}\rangle = -i\mathcal{H}|\boldsymbol{\rho}\rangle + \mathcal{R}\exp\left[\beta Z^{-1}\mathcal{H}^{(\mathrm{L})}|\mathbf{1}\rangle\langle\mathbf{1}|\right]|\boldsymbol{\rho}\rangle \tag{6.159}$$

This form has the dual advantage of being uniform and straightforward to compute —the relaxation superoperator is simply post-multiplied by an exponential of a very sparse matrix. However, like the inhomogeneous master equation in the previous section, this thermalisation is approximate – it breaks down at low temperatures and makes erroneous predictions for multi-spin correlations.

6.9 Mechanisms of Spin Relaxation

This section contains overviews and rate expressions for common spin relaxation mechanisms. With one or two exceptions, their laborious derivations are skipped because they were done using *Mathematica*—the scripts are available in the example set of *Spinach* library.

6.9.1 Empirical: Extended T_1/T_2 Model

The simplest model of spin relaxation codifies the empirical observation that, in strong magnetic fields, transverse magnetisation often decays exponentially to zero at a certain rate, and longitudinal magnetisation returns exponentially to its thermal equilibrium value, generally at a different rate. This suggests a modification to the classical precession equations for the magnetic moment $\boldsymbol{\mu}$ [94]:

$$\begin{cases} \dfrac{\mathrm{d}}{\mathrm{d}t}\mu_\mathrm{X}(t) = -\omega\mu_\mathrm{Y}(t) \\ \dfrac{\mathrm{d}}{\mathrm{d}t}\mu_\mathrm{Y}(t) = +\omega\mu_\mathrm{X}(t) \\ \dfrac{\mathrm{d}}{\mathrm{d}t}\mu_\mathrm{Z}(t) = 0 \end{cases} \Rightarrow \begin{cases} \dfrac{\mathrm{d}}{\mathrm{d}t}\mu_\mathrm{X}(t) = -\omega\mu_\mathrm{Y}(t) - \dfrac{1}{T_2}\mu_\mathrm{X} \\ \dfrac{\mathrm{d}}{\mathrm{d}t}\mu_\mathrm{Y}(t) = +\omega\mu_\mathrm{X}(t) - \dfrac{1}{T_2}\mu_\mathrm{Y} \\ \dfrac{\mathrm{d}}{\mathrm{d}t}\mu_\mathrm{Z}(t) = -\dfrac{1}{T_1}\left(\mu_\mathrm{Z} - \mu_\mathrm{eq}\right) \end{cases} \tag{6.160}$$

where T_1 became known as *longitudinal relaxation time* and T_2 as *transverse relaxation time*; these times are measured experimentally. This model covers the relaxation of single-spin states ($\mathbf{S}_\mathrm{Z}$ for longitudinal and $\mathbf{S}_\pm$ for transverse), but its original formulation [94] does not mention product states.

A simple extension is possible for non-interacting spins relaxed by mutually uncorrelated local stochastic processes. Consider a system of non-interacting spins with uncorrelated stochastic Hamiltonians:

$$\begin{gathered} \mathbf{H}(t) = \mathbf{H}_0 + \mathbf{H}_1(t) = \sum_k \left[\mathbf{H}_0^{(k)} + \mathbf{H}_1^{(k)}(t)\right] \\ \left\langle \mathbf{H}_1^{(n)}(t)\mathbf{H}_1^{(k)}(t+\tau)\right\rangle = \mathbf{0} \qquad n \neq k \end{gathered} \tag{6.161}$$

where the upper index runs over the spins in the system. Within the derivation of Redfield theory described in Sect. 6.3, the Hamiltonian autocorrelation function splits up:

$$\langle \mathbf{H}_1(t)\mathbf{H}_1(t+\tau)\rangle = \sum_k \left\langle \mathbf{H}_1^{(k)}(t)\mathbf{H}_1^{(k)}(t+\tau)\right\rangle \tag{6.162}$$

and the relaxation superoperator, therefore, acquires the following structure:

$$\begin{aligned}\mathcal{R} &= \left[\mathcal{R}^{(1)} \otimes \mathbf{1}^{(2)} \otimes \mathbf{1}^{(3)} \otimes \ldots\right] + \left[\mathbf{1}^{(1)} \otimes \mathcal{R}^{(2)} \otimes \mathbf{1}^{(3)} \otimes \ldots\right] \\ &+ \left[\mathbf{1}^{(1)} \otimes \mathbf{1}^{(2)} \otimes \mathcal{R}^{(3)} \otimes \ldots\right] + \ldots\end{aligned} \tag{6.163}$$

The relaxation rate of a product state is now the sum of the relaxation rates of its product terms:

$$\left\langle \bigotimes_k \mathbf{S}^{(k)} \middle| \mathcal{R} \middle| \bigotimes_k \mathbf{S}^{(k)} \right\rangle = \sum_k \left\langle \mathbf{S}^{(k)} \middle| \mathcal{R}^{(k)} \middle| \mathbf{S}^{(k)} \right\rangle \tag{6.164}$$

This is a rough approximation, but a useful one when detailed relaxation models are not available. This is also the reason why quantum devices do not scale – the broader the correlated state, the faster it relaxes when the noises affecting the individual parts of the device are statistically independent.

6.9.2 Coupling to Stochastic External Vectors

Some relaxation mechanisms (spin-rotation, lanthanide-induced dipolar, non-uniform magnetic field in the gas phase, scalar relaxation of the second kind, etc.) reduce mathematically to a coupling between a spin and a classical external stochastic vector with known statistics:

$$\mathbf{H}_0 = \omega_{\mathrm{S}} \mathbf{S}_{\mathrm{Z}}, \qquad \mathbf{H}_1(t) = b_{\mathrm{X}}(t)\mathbf{S}_{\mathrm{X}} + b_{\mathrm{Y}}(t)\mathbf{S}_{\mathrm{Y}} + b_{\mathrm{Z}}(t)\mathbf{S}_{\mathrm{Z}} \tag{6.165}$$

where ω_{S} is the Zeeman frequency, and the nature of the zero-mean vector $\mathbf{b}(t)$ depends on the context: random external fields have $\mathbf{b}(t) = -\gamma_{\mathrm{S}} \mathbf{B}_1(t)$, scalar relaxation of the second kind would have J-coupling multiplied by the Cartesian direction of the partner spin, etc. Equation (6.47) may be applied directly

$$\begin{aligned}\mathcal{R} &= -\left\langle \int_0^\infty \mathcal{H}_1(0) e^{-i\mathcal{H}_0 \tau} \mathcal{H}_1(\tau) e^{+i\mathcal{H}_0 \tau} d\tau \right\rangle \\ r_k &= \frac{\langle \mathbf{S}_k | \mathcal{R} | \mathbf{S}_k \rangle}{\langle \mathbf{S}_k \mid \mathbf{S}_k \rangle}, \quad k \in \{X, Y, Z\}\end{aligned} \tag{6.166}$$

with the following result for the relaxation rates of the three Cartesian projections of the magnetisation:

$$
\begin{aligned}
r_{\mathrm{X}} &= \int_0^\infty [g_{\mathrm{ZZ}}(\tau) + g_{\mathrm{YY}}(\tau)\cos(\omega_{\mathrm{L}}\tau) - g_{\mathrm{YX}}(\tau)\sin(\omega_{\mathrm{L}}\tau)]\mathrm{d}\tau \\
r_{\mathrm{Y}} &= \int_0^\infty [g_{\mathrm{ZZ}}(\tau) + g_{\mathrm{XX}}(\tau)\cos(\omega_{\mathrm{L}}\tau) + g_{\mathrm{XY}}(\tau)\sin(\omega_{\mathrm{L}}\tau)]\mathrm{d}\tau \\
r_{\mathrm{Z}} &= \int_0^\infty \{[g_{\mathrm{XX}}(\tau) + g_{\mathrm{YY}}(\tau)]\cos(\omega_{\mathrm{L}}\tau) + [g_{\mathrm{XY}}(\tau) - g_{\mathrm{YX}}(\tau)]\sin(\omega_{\mathrm{L}}\tau)\}\mathrm{d}\tau
\end{aligned}
\tag{6.167}
$$

The autocorrelation functions here refer to the Cartesian components of the random vector $\mathbf{b}\,(t)$:

$$
g_{nk}(\tau) = \langle b_n(t) b_k(t+\tau) \rangle, \qquad n,k \in \{\mathrm{X},\mathrm{Y},\mathrm{Z}\} \tag{6.168}
$$

and the usual Redfield theory validity conditions are assumed (Sect. 6.3.2): the random field component $\mathbf{H}_1(t)$ of the Hamiltonian must be a centred stationary stochastic process acting as a small perturbation to $\mathbf{H}_0$, and the resulting relaxation times must be much longer than the decay time of $g_{nk}(\tau)$. Although Eq. (6.167) does not depend on the spin quantum number, the autocorrelation function in Eq. (6.168) might—an example appears in Sect. 6.9.4 where the noise comes from the stochastic dynamics of a partner spin.

6.9.3 Scalar Relaxation: Noise in the Interaction

The situation where a spin–spin coupling is modulated in amplitude but not direction is encountered for distance-modulated dipolar interactions (scalar modulation of an anisotropic coupling) and for conformational mobility processes in organic radicals (modulation of isotropic hyperfine interaction). Conformational mobility is also a common source of *J*-coupling modulation (Fig. 6.4).

The minimal spin Hamiltonian capturing the process is

$$
\mathbf{H} = \omega_{\mathrm{L}} \mathbf{L}_{\mathrm{Z}} + \omega_{\mathrm{S}} \mathbf{S}_{\mathrm{Z}} + a(t)\mathbf{Q}, \qquad \mathbf{Q} = \mathbf{L} \cdot \mathbf{M} \cdot \mathbf{S} \tag{6.169}
$$

where $a(t)$ is a real-valued stochastic process with a zero average, and $\mathbf{M}$ is a real 3x3 matrix, not necessarily traceless or symmetric, and possibly isotropic; $\mathbf{L}$ and $\mathbf{S}$ are vectors of Cartesian spin operators. In this case, the Redfield theory integral in Eq. (6.46) is simplified because the sum only contains one term

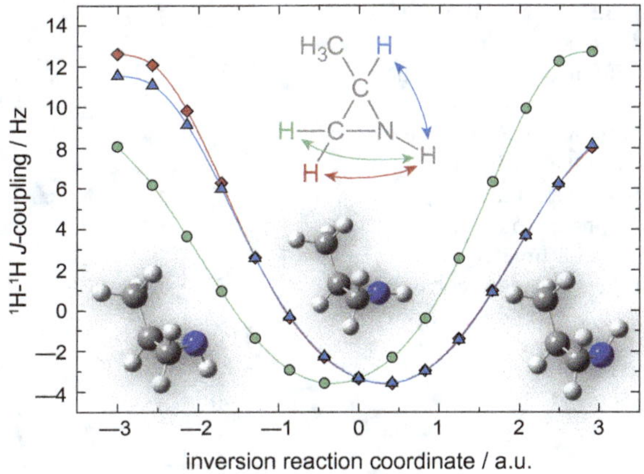

Fig. 6.4 Modulation of three-bond ^{1}H-^{1}H J-couplings to the amine proton during 2-methylaziridine nitrogen centre inversion. Reproduced from [229]

$$\mathcal{R} = -\Delta_a^2 \int_0^\infty g(\tau)\mathcal{Q}e^{-i\mathcal{H}_0\tau}\mathcal{Q}^\dagger e^{i\mathcal{H}_0\tau}d\tau, \qquad \mathcal{H}_0|\boldsymbol{\rho}\rangle = \text{vec}([\omega_L\mathbf{L}_Z + \omega_S\mathbf{S}_Z, \boldsymbol{\rho}])$$
$$\mathcal{Q}|\boldsymbol{\rho}\rangle = \text{vec}([\mathbf{L}\cdot\mathbf{M}\cdot\mathbf{S}, \boldsymbol{\rho}]), \qquad \langle a(t)a^*(t+\tau)\rangle = \Delta_a^2 g(\tau) \tag{6.170}$$

with the squared modulation depth Δ_a^2 chosen so as to have $g(0) = 1$. Relaxation rates obtained by automated symbolic processing are compiled in Table 6.2 with the following shorthands:

$$\begin{aligned} r_{L+S} &= \Delta_a^2 j(\omega_L + \omega_S)\left[(m_{XY} + m_{YX})^2 + (m_{XX} - m_{YY})^2\right] \\ r_{L-S} &= \Delta_a^2 j(\omega_L - \omega_S)\left[(m_{XY} - m_{YX})^2 + (m_{XX} + m_{YY})^2\right] \\ r_L &= \Delta_a^2 j(\omega_L)\left[m_{XZ}^2 + m_{YZ}^2\right], \quad r_0 = j(0)m_{ZZ}^2 \\ r_S &= \Delta_a^2 j(\omega_S)\left[m_{ZX}^2 + m_{ZY}^2\right], \quad j(\omega) = \text{Re}\int_0^\infty g(\tau)e^{-i\omega\tau}d\tau \end{aligned} \tag{6.171}$$

Table 6.2 Redfield theory relaxation and longitudinal cross-relaxation rates arising from scalar modulation of a general bilinear coupling. Sign convention is such that self-relaxation rates are positive

State(s)	(Cross-)relaxation rate
$\mathbf{L}_Z$	$\frac{\Delta_a^2 S(S+1)}{3}\left(r_L + \frac{1}{2}(r_{L-S} + r_{L+S})\right)$
$\mathbf{S}_Z$	$\frac{\Delta_a^2 L(L+1)}{3}\left(r_S + \frac{1}{2}(r_{L-S} + r_{L+S})\right)$
$\mathbf{L}_\pm$	$\frac{\Delta_a^2 S(S+1)}{3}\left(r_0 + \frac{1}{2}r_L + r_S + \frac{1}{4}(r_{L-S} + r_{L+S})\right)$
$\mathbf{S}_\pm$	$\frac{\Delta_a^2 L(L+1)}{3}\left(r_0 + \frac{1}{2}r_S + r_L + \frac{1}{4}(r_{L-S} + r_{L+S})\right)$
$\mathbf{L}_Z \leftrightarrow \mathbf{S}_Z$	$\frac{\Delta_a^2\sqrt{L(L+1)S(S+1)}}{6}\left(r_{L+S} - r_{L-S}\right)$

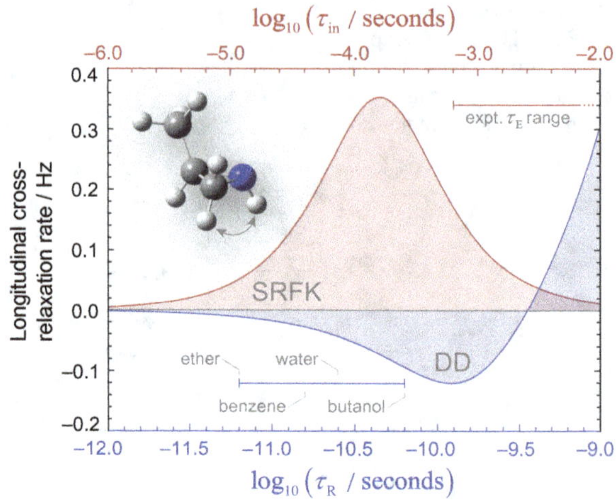

Fig. 6.5 Dipolar (blue curve) and SRFK (red curve) longitudinal cross-relaxation rates between the amine proton and the CH_2 proton trans to the methyl group in 2-methylaziridine, as functions of rotational correlation time τ_R and nitrogen inversion correlation time τ_{in}. Reproduced from [229]

When the tensor **M** is isotropic, these processes are called *scalar (cross-) relaxation of the first kind*. The full symbolic relaxation superoperator may be generated using the Mathematica worksheet supplied with the example set of Spinach library. Direct inspection indicates that there is a five-dimensional subspace of states ($\mathbf{L}_Z\mathbf{S}_Z$, $\mathbf{L}_\pm\mathbf{S}_\pm$, $\mathbf{L}_\pm\mathbf{S}_\mp$) that are immune to relaxation under scalar modulation of isotropic bilinear couplings.

Electron-nuclear cross-relaxation due to scalar modulation of the hyperfine coupling tensor occurs in dynamic nuclear polarisation experiments [230], where there are two statistically correlated terms in the interaction Hamiltonian: translationally modulated electron-nuclear dipolar coupling and collisionally modulated contact interaction (Sect. 3.1.5). Inter-nuclear scalar cross-relaxation is seldom seen [229], but can be stronger than the Nuclear Overhauser Effect (Sect. 6.9.5.2) in systems that have slow conformational exchange processes (Fig. 6.5).

6.9.4 Scalar Relaxation: Noise in the Partner Spin

Consider a two-spin system with a scalar coupling between the spins:

$$\mathbf{H} = \omega_L \mathbf{L}_Z + \omega_S \mathbf{S}_Z + a\mathbf{L}\cdot\mathbf{S} \tag{6.172}$$

When L relaxes slowly and S relaxes rapidly on the time scale of the interaction (for example, in a *J*-coupled ^{1}H-^{14}N pair in an aqueous protein solution), the dynamics of S is seen by L as an external stochastic process:

$$\mathbf{H}_0^L = \omega_L \mathbf{L}_Z, \qquad \mathbf{H}_1^L(t) = b_X(t)\mathbf{L}_X + b_Y(t)\mathbf{L}_Y + b_Z(t)\mathbf{L}_Z \tag{6.173}$$

With respect to spin L, the relaxation problem is now an instance of relaxation due to a random external field (Sect. 6.9.2). The correlation functions in Eq. (6.168) are

$$g_{nk}(\tau) = \langle b_n(t) b_k(t+\tau) \rangle = a^2 \langle \mathbf{S}_n(0) \mathbf{S}_k(\tau) \rangle, \qquad n,k \in \{\mathrm{X},\mathrm{Y},\mathrm{Z}\} \tag{6.174}$$

The ensemble average may be expressed in the Heisenberg picture as a trace with the thermal equilibrium density matrix, which at high temperature is close to a multiple of the unit matrix:

$$g_{nk}(\tau) = a^2 \mathrm{Tr}\left[\boldsymbol{\rho}_{\mathrm{eq}} \mathbf{S}_n \mathcal{P}(\tau) \mathbf{S}_k\right] = \frac{a^2}{2S+1} \mathrm{Tr}[\mathbf{S}_n \mathcal{P}(\tau) \mathbf{S}_k] \tag{6.175}$$

where $\mathcal{P}(\tau)$ is the time propagation superoperator. In its action on spin S, we will explicitly assume the mono-exponential loss of autocorrelation:

$$\begin{aligned} \mathcal{P}(\tau)\mathbf{S}_\mathrm{X} &= (\mathbf{S}_\mathrm{X} \cos(\omega_\mathrm{S}\tau) + \mathbf{S}_\mathrm{Y} \sin(\omega_\mathrm{S}\tau)) \mathrm{e}^{-\tau/T_{2\mathrm{S}}} \\ \mathcal{P}(\tau)\mathbf{S}_\mathrm{Y} &= (\mathbf{S}_\mathrm{Y} \cos(\omega_\mathrm{S}\tau) - \mathbf{S}_\mathrm{X} \sin(\omega_\mathrm{S}\tau)) \mathrm{e}^{-\tau/T_{2\mathrm{S}}} \\ \mathcal{P}(\tau)\mathbf{S}_\mathrm{Z} &= \mathbf{S}_\mathrm{Z} \mathrm{e}^{-\tau/T_{1\mathrm{S}}} \end{aligned} \tag{6.176}$$

Performing substitutions and simplifications leads us to traces of Cartesian spin operator products

$$\mathrm{Tr}[\mathbf{S}_n \mathbf{S}_k] = \delta_{nk} S(S+1)(2S+1)/3 \tag{6.177}$$

and eventually yields the following correlation functions:

$$\begin{aligned} g_\mathrm{XX}(\tau) = g_\mathrm{YY}(\tau) &= \frac{a^2 S(S+1)}{3} \cos(\omega_\mathrm{S}\tau) \mathrm{e}^{-\tau/T_{2\mathrm{S}}}, \\ g_\mathrm{ZZ}(\tau) &= \frac{a^2 S(S+1)}{3} \mathrm{e}^{-\tau/T_{1\mathrm{S}}} \\ g_\mathrm{XY}(\tau) &= \frac{a^2 S(S+1)}{3} \sin(\omega_\mathrm{S}\tau) \mathrm{e}^{-\tau/T_{2\mathrm{S}}}, \\ g_\mathrm{YX}(\tau) &= -\frac{a^2 S(S+1)}{3} \sin(\omega_\mathrm{S}\tau) \mathrm{e}^{-\tau/T_{2\mathrm{S}}} \end{aligned} \tag{6.178}$$

Substitution into Eq. (6.167) and integration produces the rates for the relaxation process that is called *scalar relaxation of the second kind* [230]:

$$\begin{aligned} R_\mathrm{Z} &= \frac{a^2 S(S+1)}{3} \frac{2T_{2\mathrm{S}}}{1+(\omega_\mathrm{L}-\omega_\mathrm{S})^2 T_{2\mathrm{S}}^2} \\ R_{\mathrm{X,Y}} &= \frac{a^2 S(S+1)}{3} \left(T_{1\mathrm{S}} + \frac{T_{2\mathrm{S}}}{1+(\omega_\mathrm{L}-\omega_\mathrm{S})^2 T_{2\mathrm{S}}^2} \right) \end{aligned} \tag{6.179}$$

The full symbolic relaxation superoperator may be generated using the *Mathematica* worksheet supplied with the example set of the *Spinach* library. Note the subtle difference with scalar relaxation of the first kind in the previous section: there, it was the interaction coefficient a that was the source of time-dependent noise that came from things like conformational mobility. Here, the coefficient a is fixed—it is only a conduit of the noise that is generated by the partner spin.

6.9.5 Isotropic Rotational Diffusion

Dominant spin relaxation mechanisms in liquid state magnetic resonance proceed from rotational diffusion and the associated stochastic modulation of anisotropic interactions. The presentation given in this section differs from previous accounts in several respects. Firstly, we break with the tradition of using [eigenvalues + orientation] specifications, and work with Cartesian forms of all interaction tensors. This leads to notational simplification, particularly for cross-correlations of rhombic tensors. Secondly, the often ignored antisymmetric (first spherical rank) anisotropies are treated with the same respect as the dominant symmetric (second spherical rank) anisotropies. This is because situations exist (*e.g.* nuclear shielding by paramagnetic metal centres, Sect. 3.1.7) where first-rank components dominate. Lastly, we replace laborious derivations with machine algebra scripts (downloadable as a part of *Spinach*) and only discuss the problem setting and the properties of the resulting relaxation rates.

For a 3×3 interaction tensors $\mathbf{A}$ and $\mathbf{B}$, the following norms and their associated (via the polarisation relation) inner products will occur repeatedly in the equations presented in this section. For the first spherical rank (antisymmetric) anisotropies:

$$\begin{gathered} \Lambda_{\mathbf{A}}^2 = (a_{\mathrm{XY}} - a_{\mathrm{YX}})^2 + (a_{\mathrm{XZ}} - a_{\mathrm{ZX}})^2 + (a_{\mathrm{YZ}} - a_{\mathrm{ZY}})^2 \\ \daleth_{\mathbf{A},\mathbf{B}} = \left(\Lambda_{\mathbf{A}+\mathbf{B}}^2 - \Lambda_{\mathbf{A}-\mathbf{B}}^2\right)/4 \end{gathered} \tag{6.180}$$

And for the second spherical rank (symmetric) anisotropies:

$$\begin{aligned} \Delta_{\mathbf{A}}^2 &= a_{\mathrm{XX}}^2 + a_{\mathrm{YY}}^2 + a_{\mathrm{ZZ}}^2 - a_{\mathrm{XX}}a_{\mathrm{YY}} - a_{\mathrm{XX}}a_{\mathrm{ZZ}} - a_{\mathrm{YY}}a_{\mathrm{ZZ}} \\ &\quad + \frac{3}{4}\left[(a_{\mathrm{XY}} + a_{\mathrm{YX}})^2 + (a_{\mathrm{XZ}} + a_{\mathrm{ZX}})^2 + (a_{\mathrm{YZ}} + a_{\mathrm{ZY}})^2\right] \\ \aleph_{\mathbf{A},\mathbf{B}} &= \left(\Delta_{\mathbf{A}+\mathbf{B}}^2 - \Delta_{\mathbf{A}-\mathbf{B}}^2\right)/4 \end{aligned} \tag{6.181}$$

These parameters follow the bilinear relationships pertaining to norms and scalar products:

$$\begin{array}{lll} \Delta_{\mathbf{A}}^{2}=\aleph_{\mathbf{A},\mathbf{A}}, & \aleph_{\mathbf{A},\mathbf{B}}+\aleph_{\mathbf{A},\mathbf{C}}=\aleph_{\mathbf{A},\mathbf{B}+\mathbf{C}}, & \aleph_{\mathbf{A},\mathbf{C}}+\aleph_{\mathbf{B},\mathbf{C}}=\aleph_{\mathbf{A}+\mathbf{B},\mathbf{C}} \\ \Lambda_{\mathbf{A}}^{2}=\daleth_{\mathbf{A},\mathbf{A}}, & \daleth_{\mathbf{A},\mathbf{B}}+\daleth_{\mathbf{A},\mathbf{C}}=\daleth_{\mathbf{A},\mathbf{B}+\mathbf{C}}, & \daleth_{\mathbf{A},\mathbf{C}}+\daleth_{\mathbf{B},\mathbf{C}}=\daleth_{\mathbf{A}+\mathbf{B},\mathbf{C}} \end{array} \tag{6.182}$$

We will explicitly account for the multiplicity of all spins. This is necessary because Frobenius inner products of Cartesian spin operators are representation-dependent:

$$\mathrm{Tr}[\mathbf{S}_n\mathbf{S}_k]=\delta_{nk}S(S+1)(2S+1)/3, \qquad n,k\in\{\mathrm{X},\mathrm{Y},\mathrm{Z}\} \tag{6.183}$$

where S is the spin quantum number. Because the discussion below does not assume any particular spin quantum numbers, the following pre-factors will make an appearance:

$$\langle\mathbf{S}^2\rangle=S(S+1), \quad m_{\mathrm{S}}^2=(2S+1)^2 \tag{6.184}$$

In tables below, spectral density functions and correlation times of spherical rank l are defined as real parts of the Fourier transforms of the correlation functions in Eq. (6.82):

$$\begin{aligned} J_l(\omega) &= \frac{\tau_{\mathrm{C}}^{(l)}}{2l+1}\left[1+\left(\tau_{\mathrm{C}}^{(l)}\omega\right)^2\right]^{-1}, \\ \tau_{\mathrm{C}}^{(l)} &= \frac{1}{l(l+1)D_{\mathrm{R}}}=\frac{8\pi\eta r^3}{l(l+1)k_{\mathrm{B}}T} \end{aligned} \tag{6.185}$$

where D_{R} is the rotational diffusion coefficient. When τ_{C} and $J(\omega)$ are used without the rank index, a historical convention implies second spherical rank.

6.9.5.1 Zeeman Interactions

Consider stochastic modulation of anisotropic Zeeman interactions (chemical shift anisotropy for nuclei, g-tensor anisotropy for electrons) by rotational diffusion in an isotropic liquid [231]. When the magnetic field is directed along the Z-axis of the laboratory frame, the Hamiltonian is

$$\mathbf{H}=\omega\mathbf{S}_{\mathrm{Z}}+\mathbf{S}\cdot\mathbf{Z}\cdot\mathbf{B}_0, \qquad \mathbf{B}_0=\begin{bmatrix}0 & 0 & B_0\end{bmatrix}^{\mathrm{T}} \tag{6.186}$$

where the first term in the sum corresponds to the isotropic part of the Zeeman interaction tensor, and the matrix $\mathbf{Z}$ in the second term is traceless. In the notational convention of NMR spectroscopy:

$$\begin{aligned} \mathbf{H}_{\mathrm{NMR}} &= -\gamma\mathbf{S}\cdot(\mathbf{1}+\boldsymbol{\delta})\cdot\mathbf{B}_0=[-\gamma(1+\delta_{\mathrm{iso}})B_0]\mathbf{S}_{\mathrm{Z}}+\mathbf{S}\cdot[-\gamma\boldsymbol{\delta}_{\mathrm{aniso}}]\cdot\mathbf{B}_0 \\ \delta_{\mathrm{iso}} &= \mathrm{Tr}(\boldsymbol{\delta})/3, \qquad \boldsymbol{\delta}_{\mathrm{aniso}}=\boldsymbol{\delta}-\mathbf{1}\delta_{\mathrm{iso}} \end{aligned} \tag{6.187}$$

Table 6.3 Redfield theory relaxation rates under isotropic rotational diffusion of antisymmetric (spherical rank 1) and symmetric (spherical rank 2) anisotropies in the Zeeman interaction tensor. The invariants are defined in Eqs. (6.179) and (6.180), spectral density functions in Eq. (6.184)

State	Antisymmetric	Symmetric
$\mathbf{L}_Z$	$\frac{1}{2}\Lambda_{\mathbf{Z}}^2 B_0^2 J_1(\omega)$	$\frac{2}{3}\Delta_{\mathbf{Z}}^2 B_0^2 J_2(\omega)$
$\mathbf{L}_\pm$	$\frac{1}{4}\Lambda_{\mathbf{Z}}^2 B_0^2 J_1(\omega)$	$\frac{1}{9}\Delta_{\mathbf{Z}}^2 B_0^2 (4J_2(0) + 3J_2(\omega))$

where γ is the magnetogyric ratio of the nucleus in question, $\boldsymbol{\delta}$ is its chemical shift tensor. This corresponds to the following definitions for the components of Eq. (2.19):

$$\omega = -\gamma(1 + \delta_{\text{iso}})B_0, \qquad \mathbf{Z} = -\gamma\boldsymbol{\delta}_{\text{aniso}} \tag{6.188}$$

In the notation convention of EPR spectroscopy:

$$\begin{gathered} \mathbf{H}_{\text{EPR}} = \frac{\mu_{\text{B}}}{\hbar}\mathbf{S}\cdot\mathbf{g}\cdot\mathbf{B} = \frac{\mu_{\text{B}} g_{\text{iso}} B_0}{\hbar}\mathbf{S}_Z + \frac{\mu_{\text{B}}}{\hbar}\mathbf{S}\cdot\mathbf{g}_{\text{aniso}}\cdot\mathbf{B} \\ g_{\text{iso}} = \text{Tr}(\mathbf{g})/3, \quad \mathbf{g}_{\text{aniso}} = \mathbf{g} - \mathbf{1}g_{\text{iso}} \end{gathered} \tag{6.189}$$

where μ_{B} is Bohr magneton and $\mathbf{g}$ is the g-tensor of the electron. This corresponds to the following definitions for the components of Eq. (6.186):

$$\omega = \frac{\mu_{\text{B}} g_{\text{iso}} B_0}{\hbar}, \qquad \mathbf{Z} = \frac{\mu_{\text{B}}}{\hbar}\mathbf{g}_{\text{aniso}} \tag{6.190}$$

Neither the chemical shift tensor nor the g-tensor is in general symmetric; Sect. 3.2.6 gives an example of strongly antisymmetric nuclear shielding tensors.

Redfield theory expressions for the longitudinal and transverse relaxation rates for the rotationally modulated CSA mechanism are given in Table 6.3. These the expressions remain the same for nuclei of different spin, and for electron shells of different spin multiplicity. Symmetric and antisymmetric parts have no cross-terms and may be added algebraically. Example plots of the relaxation rates as functions of magnet field and rotational correlation time for the ^{15}N nucleus in adenine are given in Fig. 6.6.

6.9.5.2 Bilinear Interactions

Common examples of anisotropic bilinear spin–spin couplings are inter-electron dipolar interactions (Sect. 3.2.8), dipolar parts of hyperfine couplings (Sect. 3.2.4), and dipole interactions between nuclei (Sect. 3.2.5). The minimal Hamiltonian is

$$\mathbf{H} = \omega_{\text{L}}\mathbf{L}_Z + \omega_{\text{S}}\mathbf{S}_Z + \mathbf{L}\cdot\mathbf{A}\cdot\mathbf{S} \tag{6.191}$$

where the matrix $\mathbf{A}$ is traceless, but not necessarily symmetric. Section 3.2 gives explicit expressions for this matrix in various settings. For point dipolar interaction between isotropically shielded spins:

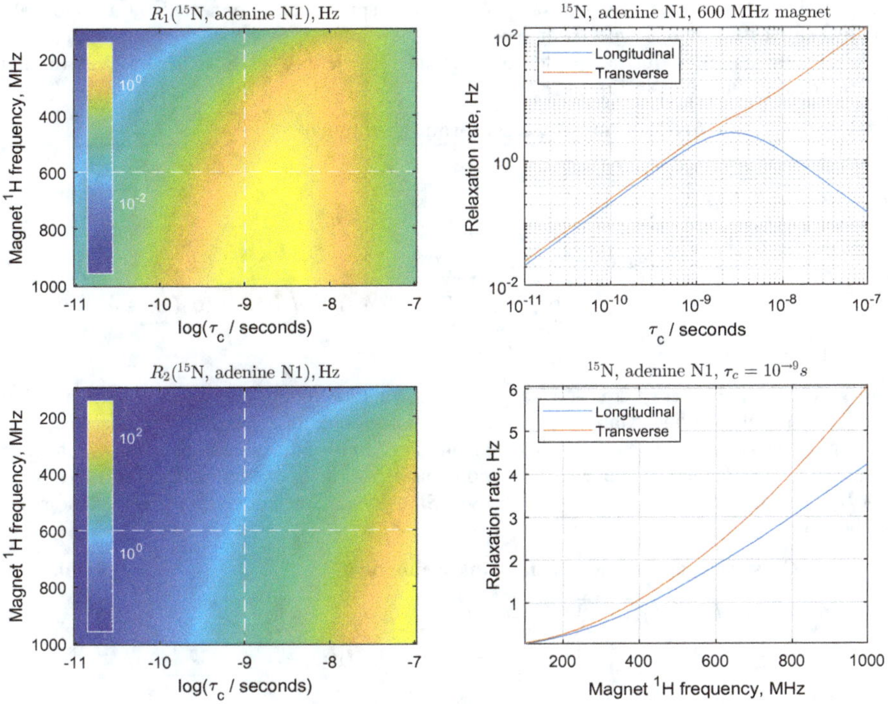

Fig. 6.6 Longitudinal and transverse CSA relaxation rates for a ^{15}N nucleus in the N1 position of the adenine ring, plotted as functions of magnet frequency and rotational correlation time

$$\Delta_{\mathbf{A}}^2 = 9\left(\frac{\mu_0}{4\pi}\right)^2 \frac{\gamma_{\mathrm{L}}^2 \gamma_{\mathrm{S}}^2 \hbar^2}{r_{\mathrm{LS}}^6}, \qquad \Lambda_{\mathbf{A}}^2 = 0 \tag{6.192}$$

Final results (obtained using *Mathematica* scripts supplied with *Spinach* [220]) for relaxation and cross-relaxation rates are given in Tables 6.4, 6.5, 6.6 and 6.7; graphical illustrations are given in Figs. 6.7 and 6.8.

A particular limit is encountered in high-field ESR spectroscopy, where the precession frequency of the electron is much greater than that of the nucleus, and thus $\omega_{\mathrm{E}} \pm \omega_{\mathrm{N}} \approx \omega_{\mathrm{E}}$. Another common limit in NMR spectroscopy of small molecules in non-viscous solvents is the *extreme narrowing limit*, where $\omega\tau_{\mathrm{C}} \ll 1$.

Dipolar cross-relaxation between $\mathbf{L}_{\mathrm{Z}}$ and $\mathbf{S}_{\mathrm{Z}}$ is called the *Overhauser effect*; it is used for distance measurement (because $\Delta_{\mathbf{A}}^2$ is proportional to r_{LS}^{-6}) and for sensitivity enhancement of low-γ nuclei. The latter may be illustrated by a steady-state calculation in the $\{\mathbf{L}_{\mathrm{Z}}, \mathbf{S}_{\mathrm{Z}}\}$ subspace. The equation of motion is

Table 6.4 Redfield theory self-relaxation rates for a spin system connected by a rotationally modulated traceless bilinear interaction, such as dipole–dipole or hyperfine coupling. L and S are spin quantum numbers, interaction tensor invariants are defined in Eqs. (6.180) and (6.191), spectral density functions in Eq. (6.184)

State	Self-relaxation rate, second-rank component
L_Z	$\frac{2S(S+1)\Delta_A^2}{27}(3J_2(\omega_L)+J_2(\omega_L-\omega_S)+6J_2(\omega_L+\omega_S))$
S_Z	$\frac{2L(L+1)\Delta_A^2}{27}(3J_2(\omega_S)+J_2(\omega_L-\omega_S)+6J_2(\omega_L+\omega_S))$
$L_\pm$	$\frac{S(S+1)\Delta_A^2}{27}(4J_2(0)+3J_2(\omega_L)+J_2(\omega_L-\omega_S)+6J_2(\omega_S)+6J_2(\omega_L+\omega_S))$
$S_\pm$	$\frac{L(L+1)\Delta_A^2}{27}(4J_2(0)+6J_2(\omega_L)+J_2(\omega_L-\omega_S)+3J_2(\omega_S)+6J_2(\omega_L+\omega_S))$

Table 6.5 Redfield theory cross-relaxation rates for a spin system connected by a rotationally modulated traceless bilinear interaction, such as dipole–dipole or hyperfine coupling. Secularity status refers to strong magnetic field directed along the Z axis of the laboratory frame of reference. L and S are spin quantum numbers, interaction tensor invariants are defined in Eqs. (6.180) and (6.191), spectral density functions in Eq. (6.184). Sign convention is such that self-relaxation rates are positive

A	B	Cross-relaxation rate, second-rank component	Secular
L_Z	S_Z	$-\frac{2\Delta_A^2}{27}\sqrt{L(L+1)S(S+1)}(J_2(\omega_L-\omega_S)-6J_2(\omega_L+\omega_S))$	Yes
$L_\pm$	$S_\pm$	$+\frac{\Delta_A^2}{27}\sqrt{L(L+1)S(S+1)}(2J_2(0)+3J_2(\omega_L)+2J_2(\omega_L-\omega_S)+3J_2(\omega_S))$	Depends

Table 6.6 Redfield theory self-relaxation rates for a spin system connected by a rotationally modulated traceless bilinear interaction, such as dipole–dipole or hyperfine coupling. L and S are spin quantum numbers, interaction tensor invariants are defined in Eqs. (6.179) and (6.191), spectral density functions in Eq. (6.184)

State	Self-relaxation rate, first-rank component
L_Z	$\frac{S(S+1)\Lambda_A^2}{6}(J_1(\omega_L)+J_1(\omega_L-\omega_S))$
S_Z	$\frac{L(L+1)\Lambda_A^2}{6}(J_1(\omega_L-\omega_S)+J_1(\omega_S))$
$L_\pm$	$\frac{S(S+1)\Lambda_A^2}{12}(J_1(\omega_L)+J_1(\omega_L-\omega_S)+2J_1(\omega_S))$
$S_\pm$	$\frac{L(L+1)\Lambda_A^2}{12}(2J_1(\omega_L)+J_1(\omega_L-\omega_S)+J_1(\omega_S))$

Table 6.7 Redfield theory cross-relaxation rates for a spin system connected by a rotationally modulated traceless bilinear interaction, such as dipole–dipole or hyperfine coupling. Secularity status refers to strong magnetic field directed along the Z axis of the laboratory frame of reference. L and S are spin quantum numbers, interaction tensor invariants are defined in Eqs. (6.179) and (6.191), spectral density functions in Eq. (6.184). Sign convention is such that self-relaxation rates are positive

A	B	Cross-relaxation rate, first-rank component	Secular
L_Z	S_Z	$-\frac{\Lambda_A^2}{6}\sqrt{L(L+1)S(S+1)}J_1(\omega_L-\omega_S)$	yes
$L_\pm$	$S_\pm$	$-\frac{\Lambda_A^2}{12}\sqrt{L(L+1)S(S+1)}(J_1(\omega_L)+J_1(\omega_S))$	depends

$$\frac{d}{dt}\begin{pmatrix} L_Z \\ S_Z \end{pmatrix} = \begin{pmatrix} -r_L & r_X \\ r_X & -r_S \end{pmatrix}\begin{pmatrix} L_Z - L_Z^{eq} \\ S_Z - S_Z^{eq} \end{pmatrix}, \quad \begin{cases} L_Z = \mathrm{Tr}(\mathbf{L}_Z\boldsymbol{\rho}) \\ S_Z = \mathrm{Tr}(\mathbf{S}_Z\boldsymbol{\rho}) \end{cases} \tag{6.193}$$

where r_L and r_S are the self-relaxation rates from Table 6.4 and r_X is the cross-relaxation rate from Table 6.5. When the L spin is continuously saturated (meaning that $L_Z = 0$) and the system is allowed to settle into the new steady state (meaning that the derivative on the left-hand side is zero), we obtain the following equation for the steady-state magnetisation S_Z^∞ of spin S

$$r_X L_Z^{eq} + r_S\left(S_Z^\infty - S_Z^{eq}\right) = 0 \tag{6.194}$$

from which it follows that $S_Z^\infty = S_Z^{eq} - (r_X/r_S)L_Z^{eq}$, meaning that the magnetisation of the spin S in the new steady state is enhanced or diminished by a fraction of the equilibrium magnetisation of spin L. When S_Z^{eq} is small (for example, ^{15}N) and L_Z^{eq} is large (for example ^{1}H or an electron), the resulting enhancement in the magnetisation of spin S can be valuable. This is illustrated in Fig. 6.8.

6.9.5.3 Quadratic Interactions

Physical origins of nuclear quadrupolar interaction and zero-field splitting are discussed in Sects. 3.2.10 and 3.2.13—electrostatic interactions manifest in the spin Hamiltonian because basis orbitals spanning the spatial part of the wavefunction become linked with specific spin states in the ground multiplet of the full electronic or nuclear structure Hamiltonian. We will confine the discussion here to the second spatial rank; higher rank NQI has not been seen experimentally yet, and higher-rank ZFS relaxation is best treated numerically using Eqs. (6.47) and (6.48). For second-rank NQI and ZFS, the Hamiltonian is

$$\mathbf{H} = \omega\mathbf{S}_Z + \mathbf{S}\cdot\mathbf{Q}\cdot\mathbf{S} \tag{6.195}$$

where the 3×3 matrix $\mathbf{Q}$ is symmetric (by the definition of electric field gradient tensor in Sect. 3.2.13) and traceless—by convention, because the trace part corresponds to $\mathbf{S}^2 = \mathbf{S}_X^2 + \mathbf{S}_Y^2 + \mathbf{S}_Z^2$ whose presence does not influence the observable dynamics because it is a multiple of a unit matrix.

Self-relaxation rates of Zeeman basis states are given in (Table 6.8); note the favourable high-field behaviour of the double-quantum and zero-quantum coherences. ZFS and NQI relaxation can be rapid enough to become sources of secondary relaxation processes; those are discussed in Sects. 6.9.5 and 6.9.7. In common magnetic resonance notations (Sect. 3.4), the amplitude multipliers are

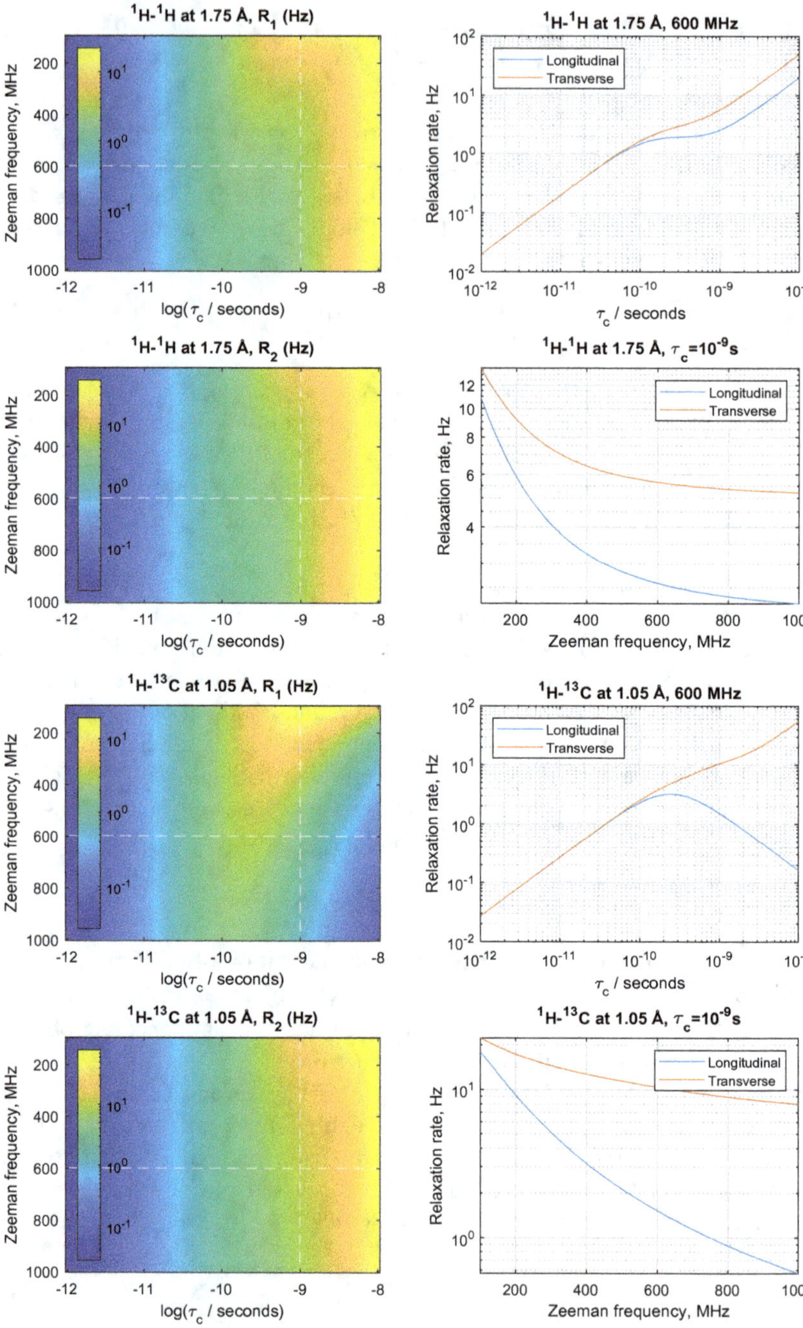

Fig. 6.7 Longitudinal and transverse dipolar relaxation rates in typical homonuclear and heteronuclear NMR spin systems, plotted as functions of magnet frequency and rotational correlation time

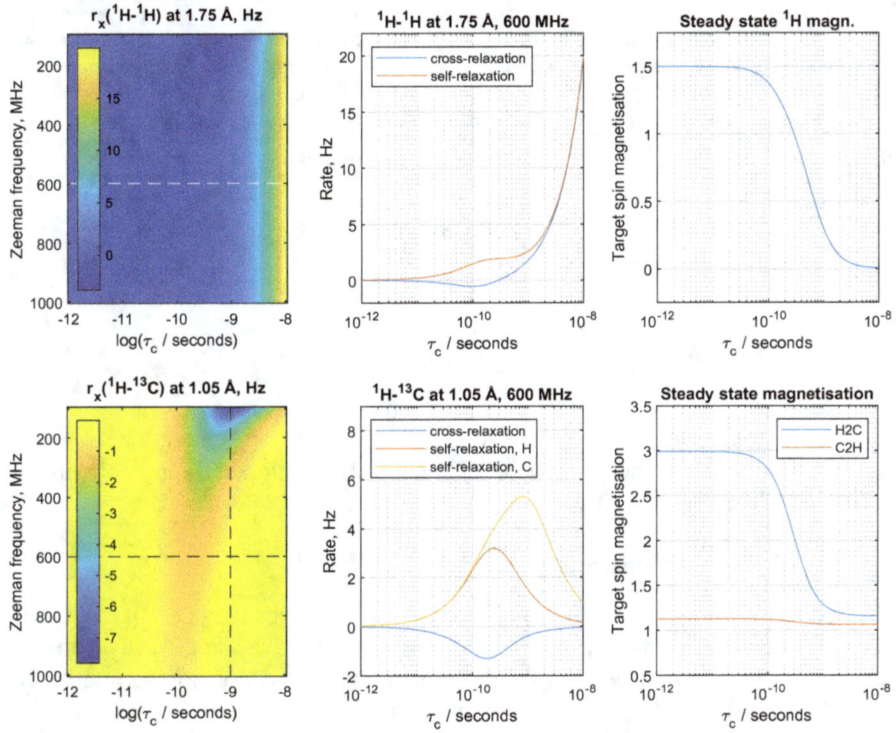

Fig. 6.8 Cross-relaxation rates as functions of proton Larmor frequency and rotational correlation time for ^{1}H-^{1}H (top left) and ^{1}H-^{13}C (bottom left) systems. Top and bottom middle panels show self- and cross-relaxation rates as functions of rotational correlation time in a popular 14.1 T (600 MHz proton Larmor frequency) magnet. Top and bottom-right panels show the steady-state magnetisation of one spin when the other is continuously saturated

Table 6.8 Redfield theory self-relaxation rates for a spin subject to a quadratic interaction, such as NQI or ZFS, modulated by isotropic rotational diffusion. S is the spin quantum number, interaction tensor invariants are defined in Eq. (6.180), and spectral density functions in Eq. (6.184)

State	Relaxation rate
S_Z	$\frac{2\Delta_Q^2}{15}(2S-1)(2S+3)(J_2(\omega_S)+4J_2(2\omega_S))$
$S_\pm$	$\frac{\Delta_Q^2}{15}(2S-1)(2S+3)(3J_2(0)+5J_2(\omega_S)+2J_2(2\omega_S))$

Table 6.9 Additional terms in the self-relaxation rates brought about by the presence of cross-correlations in a geometrically rigid system of two spin-1/2 particles (two anisotropic Zeeman interactions and a dipolar interaction) undergoing isotropic rotational diffusion relaxation. Interaction tensor invariants are defined in Eq. (6.180) and spectral density functions in Eq. (6.184) sign convention is such that pre-existing self-relaxation rates are positive

State	Additional relaxation rate
$\mathbf{L}_{\pm}\mathbf{S}_{\pm}$	$+\frac{4}{3}\aleph_{\mathbf{Z}_{\mathrm{L}},\mathbf{Z}_{\mathrm{S}}}B_0^2 J_2(0)$
$\mathbf{L}_{\pm}\mathbf{S}_{\mp}$	$-\frac{4}{3}\aleph_{\mathbf{Z}_{\mathrm{L}},\mathbf{Z}_{\mathrm{S}}}B_0^2 J_2(0)$

$$\Delta_{\mathrm{NQI}}^2 = \frac{3C^2(3+\eta^2)}{16S^2(2S-1)^2}, \qquad \Delta_{\mathrm{ZFS}}^2 = D^2 + 3E^2 \tag{6.196}$$

Systems relaxing through the zero-field splitting mechanism are rarely in the Zeeman limit and the modulation of ZFS is rarely rotational; expressions in this section should be used with care.

6.9.5.4 Cross-Correlations

Rotational diffusion does not drive *all* correlation functions to zero—in a tumbling rigid molecule, different spin interaction tensors may have fixed relative orientations: while the global rotation of the molecule is stochastic, positioning rotations of interaction tensors in the molecular frame of reference are not [232]. This observation finds much use in protein NMR spectroscopy, where the cross-correlation between dipole coupling and chemical shift anisotropy creates a line narrowing effect [233].

Because Redfield theory (Sect. 6.3) truncates the Dyson series (Sect. 4.4.3) at the second order, only pair-wise cross-correlations need to be considered. The Hamiltonian has the form obtained in Sect. 3.2.4:

$$\mathbf{H}(t) = \mathbf{H}_{\mathrm{iso}} + \sum_{lkm} \mathfrak{D}_{km}^{(l)}(t)\mathbf{Q}_{km}^{(l)} \tag{6.197}$$

where the rotational basis operators $\mathbf{Q}_{km}^{(l)}$ in the anisotropic part have internal rotations incorporated into their definitions (Sect. 3.2.4). The logistics of setting these operators up for a specific spin system are exceedingly tedious; subsequent Redfield theory processing heavier still—machine algebra or numerical methods (both available in *Spinach*) are recommended in practice.

Of the many types of cross-correlations, only one is widely used: between dipole interaction (inter-nuclear, inter-electron, and electron-nuclear) and second-rank Zeeman interaction (*g*-tensor and chemical shift tensor) anisotropy in spin-1/2 systems [233]. We, therefore, restrict our attention here to systems of spin-1/2 particles with symmetric coupling tensors and consider the following spin Hamiltonian:

$$\mathbf{H} = \mathbf{L} \cdot \mathbf{Z}_{\mathrm{L}} \cdot \mathbf{B} + \mathbf{S} \cdot \mathbf{Z}_{\mathrm{S}} \cdot \mathbf{B} + \mathbf{L} \cdot \mathbf{D} \cdot \mathbf{S} \tag{6.198}$$

where $\mathbf{Z}_{\mathrm{L,S}}$ are Zeeman tensors of the two spins, $\mathbf{D}$ is the dipolar interaction tensor, and $\mathbf{B} = [0\ 0\ B_0]^{\mathrm{T}}$ is the external magnetic field. There are two types of cross-correlations in this system: CSA-CSA (not interesting) and DD-CSA which we shall discuss in detail.

In the product basis set, the self-relaxation rates are mostly unaffected with only the flip-flip and flip-flop operators relaxing faster or slower depending on the sign of $\aleph_{\mathbf{Z}_{\mathrm{L}},\mathbf{Z}_{\mathrm{S}}}$ (Table 6.9), but multiple additional cross-relaxation terms appear (Table 6.10). Consider the difference that the corrections in Tables 6.9 and 6.10 create in the transverse relaxation rates of the left and right doublet components of the L spin when the system has a weak scalar coupling:

$$\begin{aligned} & -\frac{\langle \mathbf{L}_{+} \pm 2\mathbf{L}_{+}\mathbf{S}_{\mathrm{Z}} | \mathfrak{R} | \mathbf{L}_{+} \pm 2\mathbf{L}_{+}\mathbf{S}_{\mathrm{Z}} \rangle}{\langle \mathbf{L}_{+} \pm 2\mathbf{L}_{+}\mathbf{S}_{\mathrm{Z}} \mid \mathbf{L}_{+} \pm 2\mathbf{L}_{+}\mathbf{S}_{\mathrm{Z}} \rangle} \\ & = \frac{\Delta_{\mathbf{D}}^2 \pm 24\aleph_{\mathbf{Z}_{\mathrm{L}},\mathbf{D}} B_0 + 4\Delta_{\mathbf{Z}_{\mathrm{L}}}^2 B_0^2}{144} (4J_2(0) + 3J_2(\omega_{\mathrm{L}})) \\ & + \frac{\Delta_{\mathbf{Z}_{\mathrm{S}}}^2 B_0^2}{12} J_2(\omega_{\mathrm{S}}) + \frac{\Delta_{\mathbf{D}}^2}{144} (J_2(\omega_{\mathrm{L}} - \omega_{\mathrm{S}}) + 6J_2(\omega_{\mathrm{S}}) + 6J_2(\omega_{\mathrm{L}} + \omega_{\mathrm{S}})) \end{aligned} \tag{6.199}$$

A famously problematic regime of NMR spectroscopy is when the molecule is so large that $J_2(0) \propto \tau_{\mathrm{C}}$ terms dominate the spectral densities and the signal is broadened beyond detection. However, one of the two doublet components in Eq. (6.199) is narrower than the other. If we pick the narrow component and choose the magnetic field to be positive, the multiplier in front of the $J_2(0)$ term in Eq. (6.199) implies that there exists an optimal magnetic field that would yield the narrowest signal:

$$\begin{aligned} & \frac{\partial}{\partial B_0} \left(\Delta_{\mathbf{D}}^2 - 24 \left| \aleph_{\mathbf{Z}_{\mathrm{L}},\mathbf{D}} \right| B_0 + 4\Delta_{\mathbf{Z}_{\mathrm{L}}}^2 B_0^2 \right) = 8\Delta_{\mathbf{Z}_{\mathrm{L}}}^2 B_0 - 24 \left| \aleph_{\mathbf{Z}_{\mathrm{L}},\mathbf{D}} \right| \\ & \frac{\partial}{\partial B_0} (\cdots) = 0 \quad \Rightarrow \quad B_0^{\mathrm{opt}} = 3 \left| \aleph_{\mathbf{Z}_{\mathrm{L}},\mathbf{D}} \right| \Big/ \Delta_{\mathbf{Z}_{\mathrm{L}}}^2 \end{aligned} \tag{6.200}$$

This point is always a minimum of the transverse relaxation rate because:

$$\frac{\partial^2}{\partial B_0 \partial B_0} \left(\Delta_{\mathbf{D}}^2 \pm 24\aleph_{\mathbf{Z}_{\mathrm{L}},\mathbf{D}} B_0 + 4\Delta_{\mathbf{Z}_{\mathrm{L}}}^2 B_0^2 \right) = 8\Delta_{\mathbf{Z}_{\mathrm{L}}}^2 > 0 \tag{6.201}$$

This minimum is shown in Fig. 6.9 for the popular case of the $^{15}\mathrm{N}$–$^{1}\mathrm{H}$ spin pair in the N–H group of the peptide bond of $^{15}\mathrm{N}$-labelled proteins [234]. By a fortuitous coincidence (since neither the bond length nor the shielding tensor are in practice controllable), this minimum occurs at magnetic fields corresponding to 1.0 GHz proton Larmor frequency—the sweet spot of magnetic resonance spectroscopy magnet, electronics, and probe technologies.

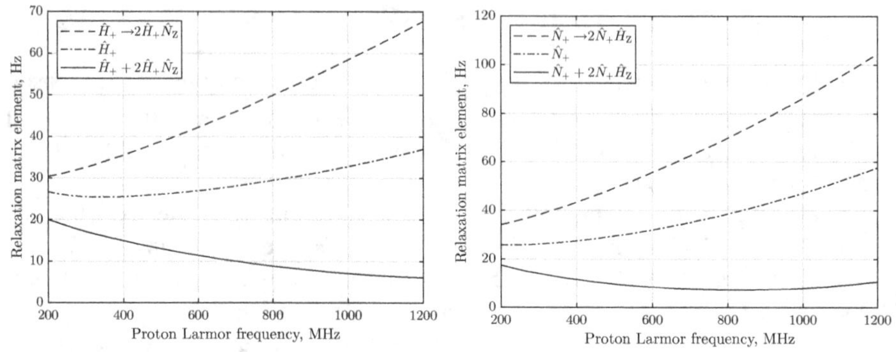

Fig. 6.9 Relaxation rates for transverse magnetisation (dot-dashed line) and two components of the *J*-coupled doublet (dashed and solid lines) in a peptide bond ^{15}N–^{1}H pair as a function of the magnet field (shown in units of proton Larmor frequency)

Table 6.10 Additional terms in the cross-relaxation rates brought about by the presence of cross-correlations in a geometrically rigid system of two spin-1/2 particles (two anisotropic Zeeman interactions and a dipolar interaction) undergoing isotropic rotational diffusion relaxation. Interaction tensor invariants are defined in Eq. (6.180) and spectral density functions in Eq. (6.184). Sign convention is such that self-relaxation rates are positive

A	B	Additional cross-relaxation rate
$\mathbf{L}_Z$	$\mathbf{L}_Z\mathbf{S}_Z$	$-\aleph_{\mathbf{Z}_L,\mathbf{D}}B_0J_2(\omega_L)$
	$\mathbf{L}_\pm\mathbf{S}_\mp$	$+\frac{1}{6}\aleph_{\mathbf{Z}_S-\mathbf{Z}_L,\mathbf{D}}B_0J_2(0)+\frac{1}{4}\aleph_{\mathbf{Z}_S+\mathbf{Z}_L,\mathbf{D}}B_0J_2(\omega_S)$
$\mathbf{S}_Z$	$\mathbf{L}_Z\mathbf{S}_Z$	$-\aleph_{\mathbf{Z}_S,\mathbf{D}}B_0J_2(\omega_S)$
	$\mathbf{L}_\pm\mathbf{S}_\mp$	$+\frac{1}{6}\aleph_{\mathbf{Z}_L-\mathbf{Z}_S,\mathbf{D}}B_0J_2(0)+\frac{1}{4}\aleph_{\mathbf{Z}_L+\mathbf{Z}_S,\mathbf{D}}B_0J_2(\omega_L)$
$\mathbf{L}_\pm$	$\mathbf{L}_Z\mathbf{S}_\pm$	$+\frac{1}{12}\aleph_{\mathbf{Z}_L,\mathbf{D}}B_0(3J_2(\omega_L)-2J_2(\omega_L-\omega_S))-\frac{1}{12}\aleph_{\mathbf{Z}_S,\mathbf{D}}B_0(2J_2(0)+3J_2(\omega_S))$
	$\mathbf{L}_\pm\mathbf{S}_Z$	$-\frac{1}{6}\aleph_{\mathbf{Z}_L,\mathbf{D}}B_0(4J_2(0)+3J_2(\omega_L))$
$\mathbf{S}_\pm$	$\mathbf{L}_\pm\mathbf{S}_Z$	$+\frac{1}{12}\aleph_{\mathbf{Z}_S,\mathbf{D}}B_0(3J_2(\omega_S)-2J_2(\omega_L-\omega_S))-\frac{1}{12}\aleph_{\mathbf{Z}_L,\mathbf{D}}B_0(2J_2(0)+3J_2(\omega_L))$
	$\mathbf{L}_Z\mathbf{S}_\pm$	$-\frac{1}{6}\aleph_{\mathbf{Z}_S,\mathbf{D}}B_0(4J_2(0)+3J_2(\omega_S))$
$\mathbf{L}_Z\mathbf{S}_\pm$	$\mathbf{L}_\pm\mathbf{S}_Z$	$-\frac{1}{2}\aleph_{\mathbf{Z}_L,\mathbf{Z}_S}B_0^2(J_2(\omega_L)+J_2(\omega_S))$
$\mathbf{L}_\pm\mathbf{S}_\mp$	$\mathbf{L}_Z\mathbf{S}_Z$	$+\frac{1}{2}\aleph_{\mathbf{Z}_L,\mathbf{Z}_S}B_0^2(J_2(\omega_L)+J_2(\omega_S))$

6.9.6 Nuclear Relaxation by Rapidly Relaxing Electrons

A common situation in paramagnetic NMR spectroscopy is the presence of a rapidly (femtoseconds to picoseconds) relaxing electron in a *d*- or *f*- orbital of a metal ion which presents itself in two ways: (1) as a magnetic susceptibility centre that shields the surrounding nuclei; (2) as a stochastic magnetic moment on the electron side of the nuclear-electron dipolar interaction. When electron spin dynamics and relaxation are both much faster than nuclear dynamics, the nuclear relaxation problem is an instance of Hubbard theory (Sect. 6.2) with the electron paying the role of the bath. The words "electron" and "spin" in such systems are loosely defined—multiple unpaired electrons may be present and "spin" is the total angular momentum of a particular multiplet (Sect. 3.2.10). The effective electron spin Hamiltonian

$$\mathbf{H}_{\mathrm{E}} = \mathbf{H}_{\mathrm{Zeeman}} + \mathbf{H}_{\mathrm{ZFS}} \tag{6.202}$$

contains a Zeeman interaction term (1–10 cm^{-1} in common NMR fields) and a zero-field splitting term (100–1000 cm^{-1}), both with stochastic components that cause electron relaxation. The large amplitude of both terms and the picosecond time scale of the resulting electron relaxation mean that the effect of nuclei on the electron spin dynamics may be ignored—this is a requirement for the bath within Hubbard theory.

The electron is then seen by the nucleus as a time-dependent magnetic moment $\boldsymbol{\mu}_{\mathrm{E}}(t)$ with the average value $\boldsymbol{\mu}^{(\mathrm{eq})}$ described by a magnetic susceptibility tensor $\boldsymbol{\chi}$ (Sect. 3.1.4), and some dynamics $\boldsymbol{\mu}(t)$ around that average, determined by the stochastic parts of Zeeman and ZFS terms in Eq. (6.202):

$$\boldsymbol{\mu}_{\mathrm{E}}(t) = \boldsymbol{\mu}^{(\mathrm{eq})} + \boldsymbol{\mu}(t), \qquad \boldsymbol{\mu}^{(\mathrm{eq})} = \boldsymbol{\chi} \cdot \mathbf{B}/\mu_0 \tag{6.203}$$

where the problem of calculating the explicit form of $\boldsymbol{\mu}(t)$ is impractically difficult because the electron Hamiltonian in Eq. (6.202) is modulated by molecular vibrations, libration of coordinated solvent and other mechanisms with unknown statistics. With Eq. (6.203) in place, the nuclear spin Hamiltonian becomes:

$$\mathbf{H}_{\mathrm{N}} = -\gamma_{\mathrm{N}} \mathbf{S} \cdot (\mathbf{1} + \boldsymbol{\delta}_0) \cdot \mathbf{B} - \gamma_{\mathrm{N}} \mu_0 \mathbf{S} \cdot \mathbf{D} \cdot \boldsymbol{\mu}_{\mathrm{E}}(t) \tag{6.204}$$

where $\mathbf{S}$ is a vector of nuclear spin operators, $\boldsymbol{\delta}_0$ is the chemical shift tensor due to the local electronic structure around the nucleus, and $\mathbf{D}$ is the dipole coupling matrix between the nucleus and the rapidly relaxing electron:

$$\mathbf{D} = \frac{1}{4\pi} \left[3 \frac{\mathbf{r} \cdot \mathbf{r}^{\mathrm{T}}}{r^5} - \frac{\mathbf{1}}{r^3} \right] \tag{6.205}$$

in which $\mathbf{r}$ is the electron-nucleus distance vector; we assume the electron to be far enough for the point dipole approximation to apply. We split $\boldsymbol{\mu}_{\mathrm{E}}(t)$ according to Eq. (6.203) and rearrange the Hamiltonian:

$$\mathbf{H}_{\mathrm{N}} = -\gamma_{\mathrm{N}} \mathbf{S} \cdot (\mathbf{1} + \boldsymbol{\delta}_0 + \mathbf{D} \cdot \boldsymbol{\chi}) \cdot \mathbf{B} - \gamma_{\mathrm{N}} \mu_0 \mathbf{S} \cdot \mathbf{D} \cdot \boldsymbol{\mu}(t) \tag{6.206}$$

The contribution to the chemical shift tensor from the magnetic susceptibility is now explicit. When electron relaxation is much faster than molecular rotation, the two terms in this Hamiltonian are statistically independent because the dynamics in $\boldsymbol{\mu}(t)$ is uncorrelated with rotational diffusion. We can, therefore, perform relaxation theory treatment for the two terms separately.

6.9.6.1 Contact Mechanism

This is an instance of scalar relaxation of the second kind (Sect. 6.9.4). When an isotropic hyperfine interaction exists between the electron L and the nucleus S:

$$\mathbf{H} = \omega_{\mathrm{L}} \mathbf{L}_{\mathrm{Z}} + \omega_{\mathrm{S}} \mathbf{S}_{\mathrm{Z}} + a \mathbf{L} \cdot \mathbf{S} \tag{6.207}$$

Exactly the same treatment as in Sect. 6.9.5 yields the following nuclear relaxation rates:

$$\begin{aligned} R_{\mathrm{Z}} &= \frac{a^2 L(L+1)}{3} \frac{2T_{2\mathrm{L}}}{1+(\omega_{\mathrm{L}} - \omega_{\mathrm{S}})^2 T_{2\mathrm{L}}^2} \\ R_{\mathrm{X,Y}} &= \frac{a^2 L(L+1)}{3} \left(T_{1\mathrm{L}} + \frac{T_{2\mathrm{L}}}{1+(\omega_{\mathrm{L}} - \omega_{\mathrm{S}})^2 T_{2\mathrm{L}}^2} \right) \end{aligned} \tag{6.208}$$

where L is the effective spin of the electron, $T_{1\mathrm{L}}$ is its longitudinal relaxation time and $T_{2\mathrm{L}}$ is the transverse relaxation time. Because contact interaction is involved (Sect. 3.1.5), this is called *contact relaxation*.

6.9.6.2 Curie Mechanism

The appearance of Eq. (6.206) suggests re-defining the chemical shift tensor as follows:

$$\boldsymbol{\delta} = \boldsymbol{\delta}_0 + \mathbf{D} \cdot \boldsymbol{\chi} \tag{6.209}$$

For a rigid molecule in a liquid, the stochastic time dependence in $\boldsymbol{\delta}_0$, $\boldsymbol{\chi}$ and $\mathbf{D}$ comes from rotational diffusion which we assume to be much faster than nuclear spin dynamics in the Zeeman rotating frame. We then have an instance of CSA relaxation (Sect. 6.9.5.1) with the chemical shift tensor now defined by Eq. (6.209). The resulting relaxation theory expressions are identical to those given in Table 6.3. This is called *Curie relaxation* [235, 236].

Compared to the diamagnetic CSA mechanism, the Curie mechanism has two notable features. Firstly, the magnetic susceptibility contribution to the chemical shift tensor

$$\mathbf{D} \cdot \boldsymbol{\chi} = \left[\mathbf{D} \cdot \chi_{\mathrm{iso}} + \mathbf{D} \cdot \boldsymbol{\chi}_{\mathrm{aniso}} \right] \tag{6.210}$$

has a significant antisymmetric component: the product of two symmetric matrices is only symmetric when they commute, which is not, in general, the case for $\mathbf{D}$ and $\boldsymbol{\chi}_{\mathrm{aniso}}$. In the absence of unpaired electrons, antisymmetric components of chemical shift tensors are usually negligible; that is not the case here. Secondly, the presence of the $\mathbf{D} \cdot \boldsymbol{\chi}_{\mathrm{aniso}}$ term means that nuclear relaxation rates depend on the relative orientation of dipolar and susceptibility tensors, and therefore on the direction of the electron-nuclear vector in the susceptibility tensor reference frame [237].

6.9.6.3 Dipolar Mechanism: Perturbative Treatment

When electron dynamics and relaxation are much faster than molecular rotation, and ZFS (which follows the molecular frame of reference) is much stronger than Zeeman interaction, the dynamics of the electron magnetic moment $\boldsymbol{\mu}(t)$ also follows the molecular frame of reference. Therefore, the dipolar term in Eq. (6.206) transforms as follows under a molecular rotation with a direction cosine matrix $\mathbf{R}$:

$$\mathbf{S}\cdot\mathbf{D}\cdot\boldsymbol{\mu}(t) \xrightarrow{\mathbf{R}} \mathbf{S}\cdot\mathbf{R}\mathbf{D}\mathbf{R}^{\mathrm{T}}\mathbf{R}\cdot\boldsymbol{\mu}(t)=\mathbf{S}\cdot\mathbf{R}\mathbf{D}\cdot\boldsymbol{\mu}(t) \tag{6.211}$$

The Hamiltonian seen by the nucleus, therefore, is

$$\mathbf{H}_0=\omega_{\mathrm{N}}\mathbf{S}_{\mathrm{Z}}, \qquad \mathbf{H}_1(t)=-\gamma_{\mathrm{N}}\mu_0\mathbf{S}\cdot\mathbf{R}(t)\mathbf{D}\cdot\boldsymbol{\mu}(t) \tag{6.212}$$

where $\mathbf{D}$ is defined in Eq. (6.205). From the point of view of the nuclear spin, the Hamiltonian in Eq. (6.212) is therefore a coupling to a stochastic external vector:

$$\begin{aligned}&\mathbf{H}_0=\omega_{\mathrm{N}}\mathbf{S}_{\mathrm{Z}}, \quad \mathbf{H}_1(t)=b_{\mathrm{X}}(t)\mathbf{S}_{\mathrm{X}}+b_{\mathrm{Y}}(t)\mathbf{S}_{\mathrm{Y}}+b_{\mathrm{Z}}(t)\mathbf{S}_{\mathrm{Z}}\\ &\mathbf{b}(t)=-\gamma_{\mathrm{N}}\mu_0\mathbf{R}(t)\mathbf{D}\cdot\boldsymbol{\mu}(t), \qquad \|\mathbf{H}_1(t)\|_2\ll\|\mathbf{H}_0\|_2\end{aligned} \tag{6.213}$$

This case was dealt with in Sect. 6.9.2; we must now find the correlation functions in Eq. (6.168). Inserting the Cartesian components of $\mathbf{b}(t)$ and opening up matrix products yields

$$\langle b_k(t)b_n(t+\tau)\rangle=\gamma_{\mathrm{N}}^2\mu_0^2\sum_{\alpha\beta\varepsilon\lambda}\langle r_{k\alpha}(t)d_{\alpha\beta}\mu_\beta(t)r_{n\varepsilon}(t+\tau)d_{\varepsilon\lambda}\mu_\lambda(t+\tau)\rangle \tag{6.214}$$

where r_{ij} are elements of $\mathbf{R}$ and d_{ij} are elements of $\mathbf{D}$. By our assumption, the dynamics of the electron magnetic moment is uncorrelated with molecular rotation, therefore the average splits:

$$\langle b_k(t)b_n(t+\tau)\rangle=\gamma_{\mathrm{N}}^2\mu_0^2\sum_{\alpha\beta\varepsilon\lambda}d_{\alpha\beta}d_{\varepsilon\lambda}\langle r_{k\alpha}(t)r_{n\varepsilon}(t+\tau)\rangle\langle\mu_\beta(t)\mu_\lambda(t+\tau)\rangle \tag{6.215}$$

Correlation functions between the elements of the rotation matrix are given in Sect. 6.4.1:

$$\langle r_{k\alpha}(t)r_{n\varepsilon}(t+\tau)\rangle=\frac{\delta_{kn}\delta_{\alpha\varepsilon}}{3}\mathrm{e}^{-\tau/\tau_{\mathrm{C}}^{(1)}} \tag{6.216}$$

where $\tau_{\mathrm{C}}^{(1)}$ is the first rank rotational correlation time defined in Eq. (6.185). Using this and the fact that $\mathbf{D}$ is a symmetric matrix simplifies the sum:

$$\langle b_k(t)b_n(t+\tau)\rangle=\frac{\gamma_{\mathrm{N}}^2\mu_0^2\delta_{kn}}{3}\mathrm{e}^{-\tau/\tau_{\mathrm{C}}^{(1)}}\sum_{\alpha\beta\varepsilon\lambda}d_{\lambda\alpha}d_{\alpha\beta}\langle\mu_\beta(t)\mu_\lambda(t+\tau)\rangle \tag{6.217}$$

With this in place, our autocorrelation functions become

$$\langle b_k(t)b_n(t+\tau)\rangle = \frac{\gamma_N^2\mu_0^2\delta_{kn}}{3}\mathrm{Tr}\left[\mathbf{D}^2\mathbf{G}(\tau)\right]\mathrm{e}^{-\tau/\tau_C^{(1)}} \tag{6.218}$$

where the autocorrelation tensor of the electron magnetic dipole vector is defined as

$$\mathbf{G}(\tau) = \left\langle \boldsymbol{\mu}(t)\boldsymbol{\mu}^{\mathrm{T}}(t+\tau)\right\rangle \tag{6.219}$$

All information about the dynamics of the electron magnetic dipole is now collected inside $\mathbf{G}(\tau)$. After writing the dipolar matrix out explicitly, we find

$$\langle b_k(t)b_n(t+\tau)\rangle = \frac{1}{3}\left(\frac{\mu_0}{4\pi}\right)^2\frac{\gamma_N^2\delta_{kn}}{r^6}\mathrm{Tr}\left[\left(3\hat{\mathbf{r}}\cdot\hat{\mathbf{r}}^{\mathrm{T}}-\mathbf{1}\right)^2\mathbf{G}(\tau)\right]\mathrm{e}^{-\tau/\tau_C^{(1)}} \tag{6.220}$$

where $\hat{\mathbf{r}}$ is the unit vector pointing in the same direction as $\mathbf{r}$. As a result, in Eq. (6.167):

$$\begin{aligned} g_{XX}(\tau) &= g_{YY}(\tau) = g_{ZZ}(\tau) = g(\tau) \\ &= \frac{1}{3}\left(\frac{\mu_0}{4\pi}\right)^2\frac{\gamma_N^2}{r^6}\mathrm{Tr}\left[\left(3\hat{\mathbf{r}}\cdot\hat{\mathbf{r}}^{\mathrm{T}}-\mathbf{1}\right)^2\mathbf{G}(\tau)\right]\mathrm{e}^{-\tau/\tau_C^{(1)}} \end{aligned} \tag{6.221}$$

and the resulting relaxation rates are

$$\begin{aligned} R_Z &= 2\int_0^\infty g(\tau)\cos(\omega_L\tau)\mathrm{d}\tau \\ R_{X,Y} &= \int_0^\infty g(\tau)[1+\cos(\omega_L\tau)]\mathrm{d}\tau \end{aligned} \tag{6.222}$$

After substitution and simplification, the following equations emerge for the relaxation rates:

$$\begin{aligned} R_1^{\mathrm{dip}} &= \frac{2}{3}\left(\frac{\mu_0}{4\pi}\right)^2\frac{\gamma_N^2}{r^6}\mathrm{Tr}\left[\left(3\hat{r}\cdot\hat{r}^{\mathrm{T}}-\mathbf{1}\right)^2\mathbf{G}(\omega_N)\right] \\ R_2^{\mathrm{dip}} &= \frac{1}{3}\left(\frac{\mu_0}{4\pi}\right)^2\frac{\gamma_N^2}{r^6}\mathrm{Tr}\left[\left(3\hat{r}\cdot\hat{r}^{\mathrm{T}}-\mathbf{1}\right)^2(\mathbf{G}(0)+\mathbf{G}(\omega_N))\right] \end{aligned} \tag{6.223}$$

where $\mathbf{G}(\omega)$ is the spectral density tensor that stores the statistics of both stochastic processes involved—the dynamics of the electron magnetic dipole and the molecular rotation:

$$\mathbf{G}(\omega) = \int_0^\infty \mathbf{G}(\tau)\cos(\omega\tau)e^{-\tau/\tau_C^{(1)}}\,d\tau \tag{6.224}$$

This matrix would be exceedingly hard to obtain from first principles; in practical calculations, $\mathbf{G}(0)$ and $\mathbf{G}(\omega_N)$ are best treated as fitting variables.

6.9.6.4 Dipolar Mechanism: Adiabatic Elimination

There are situations when both electron and nuclear spin Hamiltonians are so complicated (high-rank ZFS, significant contact coupling, significantly non-point electron-nuclear dipolar interaction, etc.) that there is no reasonable prospect of obtaining analytical theories of the kind described in the previous section. However, time scale separations mentioned there still hold: electron dynamics and relaxation are fast, molecular rotation much slower, and nuclear relaxation much slower still. The best course of action in such circumstances is adiabatic elimination (Sect. 6.1).

The assumption that electron relaxation is faster than molecular rotation allows us to remove the thermal equilibrium state from the electron equation of motion by setting $\boldsymbol{\sigma}^{(E)}(\Omega,t) = \boldsymbol{\rho}^{(E)}(\Omega,t) - \boldsymbol{\rho}_{eq}^{(E)}(\Omega)$:

$$\begin{gathered}
\frac{d}{dt}\boldsymbol{\rho}^{(E)}(\Omega,t) = -i\left[\mathcal{H}_Z(\Omega) + \mathcal{H}_{ZFS}^{(0)}(\Omega)\right]\boldsymbol{\rho}^{(E)}(\Omega,t) + \mathcal{R}\left[\boldsymbol{\rho}^{(E)}(\Omega,t) - \boldsymbol{\rho}_{eq}^{(E)}(\Omega)\right] \\
\Downarrow \\
\frac{d}{dt}\boldsymbol{\sigma}^{(E)}(\Omega,t) = -i\left[\mathcal{H}_Z(\Omega) + \mathcal{H}_{ZFS}^{(0)}(\Omega) + i\mathcal{R}(\Omega)\right]\boldsymbol{\sigma}^{(E)}(\Omega,t)
\end{gathered} \tag{6.225}$$

where $\mathcal{H}$ are Hamiltonian commutation superoperators (Zeeman and the static part of the ZFS), and $\mathcal{R}$ is the relaxation superoperator. This is an exact transformation because the definition of thermal equilibrium implies that the electron density matrix commutes with the Hamiltonian:

$$\left[\mathcal{H}_Z(\Omega) + \mathcal{H}_{ZFS}^{(0)}(\Omega)\right]\boldsymbol{\rho}_{eq}^{(E)}(\Omega) = 0 \tag{6.226}$$

The dependence of all three operators on the molecular orientation Ω is quoted to emphasize that the dynamics in Eq. (6.225) and the establishment of Eq. (6.226) are much faster than molecular rotation.

Insofar as nuclear relaxation is concerned, the effect of $\boldsymbol{\rho}_{eq}^{(E)}$ is already captured by the Curie mechanism (Sect. 6.9.6.2), which remains unchanged—the numerical scenario in this section is an alternative to the dipolar mechanism treatment. As per the adiabatic elimination procedure (Sect. 6.1), we proceed to partition the state space $\mathfrak{L}$ into the pure nuclear subspace $\mathfrak{N}$ (all states with the unit operator on the electron) and its complement $\mathfrak{L}/\mathfrak{N}$ (all states involving an electron in any way). The projector from $\mathfrak{L}$ into $\mathfrak{N}$ will be denoted $\mathcal{P}_{\mathfrak{N}}$. The complete Liouvillian of the system

$$\begin{aligned}\mathcal{L}(\Omega) = {} & \left[\mathcal{H}_{\mathrm{Z}}^{(\mathrm{E})}(\Omega) + \mathcal{H}_{\mathrm{ZFS}}^{(0)}(\Omega) + i\mathcal{R}^{(\mathrm{E})}(\Omega)\right] \otimes \mathbf{1}^{(\mathrm{N})} \\ & + \mathcal{H}_{\mathrm{HFC}}(\Omega) + \mathbf{1}^{(\mathrm{E})} \otimes \mathcal{H}^{(\mathrm{N})}(\Omega)\end{aligned} \tag{6.227}$$

is then generated using standard methods discussed in Chap. 3, the simplest way is to use *Spinach* which would simply return the required matrices. Note that the nuclear relaxation superoperator is not present at this point—but will appear once we run the adiabatic elimination of the electron degrees of freedom.

Performing projections yields the blocks required by Eq. (6.1):

$$\begin{aligned} \boldsymbol{\sigma}_0 = \mathcal{P}_{\mathfrak{N}}\boldsymbol{\sigma}, \quad & \boldsymbol{\sigma}_1 = (\mathbf{1} - \mathcal{P}_{\mathfrak{N}})\boldsymbol{\sigma} \\ \mathcal{L}_{00} = \mathcal{P}_{\mathfrak{N}}\mathcal{L}\mathcal{P}_{\mathfrak{N}}^{\dagger}, \quad & \mathcal{L}_{01} = \mathcal{P}_{\mathfrak{N}}\mathcal{L}(\mathbf{1} - \mathcal{P}_{\mathfrak{N}})^{\dagger} \\ \mathcal{L}_{01} = (\mathbf{1} - \mathcal{P}_{\mathfrak{N}})\mathcal{L}\mathcal{P}_{\mathfrak{N}}^{\dagger}, \quad & \mathcal{L}_{11} = (\mathbf{1} - \mathcal{P}_{\mathfrak{N}})\mathcal{L}(\mathbf{1} - \mathcal{P}_{\mathfrak{N}})^{\dagger} \end{aligned} \tag{6.228}$$

All assumptions made therein apply without changes; the resulting equation of motion in the nuclear subspace follows from Eq. (6.2):

$$\frac{\mathrm{d}}{\mathrm{d}t}\boldsymbol{\rho}^{(\mathrm{N})} = -i\mathcal{H}^{(\mathrm{N})}\boldsymbol{\rho}^{(\mathrm{N})} + i\mathcal{L}_{01}\mathcal{L}_{11}^{-1}\mathcal{L}_{10}\boldsymbol{\rho}^{(\mathrm{N})} \tag{6.229}$$

Because nuclear spin dynamics is much slower than molecular rotation, nuclear relaxation superoperator is obtained by averaging the dissipative $i\mathcal{L}_{01}\mathcal{L}_{11}^{-1}\mathcal{L}_{10}$ term over all orientations of the system [238].

6.9.7 Spin-Phonon Relaxation

It stands to reason that phonons introduce dynamics into molecular geometries and therefore spin Hamiltonians—the theory is essentially of a Redfield (Sect. 6.3) or Hubbard (Sect. 6.2) type with spin-displacement coupling (Sect. 5.5.3) to a phonon bath (Sect. 5.5.2). The resulting equations have some explanatory power—temperature and field dependences can be fitted—but little predictive power: neither the phonon spectrum, nor the displacement derivatives are ever known accurately enough in a realistic solid, and real-life vibrations are rarely harmonic to begin with.

At the time of writing, ab initio calculations of vibrationally driven relaxation rates yield numbers that are orders of magnitude away from the experimental data [206]. For what they are worth, this section outlines the starting points on the way to a theory of spin-phonon relaxation.

6.9.7.1 High Temperature: Redfield-type Theories

The Taylor expansion in Eq. (5.113) suggests a model wherein the time dependence in the spin Hamiltonian comes from bilinear couplings to classical lattice displacements:

$$\mathbf{H} = \mathbf{H}_0 + \sum_\alpha x_\alpha(t)\mathbf{S}_\alpha \xrightarrow{\text{Ad}} \mathcal{H} = \mathcal{H}_0 + \sum_\alpha x_\alpha(t)\mathcal{S}_\alpha \tag{6.230}$$

where $x_\alpha(t)$ are the displacements along normal modes enumerated by the index α, and $\mathbf{S}_\alpha$ are the corresponding linear response operators of the spin system.

Position autocorrelation function for the harmonic oscillator has been obtained in Sect. 5.5.1:

$$\begin{aligned} g_\alpha(\tau) &= \langle x_\alpha(0)x_\alpha(\tau)\rangle \\ &= \frac{\hbar}{2\mu_\alpha\omega_\alpha}\left[\left(e^{\hbar\omega_\alpha/kT}-1\right)^{-1}e^{+i\omega_\alpha\tau}+\left(1-e^{-\hbar\omega_\alpha/kT}\right)^{-1}e^{-i\omega_\alpha\tau}\right]e^{-\tau/\tau_\alpha} \end{aligned} \tag{6.231}$$

It is reasonable to assume uncorrelated dynamics in different normal modes:

$$\left\langle x_\alpha(0)x_{\beta\neq\alpha}(\tau)\right\rangle = 0 \tag{6.232}$$

The introduction of the phenomenological decay time τ_α is necessary from empirical evidence, and also to satisfy the validity conditions of Redfield theory (Sect. 6.3.2). How this decay appears (phonon scattering on defects, anharmonicities, mode interactions, etc.) is an open question.

Once the correlation functions are known, the relaxation superoperator is an instance of Eq. (6.46):

$$\mathcal{R} = -\sum_\alpha \mathcal{S}_\alpha \int_0^\infty e^{-i\mathcal{H}_0\tau}\mathcal{S}_\alpha e^{+i\mathcal{H}_0\tau} g_\alpha(\tau)d\tau \tag{6.233}$$

where the integral is directly computable using the auxiliary matrix relation in Eq. (6.48). The extension to the case where the Hamiltonian has significant contributions from the next order $\frac{1}{2}\sum_{\alpha\beta} x_\alpha(t)x_\beta(t)\mathbf{S}_{\alpha\beta}$ in displacements follows the same flowchart.

As all Redfield-type relaxation superoperators do, this one requires the high-temperature approximation to be applicable to the bath; it also drives the spin system to the infinite temperature state, and therefore must be thermalised or placed into an inhomogeneous master equation (Sect. 6.9) before use.

6.9.7.2 Low Temperature: Hubbard-Type Theories

In low-temperature situations, the phonon bath must be treated quantum mechanically. We can simplify the description by noting that Eq. (6.233) ended up being a sum over lattice modes, and this will be inherited by Hubbard-type theories—we need, therefore, to only look at one mode, and the sum may be taken in the end; this is equivalent to assuming that dynamics of different bath modes is uncorrelated. Within Hubbard theory, the equivalent of Eq. (6.230), therefore, is

$$\begin{gathered}\mathbf{H}=\mathbf{H}_0^{\mathrm{S}}\otimes\mathbf{1}^{\mathrm{B}}+\mathbf{1}^{\mathrm{S}}\otimes\mathbf{H}_0^{\mathrm{B}}+\mathbf{S}_\alpha\otimes\mathbf{x}_\alpha \\ \mathbf{x}_\alpha=\sqrt{\frac{2\hbar}{\mu_\alpha\omega_\alpha}}\frac{\mathbf{a}_\alpha+\mathbf{a}_\alpha^\dagger}{2},\quad \mathbf{H}_0^{\mathrm{B}}=\sum_\alpha\omega_\alpha(\mathbf{n}_\alpha+1/2)\end{gathered} \tag{6.234}$$

where $\mathbf{H}_0^{\mathrm{S}}$ is the spin Hamiltonian at the energy minimum structure, $\mathbf{H}_0^{\mathrm{B}}$ is a matrix representation (truncated at some suitable energy level to obtain finite matrices for $\mathbf{a}_\alpha$, $\mathbf{a}_\alpha^\dagger$, and $\mathbf{n}_\alpha$) of the lattice Hamiltonian in the normal mode basis, and the remaining term couples the Hermitian position response operator $\mathbf{S}_\alpha$ to the Hermitian displacement operator $\mathbf{x}_\alpha$. Direct application of Eq. (6.21) yields

$$\frac{\partial}{\partial t}\boldsymbol{\rho}^{\mathrm{S,R}}(t)=-\int_0^t\left[\begin{array}{c}\left[\mathbf{S}_\alpha^{\mathrm{R}}(t),\mathbf{S}_\alpha^{\mathrm{R}}(t')\boldsymbol{\rho}^{\mathrm{S,R}}(t)\right]\mathrm{Tr}\left(\mathbf{x}_\alpha^{\mathrm{R}}(t)\mathbf{x}_\alpha^{\mathrm{R}}(t')\boldsymbol{\rho}_{\mathrm{eq}}^{\mathrm{B}}\right)\\ -\left[\mathbf{S}_\alpha^{\mathrm{R}}(t),\boldsymbol{\rho}^{\mathrm{S,R}}(t)\mathbf{S}_\alpha^{\mathrm{R}}(t')\right]\mathrm{Tr}\left(\mathbf{x}_\alpha^{\mathrm{R}}(t')\mathbf{x}_\alpha^{\mathrm{R}}(t)\boldsymbol{\rho}_{\mathrm{eq}}^{\mathrm{B}}\right)\end{array}\right]dt' \tag{6.235}$$

where the time origin is shifted and the integration limit extended to infinity using the arguments discussed in Sect. 6.2—note that this requires the correlation functions to decay rapidly, and they would not do that unless phenomenological correlation times are again assumed for the bath:

$$\frac{\partial}{\partial t}\boldsymbol{\rho}^{\mathrm{S,R}}(t)=-\int_0^\infty\left[\begin{array}{c}\left[\mathbf{S}_\alpha^{\mathrm{R}}(t),\mathbf{S}_\alpha^{\mathrm{R}}(t-\tau)\boldsymbol{\rho}^{\mathrm{S,R}}(t)\right]\mathrm{Tr}\left(\mathbf{x}_\alpha^{\mathrm{R}}(t)\mathbf{x}_\alpha^{\mathrm{R}}(t-\tau)\boldsymbol{\rho}_{\mathrm{eq}}^{\mathrm{B}}\right)\\ -\left[\mathbf{S}_\alpha^{\mathrm{R}}(t),\boldsymbol{\rho}^{\mathrm{S,R}}(t)\mathbf{S}_\alpha^{\mathrm{R}}(t-\tau)\right]\mathrm{Tr}\left(\mathbf{x}_\alpha^{\mathrm{R}}(t-\tau)\mathbf{x}_\alpha^{\mathrm{R}}(t)\boldsymbol{\rho}_{\mathrm{eq}}^{\mathrm{B}}\right)\end{array}\right]d\tau \tag{6.236}$$

The traces are computed using the same logic as in the previous section; a tedious direct substitution of the displacement operator definition from Eq. (6.234) followed by evolution rules from Eq. (5.103) and the fact that $\boldsymbol{\rho}_{\mathrm{eq}}^{\mathrm{B}}$ commutes with $\exp\left(\pm i\mathbf{H}_0^{\mathrm{B}}t\right)$ yields

$$\begin{aligned}\operatorname{Tr}\left(\mathbf{x}_\alpha^{\mathrm{R}}(t)\mathbf{x}_\alpha^{\mathrm{R}}(t-\tau)\boldsymbol{\rho}_{\mathrm{eq}}^{\mathrm{B}}\right) &= \frac{\hbar}{2\mu_\alpha\omega_\alpha}\left(\bar{n}_\alpha \mathrm{e}^{-i\omega_\alpha\tau} + (\bar{n}_\alpha+1)\mathrm{e}^{+i\omega_\alpha\tau}\right)\mathrm{e}^{-\tau/\tau_\alpha} \\ \operatorname{Tr}\left(\mathbf{x}_\alpha^{\mathrm{R}}(t-\tau)\mathbf{x}_\alpha^{\mathrm{R}}(t)\boldsymbol{\rho}_{\mathrm{eq}}^{\mathrm{B}}\right) &= \frac{\hbar}{2\mu_\alpha\omega_\alpha}\left((\bar{n}_\alpha+1)\mathrm{e}^{-i\omega_\alpha\tau} + \bar{n}_\alpha \mathrm{e}^{+i\omega_\alpha\tau}\right)\mathrm{e}^{-\tau/\tau_\alpha}\end{aligned} \tag{6.237}$$

where the population number expectations have been obtained in Sect. 5.5.1, and we have again introduced a phenomenological decay time τ_α required by empirical observations and the validity conditions of Hubbard theory (Sect. 6.2). Just as in the previous section, this quantity depends on too many uncontrollable factors to be predictable in practice; it should be measured or fitted.

At this point, the integral in Eq. (6.236) is computable using the auxiliary matrix relation in Eq. (6.48). The extension to the case where the Hamiltonian has significant contributions from the next order $\frac{1}{2}\sum_{\alpha\beta}\mathbf{S}_{\alpha\beta}\otimes\left(\mathbf{x}_\alpha\mathbf{x}_\beta\right)$ in displacement operators follows the same flowchart.

6.9.7.3 Populations Only: Dyson-Type Theories

Redfield and Hubbard theories return complete relaxation superoperators that include cross-relaxation terms between off-diagonal elements of the spin density matrix. A simplified description in terms of energy level transition probabilities is obtained from the same Dyson perturbation theory (Sect. 4.4.3) that is responsible for the existence of the Fermi golden rule (Sect. 4.4.4). To first and second orders:

$$\begin{aligned}W_{ba}^{(1)} &= 2\pi\sum_\alpha \left|\langle b|\mathbf{S}_\alpha|a\rangle\right|^2 j_\alpha^{(1)}(\omega_{ba}) \\ W_{ba}^{(2)} &= 2\pi\sum_{\alpha\beta}\left|\sum_c \frac{\langle b|\mathbf{S}_\alpha|c\rangle\langle c|\mathbf{S}_\beta|a\rangle}{\omega_c-\omega_a\pm\omega_\beta}\right|^2 j_{\alpha\beta}^{(2)}(\omega_{ba})\end{aligned} \tag{6.238}$$

where Greek indices run over the phonon modes, Latin indices run over spin system energy levels, and the spectral densities are Fourier transforms of the autocorrelation functions that contain the corresponding—likely unpredictable on theoretical grounds—overall decay times τ_α and $\tau_{\alpha\beta}$:

$$\begin{aligned}j_\alpha^{(1)}(\omega) &= \frac{1}{\pi}\left[\frac{\tau_\alpha}{1+(\omega-\omega_\alpha)^2\tau_\alpha^2}\bar{n}_\alpha + \frac{\tau_\alpha}{1+(\omega+\omega_\alpha)^2\tau_\alpha^2}(\bar{n}_\alpha+1)\right] \\ j_{\alpha\beta}^{(2)}(\omega) &= \frac{1}{\pi}\left[\begin{aligned}&\frac{\tau_{\alpha\beta}}{1+(\omega-\omega_\beta+\omega_\alpha)^2\tau_{\alpha\beta}^2}\bar{n}_\beta(\bar{n}_\alpha+1) + \frac{\tau_{\alpha\beta}}{1+(\omega+\omega_\beta-\omega_\alpha)^2\tau_{\alpha\beta}^2}\bar{n}_\alpha(\bar{n}_\beta+1) + \dots \\ &+\frac{\tau_{\alpha\beta}}{1+(\omega-\omega_\beta-\omega_\alpha)^2\tau_{\alpha\beta}^2}\bar{n}_\beta\bar{n}_\alpha + \frac{\tau_{\alpha\beta}}{1+(\omega+\omega_\beta+\omega_\alpha)^2\tau_{\alpha\beta}^2}(\bar{n}_\alpha+1)(\bar{n}_\beta+1)\end{aligned}\right]\end{aligned} \tag{6.239}$$

and onwards to higher orders—that research is currently ongoing [206, 239]. It bears notice that exponential decay of the correlation functions is an experimentally untested assumption.

A pessimistic view here is that the unknown quantities can only be fitted and should, therefore, be packed into as few adjustable parameters as possible. For the longitudinal relaxation time, the first- and second-order contributions yield

$$\frac{1}{T_1} = \sum_n a_n^{1-\mathrm{ph}} \left(e^{\hbar\omega_n/kT} - 1 \right)^{-1} + \sum_n a_n^{2-\mathrm{ph}} e^{\hbar\omega_n/kT} \left(e^{\hbar\omega_n/kT} - 1 \right)^{-2} \tag{6.240}$$

where all unknowns are collected into the coefficients $a_n^{1-\mathrm{ph}}$ and $a_n^{2-\mathrm{ph}}$ that are fitted to experimental data. This expression still assumes that the bath remains in the thermodynamic equilibrium.

6.9.7.4 Effects of Phonon Dynamics

The assumptions made in Redfield and Hubbard theories of spin-phonon relaxation (bath in thermal equilibrium, absence of correlation between different phonon modes, rapid loss of mode autocorrelation) are not satisfied in high-quality crystals at very low temperatures: mode populations may depart from thermal equilibrium, at which point the foundation of Hubbard theory is shattered and it becomes inapplicable. The only defensible way forward in that case is to simulate lattice dynamics explicitly as a part of the system.

6.9.8 Notes on Gas-Phase Relaxation

A lengthy catalogue of gas-phase spin relaxation mechanisms may be found in Table II of [240]—every distance-dependent interaction, every contact interaction, as well as couplings to the degrees of freedom that are spatially modulated or reset by collisions can contribute to relaxation. In this book, we cover two cases with a diffusive spatial motion that lead to simple analytical expressions.

6.9.8.1 Diffusion Through Inhomogeneous Fields

When the external magnetic field is inhomogeneous, the Brownian motion creates a noisy Zeeman interaction. To first order in the Taylor expansion, this creates a coupling between position and field through the matrix Д of Cartesian coordinate derivatives of the magnetic field:

$$\mathbf{B}(t) = \mathbf{B}_0 + \text{Д} \cdot \mathbf{r}(t), \quad \text{Д}_{nk} = \left[\partial B_n / \partial r_k \right]_{\mathbf{r}=0} \tag{6.241}$$

The laboratory frame of reference will be chosen such that $\mathbf{B}_0 = \begin{bmatrix} 0 & 0 & B_0 \end{bmatrix}^{\mathrm{T}}$. For an isotropically shielded particle with a magnetogyric ratio γ, the spin Hamiltonian, therefore, is

$$\begin{aligned} \mathbf{H}(t) &= \omega_0 \mathbf{S}_{\mathrm{Z}} + \omega_{\mathrm{X}}(t)\mathbf{S}_{\mathrm{X}} + \omega_{\mathrm{Y}}(t)\mathbf{S}_{\mathrm{Y}} + \omega_{\mathrm{Z}}(t)\mathbf{S}_{\mathrm{Z}} \\ \omega_0 &= -\gamma B_0, \qquad \boldsymbol{\omega}(t) = -\gamma \mathbf{Д} \cdot \mathbf{r}(t) \end{aligned} \tag{6.242}$$

where, in the case of isotropic translational diffusion, the Cartesian components of the stochastic frequency vector $\boldsymbol{\omega}(t)$ are mutually uncorrelated, and their average is zero. This reduces the problem to an instance of relaxation through a coupling to a random external vector (Sect. 6.9.3):

$$\frac{1}{T_1} = \int_0^\infty [g_{\mathrm{XX}}(\tau) + g_{\mathrm{YY}}(\tau)] \cos(\omega_0 \tau) d\tau \tag{6.243}$$

where the Zeeman frequency autocorrelations reduce to displacement autocorrelations:

$$\begin{aligned} g_{\mathrm{XX}}(\tau) &= \langle \omega_{\mathrm{X}}(t)\omega_{\mathrm{X}}(t+\tau) \rangle = \gamma^2 \left(Д_{\mathrm{XX}}^2 + Д_{\mathrm{XY}}^2 + Д_{\mathrm{XZ}}^2 \right) \langle x(t)x(t+\tau) \rangle \\ g_{\mathrm{YY}}(\tau) &= \langle \omega_{\mathrm{Y}}(t)\omega_{\mathrm{Y}}(t+\tau) \rangle = \gamma^2 \left(Д_{\mathrm{YX}}^2 + Д_{\mathrm{YY}}^2 + Д_{\mathrm{YZ}}^2 \right) \langle y(t)y(t+\tau) \rangle \end{aligned} \tag{6.244}$$

Their Fourier transforms are a long story that falls outside the scope of this book [241]. For isotropic translational diffusion in a gas with molecular mass m and average time τ_{C} between collisions:

$$\int_0^\infty \langle x(t)x(t+\tau) \rangle \cos(\omega_0 \tau) d\tau = \frac{kT}{m} \frac{\tau_{\mathrm{C}}}{\omega_0^2 (1 + \tau_{\mathrm{C}}^2 \omega_0^2)} \tag{6.245}$$

and likewise for $g_{YY}(\tau)$. The result is the following expression for the longitudinal relaxation time:

$$\begin{aligned} \frac{1}{T_1} &= \frac{\gamma^2 kT}{m} \left(Д_{\mathrm{X}}^2 + Д_{\mathrm{Y}}^2 \right) \frac{\tau_{\mathrm{C}}}{\omega_0^2 \left(1 + \tau_{\mathrm{C}}^2 \omega_0^2 \right)} \\ Д_{\mathrm{X}}^2 &= Д_{\mathrm{XX}}^2 + Д_{\mathrm{XY}}^2 + Д_{\mathrm{XZ}}^2, \qquad Д_{\mathrm{Y}}^2 = Д_{\mathrm{YX}}^2 + Д_{\mathrm{YY}}^2 + Д_{\mathrm{YZ}}^2 \end{aligned} \tag{6.246}$$

This expression is valid in the high-pressure limit when the effect of vessel walls on particle trajectories may be ignored. Situations outside this limit must account for the vessel geometry; they are best treated numerically using the stochastic Liouville equation formalism (Sect. 6.4)

$$\begin{aligned} \frac{\partial}{\partial t} \boldsymbol{\rho}(\mathbf{r}, t) &= -i\mathcal{H}(\mathbf{r})\boldsymbol{\rho}(\mathbf{r}, t) + D\nabla^2 \boldsymbol{\rho}(\mathbf{r}, t) \\ \mathbf{H}(\mathbf{r}) &= -\gamma[B_{\mathrm{X}}(\mathbf{r})\mathbf{S}_{\mathrm{X}} + B_{\mathrm{Y}}(\mathbf{r})\mathbf{S}_{\mathrm{Y}} + B_{\mathrm{Z}}(\mathbf{r})\mathbf{S}_{\mathrm{Z}}] \end{aligned} \tag{6.247}$$

with appropriate spatial boundary conditions. SLE has the advantage of being able to handle arbitrary magnetic field directions and variations across the sample.

6.9.8.2 Collisional Relaxation: Spin-Rotation Mechanism

For an isotropically shielded nuclear spin in a rigid molecule in a gas phase, the Hamiltonian acquires a spin-rotation term derived in Sect. 5.4.1:

$$\hat{\mathbf{H}}(t) = \omega_S \mathbf{S}_Z + \mathbf{S} \cdot \mathbf{A} \cdot \hat{L}, \qquad \omega_S = -\gamma_S B_0 \tag{6.248}$$

where γ_S is the magnetogyric ratio of the spin, B_0 is the external magnetic field, $\mathbf{A}$ is the spin-rotation coupling tensor, $\mathbf{S}$ is a vector of spin operators, and $\hat{L}$ is a vector of molecular angular momentum operators. In the high-temperature and high-pressure limit, the collisions that randomise the rotational state are frequent, and the molecular angular momentum may be viewed as a noisy classical vector $\mathbf{L}(t)$:

$$\mathbf{H}(t) = \omega_S \mathbf{S}_Z + \mathbf{S} \cdot \mathbf{A}(t) \cdot \mathbf{L}(t) \tag{6.249}$$

undergoing stochastic dynamics in amplitude as well as direction. The time dependence in $\mathbf{A}(t)$ comes from molecular rotation (*atomic coordinates* are rotated), and the time dependence in $\mathbf{L}(t)$ comes from molecular collisions that alter the angular momentum (*angular velocities* are rotated). Straightforward analytical theories only exist when $\mathbf{A}(t)$ and $\mathbf{L}(t)$ are statistically uncorrelated across the ensemble; for liquids and gases, this is a reasonable assumption. For the angular momentum, the autocorrelation function is obtained from the rotational Langevin equation [242]:

$$\langle L_n(0) L_k(\tau) \rangle = \delta_{nk} I k_B T \exp(-\tau/\tau_J), \qquad n, k \in \{X,Y,Z\} \tag{6.250}$$

where I is the moment of inertia and τ_J is the characteristic time of orientational memory loss. Autocorrelation functions for the second-rank Wigner D functions occurring in the rotational expansion (Sect. 3.3.4) of the spin rotation coupling tensor $\mathbf{A}$ have been obtained in Sect. 6.4.1:

$$\left\langle \mathfrak{D}^{(2)*}_{k_1 m_1}(0) \mathfrak{D}^{(2)}_{k_2 m_2}(\tau) \right\rangle = \frac{\delta_{k_1 k_2} \delta_{m_1 m_2}}{5} \exp(-\tau/\tau_R) \tag{6.251}$$

where τ_R is the rotational correlation time. The resulting expressions have two limits—the diffusion limit (D, with $\tau_J \ll \tau_R$) in liquids and the kinetic limit (K, with $\tau_J \gg \tau_R$) in gases [243]:

$$R_1^{\mathrm{K}} = \frac{2Ik_{\mathrm{B}}T}{\hbar^2}\left(a^2 + \frac{4}{45}\Delta_{\mathbf{A}}^2\right)\tau_{\mathrm{J}}, \qquad R_1^{\mathrm{D}} = \frac{2Ik_{\mathrm{B}}T}{\hbar^2}\left(a^2 + \frac{2}{9}\Delta_{\mathbf{A}}^2\right)\tau_{\mathrm{J}} \tag{6.252}$$

where a is the isotropic part of $\mathbf{A}$ and $\Delta_{\mathbf{A}}^2$ is the second rank invariant defined in Eq. (6.181).

Incomplete Basis Sets

7

When two spin systems A and B are brought together to make a composite system A + B, their state spaces $\mathfrak{L}_{\mathrm{A}}$ and $\mathfrak{L}_{\mathrm{B}}$ combine by a Kronecker product:

$$\mathfrak{L}_{\mathrm{A+B}} = \mathfrak{L}_{\mathrm{A}} \otimes \mathfrak{L}_{\mathrm{B}} \tag{7.1}$$

because for each state in $\mathfrak{L}_{\mathrm{A}}$, the state of the other spin may be any element of $\mathfrak{L}_{\mathrm{B}}$. As discussed in Sect. 4.2.5, Hamiltonians and density matrices then combine in the following way

$$\begin{gathered} \mathbf{H}_{\mathrm{A+B}} = \mathbf{H}_{\mathrm{A}} \otimes \mathbf{1}_{\mathrm{B}} + \mathbf{1}_{\mathrm{A}} \otimes \mathbf{H}_{\mathrm{B}} + \mathbf{H}_{\mathrm{int}} \\ \boldsymbol{\rho}_{\mathrm{A+B}} = \boldsymbol{\rho}_{\mathrm{A}} \otimes \boldsymbol{\rho}_{\mathrm{B}} \end{gathered} \tag{7.2}$$

with bilinear interactions between A and B likewise expanded in Kronecker products of generators of their operator algebras:

$$\begin{gathered} \mathbf{H}_{\mathrm{int}} = \sum_n \mathbf{A}_n \otimes \mathbf{B}_n \\ \mathbf{A}_n \in \mathrm{Aut}(\mathfrak{L}_{\mathrm{A}}), \quad \mathbf{B}_n \in \mathrm{Aut}(\mathfrak{L}_{\mathrm{B}}) \end{gathered} \tag{7.3}$$

The principal problem here is the exponential growth in the matrix dimension as the spin system gets bigger—with over 20 spins, the explicit matrix representation of the Hamiltonian cannot even be stored, let alone manipulated; this situation is called *exponential scaling* wall.

Many ad hoc efficiency tweaks (sparse matrices, symmetry factorisation, etc.) can shift this wall back by a few spins, but the general problem is not solved and likely will never be. With exact solutions out of reach, a number of approximation strategies have emerged:

I. KUPROV, *Spin*,
https://doi.org/10.1007/978-3-031-05607-9_7

1. Not opening the Kronecker products in Eqs. (7.2) and (7.3)—it is possible, although logistically troublesome, to store the evolution generator **H** and/or the density matrix $\boldsymbol{\rho}$ as *tensor structures* of the following general form:

$$\begin{aligned} \mathbf{H} &= \sum_{\alpha_1 \cdots \alpha_N} \omega_{\alpha_1 \cdots \alpha_N} \mathbf{S}_{\alpha_1}^{(1)} \otimes \cdots \otimes \mathbf{S}_{\alpha_N}^{(N)} \\ \boldsymbol{\rho} &= \sum_{\alpha_1 \cdots \alpha_N} p_{\alpha_1 \cdots \alpha_N} \mathbf{S}_{\alpha_1}^{(1)} \otimes \cdots \otimes \mathbf{S}_{\alpha_N}^{(N)} \end{aligned} \tag{7.4}$$

   where $\mathbf{S}_{\alpha_k}^{(k)}$ is the α_k-th generator of the Lie algebra of the k-th spin, and the coefficient arrays can have—exactly or approximately—fewer non-zeroes than the matrix on the left-hand side. For spin Hamiltonians, the sum is always short because only unary and binary fundamental interactions appear to exist in nature.

   Depending on the constraints imposed on the combinations of indices in the coefficient arrays $\omega_{\alpha_1 \cdots \alpha_N}$ and $p_{\alpha_1 \cdots \alpha_N}$, and on the strategies for keeping those arrays manageable, these methods are called *density matrix renormalisation group* (DMRG [244]), *matrix product states and operators* (MPS and MPO [245]), *tensor trains* (TT [246]), etc. These methods will not be covered here—they are rarely a good choice for the irregular, dissipative, and densely coupled room temperature spin systems encountered in magnetic resonance, where the coefficient array $p_{\alpha_1 \cdots \alpha_N}$ quickly fills up during time evolution, and the representation stops being efficient. Although some elementary operations (matrix–matrix product, Frobenius norm, etc.) can be computed more efficiently in tensor structured representations, other important operations (notably the addition of large numbers of interaction operators when the Hamiltonian is built) become a logistical nightmare [247].

2. Finding a basis set for which some physical intuition is available (Fig. 7.1), and dropping insignificant or unpopulated states. This may be done before and/or during the simulation, based, respectively, on prior considerations and on the runtime analysis of where the system is actually going (Fig. 7.2). Typically, correlated states involving large numbers of spins and/or remote spins are not essential and can be dropped [124, 215, 248–254]. Justifications include sparsity of common spin interaction networks [253, 254], the inevitable presence of relaxation processes [83, 215], the existence of multiple non-interacting subspaces [124, 250], the presence of conservation laws [124], and simplifications brought about by the powder averaging operation [248, 252].

   Restricted state spaces are logistically straightforward and remain effective for long-range dissipative time-domain evolution—liquid-state magnetic resonance simulations of systems with hundreds of spins can now be performed routinely [253, 254] and there are encouraging signs that something similar will be possible in solid powders [252].

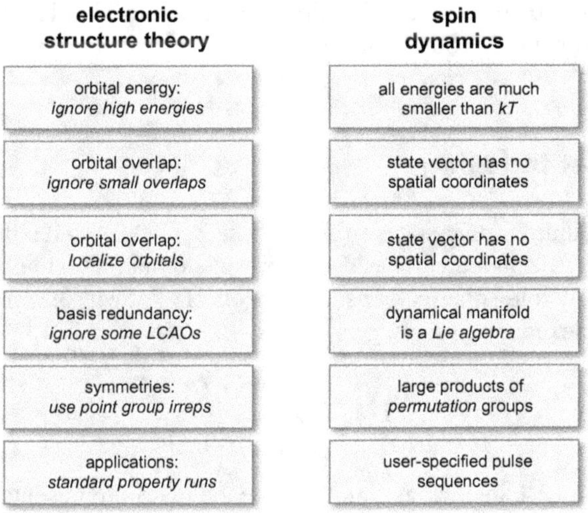

Fig. 7.1 Basis set truncation strategies used in electronic structure theory (left column) and the factors that make those strategies inapplicable to spin dynamics (right column)

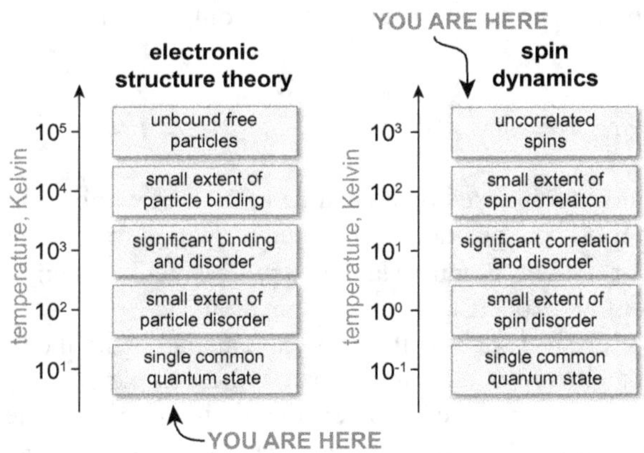

Fig. 7.2 Hierarchy of approximations in molecular electronic structure theory (left) and room-temperature spin dynamics (right). Most molecules have a single- or multi-reference electronic ground state around which rapidly convergent perturbative (Møller–Plesset) or variational (configuration interaction) expansions may be built. In that terminology, spin dynamics simulations in the direct product basis are full configuration interaction in the time domain. Thankfully, many spin systems are weakly correlated, but not in the sense that the term is used in electronic structure theory

This chapter covers Item 2—we will explore basis set truncation criteria based on relaxation rates, conservation laws and symmetries, and the interaction connectivity. For many large spin systems, after the redundant states are eliminated from the basis set, the matrix dimensions required for accurate simulations of magnetic resonance experiments become manageable.

7.1 Basis Set Indexing

In order to drop unimportant states from the basis set, a state indexing scheme must first be created. A convenient basis set for spin dynamics simulations is direct products of irreducible spherical tensor operators (IST, Sect. 3.3.2) [22, 124, 255–257] that are indexed by two integers:

$$\mathbf{T}_{lm} \quad \Leftrightarrow \quad \begin{bmatrix} l \\ m \end{bmatrix} \tag{7.5}$$

For a given spin, these integers uniquely define the matrix representation of the IST. Therefore, any product operator may be defined by an array of these integers:

$$\mathbf{T}_{l_1 m_1} \otimes \mathbf{T}_{l_2 m_2} \otimes \ldots \otimes \mathbf{T}_{l_N m_N} \quad \Leftrightarrow \quad \begin{bmatrix} l_1 & l_2 & \ldots & l_N \\ m_1 & m_2 & \ldots & m_N \end{bmatrix} \tag{7.6}$$

The array may be further transformed to require only one integer per spin:

$$\begin{gathered} \mathbf{T}_{l_1 m_1} \otimes \mathbf{T}_{l_2 m_2} \otimes \cdots \otimes \mathbf{T}_{l_N m_N} \quad \Leftrightarrow \\ \left[l_1^2 + l_1 - m_1 \quad l_2^2 + l_2 - m_2 \quad \ldots \quad l_N^2 + l_N - m_N \right], \end{gathered} \tag{7.7}$$

where the ISTs are now indexed by ascending rank l and within ranks by ascending projection number m—the position of the operator within this flattened list is given by $l^2 + l - m$. For a given product state, this representation requires exponentially less memory than the explicit matrix representation.

The number of product states still grows exponentially with the number of spins, but this indexing scheme is an improvement—the complexity scaling of at least some operations is now reduced. For example, structure coefficients c_{ijk} of the multiplicative envelope of the $\mathfrak{su}(2^N)$ Lie algebra of an N-spin system

$$\left(\bigotimes_{n=1}^{N} \mathbf{T}_{l_n^{(i)} m_n^{(i)}} \right) \left(\bigotimes_{n=1}^{N} \mathbf{T}_{l_n^{(j)} m_n^{(j)}} \right) = \sum_k c_{ijk} \left(\bigotimes_{n=1}^{N} \mathbf{T}_{l_n^{(k)} m_n^{(k)}} \right) \tag{7.8}$$

may now be computed without resort to matrix representations:

$$\begin{aligned} c_{ijk} &= \mathrm{Tr}\left[\left(\bigotimes_{n=1}^{N}\mathbf{T}_{l_n^{(i)}m_n^{(i)}}\right)\left(\bigotimes_{n=1}^{N}\mathbf{T}_{l_n^{(j)}m_n^{(j)}}\right)\left(\bigotimes_{n=1}^{N}\mathbf{T}_{l_n^{(k)}m_n^{(k)}}\right)^{\dagger}\right] \\ &= \mathrm{Tr}\left[\bigotimes_{n=1}^{N}\left(\mathbf{T}_{l_n^{(i)}m_n^{(i)}}\mathbf{T}_{l_n^{(j)}m_n^{(j)}}\mathbf{T}^{\dagger}_{l_n^{(k)}m_n^{(k)}}\right)\right] \\ &= \prod_{n=1}^{N}\mathrm{Tr}\left(\mathbf{T}_{l_n^{(i)}m_n^{(i)}}\mathbf{T}_{l_n^{(j)}m_n^{(j)}}\mathbf{T}^{\dagger}_{l_n^{(k)}m_n^{(k)}}\right) = \prod_{n=1}^{N}f_{ijk}^{(n)} \end{aligned} \tag{7.9}$$

in terms of the structure coefficients f_{ijk} of $\mathfrak{su}(2)$ algebras of individual spins, which are known and tabulated. In Eq. (7.9), the i, j, k indices run across the basis operators of $\mathfrak{su}(2^N)$, chosen to be direct products of single-spin ISTs, and the n index enumerates the spins in the system. Evaluation of a single structure coefficient using Eq. (7.9), now has linear complexity scaling with the number of spins.

7.2 Operator Representations

In order to generate a vector representation of an arbitrary state and a matrix representation of an arbitrary commutation superoperator $\mathcal{Q} = [\mathbf{Q}, \ ]$ in a given incomplete basis $\{\mathbf{O}_k\}$, it is sufficient to have matrix representations of left and right product superoperators [258] (Sect. 4.2.4), defined as

$$\mathcal{Q}^{(\mathrm{L})}\boldsymbol{\rho} = \mathbf{Q}\boldsymbol{\rho} \qquad \mathcal{Q}^{(\mathrm{R})}\boldsymbol{\rho} = \boldsymbol{\rho}\mathbf{Q} \tag{7.10}$$

A commutation superoperator is then obtained as their difference:

$$\mathcal{Q}\boldsymbol{\rho} = [\mathbf{Q}, \boldsymbol{\rho}] = \mathbf{Q}\boldsymbol{\rho} - \boldsymbol{\rho}\mathbf{Q}, \qquad \mathcal{Q} = \mathcal{Q}^{(\mathrm{L})} - \mathcal{Q}^{(\mathrm{R})}. \tag{7.11}$$

We start by expanding the density matrix $\boldsymbol{\rho}$ and the operator $\mathbf{Q}$ in our basis set:

$$\boldsymbol{\rho} = \sum_k r_k \mathbf{O}_k, \qquad \mathbf{Q} = \sum_n q_n \mathbf{O}_n, \tag{7.12}$$

The action by multiplication superoperators is then expressed via products of basis operators:

$$\mathcal{Q}^{(\mathrm{R})}\boldsymbol{\rho} = \sum_{nk} r_k q_n \mathbf{O}_k \mathbf{O}_n, \qquad \mathcal{Q}^{(\mathrm{L})}\boldsymbol{\rho} = \sum_{nk} r_k q_n \mathbf{O}_n \mathbf{O}_k. \tag{7.13}$$

Basis operators span an algebra—their products are their linear combinations:

$$\mathbf{O}_i\mathbf{O}_j = \sum_k c_{ijk}\mathbf{O}_k, \qquad c_{ijk} = \mathrm{Tr}\left(\mathbf{O}_i\mathbf{O}_j\mathbf{O}_k^{\dagger}\right), \tag{7.14}$$

where the structure coefficients are computed using the very efficient Eq. (7.9) when the destination is present in the incomplete basis, and set to zero when it is absent.

Representations of the components of the commutation superoperator then are

$$\begin{gathered}\mathcal{Q}^{(\mathrm{R})}\boldsymbol{\rho} = \sum_{nk} r_k q_n \mathbf{O}_k \mathbf{O}_n = \sum_{nkm} r_k q_n c_{knm} \mathbf{O}_m \\ \left[\mathcal{Q}^{(\mathrm{R})}\boldsymbol{\rho}\right]_m = \sum_{nk} r_k q_n c_{knm} \quad \Rightarrow \quad \left[\mathcal{Q}^{(\mathrm{R})}\right]_{mk} = \sum_n q_n c_{knm}\end{gathered} \tag{7.15}$$

and similarly for the left-side product superoperator. The representation of $\boldsymbol{\rho}$ in the same basis is the column of its expansion coefficients in Eq. (7.12). Once the representations of user-specified superoperators in the reduced basis are available, any simulation can be performed, the only logistical difference from the exact case being smaller matrix dimensions.

An important property of Eq. (7.15) is favourable computational complexity scaling with respect to the number of spins. Experimentally encountered spin interactions are at most two-particle, and the Hamiltonian is, therefore, a sum of at most two-spin operators with a known direct product structure [258]:

$$\mathbf{H}_n = \omega_n \left(\mathbf{S}_{n,1} \otimes \mathbf{S}_{n,2} \otimes \cdots \otimes \mathbf{S}_{n,N}\right) \tag{7.16}$$

where ω_n are interaction magnitudes, N is the total number of spins, and $\mathbf{S}_{n,k}$ are unit matrices or spin operators of dimension $2s_k + 1$ in which s_k is the quantum number of kth spin. When the basis set is chosen to have the same product structure:

$$\mathbf{O}_k = \bigotimes_{m=1}^{N} \mathbf{S}_{k,m} \tag{7.17}$$

the procedure described above yields

$$\left[\mathcal{H}_n^{(\mathrm{L})}\right]_{jk} = \langle \mathbf{O}_j | \mathcal{H}_n^{(\mathrm{L})} | \mathbf{O}_k \rangle = \mathrm{Tr}\left(\mathbf{O}_j^{\dagger} \mathbf{H}_n \mathbf{O}_k\right) = \ldots = \omega_n \prod_{m=1}^{N} \mathrm{Tr}\left(\mathbf{S}_{j,m}^{\dagger} \mathbf{S}_{n,m} \mathbf{S}_{k,m}\right) \tag{7.18}$$

in which the dimension of single-spin operators $\mathbf{S}_{n,k}$ is tiny and does not depend on the size of the system; the complexity of computing $\mathrm{Tr}\left(\mathbf{S}_{j,m}^{\dagger} \mathbf{S}_{n,m} \mathbf{S}_{k,m}\right)$ is, therefore, $O(1)$, and the complexity of computing one matrix element is then $O(N)$ multiplications. With $O(N^2)$ interactions in the system, this puts the worst-case complexity of building a matrix representation of the spin Hamiltonian to $O(N^3 D^2)$ multiplications, where D is the dimension of the reduced basis set.

7.3 Basis Truncation Strategies

The previous section assumed that an incomplete basis set had been chosen; here we discuss the strategies for doing that. The dominant consideration is the dissipative time-domain nature of magnetic resonance spectroscopy and imaging—basis selection strategies are radically different from those used in time-independent molecular quantum mechanics. This is illustrated in Fig. 7.1—almost nothing can be adapted.

The hierarchy of approximations is inverted relative to molecular quantum mechanics (Fig. 7.2). Instead of applying perturbation theories upwards from a small number of highly populated reference ground states, we have the system populating—only slightly unevenly—every possible excited state.

For these reasons, basis set truncation criteria in spin dynamics are based instead on reachability under coherent and/or dissipative evolution—some states may be unreachable because the necessary interactions are missing, others because relaxation drains the population on the way.

7.3.1 Correlation Order Hierarchy

Spin Hamiltonians contain at most two-particle interactions and common relaxation theories only go to second order in that Hamiltonian. Reachability would, therefore, depend on the number of non-unit matrices in the Kronecker structure $\mathbf{S}_1 \otimes \cdots \otimes \mathbf{S}_n$ of the state: the higher the correlation order (i.e. the more non-unit matrices there are), the longer it would take to reach. With a bit of luck, relaxation would take its toll, and the system would never get there.

From the computational complexity standpoint, this is an attractive proposition: dropping correlations of more than $k<n$ spins means that basis operators take the form: $\cdots \mathbf{1} \otimes \mathbf{S}_1 \otimes \cdots \otimes \mathbf{S}_2 \otimes \mathbf{1} \otimes \cdots \otimes \mathbf{S}_k \cdots$, where k spin operators are scattered among unit matrices. For a system with n spins ½, the dimension of the resulting state space then is

$$4^k \frac{n!}{k!(n-k)!} = 4^k \frac{1}{k!} n \cdot (n-1) \cdot \ldots \cdot (n-k+1) = O\left((4n)^k\right) \tag{7.19}$$

with trivial modifications if spins greater than ½ are present. Thus, the complexity scaling becomes polynomial in the total number of spins n. The procedure can of course be made adaptive, with the basis truncation level changing from one location to the next to maintain some target accuracy.

A useful by-product of basis truncation to low correlation orders is the reduction in the extreme eigenvalues of the evolution generator. This is because the highest frequencies belong to the states we are dropping. This in turn means better convergence properties for propagators (Sect. 4.9.5) and expm-times-vector operation (Sect. 4.9.6).

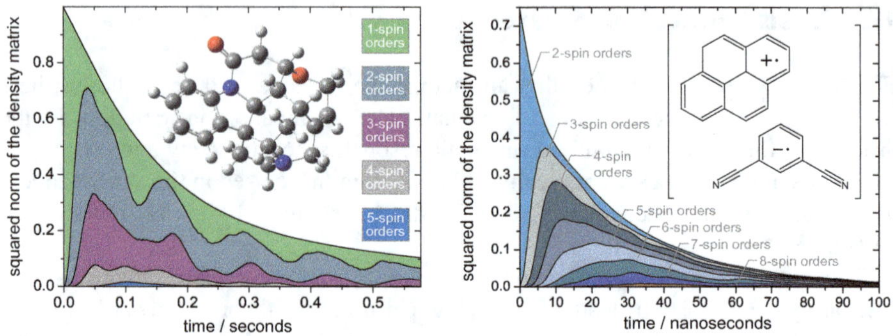

Fig. 7.3 Numerical simulation of room temperature density matrix norm dynamics during (**left**) the evolution and detection period of a pulse-acquire liquid-state NMR experiment on the 22-spin system of strychnine; (**right**) the evolution of a singlet-born pyrene-dicyanobenzene radical pair in the liquid state at a magnetic induction of 10 Gauss. Redfield relaxation superoperators were used in both cases; the thermal equilibrium state (dominated by single-spin orders at room temperature) is not shown. Adapted with permission from [215]

In liquid-state magnetic resonance, this turns out to be an excellent approximation (Fig. 7.3): empirical evidence indicates that the amplitude of high correlation orders is kept down by relaxation. It was this method that yielded polynomial complexity algorithms for liquid-state NMR spectroscopy [150, 254]. Accuracy conditions may be obtained from the analysis of population flow through the state space, which may be viewed as a direct sum of subspaces spanned by operators with different correlation orders:

$$\mathfrak{L} = \mathfrak{L}_0 \oplus \mathfrak{L}_1 \oplus \mathfrak{L}_2 \oplus \ldots \oplus \mathfrak{L}_N \tag{7.20}$$

where N is the number of spins and $\mathfrak{L}_k$ is the subspace of k-spin correlations, spanned (for example) by direct products of k irreducible spherical tensor operators each acting on its own spin. The $\mathfrak{L}_0$ subspace is spanned by the unit operator.

To obtain equations of motion for the overall population of each of the subspaces in Eq. (7.20), we split the Hamiltonian commutation superoperator into the single-spin part $\mathcal{H}_1$ (Zeeman, NQI, ZFS, etc.) and two-spin part $\mathcal{H}_2$ (all spin–spin couplings). Each subspace $\mathfrak{L}_k$ is closed under $\mathcal{H}_1$.

$$\mathcal{H}_1 \mathfrak{L}_k \subseteq \mathfrak{L}_k \tag{7.21}$$

because the number of non-unit matrices in a product operator cannot be changed by taking a commutator with a single-spin operator. $\mathfrak{L}_k$ does, however, leak into adjacent subspaces under $\mathcal{H}_2$:

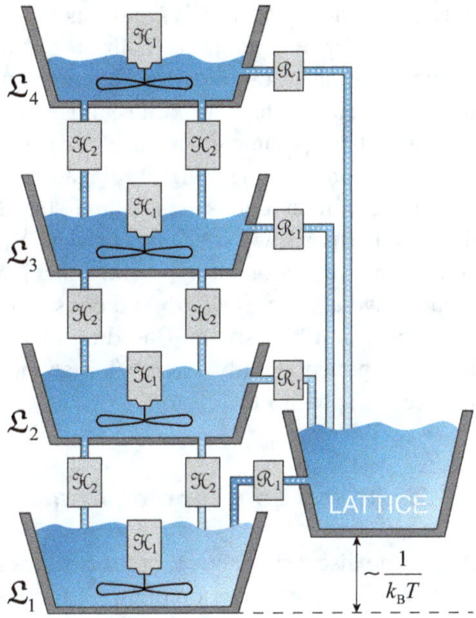

Fig. 7.4 A schematic illustration of the subspace hierarchy in Eq. Individual subspaces are invariant under commutation with the single-spin operators in $\mathcal{H}_1$, interact with their nearest neighbours under commutation with two-spin operators in $\mathcal{H}_2$, and drained by relaxation processes. High correlation levels in this subspace hierarchy are left unpopulated when relaxation is fast enough. Adapted with permission from [215]

$$\mathcal{H}_2 \mathfrak{L}_k \subseteq \mathfrak{L}_{k-1} \oplus \mathfrak{L}_k \oplus \mathfrak{L}_{k+1} \tag{7.22}$$

because a commutator of a two-spin operator with a k-spin operator can increase or reduce the correlation order by one spin, as well as leave it unchanged (Fig. 7.4).

The partitioning in Eq. (7.20) creates the corresponding partitioning of the state vector:

$$|\boldsymbol{\rho}\rangle = |\boldsymbol{\rho}_0\rangle + |\boldsymbol{\rho}_1\rangle + |\boldsymbol{\rho}_2\rangle + \ldots + |\boldsymbol{\rho}_N\rangle, \qquad |\boldsymbol{\rho}_k\rangle \in \mathfrak{L}_k \tag{7.23}$$

The extent to which the system occupies a particular subspace $\mathfrak{L}_k$ is given by the norm of the corresponding part of the state vector. Equations of motion for those norms may be obtained directly:

$$\left\{ \frac{\partial}{\partial t} \langle \boldsymbol{\rho}_k \mid \boldsymbol{\rho}_k \rangle = \left\langle \frac{\partial}{\partial t} \boldsymbol{\rho}_k \,\middle|\, \boldsymbol{\rho}_k \right\rangle + \left\langle \boldsymbol{\rho}_k \,\middle|\, \frac{\partial}{\partial t} \boldsymbol{\rho}_k \right\rangle \right._k \tag{7.24}$$

because the equation of motion for each $|\boldsymbol{\rho}_k\rangle$ is known. A protracted and technical analysis [215] of the interplay between coherent dynamics and relaxation then concludes that the fraction of $\langle \boldsymbol{\rho} \mid \boldsymbol{\rho} \rangle$ leaking outside the restricted state space does not exceed the user-specified tolerance $\xi \ll 1$ when:

$$k > 2\sqrt{\frac{h}{r}} \operatorname{erfc}^{-1}\left(\xi \operatorname{erfc}\left(\frac{1}{2}\sqrt{\frac{r}{h}} \right) \right) \tag{7.25}$$

where h is the 2-norm of $\mathbf{H}_2$ and r is the slowest single-spin relaxation rate. For a typical ^{1}H NMR simulation with an average J-coupling of 5 Hz and an average single-spin relaxation rate of 1 Hz this requires $k = 8$ for the population fraction in higher correlated states to be less than 1%.

The critical parameter in Eq. (7.25) is the interaction-relaxation ratio h/r. Systems in which the time scale of relaxation processes is comparable to the time scale of spin–spin interactions (e.g. liquid-state NMR and ESR systems) are accurately described using low-order correlations. Systems with slow relaxation and strong spin–spin interactions (e.g. solid-state NMR, particularly at low temperatures) would have a large h/r ratio and consequently require higher order correlations to be retained in the basis set. Based on this reasoning, simulation complexity scaling is truly exponential only when h/r is so big that the bound in Eq. (7.25) exceeds the number of spins in the system.

7.3.2 Interaction Topology Analysis

As per the discussion above, product states involving more than a certain number of spins and product states involving remote spins are likely to be unimportant. This section gives an example of setting up a reduced state space for liquid-state protein NMR simulations, where there are two interaction networks: J-coupling network that goes into the main Hamiltonian, and the dipole coupling network that is used to generate the relaxation superoperator (Sect. 6.3). This procedure is implemented in *Spinach* [87, 149].

1. Generate J-coupling graph (JCG) and dipolar coupling graph (DCG) from J-coupling data and Cartesian coordinates, respectively. User-specified thresholds are applied for the minimum significant J-coupling and maximum significant distance. Because spin interactions are at most two particle, the computational complexity of this procedure and the number of edges in the resulting graphs scale at most quadratically with the number of spins.
2. Use the depth-first search algorithm [259] on both JCG and DCG to generate the complete list of connected overlapping subgraphs involving a user-specified number of spins. This number controls the approximation accuracy [215] and should be specified independently for JCG and DCG. The complexity of this procedure and the number of the resulting subgraphs scale linearly with the number of edges in JCG and DCG [259].
3. For each subgraph G_k, generate a description of the complete basis set of the corresponding spin subsystem. The dimension D_k of this basis set is equal to the product of squares of multiplicities of each spin in G_k and does not depend on the size of the overall spin system. A convenient indexing scheme is the one in Eq. (7.7), where the structure of each basis operator is determined by a sequence of integers. The complete state list of a given subgraph G_k, therefore, requires nD_k integers of storage space.

4. Merge state lists of all subgraphs into a global state list, sort this list, and eliminate repetitions caused by subgraph overlap. Because the complexity scaling of each stage is polynomial in the total number of spins, the overall procedure runs in polynomial time.

This yields a basis set that contains only low orders of spin correlation (by construction, up to the size of the biggest subgraph) between spins that are proximate on JCG and DCG (by construction, because connected subgraphs were generated in Stage 2). The basis describes the entire system without gaps or cuts: once the subgraph state lists are merged and repetitions are eliminated, the result is a global list of spin operators that are expected to be populated during the spin system evolution based on the heuristics of locality and low correlation order. Accuracy is controlled by changing subgraph size in Stage 2—the limiting case of the whole system corresponds to the formally exact simulation [215] Examples of basis sets of this type implemented in *Spinach* are given in Table 7.1.

7.3.3 Zero Track Elimination

In the previous section, a reduced basis was built by analysing the interaction topology and performing what may be called *system-level pruning*—we picked the states that can potentially contribute to spin system evolution in a variety of experiments. In practical calculations, however, there is also a considerable scope for optimisations that are specific to the particular experiments and trajectories. A deeper *trajectory-level pruning* can be performed in each simulation instance to further select the states that do actually contribute to the system evolution.

Table 7.1 Examples of low correlation order basis set generation strategies for liquid-state NMR spectroscopy. The three basis set families given in this table are implemented in Spinach [87]

Basis set	Description
IK-0(n)	All spin correlations up to, and including, order n, irrespective of proximity on J-coupling or dipolar coupling graphs. Generated with a combinatorial procedure, by picking all possible groups of n spins in the current spin system and merging state spaces of those groups. Recommended for testing and debugging purposes
IK-1(n,k)	All spin correlations up to order n between directly J-coupled spins (with couplings above a user-specified threshold) and up to order k between spatially proximate spins (with distances below the user-specified threshold). Generated by coupling graph analysis as described in the text. The minimum basis set recommended for liquid state protein NMR simulations is IK-1(4,3) with a distance threshold of 4.0 Angstrom
IK-2(n)	For each spin, all of its correlations with directly J-coupled spins, and correlations up to order n with spatially proximate spins (below the user-specified distance threshold). Generated by coupling graph analysis as described in the text. Recommended for accurate simulations on large computer systems with the distance threshold of 5.0 Angstrom or greater

Strictly speaking, we want the semigroup orbit of the propagator $\mathcal{P}$ and the initial state $\boldsymbol{\rho}_0$, which is contained (due to the Taylor series relation between the propagator and the generator) in the Krylov subspace of the evolution generator $\mathcal{L}$ and the initial state:

$$\{\boldsymbol{\rho}_0, \mathcal{P}\boldsymbol{\rho}_0, \mathcal{P}^2\boldsymbol{\rho}_0, \ldots\} \in \operatorname{span}\{\boldsymbol{\rho}_0, \mathcal{L}\boldsymbol{\rho}_0, \mathcal{L}^2\boldsymbol{\rho}_0, \ldots\} \tag{7.26}$$

We cannot afford to use this definition directly—it is equivalent to running the exact simulation. We can, however, reformulate the question—instead of looking for the vectors that *do* appear in the basis of the Krylov subspace (and then projecting into that subspace), we can look for the vectors that *do not* appear (and drop them from the basis set).

Cheap screening criteria for the latter process do exist. A good example is *zero track elimination* (ZTE) in magnetic resonance (Fig. 7.5), where the initial state vectors are very sparse. That sparsity is different from the sparsity of the evolution generator—the latter refers to the general properties of the spin system, whereas the former is a property of the particular experiment through which the system evolves. It is a common observation that, in the $\{\boldsymbol{\rho}_0, \mathcal{P}\boldsymbol{\rho}_0, \mathcal{P}^2\boldsymbol{\rho}_0, \ldots\}$ sequence, many elements of the state vector stay zero throughout the calculation, exactly or approximately. Those elements, and the corresponding state-space dimensions, maybe pruned out.

Theorem 1 ***(zero track theorem)***: if the coefficient in front of the basis element $|\mathbf{b}\rangle$ in the state vector remains identically zero during the first finite step Δt of the evolution under a constant generator $\mathcal{L}$, that coefficient will stay zero during the subsequent evolution.

$$\langle\mathbf{b}|e^{-i\mathcal{L}t}|\boldsymbol{\rho}_0\rangle = 0 \quad \forall t \in [0, \Delta t] \qquad \Rightarrow \qquad \langle\mathbf{b}|e^{-i\mathcal{L}t}|\boldsymbol{\rho}_0\rangle = 0 \quad \forall t \in [0, \infty) \tag{7.27}$$

Proof: because the Taylor expansion of a continuous function is unique in any given interval, the following series, in order to be zero everywhere on $t \in [0, \Delta t]$, must have zero coefficients:

$$\langle\mathbf{b}|e^{-i\mathcal{L}t}|\boldsymbol{\rho}_0\rangle = \langle\mathbf{b}|\mathbf{1}|\boldsymbol{\rho}_0\rangle + \langle\mathbf{b}|\mathcal{L}|\boldsymbol{\rho}_0\rangle(-it) + \frac{1}{2!}\langle\mathbf{b}|\mathcal{L}^2|\boldsymbol{\rho}_0\rangle(-it)^2 + \ldots \tag{7.28}$$

meaning that

$$\langle\mathbf{b}|\mathbf{1}|\boldsymbol{\rho}_0\rangle = \langle\mathbf{b}|\mathcal{L}|\boldsymbol{\rho}_0\rangle = \langle\mathbf{b}|\mathcal{L}^2|\boldsymbol{\rho}_0\rangle = \ldots = 0 \tag{7.29}$$

i.e., that $|\mathbf{b}\rangle$ is orthogonal to the Krylov subspace generated by $\mathcal{L}$ and $|\boldsymbol{\rho}_0\rangle$. The series in Eq. (7.28) will therefore stay zero for all values of t. ■

As may be seen from the proof, ZTE detects the states that do not become populated during system evolution; those states may be dropped from the basis. This algorithm runs in reverse order compared to the Lanczos pruning procedure

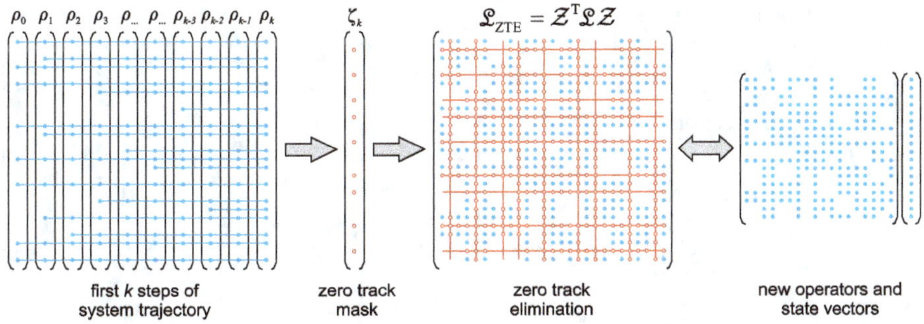

Fig. 7.5 A schematic of the implementation of zero track elimination in Spinach. The initial state vector (blue dots denote non-zeroes) is propagated forward in time for a few time steps (equal to the reciprocal upper bound for the evolution generator 2-norm) using the procedure described in Sect. 4.9.6. Zero tracks (red dots) are detected and removed from the state vector; rows and columns of the evolution generator that operate on zero tracks are also removed. The procedure is reversible (see Fig.(7.6).). Adapted with permission from [253]

[260], where the Krylov subspace is first mapped in its entirety and then projected into. One possible implementation of the process (from *Spinach* [87]) is shown schematically in Fig. 7.5. ZTE does require a few steps to be computationally affordable; the procedure for taking a step described in Sect. 4.9.6 is recommended. With more than 10 spins, system-level basis pruning must be applied first.

Theorem 1 is a good start, but the number of inner products $\langle \mathbf{b}|\mathrm{e}^{-i\mathfrak{L}t}|\boldsymbol{\rho}_0\rangle$ that would stay *identically* zero is likely to be small. A bigger set would stay *approximately* zero.

Theorem 2 *(thin track theorem):* if the absolute value of the inner product $\langle \mathbf{b}|e^{-i\mathfrak{L}t}|\boldsymbol{\rho}_0\rangle$ stays smaller than some number $\varepsilon \ll 1$ during the first finite step $\Delta t < \|\mathfrak{L}\|_2^{-1}$, removing $\langle \mathbf{b}|$ from the basis set is equivalent to restricting the simulation to a Krylov subspace generated by $\mathfrak{L}$ and $|\boldsymbol{\rho}_0\rangle$, ignoring the contributions to that subspace from high values of n in $\mathfrak{L}^n|\boldsymbol{\rho}_0\rangle$.

Proof: We observe that for $0 < t \leq \Delta t < \|\mathfrak{L}\|_2^{-1}$, the Taylor series for the inner product $\langle \mathbf{b}|e^{-i\mathfrak{L}t}|\boldsymbol{\rho}_0\rangle$ converges monotonically: the absolute value of each subsequent term is smaller than that of the preceding one. Since the absolute value of the entire series is bounded from above by ε,

$$\langle \mathbf{b}|\mathrm{e}^{-i\mathfrak{L}t}|\boldsymbol{\rho}_0\rangle = \left|\sum_n \frac{(-it)^n}{n!}\langle \mathbf{b}|\mathfrak{L}^n|\boldsymbol{\rho}_0\rangle\right| < \varepsilon \tag{7.30}$$

the high-order terms in this expansion are necessarily much smaller than ε for all values of t. That is, within some small tolerance, the contribution to the propagator from $\mathfrak{L}^n|\boldsymbol{\rho}_0\rangle$ does not contain any $|\mathbf{b}\rangle$, which is, therefore, absent to that tolerance from the Krylov subspace of interest. ∎

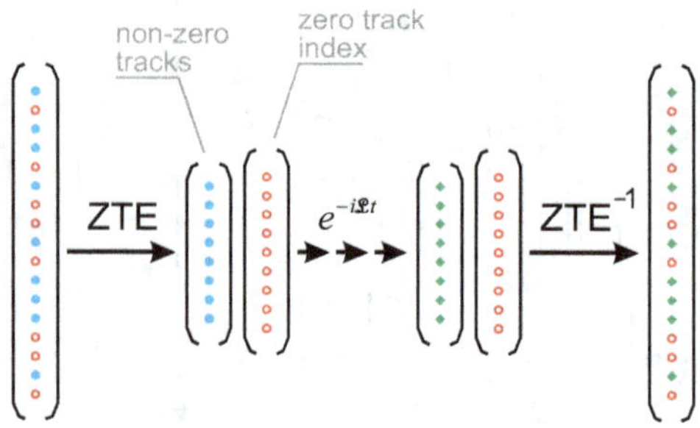

Fig. 7.6 A schematic illustration of the return to the full state space by reinstating zeroes at their original positions within the density matrix. Adapted with permission from [253]

The implementation of ZTE in *Spinach* (Fig. 7.5) has the following stages:

1. Generation of the fist K trajectory steps of duration $\Delta t \approx \|\mathfrak{L}\|_2^{-1}$ using the method described in Sect. 4.9.6 to avoid computing the matrix exponential. Efficient upper bounds on the 2-norm exist for sparse matrices (Sect. 9.3.1).
2. Detection of zero tracks in the resulting trajectory (by comparing absolute values with the user-specified tolerance δ), and generation of the zero track mask $|\zeta\rangle$.

$$\zeta_n = \begin{cases} 1 & \text{if } \frac{1}{K}\sum_{k=1}^{K}\left|\rho_n^{(k)}\right| < \delta \\ 0 & \text{otherwise} \end{cases} \tag{7.31}$$

3. Construction of the projector $\mathcal{Z}$, which is a unit matrix with columns flagged in the zero track mask $|\zeta\rangle$ taken out. The following transformation then eliminates the zero tracks:

$$\mathfrak{L}_{\mathrm{ZTE}} = \mathcal{Z}^{\mathrm{T}}\mathfrak{L}\mathcal{Z} \qquad \left|\boldsymbol{\rho}_{0,\mathrm{ZTE}}\right\rangle = \mathcal{Z}^{\mathrm{T}}|\boldsymbol{\rho}_0\rangle \tag{7.32}$$

   This equation is a formality—in practice, the computer is simply instructed to drop a number of rows and columns from $\mathfrak{L}$ and $|\boldsymbol{\rho}_0\rangle$.
4. Time-domain simulation in the resulting reduced representation. All simulation techniques described elsewhere in this book remain applicable. Calculation of observables may be done in the reduced state space.
5. Optionally, a return to the original state space by re-inserting zeroes into their original positions in the state vector (Fig. 7.6) and returning to the original Liouvillian. This may be necessary if the next time evolution stage has a different generator.

The procedure can be made more general (but more computationally expensive) if the inexpensive zero check is replaced by the inner product check prescribed by Theorems 1 and 2.

7.3.4 Conservation Law Screening

Consider a continuous unitary transformation $\mathbf{U} = \exp(-i\mathbf{G}\alpha)$ with a Hermitian generator $\mathbf{G}$ and a real parameter α that does not change the energy of any system state:

$$|\psi\rangle \to \mathbf{U}|\psi\rangle \quad \Rightarrow \quad \langle\psi|\mathbf{U}^{\dagger}\mathbf{H}\mathbf{U}|\psi\rangle = \langle\psi|\mathbf{H}|\psi\rangle \tag{7.33}$$

Because this equality holds for any wave function, it must hold for the operators between Dirac brackets. Therefore, $\mathbf{U}$ commutes with the Hamiltonian:

$$\mathbf{U}^{\dagger}\mathbf{H}\mathbf{U} = \mathbf{H} \quad \Rightarrow \quad \mathbf{U}^{-1}\mathbf{H}\mathbf{U} = \mathbf{H} \quad \Rightarrow \quad [\mathbf{H}, \mathbf{U}] = 0 \tag{7.34}$$

In particular, so does the infinitesimal transformation, and therefore the generator. This leads to a conservation law for the observable that corresponds to the generator:

$$\frac{\mathrm{d}}{\mathrm{d}t}\langle\psi|\mathbf{G}|\psi\rangle = \ldots = i\langle\psi|[\mathbf{H}, \mathbf{G}]|\psi\rangle = 0 \tag{7.35}$$

That is one of the formulations of Noether's first theorem [261]: *every differentiable symmetry of the Hamiltonian of a physical system has a corresponding conservation law*. Illustrations were provided in Chap. 2, where time translation invariance led to the conservation of energy, space translation invariance to the conservation of linear momentum, rotation invariance to the conservation of angular momentum, and the conservation law corresponding to Lorenz invariance has yielded the concept of spin. Quantities often conserved in magnetic resonance experiments are total spin $\mathbf{S}^2$ (for example, in symmetric systems and at zero field) and the total Z-axis projection of the spin $\mathbf{S}_{\mathrm{Z}}$ (for example, in a strong magnetic field directed along the notional Z-axis of the laboratory frame):

$$\mathbf{S}^2 = \mathbf{S}_{\mathrm{X}}^2 + \mathbf{S}_{\mathrm{Y}}^2 + \mathbf{S}_{\mathrm{Z}}^2, \quad \mathbf{S}_{\mathrm{X,Y,Z}} = \sum_k \mathbf{S}_{\mathrm{X,Y,Z}}^{(k)}. \tag{7.36}$$

where the sum is over the individual spins in the system. A promising trajectory level pruning method is to eliminate all states that violate conservation laws for a given initial condition. If the observable corresponding to an operator $\mathbf{A}$ is conserved:

$$\langle\mathbf{A}\rangle = \mathrm{const} \quad \Leftrightarrow \quad [\mathbf{H}, \mathbf{A}] = 0 \quad \Leftrightarrow \quad \mathcal{H}\mathbf{A} = 0 \tag{7.37}$$

then the subspaces spanned by the eigenstates of $\mathbf{A}$ that do not occur in the initial condition can be dropped from the basis set. In particular, if the initial density matrix $\boldsymbol{\rho}_0$ is an eigenstate of $\mathcal{A}$:

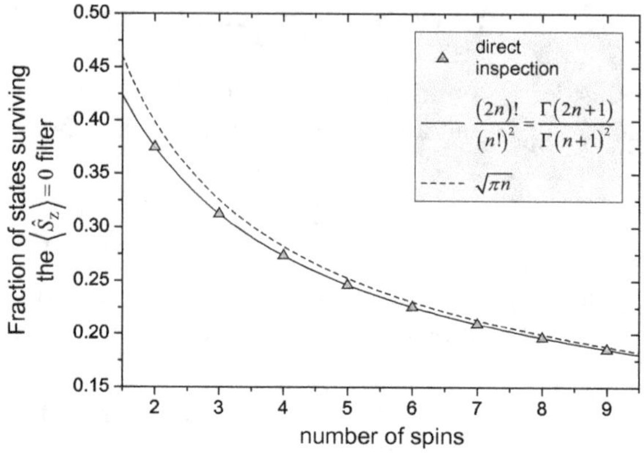

Fig. 7.7 Fraction of states surviving the $\langle S_Z \rangle = 0$ conservation filter (and therefore contributing to the spin system evolution) as a function of the total number of spin-1/2 particles in the system. Reproduced with permission from [124]

$$\mathcal{A}|\mathbf{\rho}_0\rangle = [\mathbf{A}, \mathbf{\rho}_0] = a|\mathbf{\rho}_0\rangle, \tag{7.38}$$

then the eigenvalue a may be used to index non-interacting subspaces, because

$$\mathcal{A}|\mathbf{\rho}(t)\rangle = \left[\mathbf{A}, \mathrm{e}^{-i\mathbf{H}t}\mathbf{\rho}_0\mathrm{e}^{+i\mathbf{H}t}\right] = \mathrm{e}^{-i\mathbf{H}t}[\mathbf{A}, \mathbf{\rho}_0]\mathrm{e}^{+i\mathbf{H}t} = a|\mathbf{\rho}(t)\rangle. \tag{7.39}$$

The simulation trajectory would, therefore, be confined to the subspace spanned by the states with the same eigenvalues of $\mathcal{A}$ as those that were present in the initial condition.

If the basis set is chosen to be the eigenstates of $\mathcal{A}$ (e.g. direct products of irreducible spherical tensors $\mathbf{T}_{\mathrm{lm}}$ in the case of $\mathbf{S}^2$ and $\mathbf{S}_\mathrm{Z}$), a fast basis screening procedure can be implemented because the decision to include or exclude a basis state can be made at the descriptor level: l and m indices in Eq. (7.7) are precisely the indices that make up the eigenvalues of $\mathbf{S}^2$ and $\mathbf{S}_\mathrm{Z}$.

Even for one conserved quantity, the reduction in problem dimensionality may be significant (Fig. 7.7), and many such quantities may exist—as per Eq. (7.37), their number is equal to the dimension of the null space of $\mathcal{H}$, where "null" need not refer only to exact zeros, but also to the eigenvalues $\mathcal{H}$ that are small enough to be negligible on the time scale of the simulation.

7.3.5 Generator Path Tracing

Even after state-space restriction, zero track elimination, and symmetry factorisation, the evolution generator $\mathcal{L} = \mathcal{H} + i\mathcal{R}$, can still be very sparse, with each basis state only directly connected to a few others by the off-diagonal elements (Fig. 7.8). Within that connectivity network, disconnected subnetworks may exist, corresponding to non-interacting subspaces that were not picked up by conservation law filters.

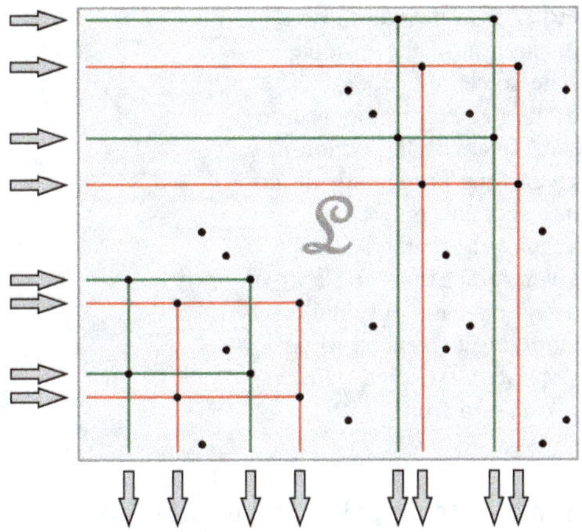

Fig. 7.8 Schematic illustration of evolution generator connectivity analysis. The generator matrix can be sparse even in densely coupled spin systems because the Hamiltonian only contains one- and two-spin operators it may be treated as the adjacency matrix of the Liouville connectivity graph (dots denote non-zero elements, lines show coherence transfer paths in an infinitesimal propagation step). Efficient procedures exist for partitioning sparse graphs into connected subgraphs, which here correspond to non-interacting subspaces. Adapted with permission from [124]

To separate those, the evolution generator may be treated as the adjacency matrix of a graph: diagonal elements correspond to nodes and off-diagonal elements to edges. Non-interacting subspaces then correspond to disjoint subgraphs and may be found in $O(\text{nnz})$ time with respect to the number of non-zeros in the generator matrix [124]. The implementation in *Spinach* uses Tarjan's graph partitioning algorithm [262, 263] on the adjacency matrix $\mathcal{J}$ obtained from the evolution generator:

$$\mathcal{J}_{nk} = \begin{cases} 1 & \text{if } |\mathfrak{L}_{nk}| > \varepsilon \\ 0 & \text{otherwise} \end{cases} \tag{7.40}$$

where ε is a user-specified tolerance. Tarjan's algorithm returns node indices for disjoint subgraphs; they correspond to state indices for non-interacting subspaces. Once the state space is partitioned into a direct sum of those, the dynamics of an observable $\langle \mathbf{A} \rangle$ is a sum over subspaces:

$$\begin{aligned} \langle \mathbf{A} | \boldsymbol{\rho}(t) \rangle &= \langle \mathbf{A} | e^{-i\mathfrak{L}t} \boldsymbol{\rho}_0 \rangle = \left\langle \bigoplus_{n=1}^{N} \mathbf{A}^{(n)} \middle| \exp\left[-i \left(\bigoplus_{n=1}^{N} \mathfrak{L}^{(n)} \right) t \right] \middle| \bigoplus_{n=1}^{N} \boldsymbol{\rho}_0^{(n)} \right\rangle \\ &= \left\langle \bigoplus_{n=1}^{N} \mathbf{A}^{(n)} \middle| \left(\bigoplus_{n=1}^{N} e^{-i\mathfrak{L}^{(n)}t} \right) \middle| \bigoplus_{n=1}^{N} \boldsymbol{\rho}_0^{(n)} \right\rangle = \sum_{n=1}^{N} \left\langle \mathbf{A}^{(n)} \middle| e^{-i\mathfrak{L}^{(n)}t} \middle| \boldsymbol{\rho}_0^{(n)} \right\rangle \end{aligned} \tag{7.41}$$

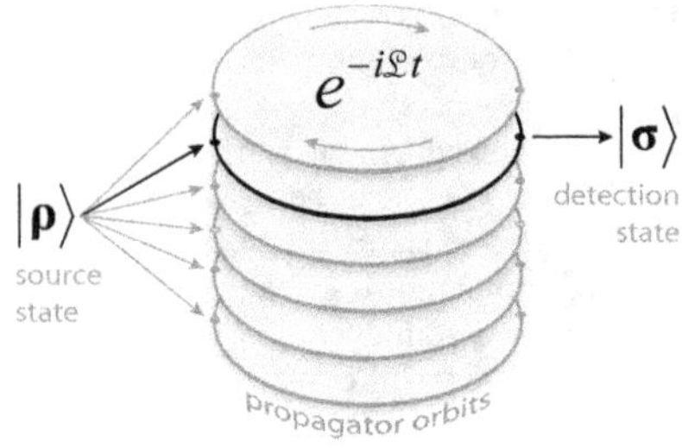

Fig. 7.9 Schematic illustration of the destination state screening process. The orbits induced in the system state space by the action of a given time propagator do not intersect and may be simulated separately. For a given initial state and a given detection state, only the orbits involving both need to be simulated. Adapted with permission from [250]

Some of the blocks may turn out to be unpopulated by the initial condition, either exactly or to some user-specified tolerance on the norm of $\boldsymbol{\rho}_0^{(n)}$. Unpopulated subspaces may be dropped.

Generator diagonalisation would, of course, achieve the same result—normal modes do not interact—but we assume that the system is so large as to render diagonalisation infeasible.

7.3.6 Destination State Screening

Any state that gets populated but never evolves into the detection state may be dropped because it will never contribute to the observable of interest. This is an instance of the reachability problem:

$$\left\langle \boldsymbol{\sigma} \mid \mathrm{e}^{-i\mathcal{H}t+\mathcal{R}t}\boldsymbol{\rho}_0 \right\rangle = \left\langle \mathrm{e}^{+i\mathcal{H}t+\mathcal{R}t}\boldsymbol{\sigma} \mid \boldsymbol{\rho}_0 \right\rangle \tag{7.42}$$

where the elements of $\boldsymbol{\rho}_0$ that are not reachable from the detection state $\boldsymbol{\sigma}$ do not contribute to the simulation result. Because $\boldsymbol{\sigma}$ can be simpler than $\boldsymbol{\rho}_0$, the dimension of the space spanned by the propagator semigroup orbit of $\boldsymbol{\sigma}$ can be smaller than that of $\boldsymbol{\rho}_0$. In particular, those of the independently evolving subspaces (Sect. 7.3.4) that do not contain the detection state can be dropped from the basis (Fig. 7.9).

7.4 Performance Illustrations

This section describes two favourable cases from liquid-state NMR spectroscopy, where the low correlation order approximation yields accurate simulations of systems beyond the exponential scaling wall of Kronecker product methods and beyond the applicability conditions of tensor network methods.

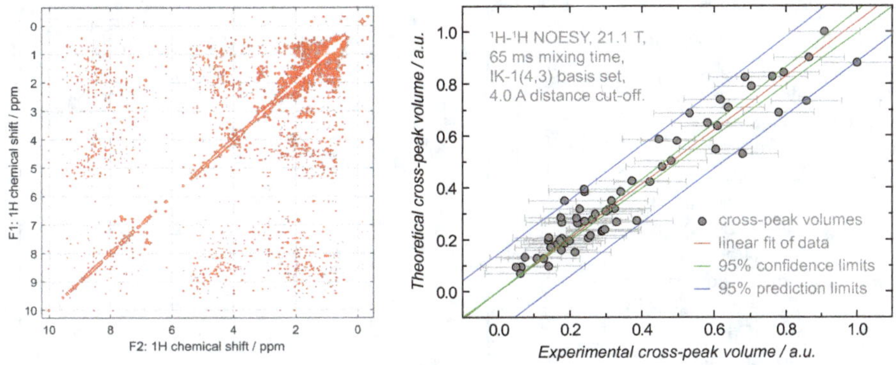

Fig. 7.10 **Left:** simulated ^{1}H-^{1}H NOESY spectrum of ubiquitin at 900 MHz proton frequency with a mixing time of 65 ms. The simulation was carried out by literal time-domain propagation through the NOESY pulse sequence using a restricted Liouville space and Redfield's relaxation superoperator computed as described in Chap. 6. Protein structure was assumed to be rigid and a single global rotational correlation time of 5 ns was used. **Right:** correlation between experimental and theoretical ^{1}H-^{1}H NOESY cross-peak volumes. Adapted with permission from [149]

Table 7.2 CPU time and memory utilisation statistics for ubiquitin ^{1}H-^{1}H NOESY simulation at different accuracy levels

Basis set for the reduced state space (see Table 7.1)	IK-1(2,2)	IK-1(3,2)	IK-1(4,2)	IK-1(4,3)
Reduced state-space dimension	29k	56k	210k	849k
Number of non-zeroes in Hamiltonian superoperator	43k	223k	1420k	2500k
Number of non-zeroes in relaxation superoperator	102k	142k	360k	1800k
Wall clock time (16 Sandy Bridge cores at 2.4 GHz)	20 min	58 min	8 h	24 h

7.4.1 ^{1}H-^{1}H NOESY Spectrum of Ubiquitin

Proteins contain hundreds of interacting (both coherently and dissipatively) nuclear spins. Their irregular three-dimensional polycyclic interaction networks are far from chain or tree topologies required by tensor network methods. Simulations of the most informative pulse sequences (e.g. NOESY and TROSY-HSQC) require long-range time evolution and treatment of relaxation processes at the Redfield superoperator level. Without the methods described in this chapter, such simulations are impractical.

Matrix dimension, storage, and CPU time statistics for a 512 $\times$ 512 point ^{1}H-^{1}H NOESY simulation of ubiquitin (573 protons, around 50,000 terms in the dipolar Hamiltonian) are given in Table 7.2. As demonstrated in Fig. 7.10, the simulation is in good agreement with the experimental data. The state-space restriction approximation reduces the Hamiltonian superoperator dimension from 4^{573} to 848,530.

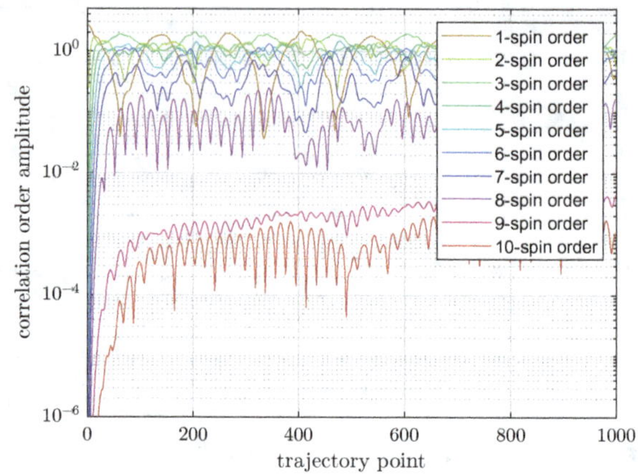

Fig. 7.11 Contributions from different orders of spin correlation to the system trajectory in the pulse-acquire ^{1}H NMR simulation of anti-3,5-difluoroheptane (16 spins). Different curves correspond to the norms of the projection of the density matrix into the subspace of one-, two-, three-, etc., spin correlations. The two traces in the lower part of the figure correspond to nine- and ten-spin correlations—there are no detectable changes in the simulated spectrum when they are dropped: only correlations of up to eight spins need to be accounted for in this system. Adapted with permission from [264]

The reduced Hamiltonian is still sparse, and therefore within reach of the methods described in Sect. 4.9.6 for time propagation and Sect. 6.3.1 for the relaxation superoperator—the simulation shown in Fig. 7.10 took a few hours on a contemporary computer with 256 GB of RAM.

The biggest challenge associated with the simulation in Fig. 7.10 is not actually spin dynamics—the algorithms described in this chapter do their job well—but collecting all the necessary interaction data, such as chemical shift tensors and *J*-couplings [149].

7.4.2 ^{19}F and ^{1}H NMR of Anti-3,5-Difluoroheptane

The full Liouville state-space dimension for the 16-spin system of anti-3,5-difluoro-*n*-heptane is $4^{16} \approx 4.3 \times 10^{9}$. The following sequence of state space reduction stages is typical for such systems.

1. *Low correlation order approximation*: as discussed in Sect. 7.3.1, high orders of spin correlation remain unpopulated in liquid-state NMR experiments. This may be confirmed by direct inspection—Fig. 7.11 shows the dynamics of the density matrix norm partitioned into contributions from subspaces with different orders of spin correlation. The amplitudes of states involving more than eight spins are three orders of magnitude smaller than the amplitude of states responsible for the transverse magnetisation. Correlations of more than eight spins can be dropped; this yields a reduction in the dimension of the Liouville space from $4.3{\cdot}10^{9}$ to 1,564,672.

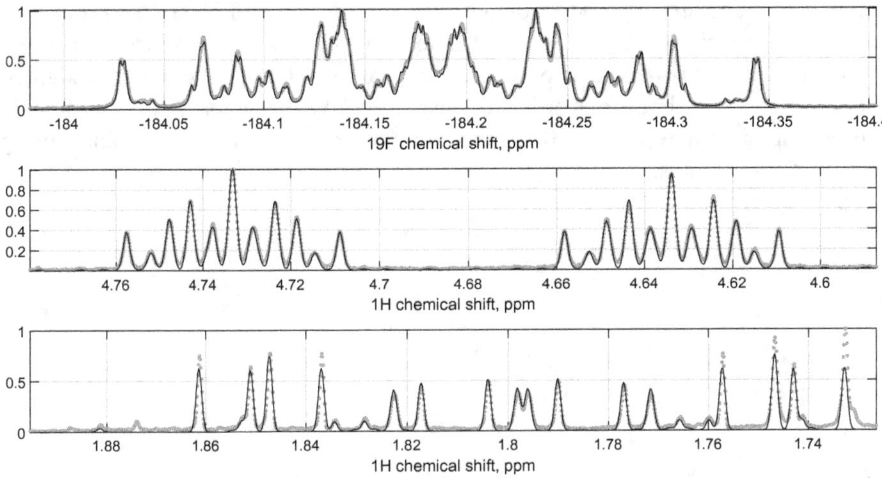

Fig. 7.12 Experimental (grey dots) and theoretical (black lines) ^{19}F and ^{1}H NMR spectra of anti-3,5-difluoroheptane in chloroform at 11.75 T magnetic induction. Experimental data kindly provided by Bruno Linclau (University of Ghent) and Neil Wells (University of Southampton)

2. *Conservation law filter*: in high-field NMR experiments, the state of the spins that are connected to the rest of the system via $\mathbf{L}_Z\mathbf{S}_Z$ operators, and not pulsed directly, stays longitudinal. A longitudinal spin order filter was therefore applied in the ^{19}F subspace for proton NMR simulations, and in the ^{1}H subspace for fluorine NMR simulations. In the former case, this yielded a further reduction in the Liouville space dimension from 1,564,672 to 520,192.
3. *Conservation law filter*: in high-field NMR experiments, the total projection quantum number of the spin system is conserved. A spin system that starts its evolution in the $\mathbf{L}_+$ state (at the beginning of the quadrature detection period) must remain in the $m_Z = +1$ subspace for the entire evolution period. Dropping other subspaces reduces the dimension further from 520,192 to 90,681.
4. *Direct product symmetry factorisation*: the protons of the two rapidly rotating methyl groups obey an S_3 permutation symmetry group *each*, meaning that the total system symmetry group is $S_3 \otimes S_3$, with 36 symmetry operations and 9 irreducible representations of dimensions 1, 1, 2, 1, 1, 2, 2, 2, and 4. As discussed in Sect. 4.6.2, only the fully symmetric irrep is populated in Liouville space. This reduces the state space dimension further from 90,681 to 58,473.

At this point, the propagation method described in Sect. 4.9.6 becomes viable. Although the final Liouville space dimension (58,473) is bigger than the achievable Hilbert space dimension (16,384 when symmetry is taken into account), a Liouville space propagation step (one matrix–vector multiplication, $58{,}473^2 \approx 3.4 \times 10^9$ flops $\approx$ 30 ms) is cheaper than a Hilbert space-time propagation step (two matrix–matrix multiplications, $2{\times}16{,}384^3 \approx 8.8 \times 10^{12}$ flops $\approx$ 900 s). Given that 4,096 propagation steps are required to obtain a spectrum with sufficient resolution, this difference is decisive and the improvement in simulation time from using a restricted symmetry-adapted Liouville space compared to the symmetry-adapted Hilbert space is by about four orders of magnitude. When sparse matrix arithmetic is used on a contemporary GPU, the simulation runs in seconds and fitting (Fig. 7.12) in minutes.

8 Optimal Control of Spin Systems

Taking a system that follows the Schrödinger equation (or an ensemble of systems that follows the Liouville–von Neumann equation) from one state to another to a specified accuracy with minimal expenditure of time and energy—with the emphasis on the word *minimal*—is increasingly important in science and engineering [266]. That is an optimisation problem: we define some measure of accuracy and some running cost, and proceed to construct a functional that is to be optimised.

We will use the Liouville space version of the Liouville–von Neumann equation in this book; less general pictures (Schrödinger equation and Hilbert space version of the LvN equation) are a notational transformation away. Popular measures of accuracy are functions of the overlap f between the final state of the system $\boldsymbol{\rho}(T)$ and the desired destination state $\boldsymbol{\delta}$:

$$f = \langle \boldsymbol{\delta} \mid \boldsymbol{\rho}(T) \rangle = \langle \boldsymbol{\delta} | \overleftarrow{\exp} \left[-i \int_0^T (\mathcal{H}(t) + i\mathcal{R}) \mathrm{d}t \right] | \boldsymbol{\rho}_0 \rangle \tag{8.1}$$

where T is the duration of the experiment, $\mathcal{H}(t)$ is the Hermitian part of the evolution generator, $\mathcal{R}$ is the relaxation superoperator (where time dependence is uncommon) and $\overleftarrow{\exp}$ denotes a time-ordered exponential (Sect. 4.1). It is common to define a Liouvillian superoperator as their sum:

$$\mathcal{L}(t) = \mathcal{H}(t) + i\mathcal{R} \tag{8.2}$$

and then to split it into the *drift part* $\mathcal{D}(t)$ that cannot be influenced and the *control part* that the instrument can vary as a function of time:

$$\mathcal{L}(t) = \mathcal{D}(t) + \sum_k c^{(k)}(t)\mathcal{C}_k \tag{8.3}$$

I. KUPROV, *Spin*,
https://doi.org/10.1007/978-3-031-05607-9_8

where $\mathcal{C}_k$ are time-independent *control* operators (e.g. radiofrequency and microwave Zeeman operators) and $c^{(k)}(t)$ are their time-dependent coefficients called *control sequences*. Maxima and minima of various combinations of the fidelity functional and the running cost with respect to control sequences are the subject of the mathematical branch of optimal control theory.

8.1 Gradient Ascent Pulse Engineering

A popular approach in spin dynamics, called *gradient ascent pulse engineering* (GRAPE [267]) is to discretise time and to work in the piecewise-constant control sequence approximation. On a finite grid of time points $0 \le t_n \le T$, the control sequence becomes a vector:

$$c^{(k)}(t) = c_n^{(k)}, \quad \mathcal{D}(t) = \mathcal{D}_n, \qquad t_{n-1} \le t < t_n \tag{8.4}$$

and the task of optimising a target functional becomes a finite-dimensional optimisation problem with respect to the elements of that vector.

Numerical optimisation is well researched [268, 269]. Because control sequence dependence in Eq. (8.1) is continuous and differentiable, gradient descent [270], conjugate gradients [271], quasi-Newton [272], and Newton–Raphson [144, 273] families of methods work well—the problem is reduced to computing first and (optionally) second derivatives of the target functional with respect to control amplitudes. Running cost functionals and their derivatives (Sect. 8.3.5) are usually uncomplicated; in the piecewise-constant approximation, the derivatives of the fidelity functional are

$$\begin{aligned} \frac{\partial f}{\partial c_n^{(k)}} &= \langle \boldsymbol{\delta} | \mathcal{P}_N \cdots \mathcal{P}_{n+1} \frac{\partial \mathcal{P}_n}{\partial c_n^{(k)}} \mathcal{P}_{n-1} \cdots \mathcal{P}_1 | \boldsymbol{\rho}_0 \rangle \\ \frac{\partial^2 f}{\partial c_n^{(k)} \partial c_m^{(j)}} &= \langle \boldsymbol{\delta} | \mathcal{P}_N \cdots \mathcal{P}_{n+1} \frac{\partial \mathcal{P}_n}{\partial c_n^{(k)}} \mathcal{P}_{n-1} \cdots \mathcal{P}_{m+1} \frac{\partial \mathcal{P}_m}{\partial c_m^{(j)}} \mathcal{P}_{m-1} \cdots \mathcal{P}_1 | \boldsymbol{\rho}_0 \rangle \end{aligned} \tag{8.5}$$

where slice propagators (Sect. 4.9.5) apply the time evolution through each time step Δt_n:

$$\mathcal{P}_n = \exp\left[-i \left(\mathcal{D}_n + \sum_k c_n^{(k)} \mathcal{C}_k \right) \Delta t_n \right] \tag{8.6}$$

The task is, therefore, reduced to calculating directional derivatives of matrix exponentials. The only term in Eq. (8.5) that depends on $c_n^{(k)}$ is $\mathcal{P}_n$. Therefore, the following scheme may be used:

$$
\begin{array}{c}
f = \langle \boldsymbol{\delta} | \mathcal{P}_N \mathcal{P}_{N-1} \cdots \mathcal{P}_{n+1} \mathcal{P}_n \overbrace{\mathcal{P}_{n-1} \cdots \mathcal{P}_2 \mathcal{P}_1 | \boldsymbol{\rho}_0 \rangle}^{\text{evolve forward from source}} \\
\uparrow \\
\left[\dfrac{\partial}{\partial c_n^{(k)}} \mathcal{P}_n \right] \\
\uparrow \\
f = \underbrace{\langle \boldsymbol{\delta} | \mathcal{P}_N \mathcal{P}_{N-1} \cdots \mathcal{P}_{n+1}}_{\text{evolve back from destination}} \mathcal{P}_n \mathcal{P}_{n-1} \cdots \mathcal{P}_2 \mathcal{P}_1 | \boldsymbol{\rho}_0 \rangle
\end{array}
\tag{8.7}
$$

Thus, the system trajectory needs to be calculated once forward from the initial condition and once backward from the destination state. Then scalar products between the pieces of the two trajectories should be taken with the propagator derivative at each step. The calculation of the fidelity gradient, therefore, requires $2N$ exponential-times-vector operations (forward and backward time propagation) and NK calculations of $\langle \mathbf{a} | \mathcal{P}' | \mathbf{b} \rangle$ inner products, where N is the number of time discretisation points and K is the number of control channels. Compared to the cost of estimating the same gradient using a finite-difference wrapper around a pre-existing simulation script (at least N^2K exponential-times-vector operations), this is a very efficient procedure and the reason for the popularity of GRAPE [267].

8.2 Derivatives of Spin Dynamics Simulations

Derivatives of matrix functions occur in multiple locations in this book, but their central role in quantum control theory warrants a detailed exposition in this chapter. The derivations—mostly skipped—for the relations reported below are all obtained using two strategies: power series expansion-differentiation-resummation and differentiation of explicitly indexed expressions. We use the *consistent numerator layout* for array derivatives: if elements of arrays $\mathbf{A}$ and $\mathbf{B}$ are indexed as $a_{ijk\ldots}^{\alpha\beta\gamma\ldots}$ and $b_{pqr\ldots}^{\kappa\mu\nu\ldots}$ (top indices for columns and bottom indices for rows), then the elements of $\mathbf{C} = \partial \mathbf{A} / \partial \mathbf{B}$ are indexed as

$$
c_{ijk\ldots\kappa\mu\nu\ldots}^{\alpha\beta\gamma\ldots pqr\ldots} = \partial a_{ijk\ldots}^{\alpha\beta\gamma\ldots} \Big/ \partial b_{pqr\ldots}^{\kappa\mu\nu\ldots}
\tag{8.8}
$$

In particular, this means that $\partial \mathbf{x} / \partial y$ is a column vector, $\partial x / \partial \mathbf{y}$ is a row vector, $\partial \mathbf{x} / \partial y$ is a matrix of the same size as $\mathbf{X}$, and $\partial x / \partial \mathbf{Y}$ is a matrix of the same size as $\mathbf{Y}^{\mathrm{T}}$.

8.2.1 Elementary Matrix Calculus

The identities collected in this section may be proven by applying scalar calculus relations to element-by-element expressions. The following vector-by-vector derivatives will be useful below:

$$\begin{aligned} &\frac{\partial \mathbf{x}}{\partial \mathbf{x}} = \mathbf{1}, && \frac{\partial}{\partial \mathbf{x}}(\mathbf{A}\mathbf{x}) = \mathbf{A}, \\ &\frac{\partial}{\partial \mathbf{x}}\left(\mathbf{x}^{\mathrm{T}}\mathbf{A}\right) = \mathbf{A}^{\mathrm{T}}, && \frac{\partial}{\partial \mathbf{x}}\mathbf{f}[\mathbf{y}(\mathbf{x})] = \frac{\partial \mathbf{f}}{\partial \mathbf{y}}\frac{\partial \mathbf{y}}{\partial \mathbf{x}} \end{aligned} \tag{8.9}$$

Scalar-by-vector product and chain rules inherit the structure from the scalar-by-scalar case:

$$\frac{\partial}{\partial \mathbf{x}}[f(\mathbf{x})g(\mathbf{x})] = f(\mathbf{x})\frac{\partial g}{\partial \mathbf{x}} + g(\mathbf{x})\frac{\partial f}{\partial \mathbf{x}}, \qquad \frac{\partial}{\partial \mathbf{x}}f[g(\mathbf{x})] = \frac{\partial f}{\partial g}\frac{\partial g}{\partial \mathbf{x}} \tag{8.10}$$

In particular, for inner products:

$$\frac{\partial}{\partial \mathbf{x}}\left(\mathbf{x}^{\mathrm{T}}\mathbf{A}\mathbf{x}\right) = \mathbf{x}^{\mathrm{T}}\left(\mathbf{A}+\mathbf{A}^{\mathrm{T}}\right), \qquad \frac{\partial^2}{\partial \mathbf{x}\partial \mathbf{x}^{\mathrm{T}}}\left(\mathbf{x}^{\mathrm{T}}\mathbf{A}\mathbf{x}\right) = \mathbf{A}+\mathbf{A}^{\mathrm{T}} \tag{8.11}$$

This includes the 2-norm (Sect. 1.4.5) as a special case:

$$\frac{\partial}{\partial \mathbf{x}}\|\mathbf{x}\|_2 = \frac{\mathbf{x}^{\mathrm{T}}}{\|\mathbf{x}\|_2}, \qquad \frac{\partial}{\partial \mathbf{x}}\|\mathbf{x}\|_2^2 = 2\mathbf{x}^{\mathrm{T}}, \qquad \frac{\partial}{\partial \mathbf{x}}\frac{\mathbf{x}}{\|\mathbf{x}\|_2} = \frac{\mathbf{1}}{\|\mathbf{x}\|_2} - \frac{\mathbf{x}\mathbf{x}^{\mathrm{T}}}{\|\mathbf{x}\|_2^3} \tag{8.12}$$

Pertinent scalar-by-matrix derivatives are those of inner products, traces and of the Frobenius norm:

$$\begin{gathered} \tfrac{\partial}{\partial \mathbf{X}}\left(\mathbf{a}^{\mathrm{T}}\mathbf{X}\mathbf{b}\right) = \mathbf{b}\mathbf{a}^{\mathrm{T}}, \qquad \tfrac{\partial}{\partial \mathbf{X}}\mathrm{Tr}[\mathbf{X}] = \mathbf{1}, \qquad \tfrac{\partial}{\partial \mathbf{X}}\mathrm{Tr}[f(\mathbf{X})] = f'(\mathbf{X}), \\ \tfrac{\partial}{\partial \mathbf{X}}\mathrm{Tr}[\mathbf{A}\mathbf{X}\mathbf{B}] = \mathbf{B}\mathbf{A}, \qquad \tfrac{\partial}{\partial \mathbf{X}}\|\mathbf{X}\|_{\mathrm{F}}^2 = 2\mathbf{X} \end{gathered} \tag{8.13}$$

where $f'(\mathbf{X})$ is the derivative of the scalar function $f(x)$ evaluated with a matrix argument $\mathbf{X}$ using the Taylor series extension:

$$f(x) = \sum_{n=0}^{\infty}\frac{f^{(n)}(0)}{n!}x^n \quad \Rightarrow \quad f(\mathbf{X}) = \sum_{n=0}^{\infty}\frac{f^{(n)}(0)}{n!}\mathbf{X}^n \tag{8.14}$$

Pertinent matrix-by-scalar relations are

$$\begin{aligned} &\frac{\partial \mathbf{X}^{-1}}{\partial a} = -\mathbf{X}^{-1}\frac{\partial \mathbf{X}}{\partial a}\mathbf{X}^{-1}, \qquad \frac{\partial}{\partial a}f(a\mathbf{X}) = \mathbf{X}f'(a\mathbf{X}) \\ &\frac{\partial^2 \mathbf{X}^{-1}}{\partial a \partial b} = \mathbf{X}^{-1}\left(\frac{\partial \mathbf{X}}{\partial a}\mathbf{X}^{-1}\frac{\partial \mathbf{X}}{\partial b} - \frac{\partial^2 \mathbf{X}}{\partial a \partial b} + \frac{\partial \mathbf{X}}{\partial b}\mathbf{X}^{-1}\frac{\partial \mathbf{X}}{\partial a}\right)\mathbf{X}^{-1} \end{aligned} \tag{8.15}$$

A large collection of other relations of this kind may be found in the Matrix Cookbook [274].

8.2.2 Eigensystem Derivatives

Hellmann–Feynman theorem was covered from the analytical perspective in Sect. 2.6.8; here we focus on its implementation in matrix representations.

8.2.2.1 Non-degenerate Eigenvalues

The definition $\mathbf{H}\mathbf{v}_k = \lambda_k \mathbf{v}_k$ of a normalised eigenvector $\mathbf{v}_k$ of a matrix $\mathbf{H}$ that depends on a parameter α may be differentiated and rearranged into

$$[\mathbf{H} - \lambda_k \mathbf{1}]\mathbf{v}'_k = -[\mathbf{H}' - \lambda'_k \mathbf{1}]\mathbf{v}_k \tag{8.16}$$

where prime indicates a derivative with respect to α and $\mathbf{1}$ is the identity matrix. Both matrices in the square brackets are singular because one eigenvalue of $\mathbf{H}$ matches λ_k and the corresponding eigenvalue of $\mathbf{H}'$ matches λ'_k. When all eigenvalues λ_k are different, the problem is regularised (following the discussion given in Sect. 2.6.8) by requiring that $\mathbf{v}_k^\dagger \mathbf{v}'_k = 0$ and augmenting Eq. (8.16) as follows:

$$\begin{pmatrix} \mathbf{H} - \lambda_k \mathbf{1} & -\mathbf{v}_k \\ \mathbf{v}_k^\dagger & 0 \end{pmatrix} \begin{pmatrix} \mathbf{v}'_k \\ \lambda'_k \end{pmatrix} = \begin{pmatrix} -\mathbf{H}'\mathbf{v}_k \\ 0 \end{pmatrix} \tag{8.17}$$

This equation may be solved using standard methods such as GMRES [151]. A convenient numerical accuracy check is to compare the resulting λ'_k to $\mathbf{v}_k^\dagger \mathbf{H}' \mathbf{v}_k$.

8.2.2.2 Degenerate Eigenvalues

When the eigensystem is degenerate, we consider each degenerate block individually:

$$\mathbf{H}\mathbf{V}_k = \mathbf{V}_k \mathbf{\Lambda}_k \tag{8.18}$$

where $\mathbf{V}_k$ is a slim matrix whose columns are normalised eigenvectors of the kth degenerate set, and $\mathbf{\Lambda}_k = \lambda_k \mathbf{1}$. Numerical diagonalisation of $\mathbf{H}$ will return some linear combination $\mathbf{V}_k \mathbf{P}_k^\dagger$ of those eigenvectors, where $\mathbf{P}_k$ is an unknown unitary matrix.

When the Hellmann–Feynman theorem (Sect. 2.6.8) is applied, the presence of $\mathbf{P}_k$ introduces an unknown unitary transformation that mixes the derivatives:

$$\mathbf{\Lambda}'_k = \mathbf{P}_k \mathbf{V}_k^\dagger \mathbf{H}' \mathbf{V}_k \mathbf{P}_k^\dagger \tag{8.19}$$

This mixing is undone by computing and diagonalising $\mathbf{V}_k^\dagger \mathbf{H}' \mathbf{V}_k$; this yields the matrix $\mathbf{P}_k$ that makes $\mathbf{\Lambda}'_k(\alpha)$ diagonal and unmixes the eigenvector block. Differentiating Eq. (8.18) then yields:

$$[\mathbf{H} - \lambda_k \mathbf{1}]\mathbf{V}'_k = \mathbf{V}_k \boldsymbol{\Lambda}'_k - \mathbf{H}'\mathbf{V}_k \tag{8.20}$$

which is analogous to Eq. (8.16) and singular for the same reason. To stabilise this equation, we once again use the normalisation condition discussed in Sect. 2.6.8:

$$\mathbf{V}_k^\dagger \mathbf{V}_k = \mathbf{1} \quad \Rightarrow \quad \mathbf{V}_k^\dagger \mathbf{V}'_k = \mathbf{0} \tag{8.21}$$

and form the block matrix equation

$$\begin{pmatrix} \mathbf{H} - \lambda_k \mathbf{1} & -\mathbf{V}_k \\ \mathbf{V}_k^\dagger & \mathbf{0} \end{pmatrix} \begin{pmatrix} \mathbf{V}'_k \\ \boldsymbol{\Lambda}'_k \end{pmatrix} = \begin{pmatrix} -\mathbf{H}'\mathbf{V}_k \\ \mathbf{0} \end{pmatrix} \tag{8.22}$$

which is solved by multiplying both sides by the inverse of the (now regular) leftmost matrix. Comparing the resulting $\boldsymbol{\Lambda}'_k$ to the one obtained in Eq. (8.19) provides a practical test of numerical accuracy.

8.2.3 Trajectory Derivatives

Diagonalisation is rarely an efficient way forward in spin dynamics simulations due to its unfavourable computational complexity and storage requirement scaling with the matrix size. Time-domain methods can be more efficient (Chap. 7); quantum control problems are also time-domain. It is, therefore, necessary to consider derivatives of system trajectories and of their Fourier transforms.

8.2.3.1 Derivative Superoperator

In Liouville space, one may seek a superoperator $\mathcal{D}_\alpha(t)$ that acts on the density operator and returns its derivative $\boldsymbol{\rho}'_\alpha(t)$ with respect to a parameter α:

$$\boldsymbol{\rho}'_\alpha(t) = \mathcal{D}_\alpha(t)\boldsymbol{\rho}(t). \tag{8.23}$$

The infinitesimal time step solution of LvN equation

$$\boldsymbol{\rho}(t+\Delta t) = \mathcal{P}\boldsymbol{\rho}(t), \qquad \mathcal{P} = \exp[-i\mathcal{L}(t)\Delta t], \tag{8.24}$$

may be differentiated with respect to the parameter:

$$\boldsymbol{\rho}'_\alpha(t+\Delta t) = \mathcal{P}'_\alpha\boldsymbol{\rho}(t) + \mathcal{P}\boldsymbol{\rho}'_\alpha(t) \tag{8.25}$$

Combining this with the definition in Eq. (8.23) and taking the limit $\Delta t \to 0$ yields

$$\frac{\partial \mathcal{D}_\alpha}{\partial t} = -i[\mathcal{L}, \mathcal{D}_\alpha] - i\mathcal{L}'_\alpha \tag{8.26}$$

where $\mathcal{L}'_\alpha$ is a Liouvillian derivative with respect to the simulation parameter in question. The initial conditions are in most cases $\mathcal{D}_\alpha(0) = \mathbf{0}$ and $\boldsymbol{\rho}'_\alpha(0) = \mathbf{0}$, reflecting the fact that the simulation starts from a state that does not depend on α. Higher order derivatives may be obtained in a similar way [342].

8.2.3.2 Derivative Co-propagation

Evolving a superoperator under Eq. (8.26) is computationally inefficient. A better approach [342] uses an equation of motion for the derivative of the density matrix. From Eq. (8.25), we already know the solution of that equation; taking the $\Delta t \to 0$ limit yields

$$\frac{\partial \boldsymbol{\rho}'_\alpha}{\partial t} = -i\mathcal{L}\boldsymbol{\rho}'_\alpha - i\mathcal{L}'_\alpha\boldsymbol{\rho} \tag{8.27}$$

Similar reasoning leads to the following equation of motion for the second derivative:

$$\frac{\partial \boldsymbol{\rho}''_{\alpha\beta}}{\partial t} = -i\mathcal{L}''_{\alpha\beta}\boldsymbol{\rho} - i\mathcal{L}'_\alpha\boldsymbol{\rho}'_\beta - i\mathcal{L}'_\beta\boldsymbol{\rho}'_\alpha - i\mathcal{L}\boldsymbol{\rho}''_{\alpha\beta} \tag{8.28}$$

and likewise for higher derivatives. Solutions are obtained by taking appropriate derivatives of the time propagation rule in Eq. (8.24). In the absence of dissipative dynamics, Hilbert space versions are

$$\begin{aligned} \frac{\partial \boldsymbol{\rho}'_\alpha}{\partial t} &= -i\left[\mathbf{H}, \boldsymbol{\rho}'_\alpha\right] - i\left[\mathbf{H}'_\alpha, \boldsymbol{\rho}\right] \\ \frac{\partial \boldsymbol{\rho}''_{\alpha\beta}}{\partial t} &= -i\left[\mathbf{H}''_{\alpha\beta}, \boldsymbol{\rho}\right] - i\left[\mathbf{H}'_\alpha, \boldsymbol{\rho}'_\beta\right] - i\left[\mathbf{H}'_\beta, \boldsymbol{\rho}'_\alpha\right] - i\left[\mathbf{H}, \boldsymbol{\rho}''_{\alpha\beta}\right] \end{aligned} \tag{8.29}$$

and likewise for higher derivatives. In the optimal control context, the evolution generator is commonly a linear function of the parameter; in that case, second and higher derivatives of $\mathcal{L}$ and $\mathbf{H}$ are zero.

8.2.3.3 Frequency-Domain Derivatives

When the evolution generator $\mathcal{L} = \mathcal{H} + i\mathcal{R}$ is time-independent and has a negative-definite definite dissipative part $\mathcal{R}$, the positive-time Fourier transform of the Liouville–von Neumann equation

$$\begin{cases} \dfrac{\partial}{\partial t}\rho(t) = -i\mathcal{L}\boldsymbol{\rho}(t) \\ \boldsymbol{\rho}(0) = \boldsymbol{\rho}_0 \end{cases} \quad \Rightarrow \quad \begin{cases} (i\omega\mathbf{1} + i\mathcal{L})\boldsymbol{\rho}(\omega) = \boldsymbol{\rho}_0 \\ \boldsymbol{\rho}(\omega) = \int\limits_0^\infty \boldsymbol{\rho}(t)\mathrm{e}^{-i\omega t}\mathrm{d}t \end{cases} \tag{8.30}$$

can be differentiated with respect to a parameter α, yielding

$$\boldsymbol{\rho}'_\alpha(\omega) = -(\omega\mathbf{1} + \mathcal{L})^{-1}\mathcal{L}'_\alpha\boldsymbol{\rho}(\omega) = i(\omega\mathbf{1} + \mathcal{L})^{-1}\mathcal{L}'_\alpha(\omega\mathbf{1} + \mathcal{L})^{-1}\boldsymbol{\rho}_0 \tag{8.31}$$

where GMRES [151] is recommended for the matrix-inverse-times-vector operations and logistical optimisations are possible because $\mathcal{L}'_\alpha$ is commonly very sparse and/or low-rank. The advantage here is that specific frequency points may be monitored without the need to simulate the entire trajectory.

Similar reasoning leads to the following equation for the second derivative

$$\begin{aligned}\boldsymbol{\rho}''_{\alpha\beta}(\omega) &= -(\omega\mathbf{1}+\mathcal{L})^{-1}\left(\mathcal{L}'_\alpha\boldsymbol{\rho}'_\beta(\omega)+\mathcal{L}'_\beta\boldsymbol{\rho}'_\alpha(\omega)+\mathcal{L}''_{\alpha\beta}\boldsymbol{\rho}(\omega)\right)\\ &= -i(\omega\mathbf{1}+\mathcal{L})^{-1}\left(\mathcal{L}'_\alpha(\omega\mathbf{1}+\mathcal{L})^{-1}\mathcal{L}'_\beta+\mathcal{L}'_\beta(\omega\mathbf{1}+\mathcal{L})^{-1}\mathcal{L}'_\alpha-\mathcal{L}''_{\alpha\beta}\right)(\omega\mathbf{1}+\mathcal{L})^{-1}\boldsymbol{\rho}_0\end{aligned} \tag{8.32}$$

with similar logistical optimisations regarding the sparsity and/or low rank of the first derivatives of the evolution generator, and the fact that its second derivatives are zero when the dependence of $\mathcal{L}$ on the parameters α and β is linear [342].

8.2.4 Matrix Exponential Derivatives

Much of this chapter assumes that we have a way of computing derivatives of matrix exponentials, specifically the directional derivatives $\exp[-i(\mathbf{A}+\alpha\mathbf{B})t]$ with respect to α. Some methods (in the order of increasing sophistication and efficiency) are discussed in this section. The definition of time slice propagator given in Eq. (8.6) will be used throughout.

8.2.4.1 Finite Differences

Standard finite difference schemes [195] remain applicable in the matrix case, for example:

$$\frac{\partial\mathcal{P}_n}{\partial c_n^{(k)}}=\frac{\mathcal{P}_n\left(\ldots,c_n^{(k)}+h,\ldots\right)-\mathcal{P}_n\left(\ldots,c_n^{(k)},\ldots\right)}{h}+O(h) \tag{8.33}$$

$$\frac{\partial\mathcal{P}_n}{\partial c_n^{(k)}}=\frac{\mathcal{P}_n\left(\ldots,c_n^{(k)}+h,\ldots\right)-\mathcal{P}_n\left(\ldots,c_n^{(k)}-h,\ldots\right)}{2h}+O\left(h^2\right) \tag{8.34}$$

where the amplitude of the kth control at the nth time point is varied by a finite amount h. Equations (8.33) and (8.34) are simple examples of a large class of numerical finite-difference approximations for the derivative [195]. The balance to be maintained in this approach is between the approximation accuracy, the numerical accuracy in finite precision arithmetic, and the computational cost [275].

For the approximation accuracy analysis, consider the worst-case scenario, where the drift dynamics is dominated by the largest singular value ω_0 of $\mathcal{D}_n+\sum_{m\neq k}c_n^{(m)}\mathcal{C}_m$ and the control dynamics is dominated by the largest singular value $c_n^{(k)}\omega_k$ of $c_n^{(k)}\mathcal{C}_k$. In that case, the fastest harmonic in the system that would dominate the finite difference error is

$$f\left(c_n^{(k)}\right) = \exp\left[-i\left(\omega_0 + c_n^{(k)}\omega_k\right)\Delta t_n\right] = \exp[-i\omega_0\Delta t]\exp\left[-ic_n^{(k)}\omega_k\Delta t_n\right] \quad (8.35)$$

where Δt_n is the time slice duration. The largest singular values in question are the definition of the matrix 2-norm (Sect. 1.4.5) for which upper bounds are computationally affordable (Sect. 9.3.1).

Consider first the forward finite difference scheme in Eq. (8.33). The approximation accuracy analysis uses the Taylor expansion of the incremented term, for which the integral form of the remainder [342] yields

$$f\left(c_n^{(k)}+h\right) = f\left(c_n^{(k)}\right) + f'\left(c_n^{(k)}\right)h + \int\limits_{c_n^{(k)}}^{c_n^{(k)}+h} f''(\alpha)\left(c_n^{(k)}+h-\alpha\right)\mathrm{d}\alpha \quad (8.36)$$

Inserting this into the forward finite difference approximation for $f'\left(c_n^{(k)}\right)$ yields the expression for the approximation error:

$$\begin{aligned} f'\left(c_n^{(k)}\right) &= \frac{f\left(c_n^{(k)}+h\right) - f\left(c_n^{(k)}\right)}{h} - R\left(c_n^{(k)},h\right) \\ R\left(c_n^{(k)},h\right) &= \frac{1}{h}\int\limits_{c_n^{(k)}}^{c_n^{(k)}+h} f''(\alpha)\left(c_n^{(k)}+h-\alpha\right)\mathrm{d}\alpha \end{aligned} \quad (8.37)$$

After substituting Eq. (8.35) and a few rounds of simplifications, the ratio of the derivative approximation error $R(x,h)$ to the derivative itself becomes

$$\left|\frac{R\left(c_n^{(k)},h\right)}{f'\left(c_n^{(k)}\right)}\right| = \frac{|h\omega_k\Delta t_n|}{2} + O\left(|h\omega_k\Delta t_n|^2\right) \quad (8.38)$$

Thus, the accuracy of the forward finite difference approximation improves linearly when a reduction is made in the stencil step size h, the 2-norm of the control operator ω_k, or the time step size Δt_n; there is no dependence on the 2-norm of the drift.

A similar calculation (where the Taylor expansion needs to be carried to the h^2 term) for the central finite difference scheme in Eq. (8.34) yields a more favourable quadratic improvement in the accuracy when the same three parameters are reduced:

$$\left|\frac{R\left(c_n^{(k)},h\right)}{f'\left(c_n^{(k)}\right)}\right| = |h\omega_k\Delta t_n|^2 + O\left(|h\omega_k\Delta t_n|^3\right) \tag{8.39}$$

The same accuracy analysis may be performed for higher order finite difference approximations. An important logistical optimisation is that, in the Liouville space formulation of GRAPE given in Eq. (8.7), only the action by the propagator derivative on a vector is needed. This means that a simple reordering of operations in Eqs. (8.33), (8.34), and their higher order extensions would reduce the problem to efficient matrix exponential-times-vector calls (Sect. 4.9.6).

8.2.4.2 Complex Step Method

In the cases where matrices $\mathbf{A}$ and $\mathbf{B}$ are real (for example, in Bloch equation models of magnetic resonance imaging), the derivative of an analytic function $f(\mathbf{A})$ in the direction $\mathbf{B}$ may be approximated by $\mathrm{Im}\,[f(\mathbf{A}+ih\mathbf{B})]/h$ for a sufficiently small value of the real increment h. Accuracy analysis along the same lines as in the previous section reveals quadratic accuracy with a more favourable overall multiplier than the central finite difference in Eq. (8.34):

$$\left|\frac{R\left(c_n^{(k)},h\right)}{f'\left(c_n^{(k)}\right)}\right| = \frac{|h\omega_k\Delta t_n|^2}{6} + O\left(|h\omega_k\Delta t_n|^3\right) \tag{8.40}$$

This method has better round-off error tolerance in finite precision arithmetic [276] and lower computational cost than Eq. (8.34).

8.2.4.3 Eigensystem Differentiation

When the propagator is computed using the (very inefficient) generator diagonalisation method:

$$\mathbf{A}=\mathbf{V}\mathbf{D}\mathbf{V}^\dagger \quad \Rightarrow \quad \mathrm{e}^{\mathbf{A}} = \mathbf{V}\mathrm{e}^{\mathbf{D}}\mathbf{V}^\dagger \tag{8.41}$$

where $\mathbf{D}$ is a diagonal matrix of eigenvalues (its exponential is, therefore, computed element-wise), the eigenvector array $\mathbf{V}$ may be re-used to calculate the derivative with respect to a parameter α

$$\frac{\partial}{\partial\alpha}\mathrm{e}^{\mathbf{A}}=\frac{\partial\mathbf{V}}{\partial\alpha}\mathrm{e}^{\mathbf{D}}\mathbf{V}^\dagger + \mathbf{V}\frac{\partial\mathbf{D}}{\partial\alpha}\mathrm{e}^{\mathbf{D}}\mathbf{V}^\dagger + \mathbf{V}\mathrm{e}^{\mathbf{D}}\frac{\partial\mathbf{V}^\dagger}{\partial\alpha} \tag{8.42}$$

with the expressions for $\partial\mathbf{V}/\partial\alpha$ and $\partial\mathbf{D}/\partial\alpha$ are given in Sect. 8.2.2 that deals with eigensystem differentiation. This method is too expensive for practical numerical calculations.

8.2.4.4 Differentiation of Power Series

Since the best practical method of computing exponentials of very sparse matrices is the Taylor series with scaling and squaring (Sect. 4.9.5), differentiating that series is a possibility:

$$\begin{aligned}&\mathfrak{L}_n = \mathfrak{D}_n + \sum_k c_n^{(k)} \mathfrak{C}_k, \qquad \frac{\partial}{\partial c_n^{(k)}} \exp(-i\mathfrak{L}_n \Delta t_n)\\ &= \frac{\partial}{\partial c_n^{(k)}} \sum_{m=0}^{\infty} \frac{(-i\Delta t_n)^m}{m!} \mathfrak{L}_n^m = \sum_{m=0}^{\infty} \frac{(-i\Delta t_n)^m}{m!} \frac{\partial}{\partial c_n^{(k)}} \underbrace{(\mathfrak{L}_n \mathfrak{L}_n \cdots \mathfrak{L}_n)}_{m \text{ times}}\\ &= \sum_{m=0}^{\infty} \frac{(-i\Delta t_n)^m}{m!} \sum_{r=1}^{m} \Bigg(\mathfrak{L}_n \mathfrak{L}_n \cdots \underbrace{\mathfrak{C}_k}_{r^{\text{th}} \text{ position}} \cdots \mathfrak{L}_n \Bigg)\end{aligned} \tag{8.43}$$

The second sum appears because $\mathfrak{L}_n$ and $\mathfrak{C}_k$ do not necessarily commute; this is inconvenient. A lengthy rearrangement [277] yields a more computationally friendly single sum:

$$\begin{aligned}&\frac{\partial}{\partial c_n^{(k)}} \exp[-i\mathfrak{L}_n \Delta t_n] = \exp[-i\mathfrak{L}_n \Delta t_n]\\ &\times \left(-i\mathfrak{C}_k \Delta t_n + \frac{\Delta t_n^2}{2} [\mathfrak{L}_n, \mathfrak{C}_k] + \frac{i\Delta t_n^3}{6} [\mathfrak{L}_n, [\mathfrak{L}_n, \mathfrak{C}_k]] - \frac{\Delta t_n^4}{24} [\mathfrak{L}_n, [\mathfrak{L}_n, [\mathfrak{L}_n, \mathfrak{C}_k]]] + \ldots \right)\end{aligned} \tag{8.44}$$

where the summation of the series is to be continued until the desired accuracy (as indicated by the residual norm) is achieved. Convergence of the Taylor series for the matrix exponential is monotonic (and round-off losses in finite precision arithmetic are well-behaved) when the 2-norm of $-i\mathfrak{L}_n \Delta t_n$ is inside [0, 1] interval. That is achieved by scaling down the time step before the exponentiation and then squaring the propagator back up to the original time step (Sect. 4.9.5). The method may be extended to derivatives using the product rule:

$$\frac{\partial}{\partial \alpha} \exp(\mathbf{A}) = \exp\left(\frac{\mathbf{A}}{2}\right) \left[\frac{\partial}{\partial \alpha} \exp\left(\frac{\mathbf{A}}{2}\right)\right] + \left[\frac{\partial}{\partial \alpha} \exp\left(\frac{\mathbf{A}}{2}\right)\right] \exp\left(\frac{\mathbf{A}}{2}\right) \tag{8.45}$$

The same efficiency caveats of matrix–vector multiplication apply—because Eq. (8.7) is ultimately a matrix–vector product, operations may be reordered accordingly (Sect. 4.9.6).

8.2.4.5 Auxiliary Matrix Method

At the time of writing, the best method for computing propagator derivatives is due to van Loan [276], who noted that matrix exponential derivatives can be expressed via integrals, and that the integrals are solutions of linear block matrix differential equations, which may be solved via block matrix exponentials—the problem of computing exponential derivatives and integrals is reduced to computing exponentials of bigger matrices. Van Loan's method was subsequently refined by Carbonell et al., who derived a convenient expression for the exponential of a block-bidiagonal *auxiliary matrix* [277]:

$$\mathbf{M}=\begin{pmatrix} \mathbf{A}_{11} & \mathbf{A}_{12} & \mathbf{0} & \mathbf{0} & \mathbf{0} \\ \mathbf{0} & \mathbf{A}_{22} & \mathbf{A}_{23} & \mathbf{0} & \mathbf{0} \\ \mathbf{0} & \mathbf{0} & \mathbf{A}_{33} & \ddots & \mathbf{0} \\ \mathbf{0} & \mathbf{0} & \mathbf{0} & \ddots & \mathbf{A}_{k-1,k} \\ \mathbf{0} & \mathbf{0} & \mathbf{0} & \mathbf{0} & \mathbf{A}_{kk} \end{pmatrix}, \qquad \exp(\mathbf{M}t)=\begin{pmatrix} \mathbf{B}_{11} & \mathbf{B}_{12} & \mathbf{B}_{13} & \cdots & \mathbf{B}_{1k} \\ \mathbf{0} & \mathbf{B}_{22} & \mathbf{B}_{23} & \cdots & \mathbf{B}_{2k} \\ \mathbf{0} & \mathbf{0} & \mathbf{B}_{33} & \ddots & \vdots \\ \mathbf{0} & \mathbf{0} & \mathbf{0} & \ddots & \mathbf{B}_{k-1,k} \\ \mathbf{0} & \mathbf{0} & \mathbf{0} & \mathbf{0} & \mathbf{B}_{kk} \end{pmatrix}$$

$$\mathbf{B}_{1k}=\int_0^t dt_1\int_0^{t_1} dt_2 \ldots \int_0^{t_{k-2}} dt_{k-1}\left\{e^{\mathbf{A}_{11}(t-t_1)}\mathbf{A}_{12}e^{\mathbf{A}_{22}(t_1-t_2)}\mathbf{A}_{23}\ldots\mathbf{A}_{k-1,k}e^{\mathbf{A}_{kk}t_{k-1}}\right\} \tag{8.46}$$

We have already seen a special case of this in Sect. 6.3.1, where it provided a way of computing Redfield's integral. Because the kth derivative $\mathbf{D}_k(t)$ of $\exp[-i(\mathbf{H}_0+\alpha\mathbf{H}_1)t]$ with respect to α at $\alpha=0$ obeys the following recurrence relation:

$$\begin{aligned} &\mathbf{D}_k(t)=ke^{\mathbf{A}t}\int_0^t e^{-\mathbf{A}\alpha}\mathbf{B}\mathbf{D}_{k-1}(\alpha)d\alpha, \\ &\mathbf{D}_0(t)=e^{\mathbf{A}t}, \qquad \mathbf{A}=-i\mathbf{H}_0, \qquad \mathbf{B}=-i\mathbf{H}_1 \end{aligned} \tag{8.47}$$

the derivatives may be computed by constructing and exponentiating a block-bidiagonal matrix:

$$\begin{aligned} &\mathbf{M}=\begin{pmatrix} \mathbf{A} & \mathbf{B} & \mathbf{0} & \mathbf{0} & \mathbf{0} \\ \mathbf{0} & \mathbf{A} & \mathbf{B} & \mathbf{0} & \mathbf{0} \\ \mathbf{0} & \mathbf{0} & \mathbf{A} & \ddots & \mathbf{0} \\ \mathbf{0} & \mathbf{0} & \mathbf{0} & \ddots & \mathbf{B} \\ \mathbf{0} & \mathbf{0} & \mathbf{0} & \mathbf{0} & \mathbf{A} \end{pmatrix} \\ &\exp(t\mathbf{M})=\begin{pmatrix} \mathbf{D}_0 & \mathbf{D}_1/1! & \mathbf{D}_2/2! & \cdots & \mathbf{D}_k/k! \\ \mathbf{0} & \mathbf{D}_0 & \mathbf{D}_1/1! & \cdots & \mathbf{D}_{k-1}/(k-1)! \\ \mathbf{0} & \mathbf{0} & \mathbf{D}_0 & \ddots & \vdots \\ \mathbf{0} & \mathbf{0} & \mathbf{0} & \ddots & \mathbf{D}_1/1! \\ \mathbf{0} & \mathbf{0} & \mathbf{0} & \mathbf{0} & \mathbf{D}_0 \end{pmatrix} \end{aligned} \tag{8.48}$$

and extracting blocks from the result. In the GRAPE context, for the first propagator derivative:

$$\begin{pmatrix} \mathcal{P}_n & \frac{\partial \mathcal{P}_n}{\partial c_n^{(k)}} \\ \mathbf{0} & \mathcal{P}_n \end{pmatrix} = \exp\left[-i \begin{pmatrix} \mathcal{L}_n & \mathcal{C}_k \\ \mathbf{0} & \mathcal{L}_n \end{pmatrix} \Delta t_n \right] \tag{8.49}$$

and similarly for higher derivatives, for example:

$$\begin{gathered} \begin{pmatrix} \mathcal{P}_n & \frac{\partial \mathcal{P}_n}{\partial c_n^{(k)}} & \mathcal{A}_n^{(kj)} \\ & \mathcal{P}_n & \frac{\partial \mathcal{P}_n}{\partial c_n^{(j)}} \\ & & \mathcal{P}_n \end{pmatrix} = \exp\left[-i \begin{pmatrix} \mathcal{L}_n & \mathcal{C}_k & \mathbf{0} \\ \mathbf{0} & \mathcal{L}_n & \mathcal{C}_j \\ \mathbf{0} & \mathbf{0} & \mathcal{L}_n \end{pmatrix} \Delta t_n \right], \\ \frac{\partial^2 \mathcal{P}_n}{\partial c_n^{(k)} \partial c_n^{(j)}} = \mathcal{A}_n^{(kj)} + \mathcal{A}_n^{(jk)} \end{gathered} \tag{8.50}$$

An advantage of auxiliary matrix techniques is logistical simplicity—assembling a block matrix, exponentiating it and extracting a block from the result is neater than summing a commutator series. The method is compatible with the efficiency savings described in Sect. 4.9.6 when only the action by the propagator derivative on a vector is needed.

8.2.5 GRAPE Derivative Translation

The efficiency of element-by-element waveform gradient calculation in GRAPE may be extended to arbitrary waveform basis sets. Let there be a real orthonormal basis set of waveforms:

$$\mathbf{W} = \begin{bmatrix} | & \cdots & | \\ \mathbf{w}_1 & \cdots & \mathbf{w}_M \\ | & \cdots & | \end{bmatrix}, \qquad \mathbf{W}^{\mathrm{T}} \mathbf{W} = \mathbf{1} \tag{8.51}$$

such that the control sequence $\mathbf{c}^{(k)}$ at kth channel has an expansion

$$\mathbf{c}^{(k)} = \sum_m \alpha_m^{(k)} \mathbf{w}_m = \mathbf{W} \alpha^{(k)} \quad \Leftrightarrow \quad c_n^{(k)} = \sum_m w_{nm} \alpha_m^{(k)} \tag{8.52}$$

Then derivatives of any function f of $\mathbf{c}^{(k)}$ vector are translated as follows into the corresponding derivatives with respect to the waveform basis expansion coefficient vector $\boldsymbol{\alpha}^{(k)}$:

$$\begin{aligned}\frac{\partial f}{\partial \alpha_m^{(k)}} = \sum_n \frac{\partial f}{\partial c_n^{(k)}} \frac{\partial c_n^{(k)}}{\partial \alpha_m^{(k)}} = \sum_n \frac{\partial f}{\partial c_n^{(k)}} w_{nm} \quad &\Leftrightarrow \quad \nabla_{\boldsymbol{\alpha}} f = \mathbf{W}^{\mathrm{T}} [\nabla_{\mathbf{c}} f] \\ \frac{\partial^2 f}{\partial \alpha_m^{(k)} \partial \alpha_{m'}^{(k')}} = \sum_{n,n'} \frac{\partial f}{\partial c_n^{(k)} \partial c_{n'}^{(k')}} w_{nm} w_{n'm'} & \end{aligned} \tag{8.53}$$

where m enumerates basis waveforms, n enumerates time points, and k enumerates control channels. These are special cases of the matrix chain rules discussed in Sect. 8.2.1; these relations connect the methods that use waveform basis sets to GRAPE algorithms [265, 275].

8.3 Optional Components

An advantage of GRAPE over other optimal control formalisms is the simplicity with which secondary considerations may be introduced. The options described in this section are implemented in *Spinach* [87].

8.3.1 Prefixes, Suffixes, and Dead Times

The definition of any fidelity functional based on the overlap between states is compatible with the presence of propagators that do not depend on the control sequence, for example:

$$f = \langle \boldsymbol{\delta} | \mathcal{P}_{\text{suff}} (\mathcal{P}_N \cdots \mathcal{P}_1) \mathcal{P}_{\text{pref}} | \boldsymbol{\rho}_0 \rangle \tag{8.54}$$

This is convenient in situations where the optimal control stage is positioned inside a larger experiment—preceding events may be packaged into the prefix propagator $\mathcal{P}_{\text{pref}}$, and the subsequent events into the suffix propagator $\mathcal{P}_{\text{suff}}$. The control problem is then reformulated as running from the initial condition $\mathcal{P}_{\text{pref}} | \boldsymbol{\rho}_0 \rangle$ with the projection onto the destination state $\langle \boldsymbol{\delta} | \mathcal{P}_{\text{suff}}$.

A special case of suffix propagator is dead time—a finite period between the last control operation and the start of the detection stage. This can be significant enough for many control problems to require pre-phased solutions: the system is prepared in a state that drifts into the destination during the specified dead time; a common example is echo-detected experiments in electron spin resonance and quadrupolar nuclear magnetic resonance (Fig. 8.1).

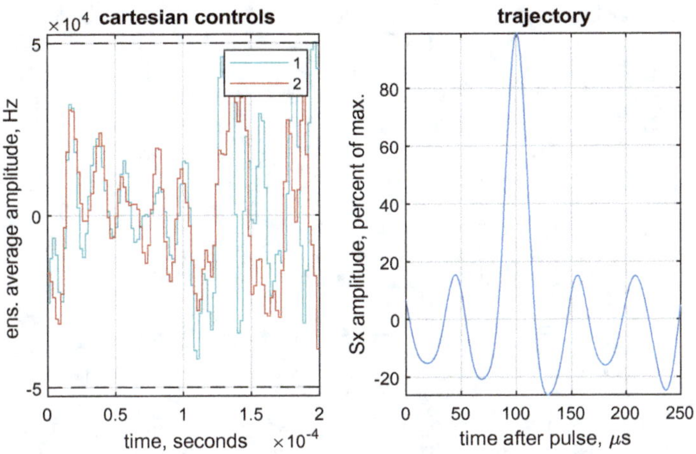

Fig. 8.1 An example of a pre-phasing optimal control pulse in quadrupolar magnetic resonance. The ensemble is $-CD_3$ alanine powder with a uniform distribution over orientations and B_1 nutation frequency from 40 to 60 kHz. A 100-point 200 μs radiofrequency pulse (left panel) is optimised to send all ensemble members (100 orientations on a finite grid) into $\mathbf{S}_X$ 100 μs after the end of the pulse (right panel)

8.3.2 Keyholes and Freeze Masks

It may be necessary to send the system or an ensemble through a specific state or a specific subspace $\mathfrak{L}$ at a particular time point. This is straightforward—one or more projection superoperators $\mathcal{W}_{\mathfrak{L}}$ onto the states or into the subspaces in question are inserted at the appropriate times:

$$f = \langle \boldsymbol{\delta} | \mathcal{P}_N \cdots \mathcal{P}_n \mathcal{W}_{\mathfrak{L}} \mathcal{P}_{n-1} \cdots \mathcal{P}_1 | \boldsymbol{\rho}_0 \rangle \tag{8.55}$$

The presence of such projectors does not affect the calculation of control sequence derivatives, but they reduce the fidelity when the system populates states outside $\mathfrak{L}$ between time slice $n-1$ and n. When, in the course of fidelity optimisation, the system eventually learns to pass through $\mathfrak{L}$, the projectors stop having an effect. A common example in magnetic resonance (Fig. 8.2) is sending the system through a particular spin correlation order at a particular stage in the experiment.

A related notion is that of a freeze mask—some points of the control sequence may be required to remain at their initial values during the optimisation (Fig. 8.3). This is accomplished by zeroing the corresponding elements of the step vector (not of the gradient, because the gradient may be used for other purposes elsewhere, Sect. 8.4) at the waveform update stages of the optimisation.

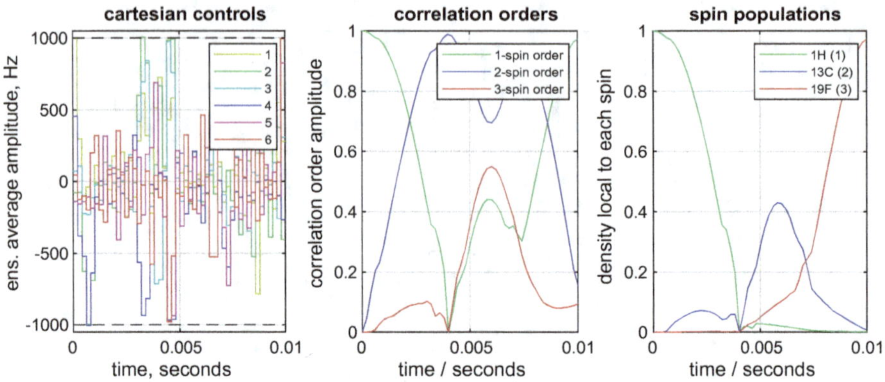

Fig. 8.2 An example of a keyholed optimal control problem in liquid state nuclear magnetic resonance. A six-channel (H_X, H_Y, C_X, C_Y, F_X, F_Y) piecewise-constant radiofrequency pulse (left panel) is optimised to transport all longitudinal magnetisation from ^{1}H to ^{19}F in a ^{1}H–^{13}C–^{19}F fragment with *J*-couplings between the spins connected by a chemical bond. A keyhole is set at 4 ms which requires the system to be in the two-spin correlation subspace at that point. At the end of the optimisation, the trajectory (middle panel) conforms to that requirement and an abrupt switch in the system dynamics (right panel) is visible at that point

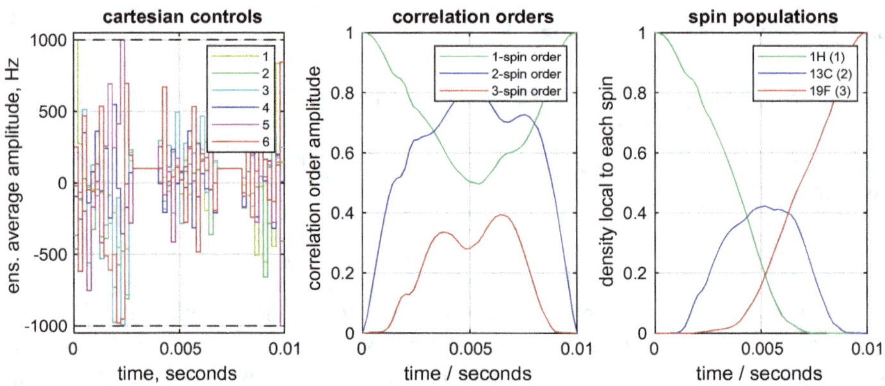

Fig. 8.3 An example of partially frozen optimal control optimisation. A six-channel (H_X, H_Y, C_X, C_Y, F_X, F_Y) piecewise-constant radiofrequency pulse (left panel) is optimised to transport all longitudinal magnetisation from ^{1}H to ^{19}F in a ^{1}H–^{13}C–^{19}F fragment with *J*-couplings between the spins connected by a chemical bond. Two regions of the pulse sequence (left panel) remain fixed at their initial value of 100 Hz (all six channels) for the duration of the optimisation

8.3.3 Multi-target and Ensemble Control

A simple extension of the above mathematics is to require multiple initial conditions $\left\{\boldsymbol{\rho}_0^{(1)}, \boldsymbol{\rho}_0^{(2)}, \ldots\right\}$ to arrive at the corresponding destination states $\left\{\boldsymbol{\delta}^{(1)}, \boldsymbol{\delta}^{(2)}, \ldots\right\}$. In the limit where the number of linearly independent sources and targets is equal to the dimension of the state space, we have a gate design or a universal rotation design problem. The fidelities may be combined in any differentiable way, the simplest example being a weighted sum

$$f = \sum_m w_m \mathrm{Re}\left\langle \boldsymbol{\delta}^{(m)} \middle| \mathcal{P}_N \cdots \mathcal{P}_1 \middle| \boldsymbol{\rho}_0^{(m)} \right\rangle \tag{8.56}$$

with the obvious parallelisation opportunity over the state pairs. Similar sums (and approximations of integrals) may be constructed over any property of the system or the control apparatus. The result of the optimisation would then be a control sequence that operates on ensembles of system-instrument combinations. Examples include distributions of drift Hamiltonian parameters, of power level multipliers and direction modifiers reflecting uneven distributions of control fields across the sample, of decoherence rates, etc. Some of those distributions may be correlated—e.g. a different initial condition in each member of the drift Hamiltonian ensemble. Once again, the fidelities may be combined in any differentiable way, the simplest example is again a weighted sum

$$f = \sum_{\alpha\beta\gamma\ldots} w_{\alpha\beta\gamma\ldots} \mathrm{Re}\left\langle \boldsymbol{\delta}^{(\alpha)} \middle| \mathcal{P}_N^{(\beta\gamma\ldots)} \ldots \mathcal{P}_1^{(\beta\gamma\ldots)} \middle| \boldsymbol{\rho}_0^{(\alpha)} \right\rangle \tag{8.57}$$

with large-scale parallelisation becoming possible over the multi-index $\{\alpha, \beta, \gamma, \ldots\}$ that runs over everything that is distributed in the ensemble. Because derivatives are easily propagated through any differentiable expressions using the chain rule, the flow of the GRAPE algorithm remains unchanged. Altering weights in Eqs. (8.56) and (8.57) allows elaborate spatial patterns, ensemble correlations, and statistical correlations between quantum mechanical states to be imprinted into the ensemble. A good example is metabolite-selective localised excitation in magnetic resonance imaging [343].

8.3.4 Cooperative Control and Phase Cycles

We have so far assumed a single contiguous control event targeting specific source–destination pairs or—in the limit—a particular state-space transformation for a system or ensemble. However, spectroscopic experiments (e.g. magnetic resonance) typically have the following structure:

$$[\text{preparation} \rightarrow \text{evolution}]_n \rightarrow \text{preparation} \rightarrow \text{detection} \tag{8.58}$$

where the preparation stages involve interaction with control fields, but only the drift Liouvillian is active during the evolution and detection stages. Optimisation of *individual* preparation stages is not necessarily the best approach because it could be possible to compensate the imperfections of one preparation stage during later preparation stages.

This is the essence of *cooperative control*, which deals with experiment-wide optimisation wherein all preparation stages are optimised simultaneously to improve some experiment-wide outcome. We will follow the magnetic resonance tradition and call each control event a pulse. Two scenarios are possible:

(a) Pulses working to compensate each other's imperfections within a single experiment—a generalisation of composite pulses originally developed in NMR spectroscopy [278].

(b) Pulses working to make sure that undesired components in the observables of interest cancel when the results of multiple experiments are combined at the data processing stage—a generalisation of phase cycles, also originally developed in NMR [279].

Situation (a) is called *single-scan cooperativity* and situation (b) *multi-scan cooperativity* [344].

8.3.4.1 Single-Scan Cooperative Pulses

As an example of the event train in Eq. (8.58), consider the MQMAS pulse sequence [282]:

$$\underbrace{\left[\mathbf{S}_{\mathrm{Z}} \to \left(\mathbf{S}_{+}^{k} \pm \mathbf{S}_{-}^{k}\right)\right]}_{\text{firstpulse}} \xrightarrow{\tau} \underbrace{\left[\left(\mathbf{S}_{+}^{k} \pm \mathbf{S}_{-}^{k}\right) \to \left(\mathbf{S}_{+} \pm \mathbf{S}_{-}\right)\right]}_{\text{secondpulse}} \to \text{detection} \tag{8.59}$$

where the first pulse converts longitudinal magnetisation into a k-quantum coherence (k depends on the isotope in question) that evolves for an incremented period τ and then gets converted by the second pulse into observable magnetisation that is detected. An idealised version of the same pulse sequence would have both pulses replaced by analytical projectors that accomplish the transfers with the theoretical maximum efficiency given by the Sørensen bound [281].

Single-scan cooperative pulse optimisation here treats the entire sequence as one pulse; the free evolution period has control amplitudes frozen at zero. Optimisation proceeds in the following stages:

1. Spin system ensemble is generated by sampling from distributions of relevant parameters (quadrupolar axiality and rhombicity, chemical shifts, spatial orientation, etc.) and calculating the corresponding weights in Eq. (8.57) or some other fidelity functional.

2. The idealised pulse sequence is simulated for each system in the ensemble (initial conditions and drift generators may be different between ensemble members). The ideal outputs are recorded into a set of destination states, one for each system in the ensemble.

3. An ensemble GRAPE optimisation is performed (for example, using the methods discussed in Sect. 8.4) for the entire pulse sequence in one go, with zero control amplitudes enforced by a freeze mask during the free evolution period.

The procedure is reminiscent of setting up a training database and performing neural network training—in fact, the gradient calculation procedure in GRAPE is substantially backpropagation [282]. Just like it happens in machine learning, the performance of the resulting pulses will depend on how well the ensemble had been set up; they are not expected to work outside that ensemble.

8.3.4.2 Multi-scan Cooperative Pulses

A complementary strategy deals with the inevitable presence of impurities in the state vector that arrives at the detection stage—those impurities cannot be eliminated altogether, but their phase may be easier to control than their amplitude. This has been known for decades in NMR spectroscopy: multiple experiments may be carried out that lead to impurities of different phases—those cancel out during post-processing when the outcomes of individual experiments are combined [279]. This can effectively create non-unitary transformations that, in the absence of relaxation, a single-scan experiment cannot accomplish.

Consider an ensemble of systems (indexed by k) on which we perform multiple experiments (indexed by n) with the initial state $\boldsymbol{\rho}_k$, target state $\boldsymbol{\delta}_k$, and whole-experiment propagators $\mathcal{P}_n^{(k)}$:

$$\left\{\boldsymbol{\rho}^{(k)} \xrightarrow{\mathcal{P}_n^{(k)}} f_n^{(k)}\boldsymbol{\delta}^{(k)} + \boldsymbol{\eta}_n^{(k)}\right._{n,k} \tag{8.60}$$

where $f_n^{(k)}$ is the fidelity achieved in the nth experiment on the kth system and $\boldsymbol{\eta}_n^{(k)}$ is the corresponding impurity. If we combine the outcomes of different experiments by the addition of some observables, we would like these impurities to cancel in each member of the ensemble when the outcomes of the experiments are added up. We, therefore, seek to minimise the following quantity:

$$\begin{aligned}
\Omega &= \sum_k \left\| \boldsymbol{\eta}_1^{(k)} + \boldsymbol{\eta}_2^{(k)} + \cdots \right\|^2 = \sum_k \left\| \left(\mathbf{1} - \mathcal{W}_{\boldsymbol{\delta}}^{(k)}\right)\left(\mathcal{P}_1^{(k)}\left|\boldsymbol{\rho}^{(k)}\right\rangle + \mathcal{P}_2^{(k)}\left|\boldsymbol{\rho}^{(k)}\right\rangle + \cdots\right)\right\|^2 \\
&= \sum_k \left\langle \boldsymbol{\rho}^{(k)}\right|\left(\mathcal{P}_1^{(k)\dagger} + \mathcal{P}_2^{(k)\dagger} + \cdots\right)\left(\mathbf{1} - \mathcal{W}_{\boldsymbol{\delta}}^{(k)}\right)\left(\mathbf{1} - \mathcal{W}_{\boldsymbol{\delta}}^{(k)}\right)\left(\mathcal{P}_1^{(k)} + \mathcal{P}_2^{(k)} + \cdots\right)\left|\boldsymbol{\rho}^{(k)}\right\rangle \\
&= \sum_k \left\langle \boldsymbol{\rho}^{(k)}\right|\left(\mathcal{P}_1^{(k)\dagger} + \mathcal{P}_2^{(k)\dagger} + \cdots\right)\left(\mathbf{1} - \mathcal{W}_{\boldsymbol{\delta}}^{(k)}\right)\left(\mathcal{P}_1^{(k)} + \mathcal{P}_2^{(k)} + \cdots\right)\left|\boldsymbol{\rho}^{(k)}\right\rangle
\end{aligned} \tag{8.61}$$

where $\mathcal{W}_{\boldsymbol{\delta}}^{(k)} = \left|\boldsymbol{\delta}^{(k)}\right\rangle\left\langle\boldsymbol{\delta}^{(k)}\right|$, and the simplification is possible because projection operators are here Hermitian and idempotent. A tedious calculation yields the following expression for the gradient:

$$\nabla\Omega = 2\sum_k \nabla \mathrm{Re}\left\langle \boldsymbol{\eta}_1^{(k)} + \boldsymbol{\eta}_2^{(k)} + \cdots \right|\left(\mathcal{P}_1^{(k)} + \mathcal{P}_2^{(k)} + \cdots\right)\left|\boldsymbol{\rho}^{(k)}\right\rangle \tag{8.62}$$

Because we seek to minimise Ω while simultaneously maximising fidelity, this functional and its gradient must be subtracted from, respectively, the GRAPE fidelity and its gradient with some coefficient $\lambda/2$ (decided by the user through trial and error) that regulates the relative importance attached to the achievement of the fidelity *versus* the cancellation of the impurities:

$$\begin{aligned}\nabla\Gamma &= \nabla f - (\lambda/2)\nabla\Omega \\ &= \nabla\mathrm{Re}\sum_k \left\langle \boldsymbol{\delta}^{(k)} - \lambda\left(\boldsymbol{\eta}_1^{(k)} + \boldsymbol{\eta}_2^{(k)} + \cdots\right)\middle|\left(\mathcal{P}_1^{(k)} + \mathcal{P}_2^{(k)} + \cdots\right)\middle|\boldsymbol{\rho}^{(k)}\right\rangle\end{aligned} \tag{8.63}$$

Calculation of this gradient proceeds in two stages:

1. Forward simulations in Eq. (8.60) are carried out; the target states $\left|\boldsymbol{\delta}^{(k)}\right\rangle$ are projected out of the final states produced by experiment n on system k to obtain their impurities $\left|\boldsymbol{\eta}_n^{(k)}\right\rangle$.
2. The target state is updated to $\left\langle \boldsymbol{\delta}^{(k)} - \lambda\left(\boldsymbol{\eta}_1^{(k)} + \boldsymbol{\eta}_2^{(k)} + \cdots\right)\right|$ and submitted to the GRAPE algorithm (Sect. 8.1) for evaluation of the fidelity and its derivatives.

Once the gradient is computed, optimisation methods discussed in Sect. 8.4 become applicable. Initial guesses for pulse waveforms may be obtained by running single-experiment GRAPE optimisations.

8.3.4.3 Conventional Phase Cycles

Electromagnetic phase symmetries arise from the rotational symmetry (Sect. 1.6.2) of free space—a rotation of the coordinate system does not affect inner products. From the instrumental point of view, anything that does not follow that symmetry is likely to be an artefact. From the optimal control point of view, an artefact removal phase cycle (such as $\{+x, +y, -x, -y\}$ in pulse-acquire NMR experiments) is an instance of the ensemble control problem, the ensemble variables being uniform increments in the phases of the initial condition, XY control pairs, and the target states—for example:

$$\Omega = \mathrm{Re}\sum_k \left\langle \boldsymbol{\delta}\left(\varphi_k^{(\boldsymbol{\delta})}\right)\middle|\mathcal{P}\left(\varphi_k^{(\mathcal{P})}\right)\middle|\boldsymbol{\rho}\left(\varphi_k^{(\boldsymbol{\rho})}\right)\right\rangle \tag{8.63}$$

where $\varphi_k^{(\boldsymbol{\rho})}$ is the phase of the initial condition at the kth step of the phase cycle, $\varphi_k^{(\boldsymbol{\rho})}$ is the phase offset (there may be several if there are multiple XY control pairs) applied to the control sequence, and $\varphi_k^{(\boldsymbol{\delta})}$ is the phase of the target state. When the figure of merit Ω is maximised, the result is a control sequence that tries to follow the specified phase cycle.

The other common use of phase cycles—system state purification by cancellation of unwanted amplitudes when the results of multiple experiments are added up [281]—is an instance of multi-scan cooperativity discussed in the previous section, a good example is the double-quantum filter [284].

8.3.5 Fidelity and Penalty Functionals

The fidelity functional need not be as simple as the real part of $\langle \boldsymbol{\delta} | \mathcal{P} | \boldsymbol{\rho} \rangle$ —GRAPE algorithm (Sect. 8.1) is compatible with a variety of other optimisation targets, for example:

1. Phase-insensitive fidelity $|\langle \boldsymbol{\delta} | \mathcal{P} | \boldsymbol{\rho} \rangle|^2$ —useful in NMR when magnetisation must be made transverse without any wishes on its direction in the transverse plane.
2. Multi-target fidelity $\mathrm{Re} \sum_k \left\langle \boldsymbol{\delta}^{(k)} | \mathcal{P} | \boldsymbol{\rho}^{(k)} \right\rangle$ when amplitudes of specific initial states must be transported to specific target spates. When the target state is specified for every initial state, the fidelity functional becomes equal to the experiment propagator overlap $\mathrm{Tr}\left[\mathcal{P}_{\mathrm{targ}}^{\dagger} \mathcal{P}\right]$ with a target propagator $\mathcal{P}_{\mathrm{targ}}$; this is called the *gate design problem.*
3. Any other fidelity measure that reduces, through the application of product and chain rules, to propagator derivatives (Sect. 8.2.4) and their actions on state vectors computed within the GRAPE procedure (Sect. 8.1).

There is also liberty in the selection of penalty functionals, which are preferable to having hard constraints due to better optimisation behaviour. Instrumental limitations are rarely hard; it is, therefore, reasonable to implement them as penalties rather than hard bounds. Examples include:

1. Amplitude penalty: the maximum instrumentally available amplitude of $c^{(k)}(t)$ may be limited, for example by sample heating tolerance. The penalty may apply to values of any magnitude (*uniform penalty*), or only to values that exceed a certain threshold (*spillout penalty*).
2. Frequency penalty: the maximum instrumentally available frequency in $c^{(k)}(t)$ may be limited because waveform synthesis hardware has a finite switching time. The penalty may apply from a certain frequency onwards (*e.g.* by indexing into the Fourier transform), or penalise some frequencies more than others (*e.g.* a norm of a derivative).
3. Running cost: a more general functional of control channel amplitudes may be necessary to capture sample-, experiment-, and instrument-specific constraints. An example of a linear functional is a weighted integral:

Table. 8.1 Three common control sequence penalty functionals, including first and second derivatives

Type	Penalty K	Gradient vector g	Hessian matrix H
Weighted norm square	$\sum_k w_k c_k^2$	$g_n = 2w_n c_n$	$H_{nm} = 2w_n \delta_{nm}$
Weighted derivative norm square	$\sum_k w_k [\mathbf{Dc}]_k^2$	$g_n = 2\sum_k w_k [\mathbf{Dc}]_k D_{kn}$	$H_{nm} = 2\sum_k w_k D_{kn} D_{km}$
Weighted spill-out norm square	$\sum_k w_k (c_k - u_k)^2 h_{c_n > u_n} + \sum_k w_k (c_k - l_k)^2 h_{c_n < l_n}$	$g_n = 2w_n (c_n - u_n) h_{c_n > u_n} + 2w_n (c_n - l_n) h_{c_n < l_n}$	$H_{nm} = 2w_n \delta_{nm} h_{c_n > u_n} + 2w_n \delta_{nm} h_{c_n < l_n}$

$$\int_0^T w(t) c^{(k)}(t) \mathrm{d}t \tag{8.65}$$

where T is experiment duration and $w(t)$ is a weight function.

For gradient-based optimisers (Sect. 8.4) to work efficiently, analytical derivatives with respect to the control sequence are required—examples are given in Table 8.1, where $\mathbf{c}$ is the control sequence vector, $\mathbf{w}$ is the weight vector, $\mathbf{u}$ is the upper bound vector, $\mathbf{l}$ is the lower bound vector, and $\mathbf{D}$ is the differentiation matrix of a suitable type and order, or any other appropriate transformation matrix.

In Table 8.1, the weighted norm square spill-out penalty is designed to only apply to the elements of the control sequence vector $\mathbf{c}$ that fall outside the bounds defined by $\mathbf{u}$ and $\mathbf{l}$ vectors. The meaning of the Heaviside and delta functions used in (Table.8.1) is

$$h_{c_n > u_n} = \begin{cases} 1 & \text{if} \quad c_n > u_n \\ 0 & \text{otherwise} \end{cases}, \quad h_{c_n < l_n} = \begin{cases} 1 & \text{if} \quad c_n < l_n \\ 0 & \text{otherwise,} \end{cases}$$
$$\delta_{nm} = \begin{cases} 1 & \text{if} \quad n = m \\ 0 & \text{otherwise} \end{cases} \tag{8.66}$$

8.3.6 Instrument Response

A common situation is when the control sequence emitted by the computer is not the sequence experienced by the sample—instrument hardware introduces distortions. A general theory is only possible when the part of the instrument between the computer and the sample may be approximated by a *linear time-invariant* (LTI) system (Sect. 1.7), for which the signal $\Phi\{x(t)\}$ that reaches the sample is a convolution of the input signal $x(t)$ with the pulse response $h(t)$ of the instrument:

$$\Phi\{x(t)\} = x(t) * h(t) = \int_{-\infty}^{\infty} x(\tau)h(\tau - t)\mathrm{d}\tau \tag{8.67}$$

When the waveform $x(t)$ is discretised on some finite time grid, this relation acquires a matrix form:

$$\Phi\{\mathbf{x}\} = \mathbf{Hx} \tag{8.68}$$

where $\mathbf{H}$ is a *response matrix*; it need not be invertible or even square because the instrument response can have tails beyond the temporal extent of the input. $\mathbf{H}$ may be measured experimentally by submitting a large number of randomly generated control sequences $\mathbf{x}_k$ from the computer and measuring the resulting electromagnetic fields $\Phi\{\mathbf{x}_k\}$ at the sample points. In situations when the instrument is not an LTI, the best practical way to proceed is to measure responses to a large library of inputs and to train a pair of artificial neural networks to perform the forward and the backward transformation. Fully connected few-layer networks with the asymptotically linear softplus activation functions are recommended. Because such networks are differentiable, they are easily integrated into the GRAPE workflow.

8.4 Optimisation Strategies

We will not discuss gradient-free methods here—algorithms like simplex, genetic optimisation, and multi-grid search. The continuous and differentiable (sometimes even convex) dependence of both the fidelity and the penalties on the control sequences means that continuous optimisation methods are in practice superior. The remarkable efficiency of the GRAPE method—machine precision gradient at a small multiple of the cost of a single time-domain simulation—makes "continuous optimisation" methods appealing.

8.4.1 Gradient Descent

The original implementation of the GRAPE method made use of gradient descent with an option to perform a line search in the descent direction [267]. The method performed well, but the decision to truncate the commutator series for the propagator derivative at the first order

$$\frac{\partial}{\partial c_n^{(k)}} \mathcal{P}_n = \mathcal{P}_n\left(-i\mathcal{C}_k \Delta t_n + O\left[\Delta t_n^2\right]\right) \tag{8.69}$$

was later found to be a drag on performance [276]—as the optimisation proceeds, it is the first term in the fidelity gradient

$$\frac{\partial}{\partial c_n^{(k)}}\langle\boldsymbol{\delta}|\mathcal{P}|\boldsymbol{\rho}_0\rangle = \langle\boldsymbol{\delta}\mathcal{P}_N\cdots\mathcal{P}_{n+1}|\mathcal{P}_n(-i\mathcal{C}_k\Delta t_n)|\mathcal{P}_{n-1}\cdots\mathcal{P}_1\boldsymbol{\rho}_0\rangle + O\left[\Delta t_n^2\right] \tag{8.70}$$

that gets reduced, and the approximation error eventually starts to dominate the gradient. The remedy—if one should insist on using gradient descent—is to take a few more terms in the propagator derivative series in Eq. (8.44) or to use the auxiliary matrix method to get machine precision gradient [102].

8.4.2 Quasi-Newton Methods

When machine precision gradients are available, gradient descent is no longer the most efficient thing to do. This is because approximations of the Hessian may be obtained from gradient history and used to construct Newton-like optimisers with better convergence. The most popular method in this class is BFGS (Broyden-Fletcher-Goldfarb-Shanno [269]):

$$\begin{gathered}\mathbf{H}_{s+1} = \mathbf{H}_s + \frac{\mathbf{g}_s\mathbf{g}_s^{\mathrm{T}}}{\mathbf{g}_s^{\mathrm{T}}\mathbf{m}_s} - \frac{(\mathbf{H}_s\mathbf{m}_s)(\mathbf{H}_s\mathbf{m}_s)^{\mathrm{T}}}{\mathbf{m}_s^{\mathrm{T}}\mathbf{H}_s\mathbf{m}_s}, \qquad \mathbf{H}_0 = \mathbf{1} \\ \mathbf{g}_s = \nabla f(\mathbf{x}_{s+1}) - \nabla f(\mathbf{x}_s), \qquad \mathbf{m}_s = \mathbf{x}_{s+1} - \mathbf{x}_s\end{gathered} \tag{8.71}$$

This approximate Hessian is used to take a Newton-type optimisation step:

$$\mathbf{x}_{s+1} = \mathbf{x}_s - \alpha_s\mathbf{H}_s^{-1}\nabla f(\mathbf{x}_s), \qquad \alpha_s > 0 \tag{8.72}$$

where α_k is the line search parameter—the well-studied line search strategies are outside the scope of this book. Because matrix inversions are expensive, it is more efficient to use the corresponding update schemes for the inverse of the Hessian:

$$\mathbf{H}_{s+1}^{-1} = \left(\mathbf{1} - \frac{\mathbf{m}_s\mathbf{g}_s^{\mathrm{T}}}{\mathbf{g}_s^{\mathrm{T}}\mathbf{m}_s}\right)^{\mathrm{T}}\mathbf{H}_s^{-1}\left(\mathbf{1} - \frac{\mathbf{m}_s\mathbf{g}_s^{\mathrm{T}}}{\mathbf{g}_s^{\mathrm{T}}\mathbf{m}_s}\right) + \frac{\mathbf{m}_s\mathbf{m}_s^{\mathrm{T}}}{\mathbf{g}_s^{\mathrm{T}}\mathbf{m}_s} \tag{8.73}$$

In the case of BFGS, a very memory-efficient procedure is available for generating the next step vector directly from the past gradient history, requiring no matrix storage. It is known as memory-limited BFGS, or *L-BFGS* [271, 285]. In the context of optimal control, the number of variables can exceed 10^4; L-BFGS is the only quasi-Newton method that can handle such problems.

Quasi-Newton methods require accurate gradients and may become stuck in the same way as gradient descent when approximations are used—this is illustrated in Fig. 8.4.

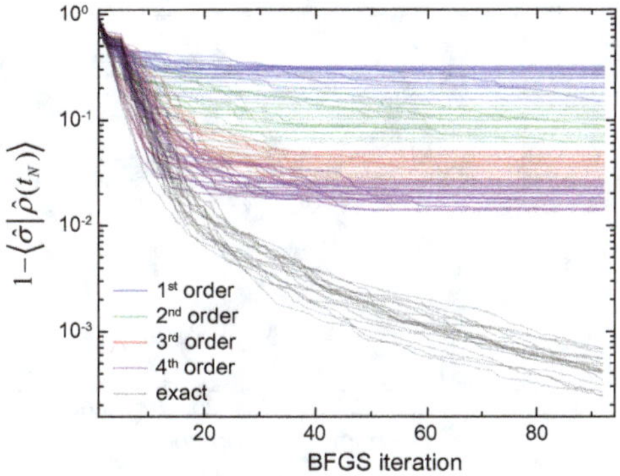

Fig. 8.4 Quality of state transfer as a function of iteration number of the BFGS algorithm. The fidelity parameter refers to the quality of magnetization inversion under a 50-point shaped radiofrequency pulse applied to a chain of 31 protons with chemical shifts spread at regular intervals over the range of 8 ppm with strong nearest neighbour J-couplings of 20 Hz in a 600 MHz magnet. Pulse duration 5 ms (100 μs per waveform step), pulse amplitude capped at 2500 Hz. State-space restriction to three-spin orders involving adjacent spins was used to reduce the matrix dimension involved in the simulation. The starting points in the optimization were set to sequences of uniformly distributed random numbers from the ± 1000 Hz interval. The "kth order" labels refer to the number of commutator series terms in Eq. (8.44), and "exact" refers to the series that has been summed to machine precision. Reproduced with permission from [277]

8.4.3 Newton–Raphson Methods

Newton–Raphson and quasi-Newton methods (minimisation is assumed here) rely on the necessary conditions for Taylor's theorem [91,287] and use a local quadratic approximation:

$$f(\mathbf{x}+\Delta\mathbf{x}) \approx f(\mathbf{x}) + \langle\nabla f(\mathbf{x}) \mid \Delta\mathbf{x}\rangle + \frac{1}{2}\langle\Delta\mathbf{x}|\nabla^2 f(\mathbf{c})|\Delta\mathbf{x}\rangle \tag{8.74}$$

The first-order necessary condition requires any minimiser $\tilde{\mathbf{x}}$ of $f(\mathbf{x})$ to be a stationary point

$$\nabla f(\tilde{\mathbf{x}}) = \mathbf{0} \tag{8.75}$$

Imposing this condition on Eq. (8.74) gives the argument update rule:

$$\mathbf{x}_{s+1} = \mathbf{x}_s - \left[\nabla^2 f(\mathbf{x}_s)\right]^{-1} \nabla f(\mathbf{x}_s) \tag{8.76}$$

The second-order necessary condition is that the Hessian $\nabla^2 f$ should be positive definite at $\tilde{\mathbf{x}}$. This is also evident from Eq. (8.76), in which a negative Hessian eigenvalue would result in a step being performed up, rather than down, the corresponding gradient direction.

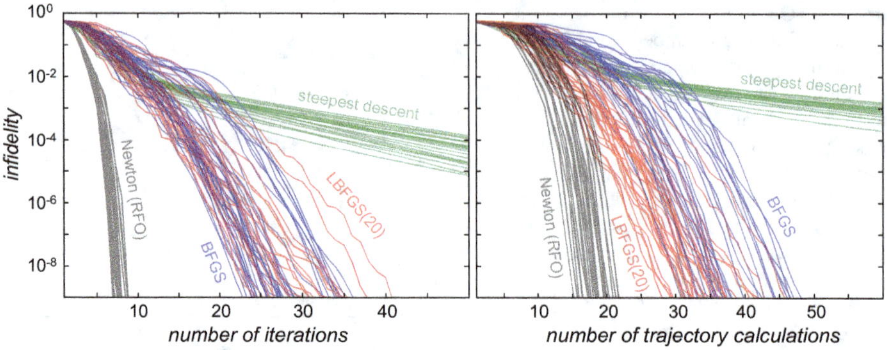

Fig. 8.5 Convergence profiles for the transfer of longitudinal magnetisation into the singlet state for the two-spin system described in the main text. The same line search method in the predicted descent direction was used in all cases. Memory time for LBFGS was set to 20 gradients. Reproduced with permission from [287]

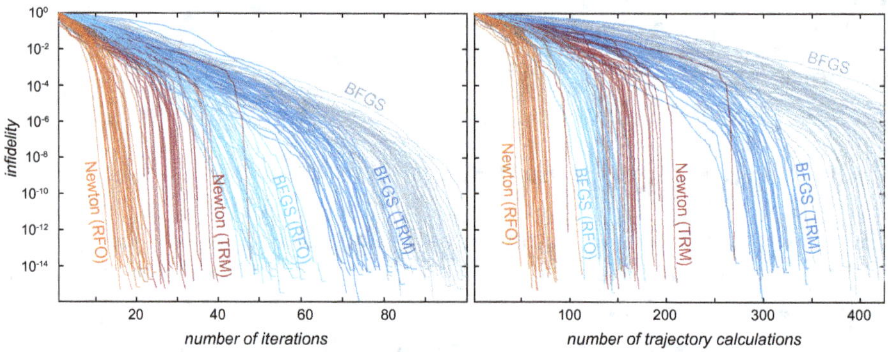

Fig. 8.6 Convergence profiles for the state transfer within the ^{1}H–^{13}C–^{19}F three-spin system described in the main text for the BFGS quasi-Newton method and the Newton–Raphson method using TRM or RFO Hessian regularization techniques. The same line search method in the predicted descent direction was used in all cases. Reproduced with permission from [287]

Second derivatives of the fidelity functional can be expensive (Sect. 8.2.4) but the performance of the resulting method as a function of iteration count is superior to both gradient descent and quasi-Newton methods (Fig. 8.5). However, the Hessian calculation becomes prohibitively expensive when the number of discretisation points in the control sequence exceeds a few hundred.

A significant problem is that, away from a minimiser $\tilde{\mathbf{x}}$, the Hessian of the figure of merit $f(\mathbf{x})$ is not guaranteed to be positive definite. Small Hessian eigenvalues are also problematic because they result in overly long steps that can be detrimental because most figures of merit are not actually quadratic. A significant amount of research has gone into modifying the Hessian in such a way as to avoid these undesired behaviours [289–299]. In the optimal control context, Hessian

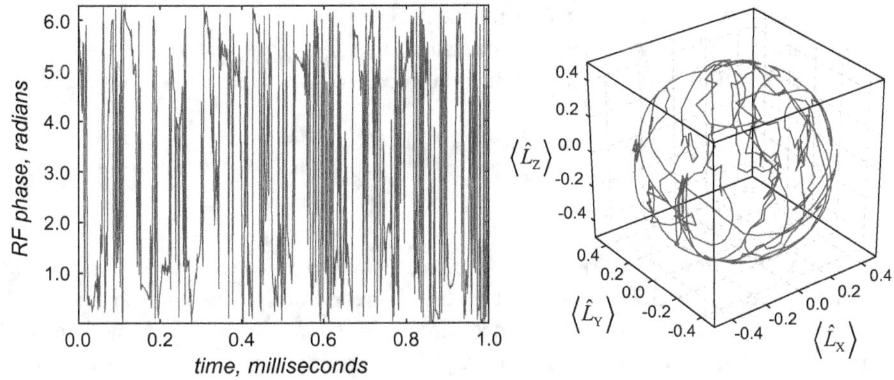

Fig. 8.7 An illustration of the fact that some optimal control pulse waveforms are not directly interpretable. **Left panel:** phase profile of a phase-modulated broadband excitation pulse that meets the following requirements: $L_z \rightarrow L_x$ excitation with at least 99% fidelity for a 50 kHz frequency range; constant RF power level of 15 kHz; tolerance for B_1 inhomogeneity of $\pm$ 30%; pulse duration 1.0 ms; 625 time discretization points. See Ref. [300] for further information on such pulses. **Right panel:** Bloch sphere representation of the dynamics of a spin that is off resonance by 250 Hz under the pulse described above. The spin eventually arrives on the X-axis with the prescribed fidelity, but its intermediate dynamics is obscure. Reproduced with permission from [300]

regularisation methods and line search procedures are outside the scope of this book; rational function optimisation with cubic line search tends to perform best (Fig. 8.6).

8.5 Trajectory Analysis

A common feature of optimal control pulses is visual randomness. As Fig. 8.7 demonstrates, plotting either the control sequence or the observables is not informative. That is an illusion—system dynamics is actually orderly—but dispelling it requires a method for high-dimensional trajectory analysis. This section reviews some recent developments in this area, with NMR pulses used as illustrations.

8.5.1 Trajectory Analysis Strategies

The dynamics of individual *observables* under optimal control pulses may be obscure, but populations of various physically relevant *subspaces* generally show interpretable dynamics [300]. Some of the heuristics used in this section are specific to magnetic resonance, but the overall product structure argument is applicable to any quantum system composed of interacting subsystems.

8.5.1.1 Correlation Order Populations

In any direct product basis set, the correlation order of a state is defined as the number of non-unit +spin operators in its direct product expression, for example:

$$\begin{array}{ll} \mathbf{S}_\mathrm{X} \otimes \mathbf{1} \otimes \mathbf{S}_+ \otimes \mathbf{1} \otimes \mathbf{1} \otimes \mathbf{1} & k = 2 \\ \mathbf{1} \otimes \mathbf{S}_\mathrm{Z} \otimes \mathbf{1} \otimes \mathbf{1} \otimes \mathbf{S}_- \otimes \mathbf{S}_\mathrm{X} & k = 3 \\ \mathbf{1} \otimes \mathbf{1} \otimes \mathbf{S}_\mathrm{Y} \otimes \mathbf{1} \otimes \mathbf{1} \otimes \mathbf{1} & k = 1 \end{array} \tag{8.77}$$

where $\mathbf{1}$ is the unit operator, $\mathbf{S}_\mathrm{XYZ}$ are Cartesian spin operators of appropriate dimension (Sect. 1.6.3.2), and $\mathbf{S}_\pm$ are the corresponding raising and lowering operators. Because magnetic resonance simulations start and get detected in low correlation orders (1 in most cases and 2 for experiments involving singlet states), correlation order populations give a measure of complexity for a given trajectory.

Classification of any state into correlation orders is always possible because the full state space $\mathfrak{L}$ of the spin system is a direct sum of correlation order subspaces $\mathfrak{L}_k$:

$$\mathfrak{L} = \mathfrak{L}_0 \oplus \mathfrak{L}_1 \oplus \ldots \oplus \mathfrak{L}_N \tag{8.78}$$

where N is the number of spins in the system and $\mathfrak{L}_0$ only contains the unit operator. The population p_k of a given correlation order k in a state $\boldsymbol{\rho}$ is

$$p_k = \| \mathcal{W}_{\mathfrak{L}_k} | \boldsymbol{\rho} \rangle \|_2 \tag{8.79}$$

where $\mathcal{W}_{\mathfrak{L}_k}$ is a projection superoperator into $\mathfrak{L}_k$. Higher correlation orders may relax faster [215] and be harder to control—a "good" control sequence would keep the population of high correlation orders low.

An example of the improvement in interpretability brought about by Eq. (8.79) is given in the middle panel of Fig. 8.8—after Eq. (8.79) is applied, the system can be seen to move very smoothly from single-spin order subspace (where the initial state lives) into two- and three-spin orders, which then fade gradually to leave a single-spin order on the destination spin. This is in contrast to the complicated appearance of the control sequence that is shown in the left panel of the same figure.

8.5.1.2 Coherence-Order Populations

Another approach is a generalization of the coherence-order diagrams [258, 301]—in spherical tensor basis sets (Sect. 3.2.2), the coherence order m of a state is defined as the sum of all projection quantum numbers in its direct product components, for example:

$$\begin{array}{ll} \mathbf{T}_{1,0} \otimes \mathbf{T}_{0,0} \otimes \mathbf{T}_{1,1} \otimes \mathbf{T}_{0,0} \otimes \mathbf{T}_{0,0} \otimes \mathbf{T}_{0,0} & m = 1 \\ \mathbf{T}_{0,0} \otimes \mathbf{T}_{2,2} \otimes \mathbf{T}_{0,0} \otimes \mathbf{T}_{0,0} \otimes \mathbf{T}_{1,-1} \otimes \mathbf{T}_{1,1} & m = 2 \\ \mathbf{T}_{0,0} \otimes \mathbf{T}_{0,0} \otimes \mathbf{T}_{2,0} \otimes \mathbf{T}_{0,0} \otimes \mathbf{T}_{0,0} \otimes \mathbf{T}_{0,0} & m = 0 \end{array} \tag{8.80}$$

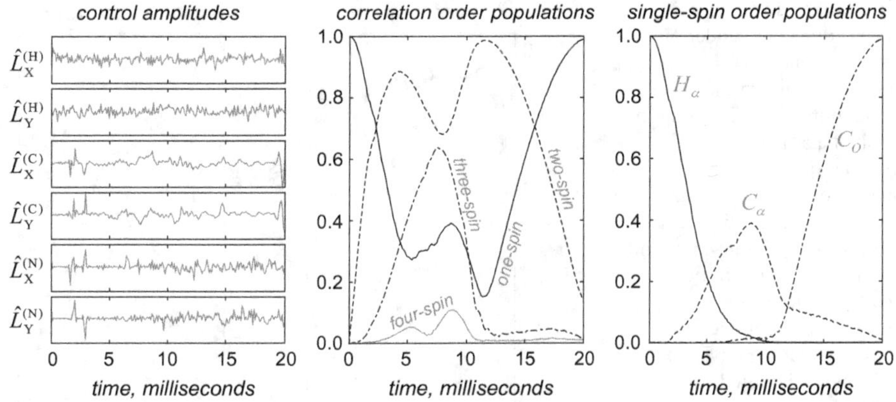

Fig. 8.8 Analysis of spin system dynamics under an optimal control pulse designed to move all magnetization from the C_α–H proton of a protein backbone fragment to the C=O carbon without leaking any magnetization to other nearby spins. **Left panel:** control operator coefficients (fractions of the nominal power level) as functions of time for the optimal solution (99% transfer fidelity). **Middle panel:** spin system dynamics, classified into spin correlation orders using Eq. (8.79). **Right panel:** further analysis of the dynamics in the single-spin order subspace using Eq. (8.84)—note the orderly transition from C_α–H proton, over to C_α carbon and onwards to the C=O carbon, in contrast with the noisy appearance of the numerically optimized pulse that is driving the system. Reproduced with permission from [300]

where $\mathbf{T}_{l,m}$ are irreducible spherical tensor operators [256,303] and $\mathbf{T}_{0,0}$ is a multiple of the unit matrix. The coherence orders also generate a partition of the state space in a way similar to Eq. (8.78):

$$\mathfrak{L} = \mathfrak{C}_{-M} \oplus \mathfrak{C}_{-M+1} \oplus \ldots \oplus \mathfrak{C}_{M-1} \oplus \mathfrak{C}_M \tag{8.81}$$

where $\mathfrak{C}_m$ is a subspace of all states with coherence order m. Coherence order may be negative and the maximum coherence order does not have to be equal to the number of spins in the system. For a given coherence order m and a given state $\boldsymbol{\rho}$ the population is given by

$$p_m = \left\| \mathfrak{P}_{\mathfrak{C}_m} |\boldsymbol{\rho}\rangle \right\|_2 \tag{8.82}$$

where $\mathfrak{P}_{\mathfrak{C}_m}$ is a projection superoperator into $\mathfrak{C}_m$. Populations of coherence-order subspaces give no indication of the complexity of dynamics (a state correlating the entire spin system can still have a zero coherence order), but they are useful in the analysis of liquid-state NMR pulse sequences because the total projection quantum number remains invariant under liquid-state NMR drift Hamiltonians and provides a convenient illustration of the sequence mechanics [258, 300]. Radiofrequency and microwave irradiation does, however, induce rotations between different coherence order subspaces and this classification is less useful in sequences involving continuous or closely spaced RF or MW events.

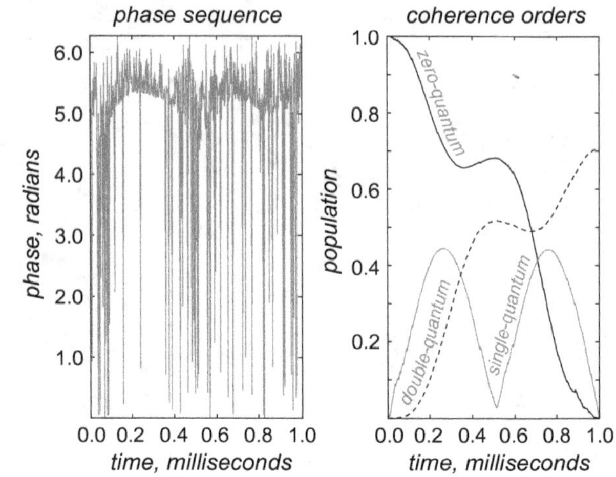

Fig. 8.9 Analysis of spin system dynamics under an optimal control pulse designed to move the population of the $\mathbf{T}_{1,0}$ state of a spin-1 particle with a rhombic quadrupolar interaction to the $\mathbf{T}_{2,2}$ state to the maximum possible extent (70.7% is the Sørensen bound in this case). **Left panel**: phase profile of the numerically optimized pulse. **Right panel**: spin system dynamics, classified into coherence orders according to Eq. (8.81). Single-quantum coherence can be seen accumulating and then fading in a very precise sequence as the system climbs into the double-quantum coherence under the influence of the control sequence. Reproduced with permission from [300]

An example of Eq. (8.82) clarifying the dynamics under an optimal control sequence is given in Fig. 8.9—a complicated numerically optimized phase-modulated pulse is seen to be driving very smooth dynamics starting at zero-quantum coherence, moving through single-quantum coherences and into the destination, which is double-quantum coherence.

8.5.1.3 Total Exclusive Population

A significant obstacle to visualization is that spin dynamics in $\mathfrak{L}_1^{(k)}$ is often obscured by fast rotations caused by the magnet and radiofrequency fields as well as quadratic interactions. The problem disappears if the *total population* of each $\mathfrak{L}_1^{(k)}$ (which is of course invariant under unitary dynamics inside $\mathfrak{L}_1^{(k)}$) is considered. We proceed by partitioning $\mathfrak{L}_1$ further into subspaces relating to individual spins:

$$\mathfrak{L}_1 = \mathfrak{L}_1^{(1)} \oplus \mathfrak{L}_1^{(2)} \oplus \ldots \oplus \mathfrak{L}_1^{(N)} \tag{8.83}$$

where the upper index enumerates spins and N is the total number of spins in the system. The populations of the individual subspaces in this direct sum are:

$$p_k = \left\| \mathcal{W}_{\mathfrak{L}_1^{(k)}} |\boldsymbol{\rho}\rangle \right\|_2 \tag{8.84}$$

where $\mathfrak{W}_{\mathfrak{L}_1^{(k)}}$ is a projection superoperator into $\mathfrak{L}_1^{(k)}$. Such analysis is useful in magnetization transfer experiments—Eq. (8.84) provides a measure of "total magnetisation" (counting both populations and coherences) on each spin in the system. An example is given in the right panel of Fig. 8.8, which reveals that the "noisy" optimal control pulse shown in the left panel is pushing the magnetization out of C_α–H proton onto C_α carbon and from there onto C=O carbon in a smooth and orderly way—something that would be contrary to intuition if only the pulse waveform were available for analysis.

Multi-spin order subspaces may be evaluated in a similar way—for example, the total population of the two-spin order subspace $\mathfrak{L}_2$ may be partitioned into contributions from individual spin pairs:

$$\mathfrak{L}_2 = \mathfrak{L}_2^{(1,2)} \oplus \mathfrak{L}_2^{(1,3)} \oplus \ldots \oplus \mathfrak{L}_2^{(1,N)} \oplus \ldots \oplus \mathfrak{L}_2^{(N,N)} \tag{8.85}$$

The norm of the projection of the density matrix into $\mathfrak{L}_2^{(n,k)}$ would then give the time dependence of the total population of all two-spin correlations between spins n and k.

8.5.1.4 Total Inclusive Population

Equations (8.83) and (8.84) only include states that are *local* to a specified spin. A complementary strategy is to examine the population of the subspace spanned by all states that *involve* the specified spin, including correlations and coherences with other spins. The system state space can be partitioned into:

$$\mathfrak{L} = \mathfrak{L}^{(k)} \oplus \left(\mathfrak{L}/\mathfrak{L}^{(k)}\right) \tag{8.86}$$

where $\mathfrak{L}^{(k)}$ is the subspace of all states that involve spin k in any way, and $\mathfrak{L}/\mathfrak{L}^{(k)}$ is the rest of $\mathfrak{L}$. The population of $\mathfrak{L}^{(k)}$ is then given by

$$p_k = \left\| \mathfrak{W}_{\mathfrak{L}^{(k)}} |\boldsymbol{\rho}\rangle \right\| \tag{8.87}$$

where $\mathfrak{W}_{\mathfrak{L}^{(k)}}$ is a projection superoperator into $\mathfrak{L}^{(k)}$. This is a broader definition than Eq. (8.84)—it gives a measure of total involvement of a given spin at a particular stage of the control sequence. Consistently low involvement levels indicate that the spin can be dropped from the simulation altogether. To that end Eq. (8.87) provides the benefit of a quantitative argument.

8.5.2 Trajectory Similarity Scores

Another property that is hard to gauge from the immediate appearance of either pulse shapes or system trajectories is the extent to which any two instances of

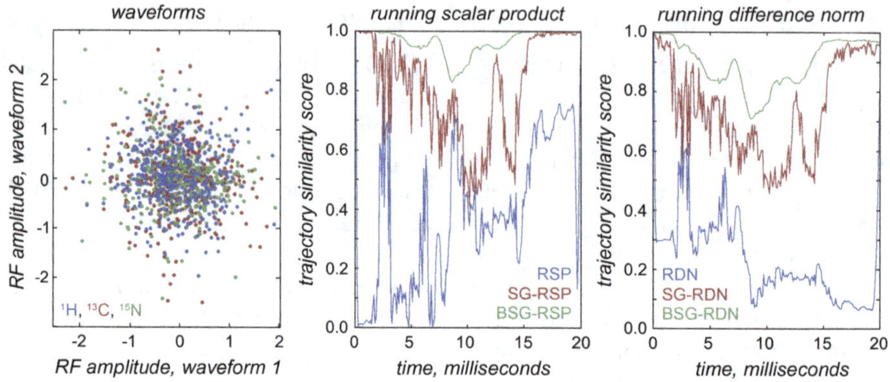

Fig. 8.10 Trajectory similarity analysis for two optimal control pulses solving the same state transfer problem (described in the caption of Fig. 8.8) to the same fidelity, but obtained from different random initial guesses. Left panel: a demonstration of the lack of direct statistical correlation between the two solutions. **Middle panel:** running scalar product similarity score for the two system trajectories without preprocessing (RSP, blue curve), with similar states grouped using Eq. (8.90) (SG-RSP, red curve) and with similar states grouped using Eq. (8.91) (BSG-RSP, green curve). **Right panel:** running difference norm similarity score for the two system trajectories without preprocessing (RDN, blue curve), with similar states grouped using Eq. (8.90) (SG-RDN, red curve) and with similar states grouped using Eq. (8.91) (BSG-RDN, green curve). Reproduced with permission from [300]

system dynamics are "similar". Optimal control solutions are not unique—a different random initial guess typically leads to a "different" pulse: the left panel of Fig. 8.10 demonstrates complete lack of direct statistical correlation between two GRAPE pulses that were obtained from different random initial guesses, but still accomplish the same goal with the same fidelity.

A more sophisticated similarity criterion is therefore required for comparing instances of spin system dynamics. From the algebraic perspective, two functions may be viewed as potentially useful:

1. Running scalar product (RSP). A step-by-step scalar product between the normalised vectors of the two trajectories

$$s_{\mathrm{AB}}(t) = \frac{\langle \boldsymbol{\rho}_{\mathrm{A}}(t) \mid \boldsymbol{\rho}_{\mathrm{B}}(t) \rangle}{\|\boldsymbol{\rho}_{\mathrm{A}}(t)\|_2 \cdot \|\boldsymbol{\rho}_{\mathrm{B}}(t)\|_2} \tag{8.88}$$

   would return 1 if the two vectors are equal, $e^{i\varphi}$ if they are different by a phase, zero if they are orthogonal and the extent and phase of their overlap if they differ in a non-trivial way.

2. Running difference norm (RDN). A step-by step norm of the difference between the corresponding vectors of the two trajectories

$$d_{\mathrm{AB}}(t) = 1 - \tfrac{1}{2}\|\boldsymbol{\rho}_{\mathrm{A}}(t) - \boldsymbol{\rho}_{\mathrm{B}}(t)\|_2 \tag{8.89}$$

would return 1 for identical vectors and zero if their tips are positioned on the opposite points of the unit ball that contains the trajectory. The choice of the norm rests with the user, but the Euclidean distance norm given in Eq (8.89) is likely the best choice.

There are situations where Eqs. (8.88) and (8.89) are too sensitive—a 90-degree difference in the phase of the magnetization vector makes the trajectories appear completely dissimilar on the RSP score ($\mathbf{S}_+$ is orthogonal to $\mathbf{S}_-$ under Frobenius inner product) and very dissimilar on the RDN score. But the physical difference is minor—the magnetisation resides on the same spin in a different phase. The definitions above, therefore, need to be updated to reflect trajectory differences in a more intuitive way.

8.5.2.1 Selective State Grouping

A significant source of cosmetic phase differences is the rapid oscillation between $\mathbf{S}_{\mathrm{X}}$ and $\mathbf{S}_{\mathrm{Y}}$ operators under Zeeman interactions. These oscillations are easy to remove from visualization by considering *the total population of the subspace* $\mathfrak{L}_{\pm}^{(k)}$ spanned by $\mathbf{S}_{+}^{(k)}$ and $\mathbf{S}_{-}^{(k)}$ operators of spin k:

$$\left\langle \mathfrak{L}_{\pm}^{(k)} \right\rangle = \sqrt{\left|\left\langle \mathbf{S}_{+}^{(k)} \right\rangle\right|^2 + \left|\left\langle \mathbf{S}_{-}^{(k)} \right\rangle\right|^2} \tag{8.90}$$

The effect this transformation has on the similarity scores is illustrated in Fig. 8.10 (centre and right panels)—two different realisations of an optimal control trajectory moving the magnetization from $^{1}\mathrm{H}_{\mathrm{C}\alpha}$ to $^{13}\mathrm{C}_{\mathrm{O}}$ in a protein backbone fragment look very dissimilar, except for the initial and the final points, on both RSP and RDN scores (blue traces). However, state grouping using Eq. (8.90) reveals that the difference is mostly in the phase of the magnetization vector—in other respects the trajectories are very similar (red traces, marked SG-RSP and SG-RDN, respectively).

It therefore appears, just as it did in the previous section, that the question of "*which subspaces does the system flow through?*" has a more interpretable answer than the same question about populations of individual states. The general approach is to identify subspaces that are invariant under uninteresting dynamics, and to interpret populations of those subspaces.

8.5.2.2 Broad State Grouping

Electromagnetic waveforms produced by optimal control methods typically engage, through Zeeman interaction, the entire dynamical Lie algebra of each spin. If the purpose of the visualization is to track magnetization transfer *between* spins, the internal dynamics within the algebra is of no interest. It may be removed from visualisation by extending Eq. (8.90) to the entire algebra of each spin, bearing in mind the note made in Sect. 2.5.5 about the dynamical lie algebra of a spin-s particle subject to external interactions being $\mathfrak{su}(2s+1)$ rather than $\mathfrak{su}(2)$. In the case of spin 1/2, we have

$$\begin{aligned} &\left\{\left\langle \mathbf{S}_{\mathrm{X}}^{(k)}\right\rangle, \left\langle \mathbf{S}_{\mathrm{Y}}^{(k)}\right\rangle, \left\langle \mathbf{S}_{\mathrm{Z}}^{(k)}\right\rangle\right\} \quad \rightarrow \\ &\left\langle \mathfrak{L}^{(k)}\right\rangle = \sqrt{\left|\left\langle \mathbf{S}_{\mathrm{X}}^{(k)}\right\rangle\right|^2 + \left|\left\langle \mathbf{S}_{\mathrm{Y}}^{(k)}\right\rangle\right|^2 + \left|\left\langle \mathbf{S}_{\mathrm{Z}}^{(k)}\right\rangle\right|^2}, \\ &\mathfrak{L}^{(k)} = \operatorname{span}\left\{\mathbf{S}_{\mathrm{X}}^{(k)}, \mathbf{S}_{\mathrm{Y}}^{(k)}, \mathbf{S}_{\mathrm{Z}}^{(k)}\right\} \end{aligned} \tag{8.91}$$

And in the case of arbitrary spin:

$$\left\langle \mathfrak{L}^{(k)}\right\rangle = \sqrt{\sum_{m=-l}^{l}\left|\left\langle \mathbf{T}_{l,m}^{(k)}\right\rangle\right|^2}, \quad \mathfrak{L}^{(k)} = \operatorname{span}\left\{\mathbf{T}_{l,-l}^{(k)}, \mathbf{T}_{l,-l+1}^{(k)}, \ldots, \mathbf{T}_{l,l}^{(k)}\right\} \tag{8.92}$$

This amounts to grouping the populations of the entire Lie sub-algebra of each individual spin—a map that may be schematically denoted as

$$\mathfrak{su}(2s_1+1)\otimes\mathfrak{su}(2s_2+1)\otimes\ldots\otimes\mathfrak{su}(2s_N+1) \quad \rightarrow \quad \mathbb{R}^1\otimes\mathbb{R}^1\otimes\ldots\otimes\mathbb{R}^1 = \mathbb{R}^N \tag{8.93}$$

where N is the number of spins in the system and $2s_k+1$ is the multiplicity of k-th spin. Equation (8.92) maps the population of each Lie algebra in the direct product into a one-dimensional subspace of a real vector space $\mathbb{R}^N$. Similarity scores computed for the trajectory image in $\mathbb{R}^N$ would only capture the *transfer* of magnetisation between spins – their internal dynamics would not be visualized.

When Eq. (8.91) is used to group populations of closely related states, the two trajectories plotted in the right panel of Fig. 8.10 turn out to be very similar—over 80% similarity throughout on RSP score and over 70% similarity on RDN score (green curves, labelled BSG-RSP and BSG-RDN respectively). This contrasts with the lack of statistical correlation for the pulse shapes themselves (Fig. 8.10, left panel).

8.6 Pulse Shape Analysis

Time-dependent perturbation theory (Sect. 4.4.3) suggests another way of interpreting optimal control waveforms: to first order in perturbation theory, a pulse is driving the transitions that match the frequencies it contains. These frequencies may themselves depend on time, and therefore time–frequency representations may be informative. A good example is Gabor transform [302]:

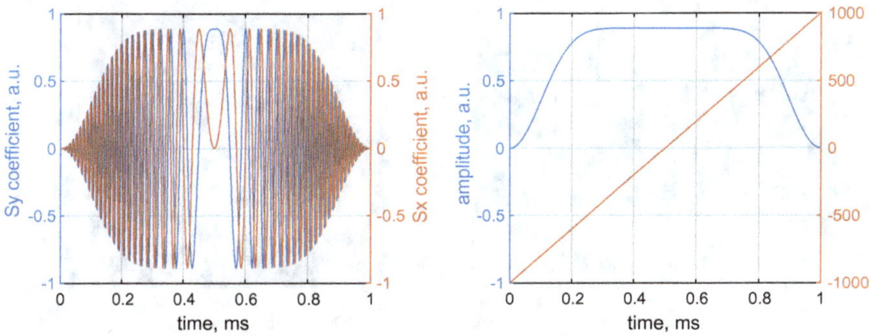

Fig. 8.11 Cartesian (left) vs. frequency-amplitude (right) representation of a frequency-swept pulse with the frequency range of 2 MHz around the centre and a duration of 1 ms. Reproduced with permission from [188]

$$f(\tau,\omega) = \int_{-\infty}^{\infty} f(t)\mathrm{e}^{-\pi(t-\tau)^2}\mathrm{e}^{-i\omega t}\mathrm{d}t \tag{8.94}$$

which is a special case of the short-time Fourier transform with a Gaussian weight. Many such representations exist [303]; we denote them all $f(\tau,\omega)$ in this section.

8.6.1 Frequency-Amplitude Plots

Many analytically derived control sequences in magnetic resonance are pairs of smooth functions corresponding to the $\mathrm{S_X}$ and $\mathrm{S_Y}$ control channels of the instrument in the rotating frame. In such cases, a transformation to the amplitude-phase representation $\{f_\mathrm{X}(t), f_\mathrm{Y}(t)\} \to \{a(t), \varphi(t)\}$ using the four-quadrant arctangent:

$$a(t) = \sqrt{f_\mathrm{X}^2(t) + f_\mathrm{Y}^2(t)}, \qquad \varphi(t) = \arctan\left[f_\mathrm{Y}(t), f_\mathrm{X}(t)\right] \tag{8.95}$$

is advantageous. In particular, frequency sweeps are simplified (Fig. 8.11) because the instantaneous frequency is the derivative of the phase.

This is useful when the pulse waveform is known analytically, but runs into trouble in other cases because the phase is 2π periodic, and the unwrapping [304] is not always a well-defined operation [305].

8.6.2 Spectrograms, Scalograms, etc.

When a short-time Fourier transform of an $\{f_\mathrm{X}(t), f_\mathrm{Y}(t)\}$ two-channel control sequence is converted into a phase-amplitude representation:

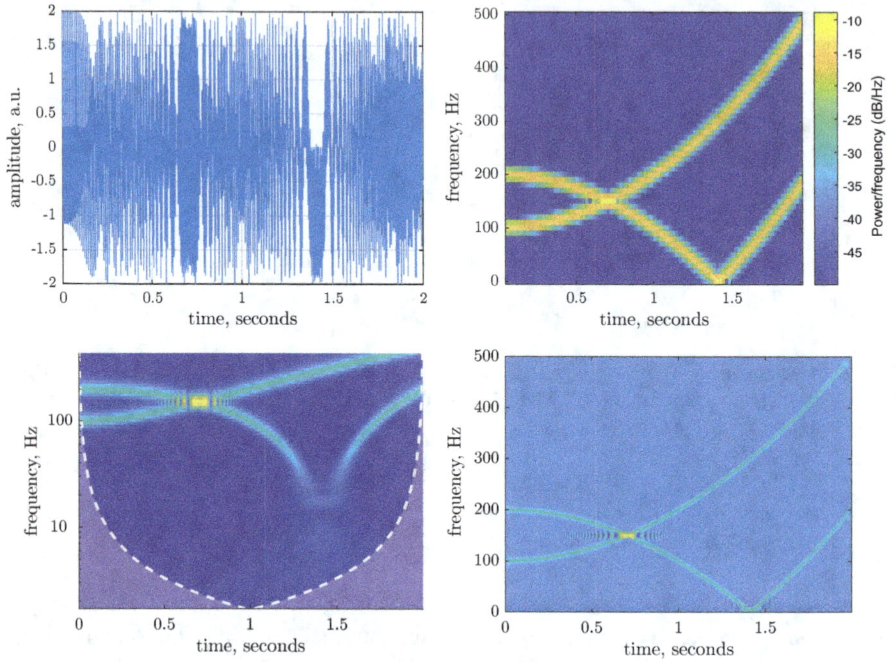

Fig. 8.12 Time-domain representation (**top left**) of a superposition of analytically generated quadratic chirp pulses, its amplitude spectrogram (**top right**), scalogram using Morse wavelets with γ = 3, time-bandwidth product of 60, and 10 voices per octave (**bottom left**), and pseudo-Wigner-Ville distribution (**bottom right**)

$$\begin{aligned} a(\tau,\omega) &= \sqrt{f_{\mathrm{X}}^2(\tau,\omega)+f_{\mathrm{Y}}^2(\tau,\omega)} \\ \varphi(\tau,\omega) &= \arctan[f_{\mathrm{Y}}(\tau,\omega), f_{\mathrm{X}}(\tau,\omega)] \end{aligned} \tag{8.96}$$

the resulting *spectrograms* may be interpretable within the Fermi golden rule framework because they show how amplitudes and phases at different frequencies change as functions of time (Fig. 8.12).

In the magnetic resonance context, it is expedient to use HSB colour space, where the amplitude is mapped into the brightness channel, phase into the hue channel, and the saturation channel value is fixed [303].

Related time–frequency representations, called *scalograms*, use the continuous wavelet transform—a generalisation of Eq. (8.94) to an arbitrary kernel function (commonly a wavelet) $w(t)$ that is shifted and scaled, with the scaling multiplier $a > 0$ playing the role of reciprocal frequency:

$$f(a,b) = \frac{1}{\sqrt{a}} \int_{-\infty}^{\infty} w^*\left(\frac{t-b}{a}\right) f(t)\,\mathrm{d}t \tag{8.97}$$

An example using Morse wavelets is shown in the bottom left panel of Fig. 8.12. A similar generalisation that also resembles a Fourier transform is the *Wigner-Ville distribution* [306]:

$$f(\omega, t) = \int_{-\infty}^{\infty} f^*\left(t - \frac{\tau}{2}\right) f\left(t + \frac{\tau}{2}\right) \mathrm{e}^{-i\omega\tau} \mathrm{d}\tau \tag{8.98}$$

When the transform is applied to the windowed version of the signal, the result is called *pseudo-Wigner-Ville distribution*; an example is shown in the bottom right panel of Fig. 8.12.

9 Notes on Software Engineering

The ongoing numerical simulation revolution has little to do with physics, but much with economics. It is driven by the increasing gap between the cost of academic brain time (over £30 per brain-hour in 2022) and the cost of CPU time (under £0.03 per core hour). The days of careful derivation, tight simplification, and thoughtful programming are gone—the game is now about dropping the problem into a matrix representation and letting a computer deal with it. Algorithmic efficiency still matters because the cost of those core-hours adds up, but the decisions are at the level of the difference between exponential and polynomial complexity methods—people no longer have the time to chase a twenty percent efficiency improvement. In the race to publish and sell, those who write good code are overtaken by those who do not.

This creates economic pressures in strange directions. One consequence is my choice of *Matlab* for *Spinach* development—this is sometimes queried on the performance compared to compiled languages. Recent versions of *Matlab* call platform-specific libraries (MKL, CUDA, etc.) for standard matrix operations; our benchmarks indicate that *Matlab* is about as fast as compiled languages. It is more convenient though: the code is platform-independent, elegant and well adapted to scientific computing—thus saving the expensive brain time. Transparent parallelisation, convenient sparse arrays, and automatic GPU support are decisive factors. With the resources realistically available from academic funding agencies, it would have been impossible to develop *Spinach* in any other language.

This chapter presents some tales from the *Spinach* [87] development road—it is not a definitive manual on how to proceed, but rather a collection of observations about what works in practice for long-range dissipative time-domain simulations of irregular spin systems.

I. KUPROV, *Spin*,
https://doi.org/10.1007/978-3-031-05607-9_9

9.1 Sparse Arrays, Polyadics, and Opia

It is a long-standing observation that spin Hamiltonians are sparse [309, 310], that they are short sums of Kronecker products of tiny matrices, and that most of those matrices are unit matrices—attractive properties begging to be used as efficiency improvement avenues. However, the same is not true of the density matrix which may fill up and become full-rank.

This section looks at what is sparse and low-rank in spin dynamics simulations, what stays sparse and low-rank, and how to use that. We deliberately skip the vast question (DMRG [245], MPO/MPS [246], TT [247], etc.) of how to keep the state vector compressed—that rarely works in irregular and dissipative magnetic resonance systems [248]—and focus instead on tensor structured representations of evolution generators. This decision was made after much trial and error [311]: we leave the state vector uncompressed and rely on reduced state space approximations (Chap. 7) to keep its dimension manageable.

9.1.1 The Need to Compress

There are two well researched limits in the numerical simulation of spin dynamics: complicated spatial dynamics of simple spin systems, and simple spatial dynamics of complicated spin systems. An example of the former is diffusion and flow MRI of mostly water [312, 313]; the latter is exemplified by spatially encoded NMR experiments [314, 315] and localised NMR spectroscopy [316]. Both cases are well covered by the existing simulation software; they are straightforward because matrix dimensions are manageable: $\mathfrak{aut}(\mathbb{R}^3) \otimes \mathfrak{su}(2)$ and $\mathbb{R}^3 \otimes \mathfrak{su}(2^N)$ are both tractable, either directly or with reasonable approximations [253, 255] for the matrix representations of $\mathfrak{su}(2^N)$, where N is the number of spins.

The simulation problem becomes numerically intractable when complicated spatial dynamics (diffusion, convection, and flow in three dimensions) is combined with complicated spin dynamics (spin–spin coupling, cross-relaxation, chemical kinetics, etc.). A well digitised human brain would have at least a hundred points in each of the three directions, meaning a dimension of at least $100^3 = 10^6$ for the spatial dynamics generator matrices. At the same time, a typical metabolite (e.g. glucose) contains upwards of ten coupled spins, meaning a Liouville space dimension of at least $4^{10} \approx 10^6$. Direct products of spin and spatial dynamics generators would then have the dimension exceeding 10^{12} even before chemical kinetics is considered—clearly an impossible proposition, even if sparse matrix arithmetic is used.

9.1.2 Sparsity of Spin Hamiltonians

It is useful to have an estimate of exactly how sparse spin Hamiltonians are. We start by noting that the dimension and the number of elements in a spin Hamiltonian are:

$$\dim(\mathbf{H}) = \prod_{k=1}^{n} (2S_k + 1), \quad \text{numel}(\mathbf{H}) = [\dim(\mathbf{H})]^2 \tag{9.1}$$

where $2S_k + 1$ is the multiplicity of the k-th spin. A system of n spins can have at most $(n^2 + 3n)/2$ interactions: $(n^2 - n)/2$ bilinear couplings between different spins, n quadrupolar interactions or zero-field splittings, and n couplings to the external magnetic field, which includes spin-rotation couplings. Each interaction has $O(1)$ spin operators in it: three for a Zeeman interaction, five for a point dipolar interaction, etc.—to a total of $O(n^2)$ operators.

It may be verified by direct inspection that Cartesian spin operators ($\mathbf{S}_\text{X}, \mathbf{S}_\text{Y}, \mathbf{S}_\text{Z}$), their binary direct products ($\mathbf{L}_\text{X}\mathbf{S}_\text{X}$, $\mathbf{L}_\text{X}\mathbf{S}_\text{Y}$, etc.), and irreducible spherical tensor operators in the Pauli basis all have $O(1)$ non-zeroes per column—usually one; the number of columns is $\dim(\mathbf{H})$. Together with the quadratic scaling of the number of interactions, this yields $O(n^2)$ non-zeroes per column, and therefore the following scaling for the number and density of non-zeroes:

$$\begin{gathered} \text{nnz}(\mathbf{H}) = O(n^2)\dim(\mathbf{H}) \\ \text{den}(\mathbf{H}) = \frac{\text{nnz}(\mathbf{H})}{\text{numel}(\mathbf{H})} = \frac{O(n^2)}{\dim(\mathbf{H})} \leq O(n^2/2^n) \end{gathered} \tag{9.2}$$

This estimate for $\text{den}(\mathbf{H})$ validates the common knowledge that spin operators are very sparse [309], but adds the important observation that the sparsity *improves exponentially* with the number of spins: in large spin systems, any physically reasonable Hamiltonian in the Pauli basis would be mostly zeroes. When a matrix representation of a spin Hamiltonian must be stored explicitly, this is a major efficiency consideration.

Exponential propagators are not necessarily sparse—non-zeroes proliferate when Hamiltonian powers are taken. This is a common problem when propagators are computed naïvely using Taylor [317], Chebyshev [317], or Padé [141] approximations; with matrix dimensions above 10^4, this is a show-stopper because of memory overflow. However, propagators may be kept sparse if the software only ever computes powers of $i\mathbf{H}\Delta t$ where $\Delta t < \|\mathbf{H}\|_2^{-1}$. That is a reasonable constraint to impose—it is the definition of the Nyquist-Shannon time step, which is required anyway to avoid signal aliasing in the frequency domain. All eigenvalues of $(i\mathbf{H}\Delta t)^n/n!$ then drop off super-exponentially with n; some of the new non-zeroes appearing after matrix multiplication are so small as to be inconsequential. They can be dropped, thereby preserving (or even improving) matrix sparsity:

$$|a_{nk}| < \varepsilon \quad \Rightarrow \quad a_{nk} \leftarrow 0 \tag{9.3}$$

where ε is a user-specified tolerance; a reasonable value is a couple of orders of magnitude lower than the reciprocal experiment duration.

This clean-up procedure is applied every time a matrix is generated or multiplied in *Spinach*. The modest overhead of examining the non-zero index for small elements is compensated by the significant reduction of memory footprint and acceleration of matrix operations. Importantly, the procedure is only effective for propagators that satisfy the above mentioned $\Delta t < \|\mathbf{H}\|_2^{-1}$ condition on the time step.

9.1.3 Tensor Structure of Liouvillians

Spin Hamiltonians are sparse, but the number of non-zeroes in Eq. (9.2) still scales exponentially with the system size: modern computers run out of memory at about twenty spins. The onset of this problem may be delayed if we observe that storing the matrix explicitly is not necessary—every spin Hamiltonian has a sum-of-direct-products structure:

$$\mathbf{H} = \sum_k \omega_k \left[\mathbf{S}_k^{(1)} \otimes \mathbf{S}_k^{(2)} \otimes \cdots \otimes \mathbf{S}_k^{(n)}\right] \tag{9.4}$$

where the number of terms (Sect. 9.1.2) is $O(n^2)$, ω_k are interaction amplitudes, and $\mathbf{S}_k^{(j)}$ are either unit matrices (need not be stored) or single-spin operators with tiny dimension and $O(1)$ non-zeroes. The total number of non-zeroes in this representation of the spin Hamiltonian is therefore $O(n^3)$—polynomial in the number of spins. The same observation applies to common relaxation superoperators (Chap. 6) which are composed of products of *tensor structures* like the one in Eq. (9.4) and therefore (because a product of krons is a kron of products) have that structure themselves. Thus, PSPACE *tensor structured matrix representations* do exist for evolution generators of large spin systems. It is only when the Kronecker products are opened that conventional matrices become impossible to store.

The same observation applies to evolution generators under a combination of spin dynamics, spatial dynamics, and chemical kinetics. They contain three principal factors: (a) spatial distributions and spatial dynamics; (b) chemical reaction dynamics; (c) spin dynamics and relaxation. These factors are in a direct product relationship—each voxel may have different concentrations and transport velocities, each chemical species may have different spin dynamics, and spin dynamics may in turn be different in each voxel due to, for example, a magnetic field gradient or a different viscosity that affects relaxation. The problem therefore has the following structure:

$$[\text{space dynamics}] \otimes [\text{reaction kinetics}] \otimes [\text{spin dynamics}] \tag{9.5}$$

The Liouville space equation of motion then necessarily has the form:

$$\frac{\mathrm{d}}{\mathrm{d}t}\boldsymbol{\rho}(t) = \left[\sum_{nmk} a_{nmk}(t)\mathbf{M}_n \otimes \mathbf{K}_m \otimes \mathbf{S}_k\right]\boldsymbol{\rho}(t) \tag{9.6}$$

where $\boldsymbol{\rho}(t)$ is the state vector, $a_{nmk}(t)$ are interaction amplitudes, $\mathbf{M}_n$ are spatial operators, $\mathbf{K}_m$ are chemical operators (possibly themselves dependent on $\boldsymbol{\rho}$ if the kinetics is non-linear), and $\mathbf{S}_k$ are (possibly dissipative) spin operators that themselves have the tensor structure shown in Eq. (9.4). Spatial operators, such as three-dimensional diffusion, are also commonly direct products of differential operators acting on the individual dimensions.

The square bracket in Eq. (9.6) is an example of a *polyadic expansion*. Efficient algorithms exist for manipulating this and other tensor structures without computing the direct products [247, 318]. Practical experience indicates that, in the context of time-domain spin dynamics, the state vector $\boldsymbol{\rho}(t)$ is best left uncompressed [311]. An important observation is that multiplication of the state vector by an exponential propagator has the same complexity scaling as the cost of multiplication by the generator. Indeed, the multiplication by the exponential propagator may be expressed as:

$$e^{-i\mathcal{L}\Delta t}\boldsymbol{\rho} = \sum_{k=0}^{\infty}\frac{(-i\mathcal{L}\Delta t)^k}{k!}\boldsymbol{\rho} = \sum_{k=0}^{\infty}\frac{(-i\Delta t)^k}{k!}(\mathcal{L}(\mathcal{L}\ldots(\mathcal{L}\boldsymbol{\rho}))) \tag{9.7}$$

When $\Delta t \le \|\mathcal{L}\|_2^{-1}$, this series converges monotonically to machine precision in about 15 iterations, meaning that only a fixed number of multiplications of $\boldsymbol{\rho}$ by the evolution generator $\mathcal{L}$ is required. An upper bound on the 2-norm of $\mathcal{L}$ only needs a few matrix–vector operations (Sect. 9.3.1). We therefore conclude that the only operation essentially required in time-domain spin dynamics simulations is kron-times-vector.

9.1.4 Kron-Times-Vector Operation

Consider a matrix–vector product where the matrix is a direct product of smaller matrices:

$$\mathbf{y} = \left[\mathbf{A}^{(1)} \otimes \mathbf{A}^{(2)} \otimes \ldots \otimes \mathbf{A}^{(N)}\right]\mathbf{x} \tag{9.8}$$

Each element of the object in square brackets is a product of the corresponding elements of $\mathbf{A}^{(k)}$:

$$[\cdots]_{i_1 j_1 i_2 j_2 \cdots i_N j_N} = a^{(1)}_{i_1 j_1} a^{(2)}_{i_2 j_2} \cdots a^{(N)}_{i_N j_N} \tag{9.9}$$

but calculating and storing the left hand side is undesirable because the number elements there is astronomical. It is more efficient to take the products on the right hand side as they become necessary. In such a scenario, the complexity is unchanged, but the memory requirements would not exceed the resources already deployed in storing the small matrices $\mathbf{A}^{(k)}$.

The procedure for computing the product $[\cdots]\mathbf{x}$ must then involve matching the linear index of the elements of $\mathbf{x}$ with the multi-index of $a_{i_1 j_1}^{(1)} a_{i_2 j_2}^{(2)} \cdots a_{i_N j_N}^{(N)}$. This is straightforward—$\mathbf{x}$ is reshaped into a multi-dimensional array whose dimensions match the row dimension of $\mathbf{A}^{(k)}$, each dimension is multiplied by the corresponding $\mathbf{A}^{(k)}$, and the result stretched back into a vector whose dimension is now the product of column dimensions of $\mathbf{A}^{(k)}$. The multiplication procedure, described by Fernandes, Plateau, and Stewart [319] and first implemented for *Matlab* by David Gleich, proceeds as follows:

1. Record row dimensions of $\mathbf{A}^{(N-k+1)}$ into c_k.
2. Reshape $\mathbf{x}$ into an N-dimensional array $\mathbf{X}$ with dimensions c_k.
3. Loop index n over the dimensions of $\mathbf{X}$

   Permute the dimensions of $\mathbf{X}$ to make its nth dimension left-most.

   Reshape $\mathbf{X}$ to into a matrix of column dimension c_n and row dimension $\prod_{m \neq n} c_m$.

   Perform reassignment $\mathbf{X} = \mathbf{A}^{(N-n+1)} \cdot \mathbf{X}$ and replace c_n by the column dimension of $\mathbf{A}^{(N-n+1)}$.

   Reshape $\mathbf{X}$ back into an N-dimensional array $\mathbf{X}$ with dimensions c_k.

   Put the dimensions of $\mathbf{X}$ back into the original order.
4. Reshape $\mathbf{X}$ back into a vector and return it as $\mathbf{y}$.

Multiplication by unit matrices may be skipped. This algorithm has complicated memory access patterns; because non-sequential memory access can be expensive, it may be viewed as a different trade-off between memory capacity and latency requirements.

9.1.5 Polyadic Objects and Opia

A considerable amount of software engineering is required before the method described in the previous section becomes useful for solving Eq. (9.6). The first significant hurdle is addition: Eq. (9.6) is a sum of direct products; it cannot be written as a single direct product. However, because matrix–vector multiplication is distributive over addition, the algorithm is easily extended to linear combinations of krons:

$$\begin{aligned} &(\alpha[\mathbf{A} \otimes \mathbf{B} \otimes \ldots] + \beta[\mathbf{C} \otimes \mathbf{D} \otimes \ldots] + \ldots)\mathbf{x} \\ &= \alpha[\mathbf{A} \otimes \mathbf{B} \otimes \ldots]\mathbf{x} + \beta[\mathbf{C} \otimes \mathbf{D} \otimes \ldots]\mathbf{x} + \ldots \end{aligned} \tag{9.10}$$

In practice, this is implemented by buffering addition: when the user adds two Kronecker products, their sum is not evaluated—the terms are simply stored in a pointer array:

$$\mathbf{A}\otimes\mathbf{B}+\mathbf{C}\otimes\mathbf{D}\otimes\mathbf{E}+\ldots \quad \Leftrightarrow \quad \{\{\mathbf{A},\mathbf{B}\},\{\mathbf{C},\mathbf{D},\mathbf{E}\},\ldots\} \tag{9.11}$$

When the time comes to multiply this object into a vector, each term in the sum is multiplied into that vector individually, and the results are added up. This is efficient because evolution generators in Eq. (9.6) are *short* sums of Kronecker products: the number of terms is much smaller than the dimension of the matrix they represent. This also offers parallelisation opportunities.

During a simulation, a polyadic representation of a matrix may be pre- or post-multiplied by a small number of other matrices. Because the only operation the entire object needs to deliver is matrix–vector product, the best strategy is again to buffer the terms into a pointer array, and apply them to the incoming vector before and after the polyadic is applied. The object structure is therefore extended as follows:

$$\begin{gathered}\mathbf{P}_1\cdot\ldots\cdot\mathbf{P}_N\cdot[\mathbf{A}\otimes\mathbf{B}+\mathbf{C}\otimes\mathbf{D}\otimes\mathbf{E}+\ldots]\cdot\mathbf{Q}_1\cdot\ldots\cdot\mathbf{Q}_M\\ \Updownarrow\\ \{\mathbf{P}_1,\ldots,\mathbf{P}_N\}\{\{\mathbf{A},\mathbf{B}\},\{\mathbf{C},\mathbf{D},\mathbf{E}\},\ldots\}\{\mathbf{Q}_1,\ldots,\mathbf{Q}_M\}\end{gathered} \tag{9.12}$$

The sequence is simply replayed from right to left every time a product into a vector is needed. Once the elements of this object are themselves allowed to be polyadic, the object can buffer an arbitrary sequence of additions, multiplications and Kronecker products.

Some matrices in Eqs. (9.4) and (9.6) are multiples of unit matrices, but automatic detection of those in finite precision arithmetic is expensive and unreliable—ideally, they should be labelled as such at the point when they are created. A convenient approach is to design an *Object Pretending It is a Unit Matrix* (OPIUM) that delivers the properties of a unit matrix (nnz, numel, conjugate-transpose, multiplications, etc.) while storing only two numbers: the coefficient and the dimension. When unit matrices in a polyadic object are replaced by opia, the efficiency is improved.

9.2 Parallelisation and Coprocessor Cards

The Moore's law curve for single thread CPU performance has plateaued around 2015, and may have partially backtracked in 2018 after mitigations were applied for speculative execution vulnerabilities discovered in major CPUs. The biggest increases to occur since have been due to the improvements in memory latency and bandwidth, and the new parallelisation opportunities offered by multi-core, cluster, and GPU architectures. Spin dynamics supports multiple parallelisation modalities: ensemble level (isotopomers, powder averages, parameter distributions), operator level (Hamiltonian and relaxation superoperator construction), trajectory level (indirect dimensions, disconnected subspaces, frequencies and fields in swept experiments), and array level (distributed matrices and vector stacks). This section contains practical advice on architectural and logistical matters. The principal problem of getting a specific calculation to run efficiently in parallel is summarised by Amdahl's Law [320]:

$$s(n) = \frac{1}{r_{\mathrm{S}} + r_{\mathrm{P}}/n} \tag{9.13}$$

where $s(n)$ is the expected speed-up from running over n parallel threads, r_{S} is the run time fraction corresponding to serial code, and r_{P} is the run time fraction corresponding to the parallel code. This function is only linear with respect to n in the unrealistic case when the serial code fraction is zero; for finite values of r_{S}, the function saturates when $n \gg r_{\mathrm{P}}/r_{\mathrm{S}}$.

Parallelisation of simulations that contain multiple independent sub-problems is straightforward. Sequential processes in monolithic spin systems are harder—the requirement for parallel scalability changes the mathematical layout of time-domain quantum mechanics. For example, matrix factorisations (diagonalisation, singular value decomposition, etc.) must now be avoided—they have large communication overheads, resulting in poor scaling even on shared-memory systems; full diagonalisation of spin Hamiltonians with dimension above 10^5 (i.e. with more than about 17 spin-1/2 particles) is currently impractical [83]. Even matrix–matrix multiplication can become hard to parallelise efficiently due to unpredictable fluctuations in the density of non-zeroes in sparse arrays [321]. For these reasons, frequency domain simulations that rely on Hamiltonian diagonalisation are hard to parallelise efficiently.

The parallelisation problem in the time domain reduces to finding a parallel algorithm for density matrix propagation. In Hilbert space, the obstacle is that propagation under the Liouville—von Neumann equation involves matrix multiplication simultaneously on the left and the right:

$$\frac{\mathrm{d}}{\mathrm{d}t}\boldsymbol{\rho}(t) = -i[\mathbf{H}(t), \boldsymbol{\rho}(t)] \qquad \Rightarrow \qquad \boldsymbol{\rho}(t+\mathrm{d}t) = \mathrm{e}^{-i\mathbf{H}(t)\mathrm{d}t}\boldsymbol{\rho}(t)\mathrm{e}^{+i\mathbf{H}(t)\mathrm{d}t} \tag{9.14}$$

Even in Liouville space, every element of $\boldsymbol{\rho}(t+\mathrm{d}t)$ depends on every element of $\boldsymbol{\rho}(t)$: when $\boldsymbol{\rho}(t+\mathrm{d}t)$ is evaluated in parallel, it must be re-sent to every worker process at each time step. This is too much communication, but fortunately Eq. (9.14) can be transformed mathematically (Sect. 9.2.3) into a form with more favourable communication requirements.

In this context, a mathematics coprocessor card (historically, a graphics processing unit, GPU) is essentially a massively parallel shared-memory device with a high internal bandwidth (better than CPU), but long data access latencies (worse than CPU). Large sets of simple sequential local operations can be performed quickly, but branching functions with non-sequential memory access (such as the above mentioned factorisations) rarely benefit from GPU processing.

9.2.1 Obvious Parallelisation Modalities

We assume the most general equation of motion in Liouville space, including coherent and dissipative dynamics in spin as well as classical degrees of freedom [231, 259]:

$$\frac{\partial \boldsymbol{\rho}}{\partial t} = -i[\mathcal{H} + i\mathcal{R} + i\mathcal{K} + i\mathcal{F} + \ldots]\boldsymbol{\rho} \tag{9.15}$$

where $\boldsymbol{\rho}$ is a state vector, $\mathcal{H}$ is a Hamiltonian commutation superoperator, $\mathcal{R}$ is a relaxation superoperator, $\mathcal{K}$ is a chemical kinetics superoperator, and $\mathcal{F}$ is a spatial dynamics generator. These quantities can depend on classical coordinates $\mathbf{x}$ and time t. Given an evolution generator $\mathcal{L}(\mathbf{x}, t)$ equal to the content of the square brackets in Eq. (9.15), the general solution is a product integral:

$$\frac{d\,\boldsymbol{\rho}(\mathbf{x}, t)}{dt} = -i\mathcal{L}(\mathbf{x}, t)\boldsymbol{\rho}(\mathbf{x}, t) \qquad \Rightarrow \qquad \boldsymbol{\rho}(\mathbf{x}, t + \mathrm{d}t) = \exp[-i\mathcal{L}(\mathbf{x}, t)\mathrm{d}t]\boldsymbol{\rho}(\mathbf{x}, t) \tag{9.16}$$

where $\boldsymbol{\rho}(\mathbf{x}, t)$ lives in the direct product of the set of all density operators (quantum degrees of freedom) and the set of all probability densities (classical degrees of freedom). There is nothing more to it—time domain spin dynamics simulations are about generating a suitable Eq. (9.16) and then persuading it to get itself computed. For sufficiently large spin systems and their ensembles, the following parallelisation opportunities present themselves:

1. **Indirect time and frequency dimensions**: any two- or higher-dimensional magnetic resonance experiment generates an array of density matrices at different time positions in the indirect dimensions. Physically, these density matrices correspond to different instances of the pulse sequence [150]; from the simulation point of view, they are independent and may be sent to different worker processes for the calculation of detection stage evolution. The same applies to slices and k-space lines of magnetic resonance imaging experiments [311]. Frequency domain simulations do by definition avoid the requirement for time order where the next time step depends on the state at the previous time step. Individual frequency points in such simulations correspond to independent calculations, although re-useable objects like linear solver preconditioners may be shared.

2. **Ensembles, isotopomers, powders, and parameter distributions**: for powder averages and voxels of MRI experiments, this parallelisation modality (essentially an outer loop over multiple independent simulations) is well covered in the existing papers and software [310, 322]. Parallelisation over ensembles (e.g. isotopomers in a natural abundance sample, or B_1 field amplitudes in an optimal control optimisation) is more subtle because multiple variables may be involved, and sometimes correlated.
3. **GRAPE gradients and Hessians**: pulse design by numerical optimisation requires derivatives of the magnetisation transfer fidelity with respect to every point of the pulse waveform [102]. While the calculation of the fidelity under any specific pulse is not easy to parallelise, its gradient and Hessian may be computed in parallel [288].
4. **Basic linear algebra**: many matrix and vector operations may be cast into a form suitable for parallel processing on a cluster, shared-memory system, or a GPU [145]. In practice, this is only efficient when the operation is performed repeatedly (*e.g.* matrix powers during the exponentiation process, or repeated matrix–vector multiplications during time propagation) in a way that requires only modest amounts of inter-thread communication [145].

9.2.2 Parallel Basis and Operator Construction

Before time evolution may be simulated, there is a considerable amount of what we will call *housekeeping*: generation of elementary operators and states, building of evolution generators, etc. Amdahl's Law does not permit us to ignore these stages; they must also—insofar as possible—be parallelised.

9.2.2.1 Basis Set Indexing

When the calculations are running in the full Hilbert or Liouville space, this stage is not necessary—the basis set is defined implicitly by the single-spin operator matrices that go into the Kronecker products. However, the more efficient polynomial complexity scaling methods (Chap. 7) in Liouville space do require a basis set to be indexed because not all possible states participate in the system dynamics.

One way to index a basis set is described in Sect. 7.1: a sparse array of integers may be used to store the direct product structure of each basis operator in the spherical tensor basis

$$\begin{gathered}[0 \quad 1 \quad 6 \quad 0 \quad 2 \quad 0 \quad \cdots] \\ \Updownarrow \\ \mathbf{T}_{0,0} \otimes \mathbf{T}_{1,1} \otimes \mathbf{T}_{2,0} \otimes \mathbf{T}_{0,0} \otimes \mathbf{T}_{1,0} \otimes \mathbf{T}_{0,0} \otimes \cdots\end{gathered} \tag{9.17}$$

where the l, m index pairs of single-spin irreducible spherical tensor operators $\mathbf{T}_{lm}$ (Sect. 3.2.2) are mapped into a single linear index as $n = l^2 + l - m$. This works because $m \in [-l, l]$. The task to be parallelised is therefore the generation of a sparse matrix (with these specifications in rows) for a given spin system and given basis state selection criteria provided by the user (Table 9.1).

In the order that the reduced basis construction stages are described in Sect. 7.3.2, the corresponding parallelisation strategies are listed in Table 9.1. All stages except parallel sorting have little to no communication requirements beyond the initial distribution of work and result collection at the end.

9.2.2.2 Hamiltonian and Relaxation Superoperator

The procedure described in Sect. 7.2 has a favourable complexity scaling even in its mathematical form, but further efficiency improvements may be made at the implementation level:

1. The structure coefficients in Eq. (7.16) may be precomputed and stored. Because high spin quantum numbers are uncommon, the arrays are small.
2. As discussed in Sect. 9.1.2, spin operators are very sparse in the Pauli basis. It is possible to set up screening criteria such that only the *bona fide* non-zeroes need to be computed.

In practice, the physically motivated list of spin interactions supplied by the user is converted into a Hamiltonian descriptor list [operator descriptor, coefficient], the list is distributed to the worker processes which compute their chunks of the total Hamiltonian and return them to the head node.

Table 9.1 Restricted state space generation stages, parallelisation avenues, and worst-case complexity scaling

Stage	Operations	Parallelisation and worst case complexity scaling
Interaction graph construction	Inspect interaction list, make [nspins x nspins] logical matrix with 1 where an interaction is significant and 0 where it is not	Over the interaction list; quadratic scaling in the number of spins
Interaction graph partitioning	A variation of Tarjan's algorithm for finding connected components of a graph	Over the node list; linear scaling in the number interactions
Subgraph basis set construction	Direct product of state lists of each spin in the subgraph	Over the subgraph list; linear scaling in the number of subgraphs
Subgraph basis set merging	Subgraph state lists are merged, repetitions are eliminated, the resulting global state list is sorted	Parallel sorting problem; slightly super-linear scaling in the number of states
Basis set pruning	States not satisfying user-specified criteria are dropped from the state list	Over the state list; linear scaling in the number of states

A general recipe covering all relaxation theories (Chap. 6) cannot be given, but the calculation of the popular Redfield superoperator is best parallelised over the summation terms in Eq. (6.46) with the individual terms computed using the auxiliary matrix method in Eq. (6.48). Screening is possible with respect to the elements of the autocorrelation function array—for isotropic rotational diffusion, most are zero. For the purely numerical version in Eq. (6.47) where the Hamiltonians come from a molecular dynamics trajectory, the parallelisation is best done over the statistically independent trajectory stripes.

A logistical problem here is summing large numbers of sparse matrices: compressed formats such as CSR and CSC make sparse matrix addition inefficient. It is best to store and communicate Hamiltonian chunks in COO sparse format, and only convert to a compressed format at the end.

9.2.3 Parallel Propagation

Time propagation commonly dominates the wall clock time of spin dynamics simulations. When the easier parallelisation routes described above are unavailable or insufficient, ways must be found to distribute the time propagation problem to the cores of a CPU or the nodes of a cluster. The main problem is communication requirements: every element $\boldsymbol{\rho}(t+\Delta t)$ does in general depend on every element of $\boldsymbol{\rho}(t)$. Efficient parallelisation avenues for time evolution do therefore attempt to break the calculation down into physically non-interacting (and therefore computationally non-communicating) parts.

9.2.3.1 Liouville Space: Generator Diagonalisation

The parallel propagation matter may superficially appear trivial—ostensibly, simply diagonalising the evolution generator $\mathcal{L}$ and moving the state vector $\boldsymbol{\rho}$ into its eigenbasis reveals independently evolving subproblems. In the simple case of a time-independent evolution generator:

$$\left[\frac{\partial}{\partial t}\boldsymbol{\rho} = -i\mathcal{L}\boldsymbol{\rho} \quad \Rightarrow \quad \left\{\mathcal{L} = \mathbf{V}\boldsymbol{\Omega}\mathbf{V}^{-1}\right\} \quad \Rightarrow \quad \underset{k}{\left\{\frac{\partial}{\partial t}\left[\mathbf{V}^{-1}\boldsymbol{\rho}\right]_k = -i\,\omega_k\left[\mathbf{V}^{-1}\boldsymbol{\rho}\right]_k\right.}\right] \tag{9.18}$$

where $\mathbf{V}$ is a matrix with eigenvectors in columns, $\boldsymbol{\Omega}$ is a diagonal matrix with eigenvalues ω_k on the diagonal, and the square brackets denote the indicated element of the vector they contain.

The problem here is the computational complexity of the diagonalisation operation (cubic with matrix dimension) and the loss of sparsity: in spin dynamics, $\mathcal{L}$ is commonly very sparse, but $\mathbf{V}$ is not. Diagonalisation operation itself does not parallelise well, and therefore this avenue is only beneficial when the diagonalisation of $\mathcal{L}$ is known a priori; that is rarely the case.

9.2.3.2 Liouville Space: Generator Block-Diagonalisation

Although full diagonalisation of the Liouvillian may be out of the question, inexpensive block-diagonalisation opportunities are plentiful [124, 254, 255]: different irreducible representations of symmetry groups (Sect. 4.6), different eigenspaces of conservation laws (Sect. 7.3.4), different propagator group orbits (Sect. 7.3.6), etc. This creates parallelisation opportunities—in liquid state NMR and ESR simulations of large spin systems there can be hundreds of independent subspaces [124, 251] and the corresponding projector matrices may be generated efficiently. Given a complete set of projectors $\{\mathcal{W}_k\}$ into independently evolving subspaces, the calculation splits up as follows:

$$\mathcal{L}_k = \mathcal{W}_k^\dagger \mathcal{L} \mathcal{W}_k, \quad \boldsymbol{\rho}_k = \mathcal{W}_k^\dagger \boldsymbol{\rho} \tag{9.19}$$

This parallelisation avenue has significant serial stages: the problem must be projected into the independent subspaces, and the state vector projected back

$$\boldsymbol{\rho} = \sum_k \mathcal{W}_k \boldsymbol{\rho}_k \tag{9.20}$$

if the evolution generator changes at the next simulation stage. When a large number of subspaces stay independent for many pulse sequence stages, this avenue is beneficial. Observables, being scalar products, may be calculated without leaving the blocked-up representation:

$$\langle \boldsymbol{\delta} \mid \boldsymbol{\rho} \rangle = \sum_k \langle \mathcal{W}_k \boldsymbol{\delta} \mid \mathcal{W}_k \boldsymbol{\rho} \rangle = \sum_k \langle \boldsymbol{\delta}_k \mid \boldsymbol{\rho}_k \rangle \tag{9.21}$$

with further screening opportunities (Sect. 7.3.6) because some $\boldsymbol{\delta}_k$ blocks may be zero.

9.2.3.3 Hilbert Space: Density Matrix Factorisation

A particular density matrix factorisation that may be known a priori (because it constitutes the definition, Sect. 4.2) is the *singular value decomposition* (SVD):

$$\boldsymbol{\rho} = \mathbf{U}\boldsymbol{\Sigma}\mathbf{V}^\dagger \tag{9.22}$$

where $\mathbf{U} = \{\mathbf{u}_1, \ldots, \mathbf{u}_p\}$ is a complex matrix with the *left singular vectors* in columns, $\mathbf{V} = \{\mathbf{v}_1, \ldots, \mathbf{v}_q\}$ is a complex matrix with the *right singular vectors* in columns, and $\boldsymbol{\Sigma}$ is a $p \times q$ diagonal matrix with positive real *singular values* $\{\sigma_1, \ldots, \sigma_n\}$ on the diagonal. All arrays are sorted in such a way as to put the singular values in decreasing order. A different way of writing Eq. (9.22) exposes the physical connection to state populations and coherences:

$$\mathbf{U\Sigma V}^{\dagger} = \sum_k \sigma_k \left(\mathbf{u}_k \otimes \mathbf{v}_k^{\dagger} \right) \equiv \sum_k \sigma_k |\mathbf{u}_k\rangle\langle\mathbf{v}_k| \tag{9.23}$$

and the mathematical connection to the rank-m Frobenius norm approximation:

$$\boldsymbol{\rho}^{(m)} = \sum_{k=1}^{m} \sigma_k \left(\mathbf{u}_k \otimes \mathbf{v}_k^{\dagger} \right), \qquad \left\| \boldsymbol{\rho} - \boldsymbol{\rho}^{(m)} \right\|_{\mathrm{F}}^{2} = \sum_{k=m+1}^{\dim(\boldsymbol{\rho})} \sigma_k^2 \tag{9.24}$$

which reflects the physical principle that small populations and weak coherences may be ignored. SVD is preferable to diagonalisation here because not all spin operators can be diagonalised.

For common density matrices (particularly those corresponding to initial and detection states), the expansion in Eq. (9.23) is known a priori and the number of terms is smaller than the dimension of $\boldsymbol{\rho}$. In that case, the double-sided propagation operation in Hilbert space (Sect. 4.2.1):

$$\begin{aligned} \boldsymbol{\rho}(t+\mathrm{d}t) &= \exp[-i\mathbf{H}(t)\mathrm{d}t]\boldsymbol{\rho}(t)\exp[+i\mathbf{H}(t)\mathrm{d}t] \\ &= \sum_k \sigma_k \exp[-i\mathbf{H}(t)\mathrm{d}t]|\mathbf{u}_k(t)\rangle\langle\mathbf{v}_k(t)|\exp[+i\mathbf{H}(t)\mathrm{d}t] \end{aligned} \tag{9.25}$$

splits into a set of independent propagation instances that may be computed in parallel [145]:

$$\begin{aligned} |\mathbf{u}_k(t+\mathrm{d}t)\rangle &= \exp[-i\mathbf{H}(t)\mathrm{d}t]|\mathbf{u}_k(t)\rangle \\ |\mathbf{v}_k(t+\mathrm{d}t)\rangle &= \exp[-i\mathbf{H}(t)\mathrm{d}t]|\mathbf{v}_k(t)\rangle \end{aligned} \tag{9.26}$$

The calculation of observables is also parallel over singular vector pairs:

$$\begin{aligned} \langle O(t)\rangle &= \mathbf{Tr}[\mathbf{O}\boldsymbol{\rho}(t)] \\ &= \sum_k \sigma_k \mathbf{Tr}[\mathbf{O}|\mathbf{u}_k(t)\rangle\langle\mathbf{v}_k(t)|] = \sum_k \sigma_k \langle\mathbf{v}_k(t)|\mathbf{O}|\mathbf{u}_k(t)\rangle \end{aligned} \tag{9.27}$$

When the SVD of $\boldsymbol{\rho}(0)$ is not known a priori, this method suffers from the same problem as the generator diagonalisation method in the previous section: the steep computational cost of singular value decomposition destroys the gains from parallelisation over singular vectors.

9.2.3.4 Hilbert Space: Factorisation-Free Observables

The factorisation that leads to the parallel split in Eq. (9.26) does not have to be SVD. If the possibility of dropping insignificant singular values is sacrificed, a suitable structure may be obtained by setting:

$$\boldsymbol{\rho}(0) = \sum_k |\boldsymbol{\rho}_k(0)\rangle\langle\boldsymbol{\upsilon}_k(0)| \tag{9.28}$$

where $|\boldsymbol{\rho}_k(0)\rangle$ is the k-th column of the initial density matrix and $|\boldsymbol{\upsilon}_k(0)\rangle$ is a vector with 1 in position k and zeros elsewhere. This formulation does not require any operations on the initial density matrix beyond sending the columns to their allocated nodes at the start of the calculation. A similar argument to Eqs. (9.23)-(9.27) yields the following propagation rules:

$$\begin{aligned} |\boldsymbol{\rho}_k(t+\mathrm{d}t)\rangle &= \exp[-i\mathbf{H}(t)\mathrm{d}t]|\boldsymbol{\rho}_k(t)\rangle \\ |\boldsymbol{\upsilon}_k(t+\mathrm{d}t)\rangle &= \exp[-i\mathbf{H}(t)\mathrm{d}t]|\boldsymbol{\upsilon}_k(t)\rangle \end{aligned} \tag{9.29}$$

and the following expression for the observables:

$$\begin{aligned} \langle O(t)\rangle &= \mathbf{Tr}[\mathbf{O}\boldsymbol{\rho}(t)] \\ &= \sum_k \mathbf{Tr}[\mathbf{O}|\boldsymbol{\rho}_k(t)\rangle\langle\boldsymbol{\upsilon}_k(t)|] = \sum_k \langle\boldsymbol{\upsilon}_k(t)|\mathbf{O}|\boldsymbol{\rho}_k(t)\rangle \end{aligned} \tag{9.30}$$

The computational complexity here is identical to that in Eqs. (9.25)-(9.27)—two sets of vectors are propagated at the worker nodes with no inter-thread communication apart from the small cost of sending the observable components $\langle\boldsymbol{\upsilon}_k(t)|\mathbf{O}|\boldsymbol{\rho}_k(t)\rangle$ back to the head node. The algorithm is:

Step 1: distribute the columns of the initial density matrix $|\boldsymbol{\rho}_k(0)\rangle$, the columns of the unit matrix $|\boldsymbol{\upsilon}_k(0)\rangle$, the Hamiltonian $\mathbf{H}$ (or the propagator if available and the Hamiltonian is time-independent), and the observable operator $\mathbf{O}$ to the worker nodes.

Step 2: on each worker node k preallocate a dense array of zeros $\mathbf{O}^{(k)}$ for the storage of the local contribution to the observable trajectory.

Step 3: on each worker node k propagate the local vectors and co-vectors using some implementation (Sect. 4.9) of Eq. (9.29): sequential matrix–vector multiplications if the exponential propagator had been supplied, and Krylov type propagation (Sect. 4.9.6) if the propagator is not available or the Hamiltonian is time-dependent. Record the local contribution to the observable dynamics using Eq. (9.30).

Step 4: collect the observable traces $\mathbf{O}^{(k)}(t)$ from every worker node and add them up element-by-element to obtain the final observable trajectory.

The resulting algorithm inherits the benefit of the method proposed by Skinner and Glaser [323]—no inter-thread communication at the propagation stage—and also removes the need to perform any kind of matrix factorisation at the problem set-up stage. The head node only needs to compute the sum of the observable traces returned by the worker nodes.

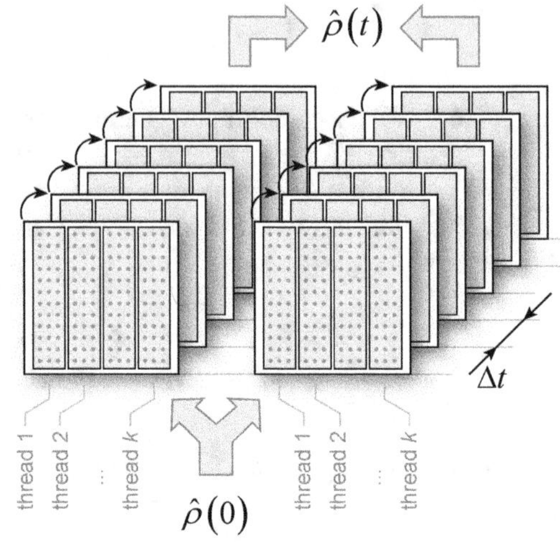

Fig. 9.1 A schematic of the split-propagate-recombine algorithm for parallel evaluation of the final density matrix in Hilbert space. Reproduced with permission from [145]

9.2.3.5 Hilbert Space: Factorisation-Free Final State

Parallel calculation of the final density matrix after an evolution period is similar to the calculation of observables discussed in the previous section:

$$\begin{aligned}\boldsymbol{\rho}(t+\mathrm{d}t) &= \exp[-i\mathbf{H}(t)\mathrm{d}t]\boldsymbol{\rho}(t)\exp[+i\mathbf{H}(t)\mathrm{d}t] \\ &= \sum_k \exp[-i\mathbf{H}(t)\mathrm{d}t]|\boldsymbol{\rho}_k(t)\rangle\langle\boldsymbol{\upsilon}_k(t)|\exp[+i\mathbf{H}(t)\mathrm{d}t]\end{aligned} \tag{9.31}$$

The individual vectors $|\boldsymbol{\rho}_k\rangle$ and $|\boldsymbol{\upsilon}_k\rangle$ may be distributed to different worker nodes for propagation and then returned to the head node where the density matrix is reassembled (Fig. 9.1):

Step 1: distribute the columns of the initial density matrix $|\boldsymbol{\rho}_k\rangle$, the columns of the unit matrix $|\boldsymbol{\delta}_k\rangle$ and the Hamiltonian $\mathbf{H}$ (or the propagator if available and the Hamiltonian is time-independent) to the worker nodes.

Step 2: on each worker node k propagate the local vectors and co-vectors through the prescribed number of time points using some implementation (Sect. 4.9) of Eq (9.29): sequential matrix-vector multiplications if the exponential propagator had been supplied, and Krylov type propagation (Sect. 4.9.6) if the propagator is not available or the Hamiltonian is time-dependent.

Step 3: collect the final vectors $|\boldsymbol{\rho}_k\rangle$ and $|\boldsymbol{\upsilon}_k\rangle$ from worker nodes and re-assemble the density matrix on the head node:

$$\boldsymbol{\rho}(t) = |\boldsymbol{\rho}_k(t)\rangle\langle\boldsymbol{\upsilon}_k(t)| \tag{9.32}$$

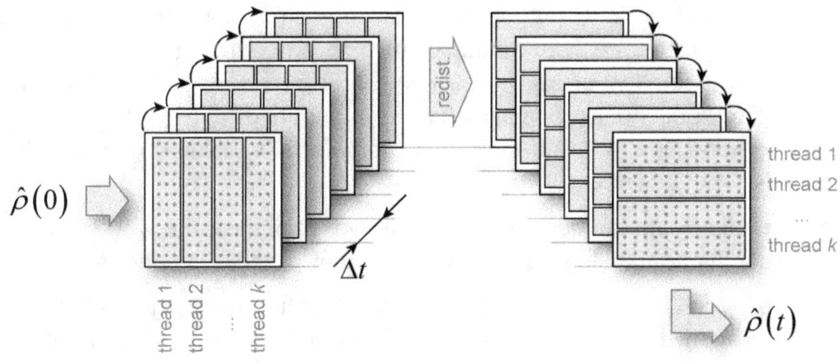

Fig. 9.2 A schematic of the propagate-redistribute-propagate algorithm for parallel evaluation of the final density matrix in Hilbert space. Reproduced with permission from [145]

This algorithm is less scalable because the head node has to do more work compared to the observable dynamics case—the cost of re-assembling the density matrix is quadratic in its dimension. It does, however, have the same advantage of zero inter-thread communication at the propagation stage.

For systems where the available communication bandwidth is large (e.g. shared-memory supercomputers) a different algorithm may be used which involves more communication but less head node processing. The idea is to reorder multiplication operations in Hilbert-space evolution calculations:

$$\boldsymbol{\rho}(t_n) = \underbrace{\mathbf{P}\cdots\mathbf{P}}_{n}\boldsymbol{\rho}(0)\underbrace{\mathbf{P}^{\dagger}\cdots\mathbf{P}^{\dagger}}_{n} = \left[\underbrace{\mathbf{P}\cdots\mathbf{P}}_{n}\left[\underbrace{\mathbf{P}\cdots\mathbf{P}}_{n}\boldsymbol{\rho}(0)\right]^{\dagger}\right]^{\dagger}, \qquad (9.33)$$
$$\mathbf{P} = \mathrm{e}^{-i\mathbf{H}\Delta t}$$

Because the density matrix is always multiplied from the left, the columns of $\boldsymbol{\rho}(0)$ may be distributed to the worker nodes and there is no inter-thread communication during the evaluation of the inner square bracket in Eq. (9.33)—the process is illustrated in Fig. 9.2.

The Hermitian conjugate is a significant communication hurdle—the density matrix that was distributed column-wise between the worker nodes, and is likely no longer sparse, has to be transposed and redistributed. After that, an identical propagation stage is carried out and the rows of the final density matrix are sent back to the head node.

Step 1: distribute the columns of the initial density matrix $|\boldsymbol{\rho}_k\rangle$ and the Hamiltonian $\mathbf{H}$ (or the propagator if available and the Hamiltonian is time-independent) to the worker nodes.

Step 2: on each node k propagate the local columns of the density matrix through the prescribed number of time points using some implementation (Sect. 4.9) of Eq. (9.29): sequential matrix–vector multiplications if the exponential propagator had been supplied, and Krylov type propagation (Sect. 4.9.6) if the propagator is not available or the Hamiltonian is time-dependent.

Step 3: re-distribute the density matrix row-wise between the worker nodes and conjugate-transpose each row on receipt to make a column.

Step 4: on each node k propagate the local columns of the density matrix through the prescribed number of time points using some implementation (Sect. 4.9) of Eq. (9.29): sequential matrix–vector multiplications if the exponential propagator had been supplied, and Krylov type propagation (Sect. 4.9.6) if the propagator is not available or the Hamiltonian is time-dependent.

Step 5: collect the columns of the density matrix on the head node and conjugate-transpose the result to obtain the final density matrix.

Step 3 in this algorithm is communication-intensive and Step 5 is potentially memory-intensive on the head node—the transpose operation requires a lot of memory access, unless the matrix is stored as a COO format sparse array [324].

9.3 Recycling and Cutting Corners

There may be a difference between the way an expression is written and the most efficient way to compute it. A good example is Frobenius inner product (Sect. 1.4.5) whose definition $\langle \mathbf{A} | \mathbf{B} \rangle = \mathrm{Tr}\left[\mathbf{A}^{\dagger}\mathbf{B}\right]$ appears to require two matrices to be multiplied—a cubic complexity operation with respect to matrix dimension. However, a simple arithmetical exercise shows that

$$\mathrm{Tr}\left[\mathbf{A}^{\dagger}\mathbf{B}\right] = \mathrm{Tot}[\mathbf{A}^{*} \odot \mathbf{B}] \tag{9.34}$$

where $\mathbf{A}^{*}$ is the element-wise complex conjugate of $\mathbf{A}$, $\odot$ denotes element-wise matrix product, and Tot stands for the total sum. The complexity of the right hand side is quadratic in the matrix dimension and all operations may be performed in-place, leading to excellent memory efficiency. Spin dynamics is rich in such efficiency tweaks, some of those are discussed in this section.

9.3.1 Efficient Norm Estimation

Spin dynamics simulations require an estimate of the largest frequency present in the system. From the physical perspective, this is necessary to avoid signal (NMR, EPR) and image (MRI) aliasing in the frequency domain. From the mathematical perspective, convergence and accuracy criteria for various series and perturbative expansions (e.g. those in Chap. 4) are formulated using the 2-norm (largest singular value, Sect. 1.4.5) of the operator. When matrix dimensions exceed 10^4, the calculation of the 2-norm becomes impractical, and even estimates become unacceptably expensive, particularly for tensor-structured objects (Sect. 9.1.5). Thankfully, an examination of the algorithms in question reveals that only an upper bound for the 2-norm is usually necessary—oversampling a trajectory (or having wider margins on an MRI image) is not a disaster. In the context of spin dynamics, the following relation is recommended:

$$\|\mathbf{A}\|_2^2 \leq \|\mathbf{A}\|_1 \|\mathbf{A}\|_\infty \tag{9.35}$$

At the time of writing, the cheapest upper bound for the 1-norm was proposed by Hager [146]. It requires a few matrix–vector products, and is therefore also compatible with tensor structured formats. Its infinity-norm version is obtained by performing vector multiplications from the other side.

9.3.2 Hash-and-Cache Recycling

Repeated evaluation of expensive matrix functions of the same argument is a common scenario in quantum dynamics simulations. In many sophisticated NMR pulse sequences, exponentials of the same matrix occur multiple times, but simply looking up a matrix in a database of previously encountered ones is impractical for the following reasons. On modern computer architectures a memory retrieval and comparison operation takes at least one CPU clock cycle. The time cost of looking up a given matrix in a sorted list of previously encountered ones is therefore at least $O(N_{\mathrm{NZ}} \log N_{\mathrm{M}})$ clocks [325], where N_{NZ} is the number of non-zero elements in the matrix, and N_{M} is the number of matrices in the database. The worst-case sorting cost for the database of previously encountered matrices is $O(N_{\mathrm{NZ}} N_{\mathrm{M}} \log N_{\mathrm{M}})$ clocks [325], which is unacceptable because the number of non-zeroes can be large.

The standard solution from database theory is to use a *hash table* [326], for which a detailed explanation is perhaps warranted. A cryptographic hash function is a function that accepts an input (called *message*) of any length and produces an output (called *digest*) with the following properties [327]:

1. The complexity of computing the digest is linear with the size of the message.
2. A single-bit modification in the message is *almost certain* to change its digest.
3. Two randomly selected messages are *almost certain* to have different digests.

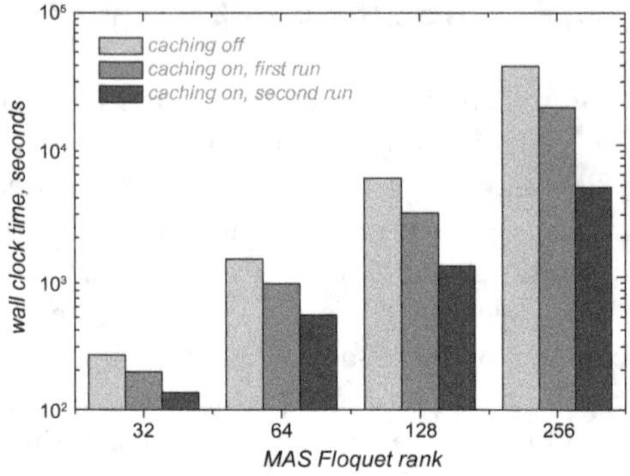

Fig. 9.3 Wall clock time consumed by the simulation of the CN2D solid state NMR experiment [330] for a ^{14}N-^{13}C spin pair in glycine with different matrix exponential caching settings in Spinach [87]. Light grey columns correspond to running with matrix caching switched off, medium grey columns are for runs with the caching switched on and a cache that is empty at the start of the simulation. Dark grey columns correspond to simulations where all required matrix exponentials are already present in the cache. Details of the NMR pulse sequence and the spin Hamiltonians involved are given in Reference [330]. Reproduced with permission from [288]

The first property offers a solution to our lookup cost problem: the complexity of computing the hash is $O(N_{\mathrm{NZ}})$—negligible compared to the cost of expensive matrix factorizations [328] and significantly smaller than the direct sorting and lookup costs discussed above. Hashing a matrix is a straightforward procedure: for full matrices, the array is typecast (in-place, to avoid making a memory copy) into UINT8 and fed into a hashing engine. For sparse matrices, the index array, the value array and the array of matrix dimensions are typecast into UINT8, concatenated and hashed as a single message. The cost of sorting the hash table is $O(N_{\mathrm{M}} \log N_{\mathrm{M}})$ and the cost of looking up a digest in the sorted list is $O(\log N_{\mathrm{M}})$ [325]. This reduces the total cost of the matrix lookup to $O(N_{\mathrm{NZ}}) + O(N_{\mathrm{M}} \log N_{\mathrm{M}}) + O(\log N_{\mathrm{M}})$. In situations where matrix operation caching is necessary, $N_{\mathrm{NZ}} \gg N_{\mathrm{M}} \log N_{\mathrm{M}} \gg \log N_{\mathrm{M}}$ (this may also be viewed as the condition under which it is sensible to use matrix operation caching)—the overall asymptotic cost of hash table matrix lookup and all the associated housekeeping is therefore $O(N_{\mathrm{NZ}})$ clocks.

The second and the third properties provide collision safety assurances: the definition of *almost certain* in this context is that one needs to calculate $2^{N/2}$ hash values (where N is the number of bits in the digest) to have a 50% probability of seeing a hash collision [329]. Even with basic 128-bit hash functions, such as MD5, this is a vanishingly rare event. If absolute certainty is required, an additional step of comparing the matrices element-by-element may be added, at the cost of extra storage, but without any changes to the asymptotic $O(N_{\mathrm{NZ}})$ complexity scaling estimate (Fig. 9.3).

The benefit derived from the caching procedure has the same caveats as the well-researched algorithms for caching disk access [331]—when the same sectors are requested repeatedly, the benefit is large, but for random access the cache can actually make the process slower. Matrix function caching should therefore only be used in situations where repeated requests for expensive functions of the same argument are likely—that situation is thankfully common.

In the quantum dynamics context, the increase in performance resulting from using matrix exponential caching is illustrated in Fig. 9.3. The CN2D NMR pulse sequence [330] is designed to correlate ^{14}N and ^{13}C NMR signals under magic angle spinning conditions. It contains multiple periods that have identical Hamiltonians—the caching algorithm identifies those automatically and avoids their recalculation.

It is in principle possible to hand-code this simulation in such a way as to avoid repeated calls to expensive functions by manually identifying the time intervals that have identical Hamiltonians. Such an approach would not, however, be scalable to more complicated experiments and to highly general and automated simulation systems, such as *Spinach* [87].

Cache destination can be a file system or a key-value store of any type—at the time of writing, an Optane card works well. In practice, the latency of the cache storage device means that for small matrices it may be faster to recalculate the function. In our practical experience, the caching procedure becomes beneficial once the dimension exceeds 512. Because matrix exponentiation dominates the numerical cost of optimal control simulations, a beneficial side effect is a rapid restart capability for the GRAPE algorithm (Sect. 8.1)—the calculation can re-trace its steps quickly.

The same applies to other expensive operations: one of the most CPU- and memory-intensive procedures in the *Spinach* kernel is Hamiltonian generation. In NMR simulations on ubiquitin [150] it required generating over 200,000 elementary spin operators from their descriptors. Because a descriptor uniquely defines an operator, it is possible to use it as a caching identifier and store the operator. Subsequent requests for that operator can be served from the cache.

9.3.3 Analytical Coherence Selection

Compared to experimental instruments, simulations have a greater measure of access to the system wavefunction, density matrix, or state vector. As a result, simulation of experiments that return combinations of signals from multiple scans may be simplified at the cost of severing the literal correspondence with what the instrument does. A simple example of this is quadrature detection, where the instrument records the real part $f_X(t) = \mathrm{Tr}[\mathbf{S}_X \boldsymbol{\rho}(t)]$ and the imaginary part $f_Y(t) = \mathrm{Tr}[\mathbf{S}_Y \boldsymbol{\rho}(t)]$ separately and then combines them into a complex signal $f_X(t) + i f_Y(t)$ prior to apodisation and Fourier transform. Because Frobenius inner product is bilinear, simulations can take a shortcut and "detect" using the non-Hermitian raising operator $\mathbf{S}_+ = \mathbf{S}_X + i\mathbf{S}_Y$, with an obvious extension to linear combinations of more than two signals—this reduces the cost of simulating phase cycles [281] in magnetic resonance spectroscopy.

In spherical tensor basis sets (Sect. 7.1) a related efficiency saving is available in high-field magnetic resonance where a common stage is coherence selection—making sure, by taking linear combinations of results of differently phased experiments, that only certain coherences and correlations (Sect. 4.2.3) are retained.

Experimentally, the procedure has combinatorial wall clock time scaling: every incremented phase at least doubles and commonly quadruples the experiment time. However, because direct products of ISTs are eigenfunctions of both coherence and correlation order operators, the simulation can proceed by simply zeroing out the unwanted elements of the state vector. The same trick applies in situations when pulsed field gradients are used to suppress undesired magnetisation transfer paths. When quadrature detection is used in multi-dimensional magnetic resonance experiments, the analytical selection results in density matrices becoming non-Hermitian. However, when the propagator sequence of the experiment is a linear operator, this has no physical consequences.

With these tricks are deployed in high-field magnetic resonance simulations, it is not uncommon to see an order of magnitude saved in the simulation time [124, 150, 251].

9.3.4 Implicit Linear Solvers

Calculation of steady states (Sect. 4.9.8) and steady orbits (Sect. 5.2.4) of dissipative spin systems leads to expressions of the following general type:

$$\boldsymbol{\rho}_{\infty} = \left(-i\mathcal{H} + \mathcal{R}\right)^{-1}\mathcal{R}\boldsymbol{\rho}_{\mathrm{eq}} \tag{9.36}$$

where calculating the matrix inverse may be impractical, or an explicit matrix representation of $\mathcal{H}$ and $\mathcal{R}$ may not be available—an example is polyadic objects (Sect. 9.1) that can only return a product into a given vector. Because $\mathcal{R}\boldsymbol{\rho}_{\mathrm{eq}}$ is also a vector, the problem is essentially about solving $\mathbf{Ax} = \mathbf{y}$ with a complex and non-Hermitian (but non-singular) square matrix $\mathbf{A}$ when only the product $\mathbf{Az}$ is available for any vector $\mathbf{z}$. A rich collection of methods exists for dealing with such cases [332]; the general idea is to iteratively minimise $\Omega(\mathbf{x}) = \|\mathbf{Ax} - \mathbf{y}\|_2^2$ with respect to $\mathbf{x}$ from some initial guess $\mathbf{x}_0$. For square and non-singular $\mathbf{A}$, the problem is continuous and convex, and therefore a unique solution exists that is reachable from any initial guess. The crucial question is about *how quickly* the solution is reached—although the gradient:

$$\nabla\Omega(\mathbf{x}) = 2\mathbf{A}^{\mathrm{T}}(\mathbf{Ax} - \mathbf{y}) \tag{9.37}$$

is available, gradient descent is impractically slow (even with line search) in situations where a machine-precision answer is required in 64-bit arithmetic. The problem gets worse when $\mathbf{A}$ is badly conditioned because the number of iterations increases with the condition number.

Of the many ways of dealing with this, we will only mention the *generalised minimal residual method* (GMRES [151]) here, which proceeds by building the Krylov subspace of the matrix $\mathbf{A}$ and the error vector, and repeatedly solving the projection of the original problem in that subspace. The projector is obtained by orthogonalising (for example, using Arnoldi process [333]) repeated actions by **A** on the error vector:

$$\mathbf{Q}(\mathbf{A},\mathbf{r}) = \mathrm{orth}\left\{\mathbf{r},\mathbf{A}\mathbf{r},\mathbf{A}^2\mathbf{r},\ldots,\mathbf{A}^{n-1}\mathbf{r}\right\}, \quad \mathbf{r} = \mathbf{y} - \mathbf{A}\mathbf{x} \tag{9.38}$$

where n is a small integer chosen by the user. If the residual minimisation step is parameterised as $\Delta\mathbf{x} = \mathbf{Q}\mathbf{a}$, where $\mathbf{a}$ is an n-dimensional vector, finding the minimum of the residual

$$\Omega(\mathbf{a}) = \|\mathbf{A}(\mathbf{x}+\mathbf{Q}\mathbf{a}) - \mathbf{y}\|_2^2 \tag{9.39}$$

is easier because the dimension of $\mathbf{a}$ is small. Once the minimiser is found, the vector $\mathbf{x}$ is updated to $\mathbf{x} \leftarrow \mathbf{x} + \mathbf{Q}\mathbf{a}$ and the procedure is repeated until convergence. After various logistical refinements, it turns out that only one multiplication of the original matrix **A** by a vector is needed per iteration, and the rest of the mathematics takes place in reduced subspaces.

References

1. F. Hausdorff, *Grundzüge der mengenlehre* (Verlag von Veit & Comp, 1914)
2. R. Dedekind, *Vorlesungen über zahlentheorie* (Friedrich Vieweg Verlag, 1894)
3. G. Peano, *Arithmetices principia nova methodo exposita* (Fratres Bocca, 1889)
4. M. Born, P. Jordan, Zur quantenmechanik. Z. Phys. **34**, 858–888 (1925)
5. G. Frobenius, Note sur la théorie des formes quadratiques à un nombre quelconque de variables. C. R. l'Académie Sci. **85**, 131–133 (1877)
6. C. Banwell, H. Primas, On the analysis of high-resolution nuclear magnetic resonance spectra, Part I: methods of calculating NMR spectra. Mol. Phys. **6**, 225–256 (1963)
7. S. Lie, *Theorie der transformationsgruppen* (BG Teubner Verlag, 1888)
8. G. Frobenius, Über gruppencharaktere, in *Sitzungsberichte der Königlich Preussischen Akademie der Wissenschaften zu Berlin* (1896), pp. 985–1021
9. H.F. Baker, Alternants and continuous groups. Proc. Lond. Math. Soc. **2**, 24–47 (1905)
10. J.E. Campbell, On a law of combination of operators (second paper). Proc. Lond. Math. Soc. 1–29, 14–32 (1897)
11. F. Hausdorff, Die symbolische exponentialformel in der gruppentheorie, in *Berichte über die Verhandlungen der Sächsischen Akademie der Wissenschaften zu Leipzig*: *Math.-Phys. Klasse* **58**, 19 (1906)
12. N. Bourbaki, Groupes de Lie et algèbres de Lie, in *Éléments d'histoire des mathématiques* (1960)
13. É. Cartan, *Sur la structure des groupes de transformations finis et continus* (Librarie Nony, 1894)
14. W. Killing, Die zusammensetzung der stetigen endlichen transformationsgruppen. Math. Ann. **33**, 1–48 (1888)
15. H.B.G. Casimir, Ph.D. Thesis: Rotation of a rigid body in quantum mechanics, University of Leiden (1931)
16. Εὐκλείδης, Στοιχεῖα, *circa* 300 BC
17. A.A. Michelson, E.W. Morley, On the relative motion of the Earth and the luminiferous ether. Am. J. Sci. **34**, 333–345 (1887)
18. H.A. Lorentz, De relatieve beweging van de aarde en den aether, in *Verslagen der Zittingen van der Wis- en Natuurkundige Afdeeling der Koninklijke Akademie van Wetenschappen* (Johannes Muller, 1892), pp. 74–79
19. H. Minkowski, Raum und zeit. Phys. Z. **10**, 104–111 (1909)
20. M. Siemens, J. Hancock, D. Siminovitch, Beyond Euler angles: exploiting the angle-axis parametrization in a multipole expansion of the rotation operator. Solid State Nucl. Magn. Reson. **31**, 35–54 (2007)
21. L. Euler, Formulae generales pro translatione quacunque corporum rigidorum, in *Novi Commentarii Academiae Scientiarum Petropolitanae* (1776), pp. 189–207
22. D.M. Brink, G.R. Satchler, *Angular Momentum*, 3rd edn. (Clarendon Press, 1993)

I. KUPROV, *Spin*,
https://doi.org/10.1007/978-3-031-05607-9

23. E. Wigner, Einige folgerungen aus der Schrödingerschen theorie für die termstrukturen. Z. Phys. **43**, 624–652 (1927)
24. I. Newton, *Opticks: or, a treatise of the reflexions, refractions, inflexions and colours of light* (Royal Society, London, 1704)
25. А.А. Морозов, А.В. Мельников, Ф.И. Скрипов, Методика свободной ядерной индукции в слабых магнитных полях в применении к некоторым задачам радиоспектроскопии высокой разрешающей силы. Известия Академии Наук СССР **22**, 1141 (1958)
26. P.A.M. Dirac, *The Principles of Quantum Mechanics* (Oxford University Press, 1981)
27. J.-B.J. Fourier, *Théorie analytique de la chaleur* (Imprimerie de Firmin Didot, 1822)
28. N. Wiener, Generalized harmonic analysis. Acta Math. **55**, 117–258 (1930)
29. D. Hilbert, *Grundzüge einer allgemeinen theorie der linearen integralgleichungen* (BG Teubner Verlag, 1912)
30. H.A. Kramers, La diffusion de la lumiere par les atomes, in *Atti del Congresso Internationale dei Fisici* (1927), pp. 545–557
31. R. de Laer Kronig, On the theory of dispersion of X-rays. J. Opt. Soc. Am. **12**, 547–557 (1926)
32. E. Schrödinger, An undulatory theory of the mechanics of atoms and molecules. Phys. Rev. **28**, 1049 (1926)
33. P.S. Laplace, *Traité de mécanique céleste*, vol. 3 (Imprimerie de Crapelet, 1802)
34. A.M. Legendre, *Recherches sur l'attraction des sphéroïdes homogènes* (Imprimerie Royale, 1785)
35. E.P. Wigner, On unitary representations of the inhomogeneous Lorentz group. Ann. Math. 149–204 (1939)
36. A. Hurwitz, Ueber die erzeugung der invarianten durch integration, Nachrichten von der Königlichen Gesellschaft der Wissenschaften zu Göttingen. Math.-Phys. Klasse 71–90 (1897)
37. M.H. Poincaré, Sur la dynamique de l'électron. Rendiconti del Circolo Matematico di Palermo (1884–1940) **21**, 129–175 (1906)
38. W. Gerlach, O. Stern, Der experimentelle nachweis der richtungsquantelung im magnetfeld. Z. Phys. **9**, 349–352 (1922)
39. G.E. Uhlenbeck, S. Goudsmit, Ersetzung der hypothese vom unmechanischen zwang durch eine forderung bezüglich des inneren verhaltens jedes einzelnen elektrons. Naturwissenschaften **13**, 953–954 (1925)
40. W. Pauli, Zur quantenmechanik des magnetischen elektrons. Z. Phys. **43**, 601–623 (1927)
41. P.A.M. Dirac, The quantum theory of the electron. Proc. R. Soc. Lond. A **117**, 610–624 (1928)
42. M. Planck, Über das gesetz der energieverteilung im normalspectrum. Ann. Phys. **309**, 553–563 (1901)
43. P.A. Gordan, *Vorlesungen über invariantentheorie* (BG Teubner Verlag, 1885)
44. E. Schrödinger, Über die kräftefreie bewegung in der relativistischen quantenmechanik, Sitzungsberichten der Preussischen Akademie der Wissenschaften. Phys.-Math. Klasse **XXIV**, 418–428 (1930)
45. J. Schwinger, On quantum-electrodynamics and the magnetic moment of the electron. Phys. Rev. **73**, 416 (1948)
46. Л.Д. Ландау, Е.М. Лифшиц, Курс теоретической физики, Том 3: Квантовая механика, нерелятивистская теория, (Физматлит, 2002)
47. E. Fermi, Über die magnetischen momente der atomkerne. Z. Phys. **60**, 320–333 (1930)
48. A. Abragam, M.H.L. Pryce, Theory of the nuclear hyperfine structure of paramagnetic resonance spectra in crystals. Proc. R. Soc. Lond. A **205**, 135–153 (1951)
49. P. Güttinger, Das verhalten von atomen im magnetischen drehfeld. Z. Phys. **73**, 169–184 (1932)
50. W. van den Heuvel, A. Soncini, NMR chemical shift as analytical derivative of the Helmholtz free energy. J. Chem. Phys. **138**, 054113 (2013)

51. A.N. Bohr, B.R. Mottelson, *Nuclear Structure* (World Scientific, 1998)
52. R.G. Parr, Y. Weitao, *Density-Functional Theory of Atoms and Molecules* (Oxford University Press, 1995)
53. A. Szabo, N.S. Ostlund, *Modern Quantum Chemistry: Introduction to Advanced Electronic Structure Theory* (Courier Corporation, 2012)
54. P.W. Atkins, R.S. Friedman, *Molecular Quantum Mechanics* (Oxford University Press, 2011)
55. R.D. Woods, D.S. Saxon, Diffuse surface optical model for nucleon-nuclei scattering. Phys. Rev. **95**, 577–578 (1954)
56. M. Włoch, D.J. Dean, J.R. Gour, M. Hjorth-Jensen, K. Kowalski, T. Papenbrock, P. Piecuch, Ab-initio coupled-cluster study of ^{16}O. Phys. Rev. Lett. **94**, 212501 (2005)
57. S.D. Bass, *The Spin Structure of the Proton* (World Scientific, 2008)
58. C. Eckart, The application of group theory to the quantum dynamics of monatomic systems. Rev. Mod. Phys. **2**, 305 (1930)
59. L.P. Gaffney, P.A. Butler, M. Scheck, A.B. Hayes, F. Wenander, M. Albers, B. Bastin, C. Bauer, A. Blazhev, S. Bönig, Studies of pear-shaped nuclei using accelerated radioactive beams. Nature **497**, 199–204 (2013)
60. K. Wolinski, J.F. Hinton, P. Pulay, Efficient implementation of the gauge-independent atomic orbital method for NMR chemical shift calculations. J. Am. Chem. Soc. **112**, 8251–8260 (1990)
61. P. Langevin, Sur la théorie du magnétisme. J. Phys. Théor. Appl. **4**, 678–693 (1905)
62. J.H. Van Vleck, On dielectric constants and magnetic susceptibilities in the new quantum mechanics, Part III: application to dia- and paramagnetism. Phys. Rev. **31**, 587–613 (1928)
63. J.H. Van Vleck, *The Theory of Electric and Magnetic Susceptibilities* (Oxford University Press, 1959)
64. N.F. Ramsey, Magnetic shielding of nuclei in molecules. Phys. Rev. **78**, 699 (1950)
65. N.F. Ramsey, Dependence of magnetic shielding of nuclei upon molecular orientation. Phys. Rev. **83**, 540 (1951)
66. N.F. Ramsey, Chemical effects in nuclear magnetic resonance and in diamagnetic susceptibility. Phys. Rev. **86**, 243 (1952)
67. R.K. Harris, E.D. Becker, S.M.C. De Menezes, P. Granger, R.E. Hoffman, K.W. Zilm, Further conventions for NMR shielding and chemical shifts. Pure Appl. Chem. **80**, 59–84 (2008)
68. H.M. McConnell, D.B. Chesnut, Theory of isotropic hyperfine interactions in π-electron radicals. J. Chem. Phys. **28**, 107–117 (1958)
69. N.F. Ramsey, E.M. Purcell, Interactions between nuclear spins in molecules. Phys. Rev. **85**, 143 (1952)
70. N.F. Ramsey, Electron coupled interactions between nuclear spins in molecules. Phys. Rev. **91**, 303 (1953)
71. W. Heisenberg, Mehrkörperproblem und resonanz in der quantenmechanik. Z. Phys. **38**, 411–426 (1926)
72. H.A. Kramers, L'interaction entre les atomes magnétogènes dans un cristal paramagnétique. Physica **1**, 182–192 (1934)
73. K. Yamaguchi, Y. Takahara, T. Fueno, Ab-initio molecular orbital studies of structure and reactivity of transition metal-oxo compounds, in *Proceedings of the Nobel Laureate Symposium on Applied Quantum Chemistry in Honor of G. Herzberg, R.S. Mulliken, K. Fukui, W. Lipscomb, R. Hoffman*, ed. by V.H. Smith, H.F. Schaefer, K. Morokuma (Springer, 1986), pp. 155–184
74. I. Dzyaloshinsky, A thermodynamic theory of "weak" ferromagnetism of antiferromagnetics. J. Phys. Chem. Solids **4**, 241–255 (1958)
75. T. Moriya, Anisotropic superexchange interaction and weak ferromagnetism. Phys. Rev. **120**, 91–98 (1960)

76. A. Abragam, B. Bleaney, *Electron Paramagnetic Resonance of Transition Ions* (Oxford University Press, 2012)
77. R. Boča, Zero-field splitting in metal complexes. Coord. Chem. Rev. **248**, 757–815 (2004)
78. D. Parker, E.A. Suturina, I. Kuprov, N.F. Chilton, How the ligand field in lanthanide coordination complexes determines magnetic susceptibility anisotropy, paramagnetic NMR shift, and relaxation behavior. Acc. Chem. Res. **53**, 1520–1534 (2020)
79. D. Ganyushin, F. Neese, First-principles calculations of zero-field splitting parameters. J. Chem. Phys. **125**, 024103 (2006)
80. O. Vahtras, O. Loboda, B. Minaev, H. Ågren, K. Ruud, Ab initio calculations of zero-field splitting parameters. Chem. Phys. **279**, 133–142 (2002)
81. A. Clebsch, *Über symbolische darstellung algebraischer formen* (De Gruyter, 1861)
82. K.W.H. Stevens, Matrix elements and operator equivalents connected with the magnetic properties of rare Earth ions. Proc. Phys. Soc. A **65**, 209–215 (1952)
83. I. Kuprov, Diagonalization-free implementation of spin relaxation theory for large spin systems. J. Magn. Reson. **209**, 31–38 (2011)
84. U. Haeberlen, High resolution NMR in solids: selective averaging, in *Advances in Magnetic Resonance, Supplement 1* (Academic Press, 1976)
85. M. Mehring, *High Resolution NMR Spectroscopy in Solids* (Springer, 2012)
86. S.D. Poisson, Mémoire sur la théorie du magnétisme en mouvement, in *Mémoires de l'Académie Royale des sciences de l'Institut Imperial de France, Paris* (1827), p. 56
87. H.J. Hogben, M. Krzystyniak, G.T.P. Charnock, P.J Hore, I. Kuprov, Spinach—a software library for simulation of spin dynamics in large spin systems. J. Magn. Reson. **208**, 179–194 (2011)
88. M.J. Frisch, G.W. Trucks, H.B. Schlegel, G.E. Scuseria, M.A. Robb, J.R. Cheeseman, G. Scalmani, V. Barone, G.A. Petersson, H. Nakatsuji, X. Li, M. Caricato, A.V. Marenich, J. Bloino, B.G. Janesko, R. Gomperts, B. Mennucci, H.P. Hratchian, J.V. Ortiz, A.F. Izmaylov, J.L. Sonnenberg, Williams, F. Ding, F. Lipparini, F. Egidi, J. Goings, B. Peng, A. Petrone, T. Henderson, D. Ranasinghe, V.G. Zakrzewski, J. Gao, N. Rega, G. Zheng, W. Liang, M. Hada, M. Ehara, K. Toyota, R. Fukuda, J. Hasegawa, M. Ishida, T. Nakajima, Y. Honda, O. Kitao, H. Nakai, T. Vreven, K. Throssell, J.A. Montgomery Jr., J.E. Peralta, F. Ogliaro, M.J. Bearpark, J.J. Heyd, E.N. Brothers, K.N. Kudin, V.N. Staroverov, T.A. Keith, R. Kobayashi, J. Normand, K. Raghavachari, A.P. Rendell, J.C. Burant, S.S. Iyengar, J. Tomasi, M. Cossi, J.M. Millam, M. Klene, C. Adamo, R. Cammi, J.W. Ochterski, R.L. Martin, K. Morokuma, O. Farkas, J.B. Foresman, D.J. Fox, Gaussian16 Revision C.01, Wallingford, CT (2016)
89. J. von Neumann, Wahrscheinlichkeitstheoretischer aufbau der quantenmechanik, Nachrichten von der Gesellschaft der Wissenschaften zu Göttingen. Math.-Phys. Klasse 245–272 (1927)
90. A.D. Bain, J.S. Martin, FT NMR of nonequilibrium states of complex spin systems, Part I: a Liouville space description. J. Magn. Reson. **29**, 125–135 (1978)
91. B. Taylor, *Methodus Incrementorum Directa & Inversa* (Impensis Gulielmi Innys, 1715)
92. F.J. Dyson, The radiation theories of Tomonaga, Schwinger, and Feynman. Phys. Rev. **75**, 486–502 (1949)
93. I.I. Rabi, Space quantization in a gyrating magnetic field. Phys. Rev. **51**, 652 (1937)
94. F. Bloch, Nuclear induction. Phys. Rev. **70**, 460–474 (1946)
95. J. Schwinger, Quantum electrodynamics, Part I: a covariant formulation. Phys. Rev. **74**, 1439–1461 (1948)
96. I.I. Rabi, N.F. Ramsey, J. Schwinger, Use of rotating coordinates in magnetic resonance problems. Rev. Mod. Phys. **26**, 167 (1954)
97. A.H. Al-Mohy, N.J. Higham, Improved inverse scaling and squaring algorithms for the matrix logarithm. SIAM J. Sci. Comput. **34**, C153–C169 (2012)
98. Е.Б. Дынкин, О представлении ряда $\log(e^{X}e^{Y})$ от некоммутирующих X и Y через коммутаторы. Математический Сборник **25**, 155–162 (1949)

99. F. Casas, A. Murua, An efficient algorithm for computing the Baker-Campbell-Hausdorff series and some of its applications. J. Math. Phys. **50**, 033513 (2009)
100. W. Magnus, On the exponential solution of differential equations for a linear operator. Commun. Pure Appl. Math. **7**, 649–673 (1954)
101. F. Casas, A. Murua, M. Nadinic, Efficient computation of the Zassenhaus formula. Comput. Phys. Commun. **183**, 2386–2391 (2012)
102. D.L. Goodwin, I. Kuprov, Auxiliary matrix formalism for interaction representation transformations, optimal control, and spin relaxation theories. J. Chem. Phys. **143**, 084113 (2015)
103. F. Fer, Resolution de l'equation matricielle U' = PU par produit infini d'exponentielles matricielles. Bull. l'Académie R. Sci., Lett. Beaux-Arts Belg. **44**, 818–829 (1958)
104. F. Casas, Sufficient conditions for the convergence of the Magnus expansion. J. Phys. A **40**, 15001–15017 (2007)
105. A. Van-Brunt, M. Visser, Special-case closed form of the Baker-Campbell-Hausdorff formula. J. Phys. A **48**, 225207 (2015)
106. U. Haeberlen, J. Waugh, Coherent averaging effects in magnetic resonance. Phys. Rev. **175**, 453 (1968)
107. J.L. Lagrange, *Recherches sur les équations séculaires des mouvemens des noeuds, et des inclinaisons des orbites des planètes* (Imprimerie de Gauthier-Villars, 1774)
108. D. Kivelson, Theory of ESR linewidths of free radicals. J. Chem. Phys. **33**, 1094–1106 (1960)
109. H.J. Bernstein, J.A. Pople, W. Schneider, The analysis of nuclear magnetic resonance spectra, Part I: Systems of two and three nuclei. Can. J. Chem. **35**, 67–83 (1957)
110. A.L. Bloom, J.N. Shoolery, Effects of perturbing radiofrequency fields on nuclear spin coupling. Phys. Rev. **97**, 1261–1265 (1955)
111. T. Gullion, J. Schaefer, Rotational-echo double-resonance NMR. J. Magn. Reson. **81**, 196–200 (1989)
112. T.G. Oas, R.G. Griffin, M.H. Levitt, Rotary resonance recoupling of dipolar interactions in solid-state nuclear magnetic resonance spectroscopy. J. Chem. Phys. **89**, 692–695 (1988)
113. M.H. Levitt, R. Freeman, Composite pulse decoupling. J. Magn. Reson. **43**, 502–507 (1981)
114. N.C. Nielsen, L.A. Strassø, A.B. Nielsen, Dipolar recoupling, in *Solid State NMR* (Springer, 2011) pp. 1–45
115. M. Suzuki, Generalized Trotter's formula and systematic approximants of exponential operators and inner derivations with applications to many-body problems. Commun. Math. Phys. **51**, 183–190 (1976)
116. N. Hatano, M. Suzuki, Finding exponential product formulas of higher orders, in *Quantum Annealing and Other Optimization Methods*. ed. by A. Das, B.K. Chakrabarti (Springer, 2005), pp. 37–68
117. E. Schrödinger, Quantisierung als eigenwertproblem. Ann. Phys. **385**, 437–490 (1926)
118. J.W. Strutt, *The Theory of Sound* (Macmillan, 1896)
119. H. Primas, Generalized perturbation theory in operator form. Rev. Mod. Phys. **35**, 710 (1963)
120. P.A.M. Dirac, The quantum theory of the emission and absorption of radiation. Proc. R. Soc. Lond. A **114**, 243–265 (1927)
121. G. Belford, R. Belford, J. Burkhalter, Eigenfields: a practical direct calculation of resonance fields and intensities for field-swept fixed-frequency spectrometers. J. Magn. Reson. **11**, 251–265 (1973)
122. S. Stoll, A. Schweiger, An adaptive method for computing resonance fields for continuous-wave EPR spectra. Chem. Phys. Lett. **380**, 464–470 (2003)
123. E.P. Wigner, *Gruppentheorie und ihre anwendung auf die quantenmechanik der atomspektren* (Springer, 1931)
124. H.J. Hogben, P.J. Hore, I. Kuprov, Strategies for state space restriction in densely coupled spin systems with applications to spin chemistry. J. Chem. Phys. **132**, 174101 (2010)

125. O. Sørensen, G. Eich, M.H. Levitt, G. Bodenhausen, R. Ernst, Product operator formalism for the description of NMR pulse experiments. Prog. Nucl. Magn. Reson. Spectrosc. **16**, 163–192 (1984)
126. P. Guntert, N. Schaefer, G. Otting, K. Wuthrich, POMA: a complete Mathematica implementation of the NMR product-operator formalism. J. Magn. Reson. **101**, 103–105 (1993)
127. E.L. Hahn, Spin echoes. Phys. Rev. **80**, 580–594 (1950)
128. G.A. Morris, R. Freeman, Enhancement of nuclear magnetic resonance signals by polarization transfer. J. Am. Chem. Soc. **101**, 760–762 (1979)
129. G. Floquet, Sur les équations différentielles linéaires à coefficients périodiques. Ann. Sci. l'École Norm. Sup. 47–88 (1883)
130. J.H. Shirley, Solution of the Schrodinger equation with a Hamiltonian periodic in time. Phys. Rev. **138**, B979–B987 (1965)
131. I. Scholz, J.D. van Beek, M. Ernst, Operator-based Floquet theory in solid-state NMR. Solid State Nucl. Magn. Reson. **37**, 39–59 (2010)
132. F. Casas, J.A. Oteo, J. Ros, Floquet theory: exponential perturbative treatment. J. Phys. A **34**, 3379 (2001)
133. C. Runge, Über die numerische auflösung von differentialgleichungen. Math. Ann. **46**, 167–178 (1895)
134. W. Kutta, Beitrag zur naherungsweisen integration totaler differentialgleichungen. Z. Math. Phys. **46**, 435–453 (1901)
135. A. Iserles, H.Z. Munthe-Kaas, S.P. Nørsett, A. Zanna, Lie-group methods. Acta Numer **9**, 215–365 (2000)
136. B. Riemann, Über die darstellbarkeit einer function durch eine trigonometrische reihe. Abh. Königlichen Ges. Wiss. Göttingen **13**, 87–132 (1867)
137. E.T. Whittaker, On the functions which are represented by the expansions of the interpolation-theory. Proc. R. Soc. Edinb. **35**, 181–194 (1915)
138. J.S. Waugh, Sensitivity in Fourier transform NMR spectroscopy of slowly relaxing systems. J. Mol. Spectrosc. **35**, 298–305 (1970)
139. D.M. Korn, Simple modification to commercial Michelson transform spectrometer for increased resolution. Rev. Sci. Instrum. **44**, 1135–1136 (1973)
140. M. Veshtort, R.G. Griffin, High-performance selective excitation pulses for solid- and liquid-state NMR spectroscopy. ChemPhysChem **5**, 834–850 (2004)
141. C. Moler, C. Van Loan, Nineteen dubious ways to compute the exponential of a matrix, twenty-five years later. SIAM Rev. **45**, 3–49 (2003)
142. П.Л. Чебышев, Математический анализ, in *Полное собрание сочинений* (Издательство Академии Наук СССР, 1948)
143. H. Padé, *Sur la représentation approchée d'une fonction par des fractions rationnelles* (Imprimerie Gauthier-Villars et Fils, 1892)
144. I. Newton, *Philosophiae Naturalis Principia Mathematica* (Royal Society Press, 1686)
145. L.J. Edwards, I. Kuprov, Parallel density matrix propagation in spin dynamics simulations. J. Chem. Phys. **136** (2012)
146. W.W. Hager, Condition estimates. SIAM J. Sci. Stat. Comput. **5**, 311–316 (1984)
147. R.B. Sidje, Expokit: a software package for computing matrix exponentials. ACM Trans. Math. Softw. **24**, 130–156 (1998)
148. L.E. Kay, M. Ikura, R. Tschudin, A. Bax, Three-dimensional triple-resonance NMR spectroscopy of isotopically enriched proteins. J. Magn. Reson. **89**, 496–514 (1990)
149. L.J. Edwards, D.V. Savostyanov, Z.T. Welderufael, D. Lee, I. Kuprov, Quantum mechanical NMR simulation algorithm for protein-size spin systems. J. Magn. Reson. **243**, 107–113 (2014)
150. Y. Saad, M.H. Schultz, GMRES: a generalized minimal residual algorithm for solving nonsymmetric linear systems. SIAM J. Sci. Stat. Comput. **7**, 856–869 (1986)
151. A.W. Overhauser, Polarization of nuclei in metals. Phys. Rev. **92**, 411 (1953)

152. T.R. Carver, C.P. Slichter, Polarization of nuclear spins in metals. Phys. Rev. **92**, 212 (1953)
153. E.N. Laguerre, Sur l'intégrale de $\exp(-x)/x$ de x à l'infini. Bull. Soc. Math. France **7**, 428–437 (1879)
154. M. Hermite, Sur un nouveau développement en série des fonctions. C. R. l'Academie Sci. **LVIII**, 93–266 (1864)
155. C.F. Gauss, *Methodvs nova integralivm valores per approximationem inveniendi* (Henricvm Dieterich, 1815)
156. В.И. Лебедев, Д.Н. Лайков, Квадратурная формула для сферы 131-го алгебраического порядка точности. Доклады Академии Наук **366**, 741–745 (1999)
157. В.И. Лебедев, О квадратурах на сфере. Журнал Вычислительной Математики и Математической Физики **16**, 293–306 (1976)
158. C. Ahrens, G. Beylkin, Rotationally invariant quadratures for the sphere. Proc. R. Soc. Lond. A **465**, 3103–3125 (2009)
159. R. Swinbank, R.J. Purser, Fibonacci grids: a novel approach to global modelling. Q. J. R. Meteorol. Soc. **132**, 1769–1793 (2006)
160. S. Stoll, Ph.D. Thesis: *Spectral Simulations in Solid-State Electron Paramagnetic Resonance* (ETH Zurich, 2003)
161. R. Hardin, N. Sloane, W. Smith, Tables of spherical codes with icosahedral symmetry (published electronically at http://neilsloane.com/icosahedral.codes), (2000)
162. D. Alderman, M.S. Solum, D.M. Grant, Methods for analyzing spectroscopic line shapes: NMR solid powder patterns. J. Chem. Phys. **84**, 3717–3725 (1986)
163. M. Bak, N.C. Nielsen, REPULSION: a novel approach to efficient powder averaging in solid-state NMR. J. Magn. Reson. **125**, 132–139 (1997)
164. M.J. Nilges, Ph.D. Thesis: *Electron Paramagnetic Resonance Studies of Low Symmetry Ni(I) and Mo(V) Complexes* (University of Illinois at Urbana-Champaign, 1979)
165. G. Voronoi, Nouvelles applications des paramètres continus à la théorie des formes quadratiques. Deuxième mémoire. Recherches sur les parallélloèdres primitifs. J. Reine Angew. Math. **1908**, 198–287 (1908)
166. G. Voronoi, Nouvelles applications des paramètres continus à la théorie des formes quadratiques. Premier mémoire. Sur quelques propriétés des formes quadratiques positives parfaits. J. Reine Angew. Math. **1908**, 97–102 (1908)
167. S.K. Zaremba, Good lattice points, discrepancy, and numerical integration. Annali di Matematica Pura ed Applicata **73**, 293–317 (1966)
168. H. Conroy, Molecular Schrödinger equation, Part VIII: a new method for the evaluation of multidimensional integrals. J. Chem. Phys. **47**, 5307–5318 (1967)
169. V.B. Cheng, H.H. Suzukawa Jr., M. Wolfsberg, Investigations of a nonrandom numerical method for multidimensional integration. J. Chem. Phys. **59**, 3992–3999 (1973)
170. C. Crăciun, Homogeneity and EPR metrics for assessment of regular grids used in CW EPR powder simulations. J. Magn. Reson. **245**, 63–78 (2014)
171. D. Wang, G.R. Hanson, A new method for simulating randomly oriented powder spectra in magnetic resonance: the Sydney opera house (SOPHE) method. J. Magn. Reson. **117**, 1–8 (1995)
172. C. Crăciun, Behaviour of twelve spherical codes in CW EPR powder simulations: uniformity and EPR properties. Stud. Univ. Babes-Bolyai—Chem. **61** (2016)
173. A. Ponti, Simulation of magnetic resonance static powder lineshapes: a quantitative assessment of spherical codes. J. Magn. Reson. **138**, 288–297 (1999)
174. M. Edén, Computer simulations in solid-state NMR, Part III: powder averaging. Concepts in Magn. Reson. A **18**, 24–55 (2003)
175. S. Fortune, A sweepline algorithm for Voronoi diagrams. Algorithmica **2**, 153–174 (1987)
176. M. Edén, M.H. Levitt, Computation of orientational averages in solid-state NMR by Gaussian spherical quadrature. J. Magn. Reson. **132**, 220–239 (1998)

177. H. Ebert, J. Abart, J. Voitländer, Simulation of quadrupole disturbed NMR field spectra by using perturbation theory and the triangle integration method. J. Chem. Phys. **79**, 4719–4723 (1983)
178. S. Stoll, A. Schweiger, EasySpin: a comprehensive software package for spectral simulation and analysis in EPR. J. Magn. Reson. **178**, 42–55 (2006)
179. C. Crăciun, Application of the SCVT orientation grid to the simulation of CW EPR powder spectra. Appl. Magn. Reson. **38**, 279–293 (2010)
180. А.Д. Милов, К.М. Салихов, М. Щиров, Применение метода двойного резонанса в электронном спиновом эхо для изучения пространственного распределения парамагнитных центров в твердых телах. Физика Твердого Тела **23**, 975–982 (1981)
181. G. Jeschke, A. Koch, U. Jonas, A. Godt, Direct conversion of EPR dipolar time evolution data to distance distributions. J. Magn. Reson. **155**, 72–82 (2002)
182. A. Fresnel, Mémoire sur la diffraction de la lumière. Ann. Chim. Phys. **1**, 239–281 (1816)
183. H.C. Torrey, Bloch equations with diffusion terms. Phys. Rev. **104**, 563 (1956)
184. E.R. Andrew, A. Bradbury, R.G. Eades, Nuclear magnetic resonance spectra from a crystal rotated at high speed. Nature **182**, 1659–1659 (1958)
185. A. Fick, Über diffusion. Ann. Phys. **170**, 59–86 (1855)
186. A.D. Fokker, Die mittlere energie rotierender elektrischer dipole im strahlungsfeld. Ann. Phys. **348**, 810–820 (1914)
187. M. Planck, Über einen satz der statistischen dynamik und seine erweiterung in der quantentheorie, in *Sitzungsberichte der Königlich Preussischen Akademie der Wissenschaften zu Berlin* (1917), pp. 324–341
188. I. Kuprov, Fokker-Planck formalism in magnetic resonance simulations. J. Magn. Reson. **270**, 124–135 (2016)
189. G. Moro, J.H. Freed, Efficient computation of magnetic resonance spectra and related correlation functions from stochastic Liouville equations. J. Phys. Chem. **84**, 2837–2840 (1980)
190. W.T. Coffey, Y.P. Kalmykov, *The Langevin Equation: with Applications to Stochastic Problems in Physics, Chemistry and Electrical Engineering* (World Scientific, 2012)
191. H. Risken, *The Fokker-Planck Equation* (Springer, 1984)
192. V. John, Higher order finite element methods and multigrid solvers in a benchmark problem for the 3D Navier-Stokes equations. Int. J. Numer. Meth. Fluids **40**, 775–798 (2002)
193. B. Gmeiner, M. Huber, L. John, U. Rüde, B. Wohlmuth, A quantitative performance study for Stokes solvers at the extreme scale. J. Comput. Sci. **17**, 509–521 (2016)
194. B. Fornberg, Generation of finite difference formulas on arbitrarily spaced grids. Math. Comput. **51**, 699–706 (1988)
195. L.N. Trefethen, *Spectral Methods in Matlab* (SIAM, 2000)
196. R. Kühne, T. Schaffhauser, A. Wokaun, R.R. Ernst, Study of transient chemical reactions by NMR: fast stopped-flow Fourier transform experiments. J. Magn. Reson. **35**, 39–67 (1979)
197. N. Singhal, C.D. Snow, V.S. Pande, Using path sampling to build better Markovian state models: predicting the folding rate and mechanism of a tryptophan zipper beta hairpin. J. Chem. Phys. **121**, 415–425 (2004)
198. P. Waage, C. Guldberg, Studier over affiniteten. Forhandlinger i Videnskabs-selskabet i Christiania **1**, 35–45 (1864)
199. R. Johnson, R. Merrifield, Effects of magnetic fields on the mutual annihilation of triplet excitons in anthracene crystals. Phys. Rev. B **1**, 896 (1970)
200. J.-B. Biot, F. Savart, Note sur le magnétisme de la pile de volta. Ann. Chim. Phys. 222–223 (1820)
201. J. Gauss, K. Ruud, T. Helgaker, Perturbation-dependent atomic orbitals for the calculation of spin-rotation constants and rotational *g*-tensors. J. Chem. Phys. **105**, 2804–2812 (1996)
202. W. Flygare, Magnetic interactions in molecules and an analysis of molecular electronic charge distribution from magnetic parameters. Chem. Rev. **74**, 653–687 (1974)

203. D. Kivelson, E. Bright Wilson Jr., Approximate treatment of the effect of centrifugal distortion on the rotational energy levels of asymmetric-rotor molecules. J. Chem. Phys. **20**, 1575–1579 (1952)
204. J. Haupt, A new effect of dynamic polarization in a solid obtained by rapid change of temperature. Phys. Lett. A **38**, 389–390 (1972)
205. S. Clough, J. Hill, Thermally induced nuclear dipolar polarization in powders. Phys. Lett. A **49**, 461–462 (1974)
206. A. Lunghi, S. Sanvito, How do phonons relax molecular spins? Sci. Adv. **5**, eaax7163 (2019)
207. W. Heisenberg, Über quantentheoretische umdeutung kinematischer und mechanischer beziehungen. Z. Phys. **33**, 879–893 (1925)
208. M.H. Devoret, Quantum fluctuations in electrical circuits, in *Proceedings of Session LXIII of the Les Houches School of Physics*, vol. 7 (1995), pp. 133–135
209. P.A.M. Dirac, The fundamental equations of quantum mechanics. Proc. R. Soc. Lond. A **109**, 642–653 (1925)
210. E.T. Jaynes, F.W. Cummings, Comparison of quantum and semiclassical radiation theories with application to the beam maser. Proc. IEEE **51**, 89–109 (1963)
211. B. Yurke, J.S. Denker, Quantum network theory. Phys. Rev. A **29**, 1419 (1984)
212. I. Solomon, Relaxation processes in a system of two spins. Phys. Rev. **99**, 559–565 (1955)
213. G. Lipari, A. Szabo, Model-free approach to the interpretation of nuclear magnetic resonance relaxation in macromolecules, Part 1: theory and range of validity. J. Am. Chem. Soc. **104**, 4546–4559 (1982)
214. R. Damadian, Tumor detection by nuclear magnetic resonance. Science **171**, 1151–1153 (1971)
215. A. Karabanov, I. Kuprov, G.T.P. Charnock, A. van der Drift, L.J. Edwards, W. Köckenberger, On the accuracy of the state space restriction approximation for spin dynamics simulations. J. Chem. Phys. **135**, 084106 (2011)
216. H. Haken, Synergetics. Phys. Bull. **28**, 412–414 (1977)
217. P.S. Hubbard, Quantum-mechanical and semiclassical forms of the density operator theory of relaxation. Rev. Mod. Phys. **33**, 249 (1961)
218. A.G. Redfield, On the theory of relaxation processes. IBM J. Res. Dev. **1**, 19–31 (1957)
219. I. Kuprov, L.C. Morris, J.N. Glushka, J.H. Prestegard, Using molecular dynamics trajectories to predict nuclear spin relaxation behaviour in large spin systems. J. Magn. Reson. **323**, 106891 (2021)
220. I. Kuprov, N. Wagner-Rundell, P.J. Hore, Bloch-redfield-wangsness theory engine implementation using symbolic processing software. J. Magn. Reson. **184**, 196–206 (2007)
221. W.E. Milne, *Numerical Calculus* (Princeton University Press, 2015)
222. P.S. Hubbard, Nuclear magnetic resonance and relaxation of four spin molecules in a liquid. Phys. Rev. **128**, 650–658 (1962)
223. R. Kubo, Generalized cumulant expansion method. J. Phys. Soc. Jpn. **17**, 1100–1120 (1962)
224. G. Lindblad, On the generators of quantum dynamical semigroups. Commun. Math. Phys. **48**, 119–130 (1976)
225. L. Boltzmann, Über die beziehung zwischen dem zweiten hauptsatze des mechanischen wärmetheorie und der wahrscheinlichkeitsrechnung, respective den sätzen über das wärmegleichgewicht, Sitzungberichte der Kaiserlichen Akademie der Wissenschaften zu Wien. Math.-Naturwiss. Classe **76**, 373–435 (1877)
226. G.W. Stewart, A Krylov-Schur algorithm for large eigenproblems. SIAM J. Matrix Anal. Appl. **23**, 601–614 (2002)
227. T.O. Levante, R.R. Ernst, Homogeneous versus inhomogeneous quantum-mechanical master equations. Chem. Phys. Lett. **241**, 73–78 (1995)
228. M.H. Levitt, L. Di Bari, Steady state in magnetic resonance pulse experiments. Phys. Rev. Lett. **69**, 3124 (1992)

229. I. Kuprov, D.M. Hodgson, J. Kloesges, C.I. Pearson, B. Odell, T.D.W. Claridge, Anomalous nuclear overhauser effects in carbon-substituted aziridines: scalar cross-relaxation of the first kind. Angew. Chem. **54**, 3697–3701 (2015)
230. A. Abragam, *The Principles of Nuclear Magnetism* (Clarendon Press, 1961)
231. H.M. McConnell, C.H. Holm, Anisotropic chemical shielding and nuclear magnetic relaxation in liquids. J. Chem. Phys. **25**, 1289–1289 (1956)
232. H.M. McConnell, Effect of anisotropic hyperfine interactions on paramagnetic relaxation in liquids. J. Chem. Phys. **25**, 709–711 (1956)
233. M. Goldman, Interference effects in the relaxation of a pair of unlike spin-1/2 nuclei. J. Magn. Reson. **60**, 437–452 (1984)
234. K. Pervushin, R. Riek, G. Wider, K. Wüthrich, Attenuated T_2 relaxation by mutual cancellation of dipole-dipole coupling and chemical shift anisotropy indicates an avenue to NMR structures of very large biological macromolecules in solution. Proc. Natl. Acad. Sci. **94**, 12366–12371 (1997)
235. M. Gueron, Nuclear relaxation in macromolecules by paramagnetic ions: a novel mechanism. J. Magn. Reson. **19**, 58–66 (1975)
236. A.J. Vega, D. Fiat, Nuclear relaxation processes of paramagnetic complexes: the slow-motion case. Mol. Phys. **31**, 347–355 (1976)
237. E.A. Suturina, K. Mason, C.F. Geraldes, N.F. Chilton, D. Parker, I. Kuprov, Lanthanide-induced relaxation anisotropy. Phys. Chem. Chem. Phys. **20**, 17676–17686 (2018)
238. J.C. Ott, E.A. Suturina, I. Kuprov, J. Nehrkorn, A. Schnegg, M. Enders, L.H. Gade, Observability of paramagnetic NMR signals at over 10,000 ppm chemical shifts. Angew. Chem. **133**, 23038–23046 (2021)
239. M. Briganti, F. Santanni, L. Tesi, F. Totti, R. Sessoli, A. Lunghi, A complete ab initio view of Orbach and Raman spin-lattice relaxation in a dysprosium coordination compound. J. Am. Chem. Soc. **143**, 13633–13645 (2021)
240. W. Happer, Optical pumping. Rev. Mod. Phys. **44**, 169 (1972)
241. D.D. McGregor, Transverse relaxation of spin-polarized ^{3}He gas due to a magnetic field gradient. Phys. Rev. A **41**, 2631 (1990)
242. P.S. Hubbard, Rotational Brownian motion. Phys. Rev. A **6**, 2421–2433 (1972)
243. P.M. Singer, D. Asthagiri, W.G. Chapman, G.J. Hirasaki, NMR spin-rotation relaxation and diffusion of methane. J. Chem. Phys. **148**, 204504 (2018)
244. S.R. White, Density matrix formulation for quantum renormalization groups. Phys. Rev. Lett. **69**, 2863–2866 (1992)
245. M. Fannes, B. Nachtergaele, R.F. Werner, Finitely correlated states on quantum spin chains. Commun. Math. Phys. **144**, 443–490 (1992)
246. I.V. Oseledets, Tensor-train decomposition. SIAM J. Sci. Comput. **33**, 2295–2317 (2011)
247. D. Savostyanov, S. Dolgov, J. Werner, I. Kuprov, Exact NMR simulation of protein-size spin systems using tensor train formalism. Phys. Rev. B **90**, 085139 (2014)
248. L.J. Edwards, D.V. Savostyanov, A.A. Nevzorov, M. Concistrè, G. Pileio, I. Kuprov, Grid-free powder averages: on the applications of the Fokker-Planck equation to solid state NMR. J. Magn. Reson. **235**, 121–129 (2013)
249. M.E. Halse, J.-N. Dumez, L. Emsley, Quasi-equilibria in reduced Liouville spaces. J. Chem. Phys. **136** (2012)
250. M. Krzystyniak, L.J. Edwards, I. Kuprov, Destination state screening of active spaces in spin dynamics simulations. J. Magn. Reson. **210**, 228–232 (2011)
251. J.-N. Dumez, M.C. Butler, L. Emsley, Numerical simulation of free evolution in solid-state nuclear magnetic resonance using low-order correlations in Liouville space. J. Chem. Phys. **133** (2010)
252. M.C. Butler, J.-N. Dumez, L. Emsley, Dynamics of large nuclear-spin systems from low-order correlations in Liouville space. Chem. Phys. Lett. **477**, 377–381 (2009)

253. I. Kuprov, Polynomially scaling spin dynamics II: further state-space compression using Krylov subspace techniques and zero track elimination. J. Magn. Reson. **195**, 45–51 (2008)
254. I. Kuprov, N. Wagner-Rundell, P. Hore, Polynomially scaling spin dynamics simulation algorithm based on adaptive state-space restriction. J. Magn. Reson. **189**, 241–250 (2007)
255. J.H. Freed, G.K. Fraenkel, Theory of linewidths in electron spin resonance spectra. J. Chem. Phys. **39**, 326 (1963)
256. B.C. Sanctuary, F.P. Temme, Multipole NMR, Part 13: multispin interactions and symmetry in Liouville space. Mol. Phys. **55**, 1049–1062 (1985)
257. D.A. Varshalovich, A.N. Moskalev, V.K. Khersonskii, *Quantum Theory of Angular Momentum* (World Scientific, 1988)
258. R.R. Ernst, G. Bodenhausen, A. Wokaun, *Principles of Nuclear Magnetic Resonance in One and Two Dimensions* (Clarendon Press, 1987)
259. S. Even, *Graph Algorithms* (Cambridge University Press, 2012)
260. G. Moro, J.H. Freed, Calculation of ESR spectra and related Fokker–Planck forms by the use of the Lanczos algorithm. J. Chem. Phys. **74**, 3757–3773 (1981)
261. E. Noether, Invariante variationsprobleme, Königlich Gesellschaft der Wissenschaften Göttingen: Nachrichten. Math.-Phys. Klasse **2**, 235–267 (1918)
262. H.N. Gabow, R.E. Tarjan, A linear-time algorithm for a special case of disjoint set union. J. Comput. Syst. Sci. **30**, 209–221 (1985)
263. R. Tarjan, Depth-first search and linear graph algorithms. SIAM J. Comput. **1**, 146–160 (1972)
264. I. Kuprov, Large-scale NMR simulations in liquid state: a tutorial. Magn. Reson. Chem. **56**, 415–437 (2018)
265. L.S. Pontryagin, V.G. Boltanskii, R.S. Gamkrelidze, E.F. Mishchenko, *The Mathematical Theory of Optimal Processes* (Pergamon, 1964)
266. N. Khaneja, T. Reiss, C. Kehlet, T. Schulte-Herbruggen, S.J. Glaser, Optimal control of coupled spin dynamics: design of NMR pulse sequences by gradient ascent algorithms. J. Magn. Reson. **172**, 296–305 (2005)
267. J. Nocedal, S.J. Wright, *Numerical Optimization* (Springer, 2006)
268. R. Fletcher, *Practical Methods of Optimization*, 2nd edn. (Wiley, 1987)
269. A. Cauchy, Méthode générale pour la résolution des systemes d'équations simultanées. C. R. l'Académie Sci. **25**, 536–538 (1847)
270. L. Lasdon, S. Mitter, A. Waren, The conjugate gradient method for optimal control problems. IEEE Trans. Autom. Control **12**, 132–138 (1967)
271. D.C. Liu, J. Nocedal, On the limited memory BFGS method for large scale optimization. Math. Program. **45**, 503–528 (1989)
272. J. Raphson, *Analysis aequationum universalis: Seu ad aequationes algebraicas resolvendas methodus generalis, & expedita, ex nova infinitarum serierum methodo, deducta ac demonstrata* (Thomas Braddyll, 1697)
273. K.B. Petersen, M.S. Pedersen, *The Matrix Cookbook* (Online Edition, 2012)
274. M.S. Vinding, D.L. Goodwin, I. Kuprov, T.E. Lund, Optimal control gradient precision trade-offs: application to fast generation of deepcontrol libraries for MRI. J. Magn. Reson. **333**, 107094 (2021)
275. A.H. Al-Mohy, N.J. Higham, The complex step approximation to the Fréchet derivative of a matrix function. Numer. Algor. **53**, 133 (2010)
276. P. de Fouquieres, S.G. Schirmer, S.J. Glaser, I. Kuprov, Second order gradient ascent pulse engineering. J. Magn. Reson. **212**, 412–417 (2011)
277. C.F. Van Loan, Computing integrals involving the matrix exponential. IEEE Trans. Autom. Control **23**, 395–404 (1978)
278. F. Carbonell, J.C. Jimenez, L.M. Pedroso, Computing multiple integrals involving matrix exponentials. J. Comput. Appl. Math. **213**, 300–305 (2008)
279. R. Freeman, S.P. Kempsell, M.H. Levitt, Radiofrequency pulse sequences which compensate their own imperfections. J. Magn. Reson. **38**, 453–479 (1980)

280. E.O. Stejskal, J. Schaefer, Data routing in quadrature FT NMR. J. Magn. Reson. **13**, 249–251 (1974)
281. A. Medek, J.S. Harwood, L. Frydman, Multiple-quantum magic-angle spinning NMR: a new method for the study of quadrupolar nuclei in solids. J. Am. Chem. Soc. **117**, 12779–12787 (1995)
282. O.W. Sørensen, A universal bound on spin dynamics. J. Magn. Reson. **86**, 435–440 (1990)
283. S. Linnainmaa, Taylor expansion of the accumulated rounding error. BIT Numer. Math. **16**, 146–160 (1976)
284. M. Rance, O. Sørensen, G. Bodenhausen, G. Wagner, R. Ernst, K. Wüthrich, Improved spectral resolution in COSY ^{1}H NMR spectra of proteins via double quantum filtering. Biochem. Biophys. Res. Commun. **117**, 479–485 (1983)
285. R.H. Byrd, J. Nocedal, R.B. Schnabel, Representations of quasi-Newton matrices and their use in limited memory methods. Math. Program. **63**, 129–156 (1994)
286. J. Gregory, *Vera circuli et hyperbolae quadratura* (Typographia Iacobi de Cadorinis, 1667)
287. D.L. Goodwin, I. Kuprov, Modified newton-raphson grape methods for optimal control of spin systems. J. Chem. Phys. **144**, 204107 (2016)
288. S.M. Goldfeld, R.E. Quandt, H.F. Trotter, Maximization by quadratic hill-climbing. Econometrica 541–551 (1966)
289. J. Greenstadt, On the relative efficiencies of gradient methods. Math. Comput. **21**, 360–367 (1967)
290. A. Banerjee, F. Grein, Convergence behavior of some multiconfiguration methods. Int. J. Quantum Chem. **10**, 123–134 (1976)
291. M.D. Hebden, *An Algorithm for Minimization Using Exact Second Derivatives*, Technical Report TP 515 (AERE Harwell Laboratory, 1973)
292. D. Goldfarb, Curvilinear path steplength algorithms for minimization which use directions of negative curvature. Math. Program. **18**, 31–40 (1980)
293. C.J. Cerjan, W.H. Miller, On finding transition states. J. Chem. Phys. **75**, 2800–2806 (1981)
294. J.J. Moré, D.C. Sorensen, *Newton's Method*, Report Number ANL-82-8 (Argonne National Lab, 1982)
295. R. Shepard, I. Shavitt, J. Simons, Comparison of the convergence characteristics of some iterative wave function optimization methods. J. Chem. Phys. **76**, 543–557 (1982)
296. J.J. Moré, D.C. Sorensen, Computing a trust region step. SIAM J. Sci. Stat. Comput. **4**, 553–572 (1983)
297. A. Banerjee, N. Adams, J. Simons, R. Shepard, Search for stationary points on surfaces. J. Phys. Chem. **89**, 52–57 (1985)
298. J. Baker, An algorithm for the location of transition states. J. Comput. Chem. **7**, 385–395 (1986)
299. T.E. Skinner, T.O. Reiss, B. Luy, N. Khaneja, S.J. Glaser, Application of optimal control theory to the design of broadband excitation pulses for high-resolution NMR. J. Magn. Reson. **163**, 8–15 (2003)
300. I. Kuprov, Spin system trajectory analysis under optimal control pulses. J. Magn. Reson. **233**, 107–112 (2013)
301. B.C. Sanctuary, Multipole operators for an arbitrary number of spins. J. Chem. Phys. **64**, 4352–4361 (1976)
302. D. Gabor, Theory of communication. Part 1: the analysis of information. J. Inst. Electr. Eng. III **93**, 429–441 (1946)
303. S. Köcher, T. Heydenreich, S. Glaser, Visualization and analysis of modulated pulses in magnetic resonance by joint time-frequency representations. J. Magn. Reson. **249**, 63–71 (2014)
304. A.V. Oppenheim, R.W. Schafer, *Digital Signal Processing* (Prentice-Hall, 1975)
305. J. Tribolet, A new phase unwrapping algorithm. IEEE Trans. Acoust. Speech Signal Process. **25**, 170–177 (1977)

306. J. Ville, Theorie et application dela notion de signal analytique. Câbles et Transmissions **2**, 61–74 (1948)
307. R.S. Dumont, S. Jain, A. Bain, Simulation of many-spin system dynamics via sparse matrix methodology. J. Chem. Phys. **106**, 5928–5936 (1997)
308. M. Veshtort, R.G. Griffin, SPINEVOLUTION: a powerful tool for the simulation of solid and liquid state NMR experiments. J. Magn. Reson. **178**, 248–282 (2006)
309. A.J. Allami, M.G. Concilio, P. Lally, I. Kuprov, Quantum mechanical MRI simulations: solving the matrix dimension problem. Sci. Adv. **5**, eaaw8962 (2019)
310. T.E. Conturo, R.C. McKinstry, J.A. Aronovitz, J.J. Neil, Diffusion MRI: precision, accuracy and flow effects. NMR Biomed. **8**, 307–332 (1995)
311. D. Le Bihan, J.F. Mangin, C. Poupon, C.A. Clark, S. Pappata, N. Molko, H. Chabriat, Diffusion tensor imaging: concepts and applications. J. Magn. Reson. Imaging **13**, 534–546 (2001)
312. L. Frydman, T. Scherf, A. Lupulescu, The acquisition of multidimensional NMR spectra within a single scan. Proc. Natl. Acad. Sci. **99**, 15858–15862 (2002)
313. K. Zangger, Pure shift NMR. Prog. Nucl. Magn. Reson. Spectrosc. **86**, 1–20 (2015)
314. P.A. Bottomley, Spatial localization in NMR spectroscopy in vivo. Ann. N. Y. Acad. Sci. **508**, 333–348 (1987)
315. H. Tal-Ezer, R. Kosloff, An accurate and efficient scheme for propagating the time-dependent Schrodinger equation. J. Chem. Phys. **81**, 3967–3971 (1984)
316. U. Schollwöck, The density-matrix renormalization group in the age of matrix product states. Ann. Phys. **326**, 96–192 (2011)
317. P. Fernandes, B. Plateau, W.J. Stewart, Efficient descriptor-vector multiplications in stochastic automata networks. J. ACM **45**, 381–414 (1998)
318. G.M. Amdahl, Validity of the single processor approach to achieving large scale computing capabilities, in *Proceedings of the ACM Joint Computer Conference*, ACM, 1967, pp. 483–485
319. A. Buluc, J.R. Gilbert, Challenges and advances in parallel sparse matrix-matrix multiplication, in *Proceedings of the 37th International Conference on Parallel Processing*, IEEE, 2008, pp. 503–510
320. Z. Tosner, R. Andersen, B. Stevensson, M. Eden, N.C. Nielsen, T. Vosegaard, Computer-intensive simulation of solid-state NMR experiments using SIMPSON. J. Magn. Reson. **246**, 79–93 (2014)
321. T. Skinner, S. Glaser, Representation of a quantum ensemble as a minimal set of pure states. Phys. Rev. A **66** (2002)
322. J.R. Gilbert, C. Moler, R. Schreiber, Sparse matrices in matlab—design and implementation. SIAM J. Matrix Anal. Appl. **13**, 333–356 (1992)
323. T.H. Cormen, *Introduction to Algorithms* (MIT Press, 2009)
324. W.D. Maurer, T.G. Lewis, Hash table methods. ACM Comput. Surv. **7**, 5–19 (1975)
325. R. Sedgewick, *Algorithms in Java* (Addison-Wesley Professional, 2002)
326. D.E. Knuth, *The Art of Computer Programming* (Addison-Wesley, 2005)
327. X. Wang, H. Yu, How to break MD5 and other hash functions, in *Advances in Cryptology—Eurocrypt 2005*, Springer, 2005, pp. 19–35
328. J.A. Jarvis, I.M. Haies, P.T.F. Williamson, M. Carravetta, An efficient NMR method for the characterisation of ^{14}N sites through indirect ^{13}C detection. Phys. Chem. Chem. Phys. **15**, 7613–7620 (2013)
329. B. Jacob, S. Ng, D. Wang, *Memory Systems: Cache, DRAM, Disk* (Morgan Kaufmann, 2010)
330. L.N. Trefethen, D. Bau III, *Numerical Linear Algebra* (SIAM, 1997)
331. W.E. Arnoldi, The principle of minimized iterations in the solution of the matrix eigenvalue problem. Q. Appl. Math. **9**, 17–29 (1951)
332. L. Euler, *Recherches sur les racines imaginaires des équations*, Histoire de l'Académie Royale des Sciences et des Belles-Lettres de Berlin, **5**, 222–288 (1751)
333. T. Levi-Civita, *Fondamenti di meccanica relativistica*, Zanichelli (1928)

334. A. Cauchy, *Sur un nouveau genre de calcul analogueau calcul infinitésimal*, in: Oeuvres Complet d'Augustin Cauchy, Gauthier-Villars, (1826)
335. R.N. Bracewell, *The Fourier transform and its applications*, McGraw-Hill, (1986)
336. C.J. Horowitz, B.D. Serot, *Self-consistent hartree description of finite nuclei in a relativistic quantum field theory*, Nucl. Phy. A. **368** 503–528 (1981)
337. J. Herzfeld, A.E. Berger, *Sideband intensities in NMR spectra ofsamples spinning at the magic angle*, J. Chem. Phys. **73** 6021–6030 (1980)
338. L. Bonacci, Liber Abaci, a manuscript, 1202
339. M. Goldman, *Formal theory of spin-lattice relaxation*, J. Magn. Reson., **149** 160–187 (2001)
340. G. Moro, J.H. Freed, *Efficient computation of magneticresonance spectra and related correlation functions from stochastic Liouville equations*, J. Phys. Chem., **84** 2837–2840 (1980)
341. G. Moro, J.H. Freed, *Calculation of ESRspectra and related Fokker–Planck forms by the use of the Lanczos algorithm*, J. Chem. Phys., **74** 3757–3773 (1981)
342. I. Kuprov, C.T. Rodgers, *Derivatives of spin dynamics simulations*, J. Chem. Phys., **131** 234108 (2009)
343. M. Holbach, J. Lambert, D. Suter, *Optimized multiple-quantum filter for robust selective excitation of metabolite signals*, J. Magn. Reson., **243** 8–16(2014)
344. M. Braun, S.J. Glaser, *Cooperative pulses*, J. Magn. Reson., **207** 114–123 (2010)

Index

I. KUPROV, *Spin*,
https://doi.org/10.1007/978-3-031-05607-9

GPSR Compliance
The European Union's (EU) General Product Safety Regulation (GPSR) is a set of rules that requires consumer products to be safe and our obligations to ensure this.

If you have any concerns about our products, you can contact us on

ProductSafety@springernature.com

In case Publisher is established outside the EU, the EU authorized representative is:

Springer Nature Customer Service Center GmbH
Europaplatz 3
69115 Heidelberg, Germany

www.ingramcontent.com/pod-product-compliance
Lightning Source LLC
Chambersburg PA
CBHW071607290626
47387CB00006B/1050
* 9 7 8 3 0 3 1 0 5 6 0 6 2 *